Physico-Chemical and Computational Approaches to Drug Discovery

RSC Drug Discovery Series

Editor-in-Chief:
Professor David Thurston, *London School of Pharmacy, UK*

Series Editors:
Dr David Fox, *Pfizer Global Research and Development, Sandwich, UK*
Professor Salvatore Guccione, *University of Catania, Italy*
Professor Ana Martinez, *Instituto de Quimica Medica-CSIC, Spain*
Professor David Rotella, *Montclair State University, USA*

Advisor to the Board:
Professor Robin Ganellin, *University College London, UK*

Titles in this Series:

How to obtain future titles on publication:
A standing order plan is available for this series. A standing order will bring delivery of each new volume immediately on publication.

For further information please contact:
Book Sales Department, Royal Society of Chemistry, Thomas Graham House, Science Park, Milton Road, Cambridge, CB4 0WF, UK
Telephone: +44 (0)1223 420066, Fax: +44 (0)1223 420247, Email: books@rsc.org
Visit our website at http://www.rsc.org/Shop/Books/

Physico-Chemical and Computational Approaches to Drug Discovery

Edited by

F. Javier Luque and Xavier Barril
Department of Physical Chemistry, University of Barcelona, Spain
Email: fjluque@ub.edu; xbarril@ub.edu

RSC Publishing

RSC Drug Discovery Series No. 23

ISBN: 978-1-84973-353-3
ISSN: 2041-3203

A catalogue record for this book is available from the British Library

© The Royal Society of Chemistry 2012

Published by The Royal Society of Chemistry,
Thomas Graham House, Science Park, Milton Road,
Cambridge CB4 0WF, UK

Registered Charity Number 207890

For further information see our web site at www.rsc.org

Printed and bound in Great Britain by Henry Ling Limited, Dorchester, DT1 1HD, UK

Preface

In a recent publication, Wild *et al.* recognized the importance of chemistry for the future of the pharmaceutical industry.[1] Starting from the widely assumed perception that the industry is now facing relevant challenges, the authors argue that only innovation will be the key to sustain the success of pharmaceutical companies, and that chemistry will continue to play a crucial role as driving force in drug discovery. Even though we have witnessed in recent years the success of therapeutic approaches based on protein drugs and antibodies, small molecules will continue playing a fundamental contribution to drug discovery. Thus, around 70% of the new chemical entities launched in 2010 were small molecules. They are well suited to interact with both extra- and intracellular targets, covering not only the binding to standard binding pockets, but also recently designed to interact with less conventional pockets (*i.e.*, allosteric sites, protein–protein binding sites). They are also amenable to chemical designs conceived to modulate the bioavailability of the drugs in the body, as well as the administration to the patient and duration of the drug action.

Since the chemical structure of a compound is the ultimate factor that determines the potency, selectivity and pharmacokinetic properties, a proper design of small molecules is crucial for developing innovative therapeutic strategies, especially keeping in mind the limited number of targets that are currently being targeted by approved drugs and the large number of potential targets that may be disclosed by the human genome project in the next years. In this scenario, computational methods appear as an extremely valuable tool to address the current challenges of pharmaceutical research, and thus contribute to abandon the rather steady time evolution of approved drugs seen in recent years, in spite of the continued increase in research and development investment by the Pharma industry.[2] As noted by Jorgensen, "*the use of computers and*

RSC Drug Discovery Series No. 23
Physico-Chemical and Computational Approaches to Drug Discovery
Edited by F. Javier Luque and Xavier Barril
© The Royal Society of Chemistry 2012
Published by the Royal Society of Chemistry, www.rsc.org

computational methods permeates all aspects of drug discovery today", and proficiency in computer-assisted drug design should therefore be advantageous for delivering new drug candidates more quickly and efficiently.[3,4]

The impact of computational methods in drug design is no longer a point of debate. Computational chemistry has made significant contributions to drug generation and optimization, including an extremely broad range of aspects such as target druggability, prediction of drug likeness, *de novo* design, fragment screening, ligand docking, estimation of binding affinity and modulation of ADMET (absorption, distribution, metabolism, excretion, toxicity) properties. Though a review of the scientific literature could allow us to pinpoint a representative number of studies where computational methods have made a decisive contribution to the success of drug discovery, it is also true that computer-based drug design has to face novel challenges posed by the continued progress in our knowledge of the factors that modulate the pharmacological action of drugs. Therefore, computer-assisted drug design must be viewed as an evolving field, with an urgent need to develop novel computational approaches able to integrate the vast amount of complex information currently available for small (bio)organic compounds, biologically relevant targets and their complexes, but also to account accurately for the thermodynamics and kinetics of protein–ligand association, the intrinsic dynamical behavior of biomolecular systems and the complexity of protein–protein networks.

This book is intended to give an overview of the recent advances in the field of computer-assisted drug design. To this end, the book covers a wide range of topics relevant to the development of drugs, thus providing a rather comprehensive description of the major methodological strategies available for drug discovery and the development of new computational tools designed to tackle the complexity of drug activity. There are two main features that distinguish this book from previous titles in the field. First, the contents are oriented to provide a physico-chemical basis of the methodological tools, as well as clear guidelines to calibrate the performance of the current methodologies used in drug design. Second, standard formalisms widely accepted in the pharmaceutical and medicinal chemistry arena are presented in conjunction with an updated review of the latest advances experienced in the field. Overall, this book should encompass an updated view of the current challenges faced by computational tools for both academic and industrial researchers and even graduate students working in modeling of biomolecular systems and drug discovery.

With this aim, the book is organized in four sections: (i) physico-chemical basis of drug activity, (ii) computational strategies in drug design, (iii) exploring diversity of drug target sites and (iv) case studies.

Physico-Chemical Basis of Drug Activity

The activity of a drug is intimately related to the interaction with the biomolecular target. The binding affinity between a drug and its target is therefore a crucial property that emerges from a delicate balance of different factors, including the intermolecular interaction between drug and target residues, (de)solvation, conformational rearrangements and entropic changes. The reliability of computational methods to predict the protein–ligand binding affinity is revisited in Chapter 1 by Llabrés, Juárez-Jiménez, Forti, Pouplana and Luque. After a brief introduction to the different contributions to the binding free energy, the authors discuss the use of classical force fields coupled to statistical mechanics methods, which provide formally rigorous approaches for the calculation of binding affinities by means of free energy calculations. Then, attention is paid to the limitations of the pairwise force field widely adopted for biomolecular simulations, including a discussion of potential improvements for ameliorating the classical description of intermolecular forces. Finally, a review of efforts made in the last years to exploit quantum mechanical-based formalisms proposed as an alternative strategy is also presented. The potential impact of these approaches is examined in light of selected studies reported recently in the literature.

A detailed analysis of the thermodynamics of ligand binding is provided in Chapter 2 by Ferenczy and Keserű. The protein–ligand affinity is determined by the free energy of binding, and its enthalpic and entropic components constitute a thermodynamic signature to extract useful information on the details of protein–ligand interaction. Knowledge of the thermodynamic signature of a compound can be exploited to optimize the binding affinity, advantageously modulating the physico-chemical properties of the compound. In this context, the authors review the basic concepts of drug binding, the enthalpic and entropic contributions and their relation to structural factors, including aspects such as the enthalpy–entropy compensation, the role of hydrophobicity, and the relation between ligand size and binding thermo-dynamics. Attention is also paid to experimental measurements of thermo-dynamic quantities, as well as to the theoretical calculation of the binding free energy and its components. With this background, the authors then discuss the impact of binding thermodynamics on medicinal chemistry optimizations, leading to distinct strategies that can be classified as enthalpy-driven or entropy-driven processes. Evaluation of binding thermodynamics at decision points such as hit prioritization, lead selection and candidate identification can then be assisted by a proper analysis of the thermodynamic signature, leading to different ligand efficiency metrics. The final part of the chapter discusses several applications where decision making was supported by thermodynamic considerations.

Solvation plays a crucial role in the intermolecular association in biological systems, and particularly in the binding of drugs to their receptors. Accounting for hydration changes during binding is therefore essential to estimate the thermodynamics of drug binding, which requires carefully calibrated solvation models. This is the main issue addressed by Sulea and Purisima in Chapter 3. In particular, the authors review the basic physico-chemical assumptions that underlie continuum solvation models, which have found a widespread application in drug discovery in recent years. To this end, the authors discuss the main features of the formalisms adopted for the evaluation of the electrostatic and non-electrostatic components of the solvation free energy, as usually adopted in continuum models. The discussion also highlights the limitations of a continuum solvation, mainly due to the lack of structural details in the first hydration shell. Finally, the chapter discusses current trends considered in the next generation of solvation models, which combine the computational efficiency of the continuum description with specific features aimed at incorporating first-shell effects in order to ensure successful large-scale applications in drug discovery.

Chapter 4 turns its attention to the bioavailability of drugs. Oral administration is the most widely used method for drug delivery into the systemic circulation. However, predicting the fraction of an administered dose of drug able to reach systemic circulation is still one of the most challenging problems. The difficulty in predicting the bioavailability of drugs stems from the balance between physico-chemical properties of compounds and physiological factors of the patient. Low bioavailability may cause lack of activity and increases the risk of high variability of response among patients. On the other hand, detection of poor bioavailability in clinical stages has an enormous economical impact. In this context, Muñoz-Muriedas reviews the development of computational tools for predicting bioavailability at early stages of drug discovery. Properties like permeability, solubility and metabolic stability can be considered the main players in controlling oral absorption. By using the data collected from studies carried out with internal datasets at GlaxoSmithKline, the author reviews the impact of these properties in the bioavailability of drugs, making particular emphasis on the nature of *in vitro* high-throughput assays for estimating these properties and to the use of *in silico* approaches based on simple physico-chemical properties. Particular attention is paid to the identification of rules-of-thumb that might be used as valuable guidelines in early stages of a drug design project.

Computational Strategies in Drug Design

This section is intended to provide a comprehensive review of current computational tools used in drug discovery, as well as of the challenges that drive their evolution toward a more accurate description of the relationships between chemical structure and pharmacological activity.

The first two chapters of this section deal with relevant aspects of ligand-based drug design. In Chapter 5, Spyrakis, Cozzini and Kellog review the molecular descriptors used in structure–activity relationships (SAR), paying particular attention to the relevance of hydrophobic properties of drugs. The chapter begins with a brief historical review of quantitative structure–activity relationship (QSAR) methods. Emphasis is placed on 3D-QSAR techniques, but attention is also paid to higher dimensionalities proposed to take into account issues such as uncertainties in ligand alignment and induced-fit effects.

Then, the main descriptors used in SAR studies are reviewed. At this point, the authors examine in more detail the information encoded in empirical hydrophobic descriptors, considering both their relationship with bioavailability properties and their contribution to protein–ligand interaction. In this latter context, the authors discuss the potential usage of hydrophobic descriptors in evaluating the *hydrophobic field* created by a molecule and its use in SAR studies. Specifically, the authors discuss the design of HINT (Hydropathic INTeractions), which is described as a "natural" non-covalent force field derived from experimental octanol/water partition coefficients that permits us to treat hydrophobicity as a 3D property. The extensive range of applications explored by HINT as an alternative approach to modeling interactions in biological systems is illustrated by selected examples.

Gillet reviews in Chapter 6 the application of pharmacophore models in drug design. Pharmacophores have traditionally been used when the 3D structure of the receptor is unknown, although more recently the use of receptor-based pharmacophore models has received a renewed interest. In this framework, the chapter reviews the assumptions underlying the definition of a pharmacophore model, leading to pharmacophoric features for active molecules that are assumed to bind in a similar way to a common receptor. Attention is paid to the proper alignment of ligands while searching for mapping between those features, as well as to the use of scoring functions in order to discern the feasibility of the pharmacophoric patterns that can arise from the multiple alignments. The application of pharmacophores in virtual screening, with the potential of retrieving compounds that belong to different chemical series (*i.e.*, scaffold hoping), is also examined. Finally, the author discusses the use of pharmacophores derived from protein binding sites, describing how the identification of potential interaction sites permits us to extract features in receptor-based pharmacophores models, which can be used as filtering techniques in hierarchical database screening protocols. The impact of pharmacophores in drug design is illustrated by the discussion of a representative selection of studies reported in the literature.

The current status of protein–ligand docking is examined in the next three chapters, which pay attention to relevant challenges in the field. A general review about docking methods and practical considerations for ligand–target docking is made by Morris in Chapter 7. The chapter starts by describing the theory that underlies docking techniques, and provides an overview about the methodologies of widely used docking tools. Emphasis is made in selected topics relevant for the success of protein–ligand docking. They include a review of search strategies adopted for the exploration of the multidimensional space for the ligand–target complex. The influence of target validation and preparation on the success of docking studies is also discussed, providing useful guidelines to validate a docking method. The author also pays attention to the selection and preparation of ligands to be used in docking screenings, and the evaluation of docking results. Finally, the chapter discusses the screening of compound libraries for their similarity to known biologically active molecules or their likelihood to bind to a target. Emphasis is made on the curation of datasets of active and inactive molecules and the effectiveness of virtual screening in identifying potential hits.

The challenging question of describing the free energy of binding in docking calculations is reviewed by Cavasotto in Chapter 8. After a brief discussion of the different (enthalpic and entropic) components of the binding free energy, the chapter discusses the problems faced by current computational methods to provide a balanced description of their contribution to the binding affinity. The implementation of those contributions into scoring functions is then discussed, paying attention to their classification into three main classes: force-field based, empirical and knowledge-based scoring functions, as well as their advantages and weaknesses. This discussion is used to review several strategies considered in the literature for the post-processing of poses in high-throughput docking. In this context, the chapter makes a review of representative studies that have been used for re-scoring of the best hits using advanced simulation techniques, including MM/PB-SA or MM/GB-SA, linear interaction energy calculations, free energy calculations and finally quantum mechanical-based approaches. The studies discussed in the chapter permit us to calibrate the accuracy and limitations of the different methodologies for pose re-ranking at the post-docking stage.

In Chapter 9, Leis and Zacharias discuss the impact of target flexibility in predicting the binding mode of ligands to their targets, which is one of the challenges for the success of virtual screening. Formation of a protein–ligand complex is often accompanied by significant conformational changes, and a proper account of target plasticity is crucial for the success of docking studies. Target flexibility involves a variety of dynamical motions, such as rearrangements of side chains, changes in secondary structural elements, deformation in loops or even alterations in the relative positions of domains. The chapter reviews the computational strategies proposed in recent years for defining representative ensembles that account for target flexibility. These strategies include, for instance, the use of backbone-dependent or independent conformational preferences of side chains, the exploration of side-chain rearrangements through molecular dynamics simulations, the identification of relevant soft degrees of freedom that may allow for an approximate inclusion of target flexibility, or the generation of ensembles representative of larger conformational alterations by exploiting loop fragments or principal components of motion. Overall, the chapter provides a basis to identify promising efforts for the inclusion of target flexibility during screening of large databases of ligands, though it also raises open questions regarding the adequacy of scoring functions to account for the inclusion of conformational changes in docking.

Morreale and Gago make an extensive review of the COMparative BInding Energy (COMBINE) strategy for analyzing the relationships between biological activities and weighted pairwise interaction energies between ligand and target residues in Chapter 10. To this end the authors first discuss the formal differences and similarities between COMBINE and 3D-QSDAR methods, such as COMFA. The main difference involves the replacement of the molecular interaction fields in the space that surrounds the ligand by the pairwise ligand–residue interaction energies, often supplemented with the desolvation term of the ligand and the receptor, based on the 3D structural data of the ligand–receptor complex. On the other hand, the similarities between COMBINE and 3D-QSDAR methods mainly affect the use of common statistical tools for the analysis of the results. At this point, the chapter discusses the basic concepts behind the statistical analysis required for derivation of the model (pretreatment of input data, partial least-squares regression, cross-validation), as well as the criteria required for selection and validation of the best model. A brief discussion of the graphical user interface called gCOMBINE is also made. Besides reporting an extensive review of selected studies that illustrate the capabilities of the method, the chapter concludes with a discussion of recent extensions of COMBINE and future challenges faced by this technique.

Finally, Rocchia, Masetti and Cavalli discuss in Chapter 11 the use of enhanced sampling methods in drug design. These techniques represent a methodological strategy aimed at escaping from a complete Boltzmann sampling while retaining correct Boltzmann statistics. In the context of drug discovery, they constitute a powerful approach to explore the mechanistic details of ligand–receptor association as well as for predicting accurate binding affinities. After a brief introduction to the basic principles of statistical mechanics and their implementation for numerical simulations of biomolecular systems, attention is paid to the binding process between a drug and its target. This is examined both in terms of the free energy change between bound and unbound states, and of the reaction coordinate chosen to follow the transition between those states. In this context, enhanced sampling techniques are presented as tools well suited to drive the sampling of rare events in drug binding. In particular, the main features of different techniques (thermodynamic integration, free energy perturbation, umbrella sampling, steered molecular dynamics and metadynamics) are discussed. Finally, the importance and usefulness of these methods is supported by the discussion of several studies devoted to protein–ligand binding, where steered molecular dynamics or metadynamics have been used to explore the ligand–receptor interaction.

Exploring the Diversity of Drug Target Sites

The aim of this section is to discuss the novel frontiers that can be explored beyond the standard concept associated with drug binding pockets. In Chapter 12, Schmidtke, Álvarez-García, Seco and Barril revise the concept of the target space. Target-based drug discovery begins with the selection of a macromolecule, the biological activity of which is sought to be modulated by a drug. Since the choice of the best target largely determines the final outcome of long and costly drug discovery projects, it is therefore crucial to ensure that the selected target offers a probability of success in line with the potential reward. In this context, the authors first discuss the distribution of druggable pockets on the proteome, identifying areas of untapped opportunities. The concept of druggability is then revisited, making a description of the properties of drug binding sites, but also opening the scenario to under-exploited molecular mechanisms of action, which can take advantage of allosteric sites, conformational trapping and cavities located in protein–protein interfaces. Next, a description of both experimental and computational methods currently available to predict druggability is provided, with especial emphasis on the advantages and limitations of each approach. Finally, the chapter concludes with an outlook about the impact of druggability prediction methods in drug discovery, not only warning about undruggable targets, but also revealing potential pharmacological intervention points that would otherwise go unnoticed.

Grimme, González-Ruiz and Gohlke examine the strategies and challenges for targeting protein–protein interactions with small molecules in Chapter 13. This contribution discusses the characteristics of protein–protein interfaces, making emphasis on the differences with binding pockets found in enzymes and receptors. The relevance of detecting "hot spots" in protein–protein interfaces is also highlighted, as they define the functional epitope associated with the primary target for binding of small molecules. The authors also examine the challenge of predicting binding sites at protein–protein interfaces from unbound protein states, as the inherent plasticity and flexibility is crucial for a proper description of the clefts that accommodate small molecules. Attention is also paid to allosteric sites as alternative targets for modulating protein–protein interactions. The final part of the chapter reviews the challenges faced by docking to protein–protein interfaces, including the description of solvent effects and the treatment of protein flexibility. Finally, the integration of experimental information, such as mutagenesis studies and NMR-derived properties, is discussed as a relevant ingredient to supplement the scoring functions and to enhance the predictive success in docking calculations.

Case Studies

The last section of the book includes two chapters chosen as case studies about the application of specific computational strategies in drug discovery. In Chapter 14, Saladino and Gervasio bring up the limitations of free energy methods based on fully atomistic Monte Carlo and molecular dynamics simulations in drug discovery, as they are usually seen as computationally very demanding and too time consuming to be routinely applied to lead discovery and optimization phases. In this scenario, enhanced sampling techniques offer a promising strategy to gain insight into the mechanistic details of protein–ligand binding. In particular, the authors concentrate the discussion on metadynamics and its ability to estimate the free energy profile along the ligand binding coordinate. After a brief description of the basic formalism of metadynamics, including the choice of the most adequate collective variables, the technique is used to explore two relevant cases: cyclin-dependent kinase 2 (CDK2) and cyclooxygenase enzymes.

Finally, in Chapter 15, Seneci, Frecer and Miertus exploit the concept of molecular diversity for the *in silico* screening of libraries. Computational methods are being increasingly used to assist the design of combinatorial libraries, in order to reduce the number of compounds that have to be synthesized, without significant decrease of the chemical diversity space coverage, as well as to increase the drug-like character of the molecules. The chapter first provides a review of the basic concepts that underlie molecular diversity and its implementation in ligand- and structure-based compound selection methods. Then, the authors illustrate the potential capabilities of these methods by discussing a selected set of successful applications of computer-assisted library design covering a broad range of therapeutic areas, including viral, bacterial and parasitic diseases.

As a final remark, we would like to thank all the contributors to this book for their fine and insightful articles that illustrate the current status and future challenges of computer-assisted drug design. We also wish to thank Gwen Jones and Juliet Binns from the Royal Society of Chemistry for their continued and invaluable assistance.

F. Javier Luque and Xavier Barril

References

1. H. O. Wild, D. Heimback and C. Huwe, *Angew. Chem., Int. Ed.*, 2011, **50**, 7452.
2. R. F. Service, *Science*, 2004, **303**, 1796.
3. W. L. Jorgensen, *Science*, 2004, **303**, 1813.
4. W. L. Jorgensen, *Acc. Chem. Res.*, 2009, **42**, 724.

Contents

RSC Drug Discovery Series No. 23
Physico-Chemical and Computational Approaches to Drug Discovery
Edited by F. Javier Luque and Xavier Barril
© The Royal Society of Chemistry 2012
Published by the Royal Society of Chemistry, www.rsc.org

Chapter 12   Expanding the Target Space: Druggability Assessments   302
Peter Schmidtke, Daniel Alvarez-Garcia, Jesus Seco and Xavier Barril

CHAPTER 1

Recognition of Ligands by Macromolecular Targets

SALOMÉ LLABRÉS, JORDI JUÁREZ, FLAVIO FORTI, RAMÓN POUPLANA* AND F. JAVIER LUQUE*

Departament de Fisicoquímica and Institut de Biomedicina (IBUB), Facultat de Farmàcia, Universitat de Barcelona, Av. Diagonal 643, E-08028 Barcelona, Spain
*E-mail: fjluque@ub.edu; rpouplana@ub.edu

1.1 Physical Basis of Ligand–Protein Binding

Molecular recognition and binding is essential in mediating a variety of processes and functions in the cell. Enzyme catalysis, receptor signalling, storage of small molecules, transport through membranes and immunological response are examples that illustrate the relevance of recognition and binding between biomolecules. The rules of physics provide the basic principles to understand molecular association, and the affinity between interacting partners can be related to macroscopic observables through the laws of thermodynamics.

Under thermodynamic equilibrium conditions, the noncovalent, reversible binding of a small molecule (*ligand*; L) to a given protein (*target*; R) is determined by the standard Gibbs free energy (ΔG^o; Eq. 1.1), which is composed of an enthalpic (ΔH^o) and an entropic ($-T\Delta S^o$) term.

$$R_{aq} + L_{aq} \Leftrightarrow R'L'_{aq}$$
$$\Delta G^o = \Delta H^o - T\Delta S^o$$

(1.1)

RSC Drug Discovery Series No. 23
Physico-Chemical and Computational Approaches to Drug Discovery
Edited by F. Javier Luque and Xavier Barril
© The Royal Society of Chemistry 2012
Published by the Royal Society of Chemistry, www.rsc.org

The binding affinity can be expressed either in terms of the standard free energy difference between bound and unbound states, or alternatively in terms of the equilibrium constant (K) for the formation of the complex between the interacting partners (Eq. 1.2).

$$\Delta G^o = -RT\ln K \qquad\qquad\qquad (1.2)$$

where R is the gas constant and T is the temperature.

The binding affinity between a ligand and its macromolecular target reflects a subtle balance between enthalpic and entropic contributions,[1] which are generally interpreted by decomposing the protein–ligand binding into a number of separate contributions. The structural (shape and size) and chemical (nature and spatial distribution of functional groups) complementarity between the ligand and the residues that are present in the binding pocket modulates the binding affinity through a variety of intermolecular interactions (Figure 1.1).[2,3] They include electrostatic interactions between the permanent charge distribution of the molecules, the induction of changes in the charge distribution due to the interaction between partners, the stabilizing contribution arising from dispersion forces, and the repulsion between electron clouds at close distances. These energy contributions contribute to typical interactions such as salt bridges, standard hydrogen bonds (where a hydrogen bond bound to an electronegative atom X forms an attractive interaction with another electronegative atom Y: X–H$\cdots$Y) and van der Waals forces. However, there has been an enrichment in the number and nature of intermolecular interactions, including interactions such as cation–π or anion–π complexes,[4–7] non-standard hydrogen bonds (C–H$\cdots$X, C–H$\cdots$$\pi$, blue-shifting hydrogen bonds)[8,9] and halogen bonding.[10,11]

The net stabilizing energy due to ligand–protein interactions compensates for unfavorable contributions to the binding. Thus, molecular association implies dehydration of the complementary surfaces of both ligand and target and reorganization of water molecules around the ligand–protein complex. Therefore, the energy gain due to the seemingly favorable interactions formed in the complex must counterbalance the cost due to breaking interactions of the separate partners with hydrating waters.[13,14] For simple neutral organic compounds the hydration free energies generally lie in a relatively narrow range (for instance, the experimental values for the transfer of ethane and acetamide from the gas phase to water amount to +1.8 and −9.7 kcal mol^{-1}, respectively).[15] However, the hydration free energy of charged compounds is much larger, which reflects the strengthening of the interactions with water molecules (as an example, hydration free energies of −73 and −77 kcal mol^{-1} have been determined for the ethylammonium cation and acetate anion).[16] Accordingly, there must be a sizable compensation between the energy cost of dehydrating both ligand and binding site residues and the energy gain due to the protein–ligand interactions upon burial of the ligand in the binding pocket.

Protein–ligand binding is often accompanied by conformational changes in the interacting partners. Beyond the rigid "lock-and-key" model, binding

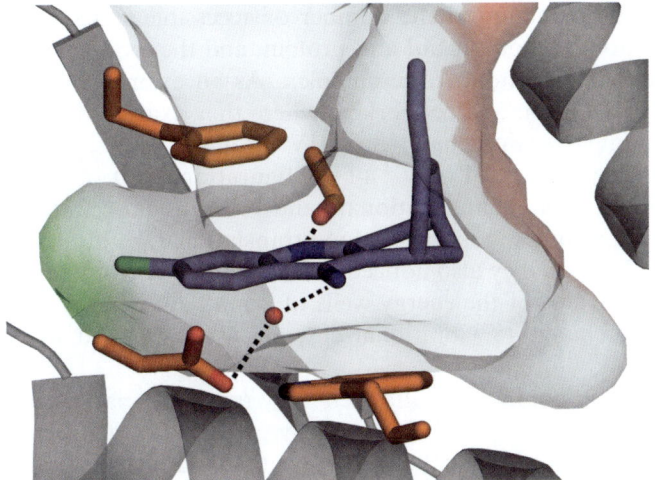

Figure 1.1 Representation of the main interactions for the binding of huprine X to the catalytic binding site of acetylcholinesterase (AChE; taken from the PDB structure 1E66).[12] The gray isocontour represents the accessible volume of the binding pocket and reveals a large degree of complementarity in shape and size with the ligand. Huprine X (shown in blue sticks) forms a variety of interactions with residues in the binding pocket (in orange sticks; for the sake of clarity, only specific fragments are shown). The nitrogen atoms of the protonated aminoacridine ring of huprine X forms a direct hydrogen bond with the carbonyl group of His440 and a water-mediated interaction with the negatively charged Asp72. In addition, there is stacking with the benzene ring of Phe330 and Trp84. Owing to the positive charge of huprine X, these interactions are reinforced by the stabilizing effect of the cation–π interaction. The chlorine atom fills a hydrophobic pocket formed by Met435, Ile439 and Trp432 (not shown). Finally, the ethyl group fills a cavity formed by Tyr121, Phe290, Phe330 and Phe331 (not shown). The tight structural and chemical complementarity explains the potent inhibition, which is reflected in an inhibition constant, K_i, of 26 pM.

events include a broader range of potential scenarios, such as the popular "induced fit" mechanism, the alternative "conformational selection" process, or even more complex models that combine the selection of specific conformations with the induction of structural readjustments by the binding partner.[17–19] However, even predicting the energy cost associated with conformational changes in the ligand has proved to be very challenging, as noted by the uncertainties associated with the choice of the level of theory used to determine the cost of selecting the bioactive conformation.[20–22] As an example, we simply quote that Tirado-Rives and Jorgensen concluded that the uncertainty in determining the conformer-focusing penalty can be anticipated conservatively to be in the 5–10 kcal mol^{-1} range.[21]

Finally, one must take into account the entropy changes, which include the loss of translational and rotational degrees of freedom upon molecular association, the reduction in the number of accessible states associated with internal rotations of both ligand and protein, and the reorganization of water molecules upon formation of the complex. As an example, we quote here a recent study by Gilson and co-workers where they examined the entropy loss for the binding of amprenavir to HIV protease.[23] They estimated that amprenavir loses 26.4 kcal mol^{-1} of configurational entropy upon binding, including both the conformational and vibrational contributions and accounting for changes in mobility along translational, rotational and internal coordinates. Finally, they also noticed that the loss of entropy results primarily from the narrowness of the energy wells of bound amprenavir relative to free ligand, as the change in vibrational entropy was estimated to be 24.6 kcal mol^{-1}.

Ligand–protein binding affinities generally fall into a narrow range varying between 10^{-2} and 10^{-12} M.[1] Remarkably, at 298 K an uncertainty in the binding free energy of 1.36 kcal mol^{-1} alters the binding constant by one order of magnitude, which highlights the need to estimate accurately the binding affinity. Nevertheless, the difficulty in predicting the binding affinity stems from the fact that the relatively narrow range of binding free energies is the result of compensation between generally large enthalpic and entropic terms,[24,25] so that small changes in the binding free energy can mask sizable and mutually compensating changes in both enthalpy and entropy. Since the enthalpic and entropic components comprise useful information on the details of protein–ligand interaction, monitoring binding thermodynamics and discriminating between enthalpy-driven and entropy-driven optimizations can be relevant for the success of drug discovery programs[26–29] (see also Chapter 2 for a detailed review). Therefore, not only the binding free energy but also the thermodynamic signature encoded by its enthalpic and entropic components are valuable for guiding lead discovery and optimization.

Predicting accurately the binding free energy is a formidable challenge to current computational methods, due to the large magnitude of the separate contributions to the binding free energy and the compensation between enthalpic and entropic terms. However, this is a fundamental ingredient for the success of drug discovery, especially when one realizes that the maximal free energy contribution per non-hydrogen atom in a drug-like ligand amounts to *ca.* −1.5 kcal mol^{-1}.[30] Therefore, maximizing the structural and chemical complementarity between a ligand and its target, and enhancing the synergistic cooperativity between ligand–protein interactions, should be effective guidelines for the success of drug discovery.

1.2 Prediction of Binding Affinities: Free Energy Calculations

A qualitative understanding of the physico-chemical features that contribute to drug binding is valuable for the analysis of structure–activity relationships encoded in pharmacophoric models, as the pharmacophore represents the ensemble of steric and electronic features necessary to ensure the optimal supramolecular interaction with a specific biological target.[31,32] However, a quantitatively accurate estimate of the binding affinity is required in other instances, such as lead optimization. The use of classical simulations in conjunction with free energy calculations have proved to be very valuable for predicting relative binding affinities arising from small chemical differences between structurally related compounds.[33–37] The most popular methods are free energy perturbation and thermodynamic integration.

In a general context, the free energy difference between systems A and B, which might represent the two ligands that bind to a common target or a mutation of a specific residue in the binding site of the target, can be expressed as indicated in eqn (1.3):

$$\Delta G = -RT \ln \langle e^{-\Delta H/RT} \rangle_A \qquad (1.3)$$

where $\Delta H = H_B - H_A$ and $\langle \ \rangle_A$ stands for the ensemble average over a system described by Hamiltonian H_A.

When systems A and B differ in a significant way, the practical solution of eqn (1.3) requires the decomposition of the alchemical transformation A→B into a number of successive steps, which is accomplished by defining a coupling parameter (λ) that controls the smooth change between initial and final states. Thus, at each intermediate step, one can define the Hamiltonian $H(\lambda)$ as indicated in eqn (1.4):

$$H(\lambda) = \lambda H_B + (1 - \lambda) H_A \qquad (1.4)$$

In free energy perturbation calculations, the free energy change for the transformation of ligand L_1 into L_2, either free in solution or in the protein–ligand complex, can be determined by the addition of the free energy changes for each of the distinct windows leading from the initial ($\lambda = 0$) to the final ($\lambda = 1$) states, as noted in eqn (1.5):

$$\Delta G = -\sum_{\lambda} RT \ln \langle e^{-\Delta H_\lambda /RT} \rangle_\lambda \qquad (1.5)$$

where $\Delta H_\lambda = H_{\lambda + \Delta \lambda} - H_\lambda$.

Thermodynamic integration provides an alternative solution, where the change in free energy between the initial and final states can be determined as indicated in eqn (1.6), where one has to evaluate the ensemble average of the derivative of the Hamiltonian with respect to the coupling parameter λ:

$$\Delta G = \int_0^1 \langle \frac{\partial H}{\partial \lambda} \rangle_\lambda \, d\lambda \qquad (1.6)$$

These techniques rely on a rigorous formalism that permits estimation of the absolute and relative binding affinities. Computation of the absolute binding affinity can be achieved by using the double decoupling methodology,[38,39] where the interaction of the ligand with the molecular environment (*i.e.*, water molecules in the free state, and the hydrated complex in the bound state) is turned on/off in different stages. To this end, the ligand is converted from a fully interacting state in the bound complex to an ideally constrained state, where interactions with the environment are turned off but the ligand is constrained to stay in the vicinity of the protein. This is convenient in order to avoid the ligand having to explore the entire simulation box, which would be extremely demanding to achieve convergence in the computed free energy change. When the ligand is fully decoupled, it is a molecule of ideal gas still constrained to occupy a given region, and release of the restraining potential (affording the ligand to occupy the whole volume and to rotate freely) provides a correction term to the free energy. Finally, the binding free energy is determined upon addition of the free energy required for removing the ligand from the bulk solvent to the gas phase.

Prediction of relative binding free energies have larger practical interest, as the relative binding affinity between two ligands can be related to specific chemical modifications introduced in a drug candidate during lead optimization. As shown in Figure 1.2, this can be determined by alchemical mutations that convert the two ligands (L_1 and L_2) in the unbound and bound states, allowing for an extensive sampling of the protein–ligand complex (and the free ligand in solution) in a realistic environment.

If evaluated accurately, the free energy change should be independent of path and simulation protocol. Nevertheless, a number of practical considerations must be taken into account.[40,41] First, the use of the thermodynamic cycle shown in Figure 1.2 is convenient because (i) it permits the simulation of non-chemical processes in order to calculate the relative binding affinity between ligands L_1 and L_2, avoiding the difficulty to carry out in a reversible way the direct association between each ligand with the target, and (ii) the comparison of the free energy changes for the alchemical transformation between ligands in solution and in the complex might benefit from the cancellation of errors in the simulations in unbound and bound states. However, the accuracy of the free energy changes is limited by different factors, such as the assumption of the simulated systems to be in equilibrium, or the need to include all relevant configurations in the ensemble. Even in the absence of large conformational changes that mediate ligand binding, which would require enhanced sampling techniques (see Chapter 11 for a review),[42] the accuracy of the results is mainly limited by the quality of the force fields, which generally rely on a pairwise description of intermolecular interactions.

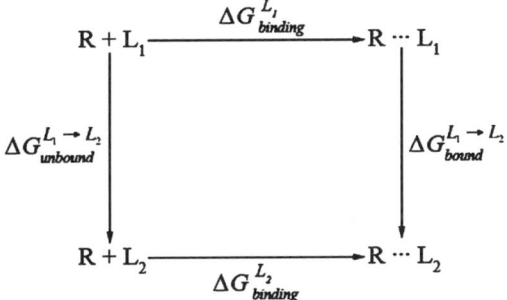

$$\Delta\Delta G = \Delta G_{binding}^{L_2} - \Delta G_{binding}^{L_1} = \Delta G_{bound}^{L_1 \rightarrow L_2} - \Delta G_{unbound}^{L_1 \rightarrow L_2}$$

Figure 1.2 Thermodynamic cycle used for the computation of relative binding affinities between two ligands to a common target. The relative free energy difference given by the experimentally available free energy changes for the binding of ligands L_1 and L_2 to the receptor R (horizontal processes) can be determined by the alchemical transformations between the ligands in the bound and unbound states (vertical processes).

1.3 Quantum Mechanical-Guided Refinements in Interaction Energy Potentials

Classical force fields only represent approximately the intermolecular interactions that mediate the recognition between ligands and proteins. Although the use of simplified expressions is understandable in terms of providing an efficient sampling, as well as of facilitating the parametrization of the large number of functional groups that can be incorporated in drug-like molecules, there is concern regarding the suitability of current biomolecular force fields to provide accurate estimates of free energy differences. As noted by Michel and Essex,[43] it seems reasonable to expect that free energy calculations cannot predict binding free energies more accurately than they can predict solvation free energies, where the uncertainties obtained for small organic compounds can be estimated to be around 1 kcal mol^{-1}. Larger deviations can be expected for more complex, flexible polyfunctional drug-like molecules.

Current efforts are being conducted toward the improvement of biomolecular force fields. A significant example of this effort is the development of the AMOEBA polarizable force field,[44] which includes a number of differences with regard to "standard" biomolecular potentials. Thus, this force field implements elaborate expressions for intramolecular energy terms that account for the inclusion of anharmonicity effects, the coupling between stretching and bending and the decomposition of angle bending into in-plane and out-of-plane components, whereas non-bonded terms include a buffered 14-7 functional for van der Waals interactions, a multipolar expansion of the permanent electrostatic distribution at each atomic center (up to quadrupole moments) and the treatment of polarization effects *via* inducible point dipole moments. Validation of the force field also involves a careful calibration against quantum mechanical (QM) data mainly derived at the second-order Møller–Plesset (MP2) level using Dunning augmented correlation consistent basis sets, as well as checking a number of molecular systems and properties.

The impact of high-quality *ab initio* QM data on the refinement of force fields is also exemplified by the quantum mechanical polarizable force field (QMPFF),[45] which was devised in order to enhance the physical basis of the energy components that were fitted to their QM counterparts. In the original publication the internal geometry was taken as rigid, but later refinements also included bonded contributions.[46] Non-bonded interactions, which are decomposed in electrostatic, exchange, induction and dispersion components, contribute to the potential energy. In QMPFF the charge distribution of an atom is treated as a core atomic point charge and a diffuse electron density, which is represented by an isotropic, exponential distribution centered on the core for an isolated atom. An exponential expression is used to reproduce the decay of exchange repulsion with the distance, whereas induction is simulated by means of a floating electron cloud subject to a nonharmonic restraint potential that constrains the cloud to remain close to a reference position, and the dispersion term follows a simple buffered r^{-6} dependence. Finally, parametrization of the energy components was performed without resorting to any experimental data, mainly using QM data derived from MP2/aTZ(-hp) calculations. Remarkably, though the force field was parametrized against gas phase QM data, it been shown to reproduce molecular properties in simulations of condensed phases.[47,48]

As a final example of the effort carried out for refining the interaction energy potential, we quote here the SIBFA procedure.[49,50] An essential feature of the SIBFA procedure is the ability to separate the interaction energy into five distinct contributions: multipolar, short-range repulsion, polarization, charge transfer and dispersion. The first term includes multipoles (up to quadrupoles) that are distributed on the atoms and bond barycenters, supplemented by an additional term that accounts for penetration effects. Short-range repulsion accounts for bond–bond, bond–lone pair and lone pair–lone pair interactions, polarization is modeled *via* induced dipoles subject to a Gaussian screening of the field created by multipoles, and charge transfer is accounted by combining the ionization potential of the electron donor and the electron affinity and "self-potential" of the electron acceptor. Finally, dispersion and exchange-dispersion components are computed as a sum of r^{-6}, r^{-8} and r^{-10} terms. QM energy decomposition schemes have been used to calibrate the different components of the SIBFA potential, which has been shown to yield accurate results for complex biomolecular systems such as metalloenzymes (for a recent review, see ref. 51).

The preceding examples reflect the interest in alleviating the deficiencies of pairwise force fields, such as an oversimplified treatment of bonded interactions or the approximation of charge distributions by point charges with consequent neglect of charge penetration effects. However, possibly the most serious defect of additive force fields is the failure to incorporate electronic polarization. In particular, since they treat implicitly the average many-body polarization effect, their reliability to model molecular systems in different environments (*e.g.* ligands in solvent-exposed regions or inner hydrophobic pockets of proteins, or the transport of molecules through membranes) is questionable. This justifies the intense effort spent by several research groups in developing efficient formalisms for the description of induction effects in biomolecular force fields, such as AMBER,[52,53] CHARMM,[54–56] OPLS[57] and GROMOS.[58]

Inclusion of polarization effects in empirical force fields is linked to a number of challenging questions, such as the choice of the physical model adopted to account for induction effects and its mathematical implementation.[59–61] Three main strategies have been adopted to account explicitly for polarization in classical force fields: fluctuating charge, induced dipoles and Drude oscillators. In the fluctuating charge method, the partial charges located on atomic sites can flow among the atoms in response to the electric field created by the environment. This is achieved by coupling the charges to the molecular environment based on electronegativity equalization, whereby charges flow between atoms until the instantaneous atomic electronegativities are equated. The induced dipole model assigns point inducible dipoles to a set of sites distributed across the molecular system. The induced dipole moment at a given site is proportional by virtue of the linear response approximation to the total local electric field, and the proportionality constant is the polarizability tensor. Finally, in the Drude oscillator approach (also known as core-shell model), polarization of an atomic site is accounted for by attaching a mobile massless particle carrying a given charge to the polarizable atom (whose net charge must be preserved) through a harmonic spring. The electrically neutral atom–Drude particle pair form the Drude oscillator, which can be polarized by an external field, so that the fluctuations in the electric field lead to oscillations of the Drude particle around the atomic site.

The implementation of explicit polarization schemes raises the question about the derivation of the parameters involved in the response model. The partitioning scheme put forward by Stone,[62] which relies on the susceptibility function of the charge density, provides distributed polarizabilities from QM calculations of the response of an isolated molecule to an external perturbation. In a different strategy, Applequist derived a heuristic approach to derive atomic polarizabilities by minimizing the deviation between calculated and experimental molecular polarizabilities.[63] Alternative schemes rely upon atomic hybrid, bond or group polarizabilities.[64] Finally, other strategies have fitted molecular polarizabilities or grids of induction energies.[65–67]

Finally, the implementation of induction schemes makes necessary the re-adjustment of the distinct energy components required to maintain the proper contribution to the total energy of the system. For the application to biomolecular simulations, it is desirable to keep the force field as simple as possible to satisfy the criteria of tractability and computational efficiency. Nevertheless, a balanced representation of the energy (electrostatic, induction and van der Waals) components is required to retain the underlying physics of intermolecular interactions. These conditions can be satisfied by calibration against the corresponding terms derived from QM energy decomposition schemes.[68-72] As an example, we limit ourselves to the Symmetry-Adapted Perturbation Theory (SAPT) approach,[73] which relies on the symmetrized Rayleigh–Schrödinger theory. Within the SAPT2 computational scheme, the interaction energy, U_{tot}^{SAPT}, which compares with the BSSE-corrected interaction energy determined at the MP2 level, is decomposed into a set of contributions with a clear physical meaning, as noted in eqn (1.7). SAPT has been used to analyze the energy components in a variety of complexes that involve different intermolecular interactions,[74,75] such as stacking and hydrogen bond complexes,[76,77] cation–π interactions[78,79] and halogen bonding.[80]

$$U_{tot}^{SAPT} = U_{ele} + U_{ind} + U_{exc} + U_{dis} + U_{exc-ind} + U_{exc-dis} + \Delta HF \qquad (1.7)$$

where the terms on the right-hand side stand for the electrostatic (U_{ele}), induction (U_{ind}), exchange (U_{exc}) and dispersion (U_{dis}) components, the coupling between exchange and induction ($U_{exc-ind}$) and between exchange and dispersion ($U_{exc-dis}$), and finally a collection of higher-order induction and exchange-induction terms ($_{\delta HF}$).

Some applications of such polarizable force fields to ligand binding studies have been reported in the literature. For instance, the AMOEBA force field has been used to compute the absolute binding free energy between trypsin and benzamidine by decoupling electrostatic and van der Waals interactions between ligand and protein, leading to a binding free energy of −6.7 kcal mol^{-1}, which compares with experimental values ranging from −6.3 to −7.3 kcal mol^{-1}.[81] By doing additional calculations "tuning off" the dipole induction term, the authors concluded that explicit treatment of polarization is critical to achieve chemical accuracy in predicting the binding affinity of charged systems. Moreover, calculation of relative binding affinities for a series five benzamidine-like ligands yielded a root-mean-square error for the computed binding free energy of 0.4 kcal mol^{-1}, the largest error being 0.7 kcal mol^{-1}.[82] The latest version of QMPFF (QMPFF3) has been used to determine the affinities of a set of five related ligands to three serine proteases: trypsin, thrombin and urokinase-type plasminogen activator.[83] Although the protein binding sites are similar and relatively rigid, the binding affinities are diverse. The calculated results were found to be in quantitative agreement with experimental data, as noted in a rmsd of 1.0 kcal mol^{-1} and a correlation between calculated and experimental data of 0.90, which were substantially better compared to the results obtained with MMFF force field (rmsd of 3.2 kcal mol^{-1}). Likewise, Kaminski *et al.* have recently used the polarizable force field (PFF)[84] to simulate the interaction between the X-linked inhibitor of apoptosis (XIAP), which is known to inhibit caspase proteins, with caspase and several XIAP–antagonist complexes, and the results were compared to those obtained with the OPLS-AA force field.[85] The authors concluded that the polarizable force field employed is not only adequate in simulating protein–ligand complexes in solutions, but also provides a prediction of a stronger success of the antagonist to the caspase–XIAP interactions.

Although the preceding examples illustrate the relevance of the explicit treatment of polarization in predicting ligand binding affinities, it is clear that the use of a polarizable force field demands more extensive testing, including diverse sets of protein–ligand complexes. On the other hand, a more complex and delicate parametrization of ligands is also expected due to the chemical diversity of functional groups encountered in drug-like compounds.

1.4 Quantum Mechanical Methods in Ligand–Protein Interactions

QM methods are nowadays a fundamental ingredient not only in the parametrization of force fields, but also in the development of more elaborate formalisms for the calculation of the interaction energy potential. The continued increase in accuracy achieved by QM methods has also stimulated the implementation and usage of QM-based techniques for different applications in the study of ligand–protein complexes. Most of these applications follow the hybrid quantum mechanical/molecular mechanical (QM/MM) computational scheme,[85–90] where the hamiltonian of the whole system can be defined as the sum of three terms (eqn 1.8) corresponding to the QM subsystem ($\hat{H}_{QM}$), the MM subsystem ($\hat{H}_{MM}$) and the coupling between the QM and MM regions ($\hat{H}_{QM/MM}$):

$$\hat{H} = \hat{H}_{QM} + \hat{H}_{MM} + \hat{H}_{QM/MM} \tag{1.8}$$

where the coupling term includes both electrostatic (eqn 1.9) and non-electrostatic (eqn 1.10) terms:

$$\hat{H}^{ele}_{QM/MM} = \sum_s \sum_x \frac{Z_s Q_x}{|r_s - r_x|} - \sum_n \sum_x \frac{Q_x}{|r - r_x|} \tag{1.9}$$

$$\hat{H}^{vW}_{QM/MM} = \sum_s \sum_x \left[\left(\frac{A_{sx}}{r_{sx}^{12}} \right) - \left(\frac{B_{sx}}{r_{sx}^6} \right) \right] \tag{1.10}$$

where s and x denote the interaction sites in the QM and MM subsystems, n runs over the number of electrons in the QM region, Z_s is the nuclear charge of a QM site, Q_x stands for the charge distribution in the MM region, A_{sx} and B_{sx}

are the van der Waals parameters for repulsive and attractive components, and $r_{sx} = |r_s - r_x|$.

Although QM/MM methods have been mainly used for the study of chemical reactive processes (*i.e.*, enzyme catalysis), they have found a wider range of applications in drug discovery in recent years, including (i) the improvement of the predictive reliability of docking calculations, (ii) the accurate calculation of the ligand–protein interaction energy and the analysis of the energy components to rationalize ligand binding, and (iii) the development of computational schemes for predicting the binding affinity.

An example of the applications of QM/MM methods in assisting the docking of ligands is the study by Friesner and co-workers, where they explored the effect of including the polarization exerted by the protein environment in the atomic charges of the ligand on the predictive capability of docking calculations.[91] To this end, fixed charges of ligands were replaced by charges derived from QM/MM calculations treating the ligand as the QM subsystem in the protein (MM) environment. Tests performed for 40 co-crystallized structures showed that, following an iterative protocol, the algorithm was in most cases able to promote the selection of correct hydrogen-bonding patterns and to converge to a native-like structure, leading to a notable improvement compared with the results derived from docking calculations based on standard fixed charges. In addition, an extension of the QM/MM docking method to metalloproteins has been recently reported.[92]

In line with the preceding findings, Reynolds and co-workers have described a strategy for including ligand and protein polarization in docking calculations *via* QM/MM calculations.[93] This strategy is based on the conversion of induced dipoles to induced charges, which can be readily implemented in docking programs. Thus, the ligand is treated quantum mechanically and is polarized by the point charges of the target protein, and the induced dipole at a given target atom is then reformulated as an induced charge in an iterative manner. The final set of polarized charges was evaluated for 12 protein–ligand systems. Although inclusion of polarization does not always led to the lowest energy pose having a lower rmsd, it was found that the polarized charges resulted in an increased cluster size and a concomitant decrease in the rmsd of the docked pose with regard to the crystallographic solution.

In addition to the structural information, QM/MM methods have also been explored as a scoring tool (*i.e.*, QM scoring).[94] The key idea is to generate diverse poses in an initial stage using standard docking methods. QM/MM calculations were subsequently performed for the subset of best scored ligand poses, allowing for a limited geometry relaxation (up to five iterations in order to limit the computational cost). Then, the pose with the lowest QM/MM energy is chosen as the top scoring one. The QM scoring approach was better than force field-based methods in modeling the binding of ligands to primarily hydrophobic sites, where structural motifs with π–π interactions are often found.

The usage of QM/MM calculations as a tool to gain insight into and rationalize the factors that determine the binding of ligands is illustrated by the study reported by Gao, Höltje and co-workers, who used a combined QM/MM approach to determine electrostatic and polarization interactions in three high-affinity inhibitors (nelfinavir, mozenavir and tipranavir) of HIV-1 protease.[95] The results highlighted that polarization was particularly important, as it contributes as much as one-third of the total electrostatic interaction energy. In addition, it was found that the 4-hydroxydihydropyrone substructure of tipranavir was the most effective structural arrangement for enhancing polarization effects.

In a similar spirit, Dubey *et al.* have used QM/MM computations to assist the docking of six kinase inhibitors active for Dengue virus, and the results showed that polarization of ligands was also relevant in defining the binding of the inhibitors.[96] Raha *et al.* have taken advantage of the pairwise decomposition of the interaction energy to explore the binding of *N*-(4-sulfamylbenzoyl)benzylamine derivatives to human carbonic anhydrase II.[97] As a final example, the factors that contribute to the interaction energy of diketo acids such as L-731,988 and S-1369 to the HIV-1 integrase have been examined by Alves *et al.* by combining molecular dynamics and QM/MM techniques. Furthermore, an energy decomposition analysis was used to examine the contributions of individual residues to the enzyme–inhibitor interactions on average structures derived from the conformational sampling. According to this analysis, specific residues (Asn155, Lys156 and Lys159) and the Mg^{2+} cation are found to be crucial for the activity of these inhibitors.[98]

As noted above, QM methods have also been used to estimate the binding affinities of ligand–protein complexes. At this point, Balaz and co-workers proposed a four-step strategy for the study of ligand–metalloprotein complexes consisting of (1) docking of ligands, (2) optimization of the complex by QM/MM methods, (3) conformational sampling on the complex with constrained metal bonds by MD simulations and (4) a single-point QM/MM energy calculation for the time-averaged structure.[99] Then, the QM/MM interaction energy was combined with a desolvation term in order to determine the binding free energy ($\Delta G_{binding}$; eqn 1.11). As in linear interaction methods, the parameters α, γ and κ were determined upon correlation with experimental binding free energies for a set of 28 hydroxamate inhibitors binding to zinc-dependent matrix metalloproteinase 9. The two descriptors included in eqn (1.11) were found to explain 90% of variance in the inhibition constants.

$$\Delta G_{binding} = \alpha\Delta\langle E_{QM/MM}\rangle + \gamma\Delta\langle SASA\rangle + \kappa \tag{1.11}$$

where $\Delta\langle E_{QM/MM}\rangle$ denotes the QM/MM interaction between ligand and protein for time-averaged structures and $\Delta\langle SASA\rangle$ denotes the change in solvent-accessible surface upon complexation.

Another approach has been adopted by Das *et al.* in order to examine the effect of using protein-polarized QM charges in GBSA calculations for nine protease inhibitors.[100] Here the general expression of a GBSA model was adopted (eqn 1.12), but attention was paid to the use of charges on the ligand, which were described by assigning MM charges or the protein-polarized ones as derived from QM/MM calculations, as well as to the influence of bridging water molecules mediating hydrogen bonding with the inhibitors. The results showed that the binding free energies determined by using those polarized charges (and specific water molecules) showed higher correlation with antiviral IC_{50} data.

$$\Delta G_{binding} = \Delta E + \Delta G_{solv} + \Delta G_{SA} \qquad (1.12)$$

where ΔE stands for the difference between the minimized energies of the protein–ligand complex, the protein and the inhibitor, ΔG_{solv} denotes the change in solvation free energy between the interacting partners and ΔG_{SA} accounts for the free energy term related to the change in solvent exposure.

In addition to the hybrid QM/MM scheme, other QM-based computational approaches have been conceived for the prediction of protein–ligand binding affinities. An elaborate scheme was reported by Raha and Merz in 2005 with the aim to perform a large-scale validation of a QM-based scoring function for predicting the binding affinity of a diverse set of ligands.[101] To this end, the binding affinity ($\Delta G_{binding}$) was determined as given in eqn (1.13), where it is decomposed into a gas-phase interaction energy (ΔG_b^{gas}) and the change in solvation free energy ($\Delta \Delta G_{solv}$) of the complex (ΔG_{solv}^{PL}) relative to the protein (ΔG_{solv}^{P}) and ligand (ΔG_{solv}^{L}):

$$\Delta G_{binding} = \Delta G_b^{gas} + \Delta \Delta G_{solv} = \Delta G_b^{gas} + \Delta G_{solv}^{PL} - \Delta G_{solv}^{P} - \Delta G_{solv}^{L} \qquad (1.13)$$

The gas-phase interaction free energy was expressed as a sum of enthalpic (ΔH_b^{gas}) and entropic ($-T\Delta S_b^{gas}$) contributions (eqn 1.14):

$$\Delta G_b^{gas} = \Delta H_b^{gas} - T\Delta S_b^{gas} \qquad (1.14)$$

The gas-phase enthalpy was determined as a sum of electrostatic and nonpolar interaction energies. The former was calculated using the divide-and-conquer method and the semiempirical AM1 or PM3 hamiltonians, so that the heat of interaction (ΔH_I) between the protein and the ligand was determined as (eqn 1.15):

$$\Delta H_I = \Delta H_f^{PL} - \Delta H_f^{P} - \Delta H_f^{L} \qquad (1.15)$$

where ΔH_f^{PL}, ΔH_f^{P} and ΔH_f^{L} denote the heat of formation of the protein–ligand complex, protein and ligand, respectively.

The nonpolar enthalpic term was determined with the classical attractive component of the Lennard-Jones interaction potential (ΔLJ_6). With regard to the entropic term, it was expressed as the addition of conformational (ΔS_{conf})

and solvent (ΔS_{solv}) entropy components. The former was estimated by considering a conformational penalty of 1 kcal mol^{-1} for each rotatable bond of the ligand and in the protein side-chains frozen upon formation of the complex. The solvent entropy term accounts for the entropy gained by release of water molecules upon binding, and it was calculated from the buried surface area resulting upon complexation. Finally, the solvation free-energy term ($\Delta\Delta G_{\text{solv}}$) was determined using QM self-consistent reaction field calculations for the complex, ligand and protein. Overall, the preceding strategy leads to eqn (1.16), where the weights of the different components were adjusted by fitting to experimental binding free energies:

$$\Delta G_{\text{binding}} = \Delta H_{\text{I}} + \Delta LJ_6 + \Delta\Delta G_{\text{solv}} + \Delta S_{\text{solv}} + \Delta S_{\text{conf}} \tag{1.16}$$

The method was shown to be effective as a scoring function for predicting ligand poses docked to a protein target and for discriminating between native and decoy poses. It has also shown good performance for the study of ligands bound to metalloenzymes, where significant and variable metal–ligand charge transfer was found.[102]

Hobza and co-workers have reported a related QM-based scheme for computation of the binding free energy. It is based on the semiempirical QM PM6-DH2 method, which includes corrections to the PM6 hamiltonian for improving the treatment of dispersion energy and hydrogen bonds.[103] The total score is determined by adding the PM6-DH2 interaction enthalpy evaluated in a continuum water environment using the COSMO model. The desolvation of the ligand was further refined by means of SMD continuum calculations. In addition, the deformation contribution due to changes in protein and ligand upon binding was also considered, though using different computational methods: for the ligand, it was estimated with the PM6-DH2 method augmented with the solvation contribution determined with the SMD model, and for the protein the AMBER potential was used. Finally, an entropic term was also considered, it being determined based on the normal mode analysis as implemented in the AMBER force field. Implementation of the method involved two steps. First, docking of the ligands (using DOCK) and then rescoring based on the binding free energies determined with the PM6-DH2 computational scheme. The method was successful in ranking 22 ligands binding to HIV-1 protease. Another study that focused on the binding of 15 structurally diverse inhibitors to CDK2 revealed a good correlation between the bare interaction enthalpy, and even the interaction enthalpy corrected for ligand desolvation and deformation energies (explaining between 77 and 87% of variance in inhibition constants), but a worsening of the results upon addition of entropic corrections (as noted in a reduction of the variance to 52%).[104] Recently, the method has been extended to treat halogen bonding, it being denoted PM6-DH2X.[105]

Anisimov and Cavasotto have adopted another strategy to evaluate the binding affinity of phosphopeptide inhibitors of the Lck SH2 domain.[106] In particular, they propose the MM/QM-COSMO strategy, which is based on a linear scaling QM-based end-point calculation. Starting from MD trajectories of the complex, a QM post-processing was made for a selection of representative snapshots, which were first refined by QM energy minimization (100 cycles) using the PM3 hamiltonian and the COSMO continuum solvent model. The binding free energy was then determined as given in eqn (1.17), where the first term on the right-hand side was determined using eqn (1.18) and the entropic term included both the changes in translational and rotational rigid body components (ΔS^{RB}) and the change in vibrational entropy (ΔS^{int}), which was determined from normal mode computations:

$$\Delta G_{binding} = \Delta\langle H^{COSMO}\rangle - T\Delta S^{RB} - T\Delta S^{int} \tag{1.17}$$

$$H^{COSMO} = E^{COSMO} + G_{np}^{solv} \tag{1.18}$$

Here E^{COSMO} represents the PM3 QM energy (including vacuum and solvation energy components) and the nonpolar contribution (G_{np}^{solv}) is determined using a relationship with the change in the solvent-accessible surface (eqn 1.19):

$$\Delta G_{np}^{solv} = \gamma\Delta SASA \tag{1.19}$$

where γ was assigned a value of 0.0020 kcal mol^{-1} Å^{-2}.

The binding affinities derived from MM/QM-COSMO calculations were compared with the results determined by MM/PBSA and MM/GBSA, as well as the Solvent Interaction Energy (SIE) method.[14] The MM/QM-COSMO method showed the best agreement, both for absolute (average unsigned error of 0.7 kcal mol^{-1}) and relative binding free energies.

The preceding discussion shows the increasing interest in exploring the potential impact of QM-based strategies in lead optimization. Clearly, the main limitation is the expensiveness of these approaches, which explains the fact that current applications tend to consider configurational sampling in an approximate way. In turn, these limitations also encourage the search for more effective, yet accurate strategies. For instance, Ryde and co-workers have recently reported a novel approach for calculating the interaction energy between a ligand and its protein target based on the Polarizable Multipole Interaction with Supermolecular Pairs (PMISP) method.[107,108] This technique treats electrostatic interaction by multipoles (up to quadrupoles) and induction (*via* anisotropic polarizabilities), whereas non-classical interactions are described by explicit QM calculations using a fragmentation approach. For the protein, electrostatics and induction are treated in the same way, supplemented by a Lennard-Jones term from a standard MM force field for non-classical terms outside a certain distance from the ligand (4–7 Å). The computational scheme is exemplified by calculating the interaction energies of biotin and BTN7 (*i.e.*, a neutral biotin analog) bound to avidin by combining properties derived at the B3LYP/aug-cc-pVTZ level and calculations at the MP2/aug-cc-pVTZ one, with an expected error of 1–2%.

A different strategy has been reported by Essex and co-workers,[109] as they present a molecular simulation protocol to compute free energies of binding that combines a QM/MM correction term with rigorous classical free energy techniques, thereby accounting for electronic polarization effects. In this computational scheme, relative free energies of binding are first computed using classical force fields, configurational sampling and replica exchange thermodynamic integration. Snapshots of the configurations at the end points of the perturbation are then subjected to DFT-QM/MM single-point calculations. Finally, the resulting QM energies are then processed using the Zwanzig equation to give free energies incorporating electronic polarization.

Finally, it is worth noting that the current efforts made for redesigning quantum chemistry codes,[110–114] making more efficient algorithms and implementing them in more powerful computational resources (*i.e.*, graphical processing units), can be relevant to alleviate the computational requirements of QM-based strategies, thus enabling the successful application of these techniques to drug discovery.

1.5 Summary and Outlook

An accurate prediction of receptor–ligand binding free energies is crucial in drug discovery, as the ability to compute *a priori* this magnitude with computer-assisted techniques would significantly reduce the amount of required experimental testing and the time needed to develop new bioactive compounds. Prediction of the binding free energy still remains a major challenge for computational chemistry due to the involvement of different enthalpic and entropic components, each playing a significant contribution, and to the important compensation between these thermodynamic quantities. The protein–ligand interaction energy is the main stabilizing contribution to the binding affinity, as it must compensate for unfavorable desolvation, conformational rearrangement and entropy loss upon formation of the ligand–protein complex. Accordingly, a quantitative understanding of the stabilization afforded by the variety of interactions formed between a ligand and its target, taking a proper accounting of electrostatic, induction, charge transfer and dispersion effects, is fundamental for rationalizing the differences in biological activity. For methods based on first principles, the path is fairly obvious: increase the level of theory to reduce the errors due to the underlying approximations adopted in simpler approaches. However, this option is seriously limited by the huge computational cost of high-level QM computations, which explains why most of the QM-based strategies devised for the study of ligand–protein complexes rely on semiempirical methods, even though their adequacy to describe certain types of interactions is questionable and demands the incorporation of correction terms. Furthermore, since biomolecules are dynamical entities, a precise knowledge of the structural plasticity of the target and its implication in mediating the binding of ligands is necessary to assist the development of more potent, selective compounds in structure-based drug design. At this stage, the adoption of statistical thermodynamics approaches is necessary to obtain a representation of the configurational ensemble of states for ligand, protein and the ligand–protein complex.

Based on the preceding discussion, QM methods can contribute to improve the accuracy of computer-assisted tools in two ways. First, by assisting the development and refinement of biomolecular force fields, which can be improved in different ways (for instance, increasing the flexibility of the fixed partial charge model or the accuracy of torsional parameters), specially regarding the inclusion of polarization effects, and keeping the balance required for the different energy components in the interaction energy potential. At some future date, the incorporation of polarization into the modeling of proteins and protein–ligand complexes will be considered a routine aspect of the calculation. However, QM-based methods used directly for the modeled structures of ligand–protein complexes or in the framework of end-point sampling techniques represent a promising alternative as a tool to develop and calibrate novel computational strategies designed to provide accurate estimates of binding affinities. At the present stage, it is clear that this effort will undoubtedly take a number of years before substantial progress is made. However, advances in both computing power and algorithms will facilitate the transition toward diverse applications in the study of protein–ligand complexes.

Acknowledgements

This work was supported by the Spanish Ministerio de Innovación y Ciencia (SAF2008-05595) and the Generalitat de Catalunya (2009-SGR00298).

References

1. H. Gohlke and G. Klebe, *Angew. Chem., Int. Ed.*, 2002, **41**, 2644.
2. C. Bissantz, B. Kuhn and M. Stahl, *J. Med. Chem.*, 2010, **53**, 5061.
3. C. A. Hunter, *Angew. Chem., Int. Ed.*, 2004, **43**, 5310.
4. J. C. Ma and D. A. Dougherty, *Chem. Rev.*, 1997, **97**, 1303.
5. A. Frontera, D. Quiñonero and P. M. Deyà, *WIRES Comput. Mol. Sci.*, 2011, **1**, 440.
6. E. Cubero, F. J. Luque and M. Orozco, *Proc. Natl. Acad. Sci. U. S. A.*, 1998, **95**, 5976.
7. A. Frontera, P. Gamez, M. Mascal, T. J. Moolbroek and J. Reedijk, *Angew. Chem., Int. Ed.*, 2011, **50**, 9564.
8. I. Alkorta and J. Elguero, *Chem. Soc. Rev.*, 1998, **27**, 163.
9. P. Hobza and Z. Havlas, *Chem. Rev.*, 2000, **100**, 4253.
10. H. L. Nguyen, P. N. Horton, M. B. Hursthouse, A. C. Legon and D. W. Bruce, *J. Am. Chem. Soc.*, 2004, **126**, 16.

11. M. G. Sarwar, B. Dragisic, L. J. Salsberg, C. Gouliaras and M. S. Taylor, *J. Am. Chem. Soc.*, 2010, **132**, 1646.
12. H. Dvir, D. M. Wong, M. Harel, X. Barril, M. Orozco, F. J. Luque, D. Muñoz-Torrero, P. Camps, T. L. Rosenberry, I. Silman and J. L. Sussman, *Biochemistry*, 2002, **41**, 2970.
13. A. M. Davis and S. J. Teague, *Angew. Chem., Int. Ed.*, 1999, **38**, 737.
14. M. Naïm, S. Bhat, K. N. Rankin, S. Dennis, S. F. Chowdhury, I. Siddiqi, P. Drabik, T. Sulea, C. Bayly, A. Jakalian and E. O. Purisima, *J. Chem. Inf. Model.*, 2007, **47**, 122.
15. S. Cabani, P. Gianni, V. Mollica and L. Lepori, *J. Solution Chem.*, 1981, **10**, 563.
16. J. R. Pliego, Jr. and J. M. Riveros, *Phys. Chem. Chem. Phys.*, 2002, **4**, 1622.
17. P. Csermely, R. Palotai and R. Nussinov, *Trends Biochem. Sci.*, 2010, **35**, 539.
18. F. Spyrakis, A. Bidon-Chanal, X. Barril and F. J. Luque, *Curr. Topics Med. Chem.*, 2011, **11**, 192.
19. D. Bucher, B. J. Grant and J. A. McCammon, *Biochemistry*, 2011, **50**, 10530.
20. E. Perola and P. S. Charifson, *J. Med. Chem.*, 2004, **47**, 2499.
21. J. Tirado-Rives and W. L. Jorgensen, *J. Med. Chem.*, 2006, **49**, 5880.
22. K. T. Butler, F. J. Luque and X. Barril, *J. Comput. Chem.*, 2009, **30**, 601.
23. C. A. Chang, W. Chen and M. K. Gilson, *Proc. Natl. Acad. Sci. U. S. A.*, 2007, **104**, 1534.
24. D. H. Williams, E. Stephens, D. P. O'Brien and M. Zhou, *Angew. Chem., Int. Ed.*, 2004, **43**, 6596.
25. C. H. Reynolds and M. K. Holloway, *ACS Med. Chem. Lett.*, 2011, **2**, 433.
26. E. Freire, *Drug Discovery Today*, 2007, **2**, 469.
27. J. E. Ladbury, G. Klebe and E. Freire, *Nat. Rev. Drug Discovery*, 2010, **9**, 23.
28. G. G. Ferenczy and G. M. Keseru, *Drug. Discovery Today*, 2010, **15**, 919.
29. G. G. Ferenczy and G. M. Keseru, *J. Chem. Inf. Model.*, 2010, **50**, 1536.
30. I. D. Kuntz, K. Chen, K. A. Sharp and P. A. Kollman, *Proc. Natl. Acad. Sci. U. S. A.*, 1999, **96**, 9997.
31. *Pharmacophores and Pharmacophore Searches*, ed. T. Langer and R. D. Hoffmann, Wiley-VCH, Weinheim, 2006.
32. See this publication, ch. 6.
33. P. A. Kollman, *Chem. Rev.*, 1993, **93**, 2395.
34. W. L. Jorgensen, *Acc. Chem. Res.*, 2009, **42**, 724.
35. *Free Energy Calculations. Theory and Applications in Chemistry and Biology*, ed. C. Chipot and A. Pohorille, Springer Series in Chemical Physics, vol. 86, Springer, Berlin, 2007.
36. A. de Ruiter and C. Oostenbrink, *Curr. Opin. Chem. Biol.*, 2011, **15**, 547.

37. J. D. Chodera, D. L. Mobley, M. R. Shirts, R. W. Dixon, K. Branson and V. S. Pande, *Curr. Opin. Struct. Biol.*, 2011, **21**, 150.
38. M. K. Gilson, J. A. Given, B. L. Bush and J. A. McCammon, *Biophys. J.*, 1997, **72**, 1047.
39. D. Hamelberg and J. A. McCammon, *J. Am. Chem. Soc.*, 2004, **126**, 7683.
40. W. F. van Gunsteren and A. E. Mark, *Eur. J. Biochem.*, 1992, **204**, 947.
41. C. D. Christ, A. E. Mark and W. F. van Gunsteren, *J. Comput. Chem.*, 2010, **31**, 1569.
42. See this publication, ch. 11.
43. J. Michel and J. W. Essex, *J. Comput. Aided Mol. Des.*, 2010, **24**, 639.
44. J. W. Ponder, C. Wu, P. Ren, V. S. Pande, J. D. Chodera, M. J. Schnieders, I. Haque, D. L. Mobley, D. S. Lambrecht, R. A. DiStasio, Jr., M. Head-Gordon, G. N. I. Clark, M. E. Johnson and T. Head-Gordon, *J. Phys. Chem. B*, 2010, **114**, 2549.
45. A. G. Donchev, V. D. Ozrin, M. V. Subbotin, O. V. Tarasov and V. I. Tarasov, *Proc. Natl. Acad. Sci. U. S. A.*, 2005, **102**, 7829.
46. A. G. Donchev, N. G. Galkin, A. A. Illarionov, O. V. Khoruzhii, M. A. Olevanov, V. D. Ozrin, M. V. Subbotin and V. I. Tarasov, *Proc. Natl. Acad. Sci. U. S. A.*, 2006, **103**, 8613.
47. A. G. Donchev, N. G. Galkin, L. B. Pereyaslavets and V. I. Tarasov, *J. Chem. Phys.*, 2007, **125**, 244107.
48. A. G. Donchev, M. G. Galkin, A. A. Illarionov, O. V. Khoruzil, M. A. Olevanov, V. D. Ozrin, L. B. Pereyaslavets and V. I. Tarasov, *J. Comput. Chem.*, 2008, **29**, 1242.
49. N. Gresh, *J. Comput. Chem.*, 1995, **16**, 856.
50. N. Gresh, J.-P. Piquemal and M. Krauss, *J. Comput. Chem.*, 2005, **26**, 1113.
51. N. Gresh, G. A. Cisneros, T. A. Darden and J.-P- Piquemal, , *J. Chem. Theory Comput.*, 2007, **3**, 1960.
52. P. Cieplak, J. Caldwell and P. A. Kollman, *J. Comput. Chem.*, 2001, **22**, 1048.
53. Z. X. Wang, W. Zhang, C. Wu, H. Lei, P. Cieplak and Y. Duan, *J. Comput. Chem.*, 2006, **27**, 781.
54. G. Lamoureux and B. Roux, *J. Chem. Phys.*, 2003, **119**, 3025.
55. G. Lamoureux, A. D. MacKerell, Jr. and B. Roux, *J. Chem. Phys.*, 2003, **119**, 5185.
56. S. Patel, A. D. Mackerell, Jr. and C. L. Brooks, III, *J. Comput. Chem.*, 2004, **25**, 1504.
57. W. L. Jorgensen, K. P. Jensen and A. N. Alexandrova, *J. Chem. Theory Comput.*, 2007, **3**, 1987.
58. D. P. Geerke and W. F. Van Gunsteren, *J. Chem. Theory Comput.*, 2007, **3**, 2128.
59. H. Yu and W. F. van Gunsteren, *Comput. Phys. Commun.*, 2005, **172**, 69.
60. P. Cieplak, F.-Y. Dupradeau, Y. Duan and J. Wang, *J. Phys.: Condens. Matter*, 2009, **21**, 333102.

61. F. J. Luque, F. Dehez, C. Chipot and M. Orozco, *WIRES Comput. Mol. Sci.*, 2011, **1**, 844.
62. A. J. Stone, *The Theory of Intermolecular Forces*, Clarendon Press, Oxford, 1996.
63. J. Applequist, J. R. Carl and K.-K. Fung, *J. Am. Chem. Soc.*, 1974, **92**, 2952.
64. K. J. Miller, *J. Am. Chem. Soc.*, 1990, **112**, 8533.
65. J. M. Strout and C. E. Dykstra, *J. Am. Chem. Soc.*, 1995, **117**, 5127.
66. F. Dehez, C. Chipot, C. Millot and J. G. Angyan, *Chem. Phys. Lett.*, 2001, **338**, 180.
67. I. Soteras, C. Curutchet, A. Bidon-Chanal, F. Dehez, J. G. Angyan, M. Orozco, C. Chipot and F. J. Luque, *J. Chem. Theory Comput.*, 2007, **3**, 1901.
68. K. Kitaura and K. Morokuma, *Int. J. Quantum Chem.*, 1976, **10**, 325.
69. P. S. Bagus, K. Hermann and C. W. Bauschlicher, Jr., *J. Chem. Phys.*, 1984, **80**, 4378.
70. W. J. Stevens and W. H. Fink, *Chem. Phys. Lett.*, 1987, **139**, 15.
71. E. Glendening, *J. Phys. Chem. A*, 2005, **109**, 11936.
72. D. G. Fedorov and K. Kitaura, *J. Comput. Chem.*, 2006, **28**, 222.
73. B. Jeziorski, R. Moszynski and K. Szalewicz, *Chem. Rev.*, 1994, **94**, 1887.
74. A. J. Misquitta and A. J. Stone, *J. Chem. Theory Comput.*, 2008, **4**, 7.
75. A. J. Misquitta, A. J. Stone and S. L. Price, *J. Chem. Theory Comput.*, 2008, **4**, 19.
76. A. Fiethen, G. Jansen, A. Hesselmann and M. Schütz, *J. Am. Chem. Soc.*, 2008, **130**, 1802.
77. H. Szatylowicz, T. M. Krygowski, J. J. Panek and A. Jezierska, *J. Phys. Chem. A*, 2008, **112**, 9895.
78. I. Soteras, M. Orozco and F. J. Luque, *Phys. Chem. Chem. Phys.*, 2008, **10**, 2616.
79. F. Archambault, C. Chipot, I. Soteras, F. J. Luque, K. Schulten and F. Dehez, *J. Chem. Theory Comput.*, 2009, **5**, 3022.
80. K. E. Riley and P. Hobza, *J. Chem. Theory Comput.*, 2008, **4**, 232.
81. D. Jiao, P. A. Golubkov, T. A.Darden and P. Ren, *Proc. Natl. Acad. Sci. U. S. A.*, 2008, **105**, 6290.
82. D. Jiao, J. Zhang, R. E. Duke, G. Li, M. J. Schnieders and P. Ren, *J. Comput. Chem.*, 2009, **30**, 1701.
83. O. Khoruzhii, A. G. Donchev, N. Galkin, A. Illarionov, M. Olevanov, V. Ozirin, C. Queen and V. Tarasov, *Proc. Natl. Acad. Sci. U. S. A.*, 2008, **105**, 10378.
84. J. R. Maple, Y. Cao, W. Damm, T. A. Halgren, G. A. Kaminski, L. Y. Zhang and R. A. Friesner, *J. Chem. Theory Comput.*, 2005, **1**, 694.
85. J. Gao, *Rev. Comput. Chem.*, 1996, **7**, 119.
86. G. Monard and K. M. Merz, Jr., *Acc. Chem. Res.*, 1999, **32**, 904.
87. M. Orozco and F. J. Luque, *Chem. Rev.*, 2000, **100**, 4187.
88. A. Warshel, *Annu. Rev. Biophys. Biomol. Struct.*, 2003, **32**, 425.
89. R. A. Friesner and V. Guallar, *Annu. Rev. Phys. Chem.*, 2005, **56**, 389.

90. A. Lodola, C. J. Woods and A. J. Mulholland, *Annu. Rep. Comput. Chem.*, 2008, **4**, 155.
91. A. E. Cho, V. Guallar, B. J. Berne and R. Friesner, *J. Comput. Chem.*, 2005, **26**, 915.
92. A. E. Cho and D. Rinaldo, *J. Comput. Chem.*, 2009, **30**, 2609.
93. C. J. R. Illingworth, G. M. Morris, K. E. B. Parkes, C. R. Snell and C. A. Reynolds, *J. Phys. Chem. A*, 2008, **112**, 12157.
94. A. E. Cho, J. Y. Chung, M. Kim and K. Park, *J. Chem. Phys.*, 2009, **131**, 134108.
95. C. Hensen, J. C. Hermann, K. Nam, S. Ma, J. Gao and H.-D. Höltje, *J. Med. Chem.*, 2004, **47**, 6673.
96. K. D. Dubey, A. K. Chaubey and R. P. Ojha, *Med. Chem. Res.*, 2011, in press; DOI: 10.1007/s00044-011-9617-1.
97. K. Raha, A. J. Van der Vaart, K. E. Riley, M. B. Peters, L. M. Westerhoff, H. Kim and K. M. Merz, Jr., *J. Am. Chem. Soc.*, 2005, **127**, 6583.
98. C. N. Alves, S. Martí, R. Castillo, J. Andrés, V. Moliner, I. Tuñón and E. Silla, *Chem. Eur. J.*, 2007, **13**, 7715.
99. A. Khandelwal, V. Lukacova, D. Comez, D. M. Kroll, S. Raha and S. Balaz, *J. Med. Chem.*, 2005, **48**, 5437.
100. D. Das, Y. Koh, Y. Tojo, A. K. Gosh and H. Mitsuya, *J. Chem. Inf. Model.*, 2009, **49**, 2851.
101. K. Raha and K. M. Merz, Jr., *J. Med. Chem.*, 2005, **48**, 4558.
102. K. Raha and K. M. Merz, Jr., *J. Am. Chem. Soc.*, 2004, **126**, 1020.
103. J. Fanfrlik, A. K. Bronowska, J. Rezac, O. Prenosil, J. Kovalinka and P. Hobza, *J. Phys. Chem. B*, 2010, **114**, 12666.
104. P. Dobes, J. Fanfrlik, J. Rezac, M. Otypeka and P. Hobza, *J. Comput. Aided Mol. Des.*, 2011, **25**, 223.
105. P. Dobes, J. Rezac, J. Fanfrlik, M. Otyepka and P. Hobza, *J. Phys. Chem. B*, 2011, **115**, 8581.
106. V. M. Anisimov and C. N. Cavasotto, *J. Comput. Chem.*, 2011, **32**, 2254.
107. P. Soderhjelm and U. Ryde, *J. Phys. Chem. A*, 2009, **113**, 617.
108. P. Soderhjelm, F. Aquilante and U. Ryde, *J. Phys. Chem. B*, 2009, **113**, 11085.
109. F. R. Beierlein, J. Michel and J. W. Essex, *J. Phys. Chem. B*, 2011, **115**, 4911.
110. I. S. Ufimtsev and T. J. Martínez, *J. Chem. Theory Comput.*, 2008, **4**, 222.
111. I. S. Ufimtsev and T. J. Martínez, *J. Chem. Theory Comput.*, 2009, **5**, 1004.
112. I. S. Ufimtsev and T. J. Martínez, *J. Chem. Theory Comput.*, 2009, **5**, 2619.
113. N. Luehr, I. S. Ufimtsev and T. J. Martínez, *J. Chem. Theory Comput.*, 2011, **7**, 949.
114. Y. Uejima, T. Terashima and R. Maezomo, *J. Comput. Chem.*, 2011, **32**, 2264.

CHAPTER 2

Thermodynamics of Ligand Binding

GYÖRGY G. FERENCZY*[a] AND GYÖRGY M. KESERŰ*[b]

[a] Department of Inorganic and Analytical Chemistry, Budapest University of Technology and Economics, 4, Szt. Gellért tér, H-1111 Budapest, Hungary; [b] Gedeon Richter Plc., 19-21 Gyömrői út, H-1103 Budapest, Hungary
*E-mail: gyorgy.ferenczy@gmail.com; gy.keseru@richter.hu

2.1 Introduction

Ligand binding is a fundamental process in many biological phenomena. Restricting our discussion to drug discovery, ligand binding includes the interaction of small molecules with their target, which is typically a protein. Binding events contribute to further properties of small molecules, whose interactions with other proteins (often identified as anti-targets) determine, for example, selectivity, metabolism and toxicity. One of the main goals of drug discovery is to realize balanced interactions of a compound with its target *versus* other proteins. Although the optimization of binding affinity is usually restricted to the target, interactions with anti-targets are largely controlled by the physico-chemical properties. These properties, on the other hand, are the critical determinants of the pharmacokinetic profile. The affinity of a ligand to a protein can be characterized by the free energy of binding. It has been recognized that the enthalpic and entropic components (thermodynamic signature) of the binding free energy comprise useful information on the details of protein–ligand interactions. A knowledge of the thermodynamic signature can be exploited to improve the binding affinity and to advantageously affect the physico-chemical properties of the compounds. In the subsequent discussion we elaborate on the

RSC Drug Discovery Series No. 23
Physico-Chemical and Computational Approaches to Drug Discovery
Edited by F. Javier Luque and Xavier Barril
© The Royal Society of Chemistry 2012
Published by the Royal Society of Chemistry, www.rsc.org

thermodynamics of ligand–protein binding, the enthalpic and entropic contributions and their relations to structural factors. The relevance of thermodynamics in drug discovery, in particular in hit selection, hit-to-lead and lead optimization processes, will be examined and guidelines will be given to apply thermodynamics-based considerations in preclinical discovery settings.

Our discussion is restricted to small molecule ligands. Although biopolymers can act as protein ligands and are increasingly important therapeutic agents, the drug discovery aspects of their binding thermodynamics are less explored and are not considered here.

2.2   Theoretical Background

2.2.1   Basic Equations of Ligand–Protein Binding Thermodynamics

The binding affinity of a ligand (L) to a protein (P) can be characterized by the dissociation constant, K_d:

$$K_d = \frac{[L][P]}{[LP]} \tag{2.1}$$

corresponding to the process

$$LP \leftrightarrow L + P \tag{2.2}$$

The logarithm of the dissociation constant is proportional to the Gibbs free energy of binding (ΔG_{bind}):

$$\Delta G_{bind} = RT \times \ln K_d \tag{2.3}$$

where R is the universal gas constant and T is the absolute temperature. ΔG_{bind} is a function of the binding enthalpy (ΔH_{bind}) and the binding entropy (ΔS_{bind}):

$$G_{bind} = H_{bind} - TS_{bind} \tag{2.4}$$

The above equations show that an improved binding affinity, *i.e.* a decreased K_d, is equivalent to a decreased (more negative) binding free energy. This can be achieved with a more negative enthalpy and with increased entropy.

2.2.2   Details of the Ligand–Receptor Binding Process; the Role of the Solvent

2.2.2.1   *Decomposition of the Binding Process*

Ligand binding signifies the process in which the originally separated and solvated ligand and protein form the solvated ligand–protein complex. This process is often decomposed into several steps. The decomposition is useful as far as it helps in the understanding and in the interpretation of the process and its thermodynamic consequences. On the other hand, it has to be kept in mind that the total binding free energy is the sum of the free energy changes of the elementary steps only if the latter link the initial and final states and all enthalpy and entropy changes are correctly taken into account. A usual decomposition includes desolvation of the ligand and the binding site, changing the conformation of both the ligand and the protein and forming interactions between them. Desolvation restructures water around the ligand, which results in a significant entropic reward. Replacement of water from the binding site may have different enthalpic and entropic consequences, depending on the binding interactions of the replaced water molecules. Binding is usually accompanied by conformational rearrangement of both the ligand and the receptor and this represents an enthalpic penalty in most cases. Formation of the ligand–receptor complex is typically coupled to forming new interactions between the ligand and its binding site that are enthalpically beneficial. Molecular recognition of the ligand, however, limits its external rotational and translational freedom as well as ligand and protein flexibility and therefore represents an entropic penalty. Although the thermodynamic impact of long-range effects is usually neglected, they could also contribute to ligand binding.

2.2.2.2 Enthalpy–Entropy Compensation

It is observed for a great variety of systems that structural variations resulting in small changes in ΔG imply more significant changes in its components, ΔH and $T\Delta S$. The origin and even the presence of enthalpy–entropy compensation have long been debated (see *e.g.* refs. 1–3). It was suspected to originate from errors associated with the determination of ΔH and $T\Delta S$ using the van't Hoff equation.[4,5] It was shown that the high correlation between enthalpy and entropy is often due to a propagation of experimental errors. However, recent calorimetric determination of thermodynamic quantities has not removed this compensation.[6,7] An interpretation of the compensation hypothesizes that an enthalpically more favorable binding imposes a more severe restriction to the motion of the interacting partners and thus a more significant unfavorable entropic change (see *e.g.* ref. 8). Water models that are able to account for the enthalpy–entropy compensation of aqueous processes have also been proposed.[9] It is worth mentioning that although the thermodynamics of host–guest complexes are different in water and in organic solvents, the enthalpy–entropy compensation is not restricted to aqueous systems.[10] Nevertheless, a detailed understanding of water properties, hydration and hydrophobicity is essential to the rationalization of binding thermodynamics in biological systems.

2.2.2.3 Hydrophobic Effect and Hydrophobic Hydration

Water is a highly complex liquid. No comprehensive theory is available to explain all experimental observations and to adequately describe aqueous processes at the level of molecular detail. A particularly intriguing phenomenon is the hydrophobic effect that refers to the transfer of apolar compounds, either from their liquid state or from a solution in an apolar solvent, to water.[11] This is a process that includes the disruption of interactions between the apolar compound and its apolar environment, refilling the vacancy in the apolar medium, cavity formation in liquid water, the insertion of the nonpolar solute, the onset of the solute–solvent interactions and the reordering of the water molecules in the close proximity of the solute. This process is accompanied by a free energy increase. At room temperature, both the enthalpy and the entropy are negative and the latter dominates. On increasing the temperature the free energy hardly changes, but the high and positive heat capacity of hydration implies that the entropy-driven process at room temperature becomes enthalpy-driven at higher temperature. Theories of the hydrophobic effect at the level of molecular structure usually concentrate on those steps that involve water (*i.e.* cavity formation in water, placement of the solute into the cavity and the restructuring of the water around the solute) and are termed hydrophobic hydration. These theories have been elaborated basically along two lines. One argues that the small size of water molecules is a key feature in producing negative entropy in opening up a cavity for the solute molecule.[12–14] The other is based on the hydrogen bonding properties of water and assumes different hydrogen bonding in the bulk than in contact with a solute (mixture or two-state water models; see ref. 11 and references therein). A combination of these factors was also proposed as the origin of the hydrophobic effect.[15] On the other hand, Monte Carlo simulations of solvation in water and in model fluids showed that the strong cohesive energy coming from the hydrogen bonds is a more important factor than the entropy price of cavity formation due to the small size of the water molecules.[16,17] This is in line with the classical view of the hydrophobic effect that assigns a structure promoting effect to the hydrophobic solute; at room temperature, water forms hydrogen-bonded structured clusters (icebergs) around the solute and these clusters have more favorable enthalpy and less favorable entropy components than does the bulk water. As the temperature increases, the hydrogen bonds of the clusters are sacrificed for increased entropy.

The effect of the solute on the hydrogen bonding state of water molecules is also a key element of Muller's two-state model.[18] It assumes that hydrogen atoms participate in hydrogen bonds in a temperature-dependent manner and this is affected by the presence of solute molecules. This model is not only able to account for the observed temperature dependence of the heat capacity of hydrophobic hydration, but also the temperature dependence of proton chemical shifts. Moreover, with appropriate parameterization it can well describe enthalpy–entropy compensation.[19] According to this model, hydrogen bonds are stronger and (in contrast to the iceberg model) more broken near to a hydrophobic solute than in bulk water.[20] The dependence of the number of hydrogen bonds of water molecules on the distance from the hydrophobic surface was used to set up analytic models for the hydration of hydrophobic particles. The models were also used to examine the solvent-mediated interactions of hydrophobic particles, including their enthalpy and entropy components together with their temperature dependence.[21]

2.2.2.4 Protein–Ligand Binding in Water

Protein–ligand complex formation in water is a related but more complex process than hydrophobic hydration; the latter can serve as a model for certain aspects of the former. Ligand binding shows resemblance to micelle formation in the sense that both processes are associated with the coalescence of solutes and thus a decrease of cavity size and a release of solvating water molecules.[22–24] Thus, ligand binding is typically accompanied by desolvation of hydrophobic groups with the corresponding thermodynamic signatures. These include entropy increase and a negative heat capacity at constant pressure, the contributing factors to the latter being debated.[25] Shape and surface to volume ratio of the interacting partners and those of the complex also influence the thermodynamics of the process[22] and may result in enthalpy-driven binding while the usual negative heat capacity feature is kept.[26] The origin of this "nonclassical" hydrophobic interaction is proposed to be the insufficient protein solvation and thus the decreased contribution of protein desolvation upon binding. In this way, the entropy gain that would normally accompany protein desolvation at ligand binding is not observed and the enthalpy gain originating from ligand–protein dispersion interactions dominates.[27]

Ligand binding to proteins is usually accompanied by conformational changes of both partners. These changes are associated with unfavorable free energy that is counterbalanced by favorable contributions of the binding. Structural changes of proteins upon ligand binding are often identified, *e.g.* by comparison of the X-ray structures of the ligand free (apo) protein with that found in the complex. However, the observed crystal structures give no direct information for the free-energy change of the complex formation for several reasons. The protein is only a part of the whole system and the examination of the free-energy consequence of its conformational change has limited significance. Moreover, X-rays produce crystal packing biased snapshots of dynamic structures and these (in the best case) may give estimates of the enthalpy but not of the free-energy change. Similarly, the ligand often binds in a conformation with higher energy (enthalpy) than that of the minimum energy solvated molecule.

The ligand and the protein form new interactions upon complex formation. It is important to realize that when the ligand binds to the protein, then ligand–water and protein–water contacts are replaced by ligand–protein and water–water contacts. The latter are formed by some water molecules participating in solvation before binding, and becoming part of the bulk solvent after binding. In this way, the newly formed ligand–protein and water–water interactions replace those that existed before binding. Thus a net free energy gain can only be achieved when good steric and electrostatic complementarity between the ligand and the protein is realized.

2.2.3 Ligand Properties and Binding Thermodynamics

The free-energy of ligand binding (ΔG) determines the dissociation constant of the complex K_d according to eqn (2.3). In many cases we characterize the binding with the affinity of the ligand to the protein and it can be described with ΔG, or equivalently with K_d. Although high affinity and thus a low binding free-energy is a target property in medicinal chemistry optimizations, there are at least two reasons to monitor not only the free energy itself but also its enthalpic and entropic components. One reason is that the knowledge of these components and their changes with structural modifications of the ligand allow a better understanding of the structural requirement of high affinity binding. The other reason is that the improvement of affinity is only one of the objectives of optimization and it has been recognized that some other objectives, like achieving selectivity and avoiding toxicity, can also be related to the free-energy components of the binding to the target protein. These considerations raise interest in the measurement and interpretation of binding enthalpy and entropy.

2.2.3.1 Enthalpic and Entropic Components of Dissociation Constants

The dissociation constant of a complex (K_d) is in a direct relation with the free energy change of complex formation (eqn 2.3), which in turn can be decomposed into enthalpic and entropic contributions (eqn 2.4). Based on these relations, the negative logarithm of the dissociation constant (pK_d) can also be decomposed into enthalpic and entropic components. Thus, a combination of eqns (2.3) and (2.4) gives:

$$H - T \times S = RT \times \ln K_d \tag{2.5}$$

which can be transformed to yield:

$$\left[\frac{-H}{2.303 \times RT}\right] + \left[\frac{S}{2.303 \times R}\right] = pK_d \tag{2.6}$$

Then, we define pK_H, the enthalpic component as:

$$\left[\frac{-H}{2.303 \times RT}\right] = pK_d^{enthalpy} = pK_H \tag{2.7}$$

and pK_S, the entropic component of pK_d, as:

$$\frac{S}{2.303 \times R} = pK_d^{entropy} = pK_S \qquad (2.8)$$

Thus we can write:

$$pK_H + pK_S = pK_d \qquad (2.9)$$

These components are useful in characterizing the thermodynamics of ligand binding. In the forthcoming discussion we analyze pK_H, which is a measure of the enthalpic component of binding. It can be derived directly from ΔH. Note that a more negative enthalpy (ΔH) corresponds to more favorable binding and to a higher pK_H.

2.2.3.2 Relation Between Ligand Size and Binding Thermodynamics

The separation of pK_d into pK_H and pK_S creates a convenient framework for a comparative analysis of affinity and its enthalpic and entropic components. In the present section, affinity and enthalpy are investigated as functions of ligand size. Figure 2.1(a) shows calorimetrical pK_d values as a function of the number of ligand heavy atoms (HA). Data for 812 ligand–protein complexes were taken from three databases: BindingDB,[28] PDBcal[29] and Scorpio.[30] The maximum pK_d values of the HA vs. pK_d plot in Figure 2.1(a) start at about $pK_d = 8$ and reach a maximum at about $pK_d = 12$ at $HA = 40$. We fitted a function to HA vs. pK_{dmax}, as shown in Figure 2.1(b). The pK_{dmax} values were extracted from the HA vs. pK_d function by dividing HA into intervals of $HA = 5$ and by picking the largest pK_d values from each interval. Seeking for a relationship between HA and pK_{dmax} in the form of:

$$\ln(pK_{dmax}) = a \times \ln(HA) + b \qquad (2.10)$$

we obtain $a = 0.3$ and $b = 1.27 \approx \ln(3.6)$ (Figure 2.1b).[31] This can be written as:

$$pK_{dmax} = 3.6 \times HA^{0.3} \qquad (2.11)$$

This relationship is highly similar to those reported by others[32,33] using IC_{50} and K_i data rather than our K_d data set obtained from thermodynamic measurements.

The pK_H data set exhibits a significantly different behavior. Figure 2.2(a) shows the HA vs. pK_H plot and Figure 2.2(b) the $\ln(HA)$ vs. $\ln(pK_{Hmax})$ plot (by analogy to the highest affinity compounds of pK_{dmax} we identify compounds of pK_{Hmax} as the highest enthalpy compounds). pK_H covers a wide range, with extremes at -13 and $+20$. This is in contrast to pK_d, which is

a

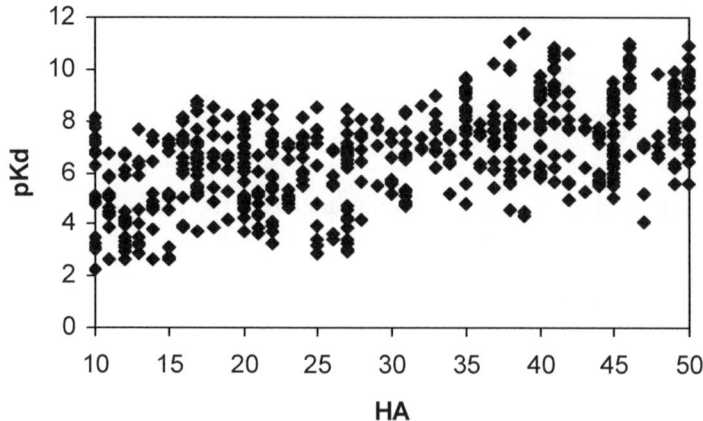

b

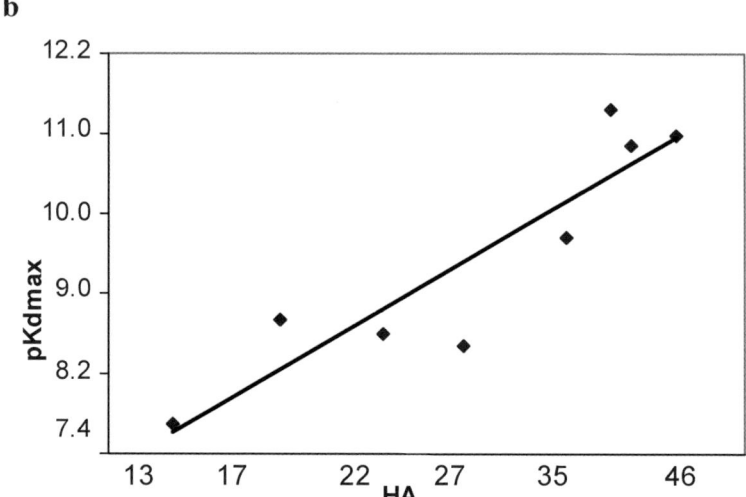

Figure 2.1   (a) Plot of pK_d *versus* number of heavy atoms (*HA*). (b) Logarithmic plot of pK_d for the most potent ligands *versus* the number of heavy atoms (taken with permission from ref. 31).

always positive and extends up to about 12 [*cf.* Figure 2.1(a) and Figure 2.2(a)]. The heavy atom dependence of pK_{Hmax} is also different from that of pK_{dmax} [*cf.* Figure 2.1(b) and Figure 2.2(b); note the different scale of the vertical axes]. While pK_{dmax} shows an increasing trend up to about 40 heavy atoms, pK_{Hmax} exhibits a decreasing trend, *i.e.* the achievable maximal binding enthalpy decreases with increasing molecular size. Replacing pK_{dmax} by pK_{Hmax} in eqn (2.10) and fitting the parameters to the points in Figure 2.2(b) results in $a = -0.3$ and $b = \ln(40)$, and thus:

a

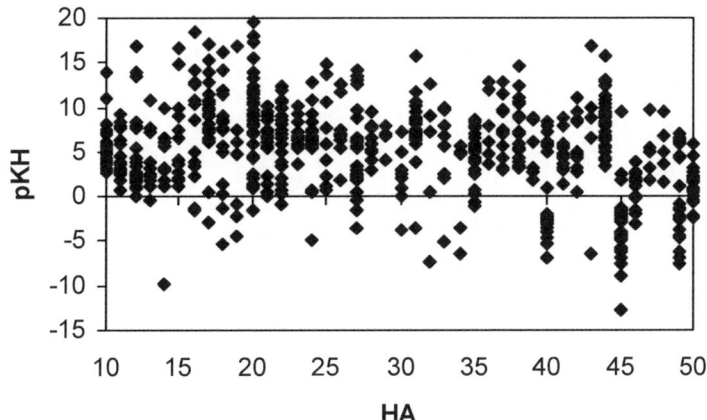

b

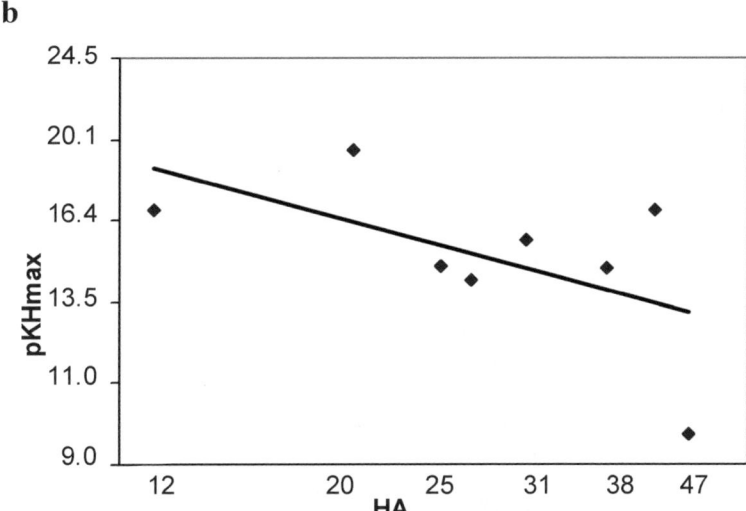

Figure 2.2 (a) Plot of pK_H *versus* the number of heavy atoms (HA). (b) Logarithmic plot of maximal pK_H *versus* the number of heavy atoms (taken with permission from ref. 31).

$$pK_{Hmax} = 40 \times HA^{-0.3} \tag{2.12}$$

Note the opposite sign of the power of HA in eqns (2.11) and (2.12).

The enthalpic content of binding is a useful parameter in assessing the quality of compounds in a drug discovery program. Compounds having a more favorable binding enthalpy are considered better starting points for further optimizations (*vide infra*). In this respect, eqn (2.12) gives guidance on how close a compound is to the available optimum. This equation also shows that the optimum depends on the compound size and that pK_H is not the optimal measure when different size compounds are to be compared. Therefore, another measure, the size-independent enthalpic efficiency (*SIHE*), is defined as:

$$SIHE = \frac{pK_H}{40} \times HA^{0.3} \tag{2.13}$$

The *HA vs. SIHE* plot is shown in Figure 2.3. The highest enthalpy compounds have *SIHE* values around 1, independently of their size. Thus *SIHE* is a good measure of the enthalpic efficiency: a higher *SIHE* is better, its limit is around 1 and the comparison of molecules is not biased by their differing size. *SIHE* can find use in drug discovery, as is discussed in Section 2.5.

The relative standard error of the parameters in eqns (2.12) and (2.13) is ~50% (*cf.* with Figures 2.2 and 2.3) and their fitted values may change when further thermodynamic data become available. However, the qualitative difference in the size dependences of affinity and enthalpy is not expected to be affected. Another factor to be taken into account when applying the *SIHE* concept is that the maximal available enthalpy may be target dependent, as was proposed for the maximal available affinity.[34]

It is worth comparing the size-independent enthalpic efficiency, *SIHE*, of eqn (2.13) with a definition of size-independent ligand efficiency, *SILE*. Nissink[33] proposed the formula:

$$SILE = pK_i \times HA^{-0.3} \tag{2.14}$$

based on the pK_i data set of Reynolds *et al.*[32] (note that in contrast to our definition of *SIHE*, this *SILE* is not normalized to 1). The different powers of

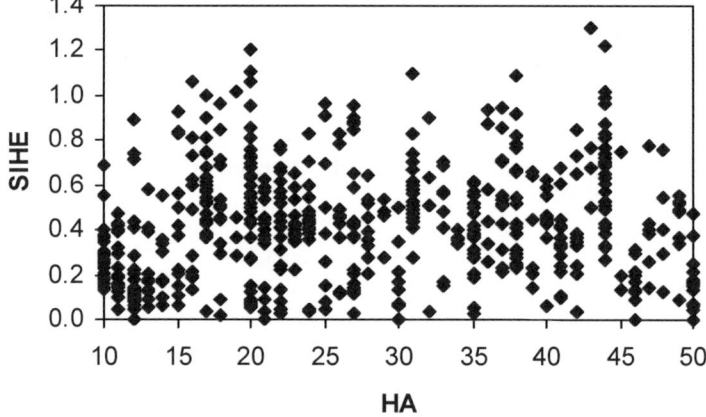

Figure 2.3   Plot of size-independent enthalpic efficiency (*SIHE*) *versus* number of heavy atoms (*HA*) (taken with permission from ref. 31).

HA in eqns (2.13) and (2.14) well reflect the different size dependence of enthalpy (pK_H) and affinity (pK_d). The positive sign in the power of *HA* in *SIHE* results in a factor $HA^{0.3}$ that increases with increasing molecular size, as is required to compensate for the decrease of pK_{Hmax}, the maximal available enthalpy. On the other hand, the opposite is true for SILE, where *HA* appears with a negative power.

2.2.3.3 Interpretation of the Size Dependency of Ligand Binding Enthalpy

It was shown above that the maximal available binding enthalpy, pK_{Hmax}, decreases with increasing ligand size. A rationalization of this finding is started by recalling that the enthalpy component of the free energy change accompanied by the binding process is dominated by polar interactions, including hydrogen bonds. A gain in binding enthalpy is realized when interactions between the ligand and the protein, and also those between the released water molecules, are more beneficial than the interactions present before the protein–ligand binding. As polar interactions are sensitive to the relative position of the interacting partners, optimal interactions can only be achieved when ligand and protein structures fit well. To appreciate this sensitivity we quote that the interaction energy of a hydrogen-bonded water dimer near to a stationary point on the potential energy surface changes over 23 kJ mol^{-1} upon $\sim$0.5 Å shrinkage of the O–O distance.[35] The high sensitivity of the enthalpy to the relative positions of the interacting molecules poses a strict restriction on the possible ligand structures that can interact with the protein with a favorable enthalpy. Furthermore, as the size of the ligand increases it is less likely that a good fit can be realized. Indeed, it was demonstrated by simple models that increasing the complexity of the system dramatically decreases the chance of useful interactions with a randomly chosen ligand.[36] On the other hand, increased complexity is required for high affinity binding, but then the entropic contribution is likely to prevail.[37,38]

The existence of a single distinguished binding site (as proposed in ref. 39) is also in favor of ligands with limited size. It is argued that the loss of rigid body entropy upon binding represents a significant barrier to binding and in most targets it can be overcome only at a single site. These critical binding regions (hot spots) are sensitive to structural changes and they provide an indispensable contribution to binding. We assume that enthalpy dominates such binding events. Note that the rigid body entropy loss has only small size dependence[39,40] and this, together with the limited entropy gain coming from the water release of the binding of small ligands, rationalizes that fragment binding is often enthalpy driven.[41] The extension of ligands to adjacent sites gives a relatively less important contribution to the binding and this is in line with the observed decrease of binding enthalpy of larger ligands. Rejto and Verkhivker[42] also proposed the existence of a unique site of binding that is characterized by an unfrustrated energy landscape and thus a unique binding mode of small core fragments. Although computational docking did not find outstanding scores for these sites,[42] it can be assumed that they correspond to the critical binding regions of ref. 36 and are associated with important enthalpy contributions.

It is instructive to decompose pK_{dmax} into its pK_H and pK_S components. We emphasize that this enthalpy component of the highest affinity compounds does not correspond to the maximal achievable binding enthalpy (pK_{Hmax}). In order to remark this distinction we will designate the enthalpy component of pK_{dmax} as pK_{H_dmax}. Note that pK_{H_dmax} values were extracted from our data set in the following way: the five highest affinity compounds for a given HA were taken and the one with the highest pK_H was selected as pK_{H_dmax}. In this way, pK_{H_dmax} represents the maximal enthalpy among the maximal affinity compounds. Figure 2.4 shows pK_{Hmax}, pK_{H_dmax} and pK_{dmax} together as functions of the number of heavy atoms. The enthalpy content of maximal affinity compounds (pK_{H_max}, red bars) decreases with increasing compound size; while pK_{H_dmax} exceeds pK_{dmax} for smaller ligands, it is below pK_{dmax} for larger ligands (*cf.* red and green bars). In other words, binding of the highest affinity ligands is basically enthalpy driven when the ligands are small and it becomes entropy driven when the ligands are large. The enthalpy dominated binding of small size ligands is also supported by thermodynamic data of fragment screening hits.[41]

The enthalpy of the highest enthalpy ligands (pK_{Hmax}) is consistently higher than pK_{dmax} (*cf.* blue and green bars). This shows that high binding enthalpy can be achieved even for large compounds and this poses the question of why the high enthalpy is not accompanied with high affinity for these large compounds. A structural comparison of large-size high affinity and high enthalpy compounds reveals that the latter are significantly more polar. This

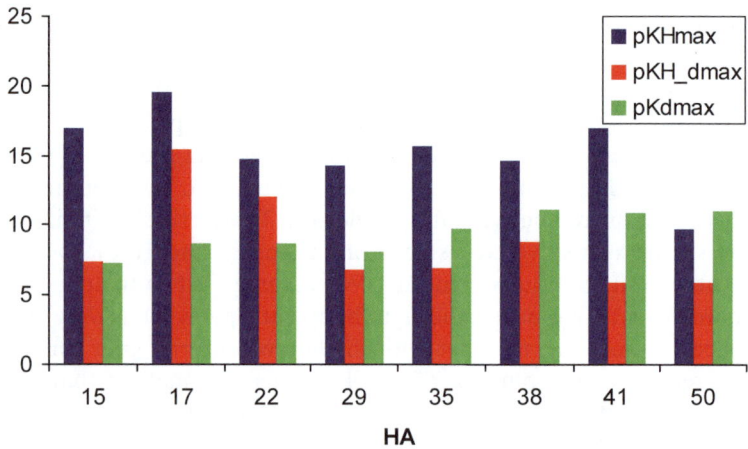

Figure 2.4 pK_H of the highest enthalpy compounds (pK_{Hmax}) and pK_H and pK_d of the highest affinity compounds (pK_{H_dmax} and pK_{dmax}) *versus* the number of heavy atoms (HA) (taken with permission from ref. 31).

point will be discussed at some detail in Section 2.5.3. Here we only note that this is in line with the observation that the transfer of polar ligands from water into the crystalline phase (the latter is assumed to have interactions similar to ligand–protein complexes) is enthalpically favorable and is accompanied by an entropic penalty.[43]

Summarizing the analysis of Figure 2.4, it can be concluded that the enthalpy content of high affinity binding decreases with increasing ligand size and that high binding enthalpy can be achieved with increased polarity that tends to decrease the affinity for larger compounds. These findings have important implications for compound selection and optimization in drug discovery programs, as it is discussed in Section 2.5.3.

2.3 Measuring Binding Thermodynamics

Isothermal titration calorimetry (ITC) and the van't Hoff analysis are the most common experimental approaches that give information on thermodynamic quantities of ligand–protein binding. While ITC refers to the experimental technique for the direct measurement of ΔH and K_d, the van't Hoff analysis signifies the procedure that derives binding free energy from binding constants measured at different temperatures with any experimental method. An overview of the principal characteristics of these two approaches is presented with a comparison of the utility of their results.

2.3.1 Isothermal Titration Calorimetry

ITC is considered to be the preferred method for thermodynamic character-ization of binding owing to its ability to directly measure ΔH, K_d and stoichiometry. Then a direct calculation provides us with ΔG and ΔS. Moreover, performing the measurements at several temperatures allows the evaluation of the heat capacity change, ΔC_p.

The ITC experiment is performed by the stepwise addition of one of the components (usually the ligand) to the other (usually the protein). The heat absorbed or released is measured in each step (see ref. 44 and Figure 2.5). At the beginning of the titration the whole amount of the added titrate forms a complex due to the large excess of the titrant and gives a measure of ΔH. The heat change diminishes during the titration and the integrated heat of binding plot makes it possible to estimate K_d. This estimate depends on the binding model and it requires a correction for other heat effects during titration, *e.g.* the dilution of the components.[45] The range of binding constant that can be determined with a usual experimental setup is $\sim 10^4$–10^8 M^{-1}. The limit for strong binders arises from the low heat change that accompanies the binding at low concentration, while the limit for weak binders is the consequence of solubility and stability issues.[45] Both limits can be extended by displacement experiments in which the competition with ligands having less extreme binding affinity is exploited to shift heat changes in the appropriate range.[46–48] It is

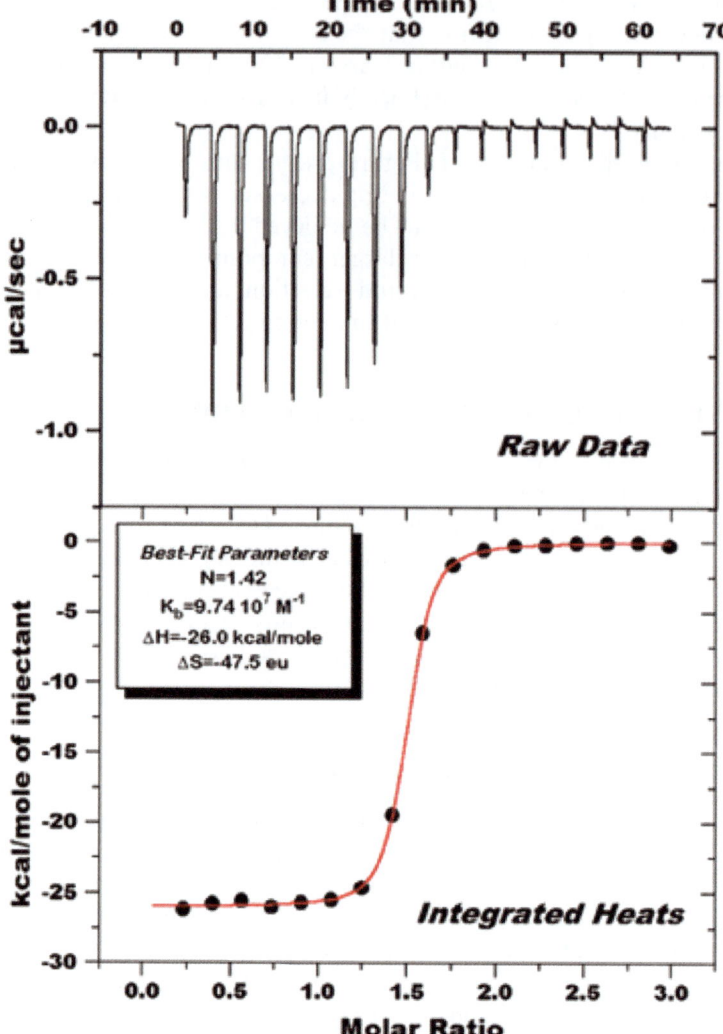

Figure 2.5   Typical ITC data. *Top*: raw ITC data; *bottom*: binding isotherm.[44]

worth noting that differential scanning calorimetry (DSC; an otherwise less preferred alternative to ITC) can be advantageously used in the case of high binding constants.[49,50]

When protonation events are linked to binding, then extra protons released or taken by the ligand–protein complex will interact with the buffer. In such cases, the measured heat depends on the ionization enthalpy of the buffer. Then, using various buffers with different ionization enthalpies makes it possible to determine the number of protons taken or released by the buffer and to calculate an intrinsic binding enthalpy that does not include the heat of protonation or deprotonation of the buffer. Having determined the change in the protonation state upon complex formation and performing calorimetric experiments at various pH values gives information on the pK_a values of the partners before and after complex formation.[51,52] Another utility of buffer variation in binding experiments coupled with protonation events is that it can shift heat changes in the range accessible for ITC.

The principal limitations of ITC experiments are high protein consumption and low throughput. Recent ITC instruments achieve more favorable limits by downsizing the sample size and increasing the sensitivity of the measurement, allowing a sample throughput of 50 samples per day or above.[53] Another recent development is the introduction of chip-based array calorimetry. Their protein consumption is reduced by several orders of magnitude and throughput is increased; however, their sensitivity requires improvement (see ref. 41 and references therein).

As a final note on ITC, the annual survey of its applications includes, among others, those relevant in drug discovery. The latest survey at the time of the preparation of the present manuscript is ref. 54.

2.3.2 Van't Hoff Analysis

A form of the van't Hoff equation well suited for our present discussion reads as:

$$\ln K_d = \frac{H}{RT} - \frac{S}{R} \tag{2.15}$$

This can be derived directly from eqns (2.3) and (2.4). It shows that measuring K_d at different temperatures and plotting $\ln K_d$ as a function of $\frac{1}{T}$ gives $\frac{H}{R}$ as the slope and $\frac{S}{R}$ as the intercept of a straight line. A complication that can arise is the nonlinearity of the van't Hoff plot that is the manifestation of the temperature dependence of ΔH. In such cases, a larger number of measurements, the inclusion of $C_p = \frac{dH}{dT}$ into the model together with an involved statistical data analysis[55] are required.

Any method that is able to measure K_d over a temperature range can be appropriate for the application of the van't Hoff method. Radioligand binding assays have been commonly used and provide data for receptors. Measured data supported the interpretation of phenomena like allostery[56] and agonist–antagonist differentiation.[57] Membrane proteins are particularly advantageous targets for the application of the van't Hoff method as they show small ΔH variations with temperature.[58] Other techniques suitable for measuring K_d at different temperatures include surface plasmon resonance (SPR), NMR, mass spectrometry and chromatographic methods.[59] SPR is especially attractive owing to its low protein requirement and relatively high throughput, although it has the disadvantage that it requires the immobilization of one of the partners.[60]

2.3.3 Further Notes on Experimental Approaches

Comparisons of the thermodynamic quantities determined by ITC and by van't Hoff's methods are hampered by the effect of varying experimental conditions to the derived quantities. Nevertheless, recent comparisons show the consistency of the data[61–63] that permits the selection of the technique most appropriate to a given system. The instrumental development holds the promise that these techniques will play an increasingly important role in the drug discovery process; in particular, they may appear as screening tools primarily in fragment-based discovery (FBDD).[41,60] FBDD is well suited for thermodynamic screening as it requires moderate throughput and thermodynamic profiling represents significant support for hit selection and optimization.

The increase of available thermodynamic data prompted the creation of databases, namely BindingDB,[64] PDBCal[65] and Scorpio,[66] that include primarily ligand–protein complexes. As thermodynamic profiling is most useful when corroborated by structural data, these databases also refer to atomic level structural information. The contents of these databases are overlapping. All together they include data for about 1000 systems and they allow the systematic analysis of various properties and the recognition of trends characteristic to ligand–protein binding (see *e.g.* refs. 31, 67–69).

2.4 Calculation of the Binding Free Energy and its Components

The calculation of ligand–protein binding free energy is a major goal of computational drug discovery. A reasonably accurate and fast prediction of free energy and thus ligand affinity of design compounds would have a fundamental impact on drug research. Concerning accuracy, we recall that according to eqn (2.3) a 1.4 kcal mol^{-1} change in the binding free energy corresponds to a 10-fold difference in the dissociation constant and this represents an approximate upper bound to the accuracy required for a useful prediction. Various techniques have been elaborated for the calculation of binding free energies; however, none can be considered as a routinely applicable accurate enough solution to the problem. In the forthcoming discussion we will survey the most popular computational techniques used for estimating the total free energy of ligand binding or the free energy difference of the binding of two similar systems. We also present methods aimed at calculating the free energy components, namely enthalpy and entropy.

2.4.1 Molecular Simulation-Based Computational Methods

These methods assume that the Gibbs free energy is well approximated by the Helmholtz free energy (F), expressed as:

$$F = -kT \times \ln Z \tag{2.16}$$

where the partition function

$$Z = \sum_i \exp\left(-\frac{E_i}{kT}\right) \tag{2.17}$$

includes a sum over all configurations of the system. Here k is the Boltzmann constant, T is the absolute temperature and E_i is the energy of the i-th state.

The evaluation of the partition function for systems as complex as those in ligand–protein binding is not feasible. Even its approximation with a set of Boltzmann-weighted snapshots [generated from Monte Carlo (MC) or molecular dynamics (MD) simulations] is highly inaccurate and therefore various methods have been worked out to estimate free energies or free energy differences without the evaluation of Z.

Binding free energy differences can be obtained with a reasonable amount of computational work by alchemical transformations. An example is presented in Figure 2.6. The free energy change in this thermodynamic cycle is zero. Thus the difference of the binding free energies of ligands A and B can be written as:

$$G = G_{\text{bind}}^{A} - G_{\text{bind}}^{B} = G_{\text{transform}}^{\text{ligand}A \to B} - G_{\text{transform}}^{\text{complex}A \to B} \tag{2.18}$$

This equation shows that the binding free energy difference of ligands can be calculated as the difference of the free energies of two alchemical transformations: one that transforms the unbound solvated ligand A into B, and another that transforms the solvated protein–ligand A complex into protein–ligand B complex. The advantage of treating these alchemical transformations is that they connect systems whose free energy difference can be calculated with improved efficiency.

The two most widely used methods for calculating free energy differences of alchemical transformations are thermodynamic integration (TI) and free energy perturbation (FEP). The TI equation has a particularly simple form when the potential functions of the two states are linear in a parameter λ:

$$G = G_2 - G_1 = \int_0^1 <V_2 - V_1>_\lambda \, \mathrm{d}\lambda \tag{2.19}$$

where V_i is the energy of the i-th system and the $<>_\lambda$ brackets indicate an ensemble average over the λ distribution of states.

The derivation of this equation, together with that of other thermo-dynamics-based methods, is given in several papers and reviews (see *e.g.* ref. 69). TI calculations include multiple simulations with different λ values and a numerical integration over λ to obtain the free energy difference.

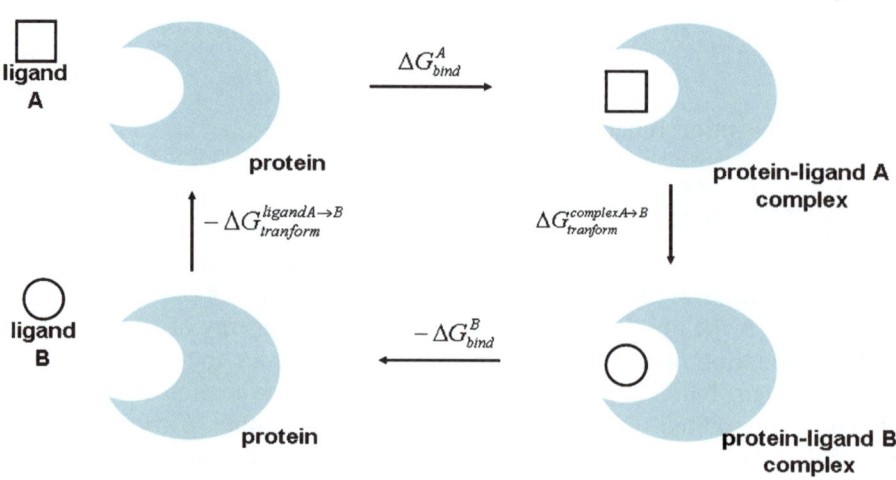

Figure 2.6   Alchemical transformation for calculating the free energy difference of
binding of two ligands to the same protein.

The basic equation of FEP is:

$$G = G_2 - G_1 = -kT \ln < \exp\left(-\frac{V_2 - V_1}{kT}\right) >_i \qquad (2.20)$$

where the $<>_i$ brackets indicate an ensemble average over system i (1 or 2).

A FEP calculation includes the evaluation of the energy difference between the two states and the ensemble average is taken over the first state. Performing also a backward simulation, *i.e.* taking an average over the second state, allows an estimation of the convergence. Improved accuracy *via* a better sampling can be achieved by dividing the transition between the two end states into several steps.

Non-equilibrium work methods[70,71] are related approaches based on the equality of the work associated with the non-equilibrium switch between two states and the free energy difference of these states:

$$G = G_2 - G_1 = -kT \ln < \exp\left(-\frac{W}{kT}\right) > \qquad (2.21)$$

where W is the external work performed on the system and the average is taken along the possible trajectories.

The above techniques can also be used to calculate the potential of mean force (PMF) or free energy profiles typically along a physical coordinate.

The double-decoupling method[72] deserves special attention as it is able to calculate standard binding free energies. The thermodynamic basis of the methods is shown in Figure 2.7. Double decoupling includes two simulations: one with the ligand in solution and another with the ligand together with the

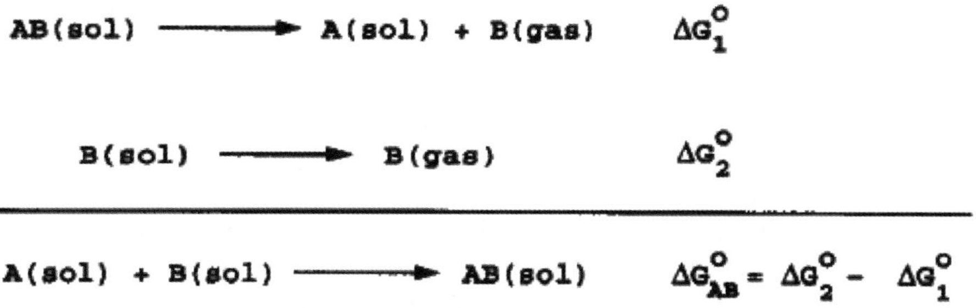

$$AB(sol) \longrightarrow A(sol) + B(gas) \qquad \Delta G_1^o$$

$$B(sol) \longrightarrow B(gas) \qquad \Delta G_2^o$$

_____

$$A(sol) + B(sol) \longrightarrow AB(sol) \qquad \Delta G_{AB}^o = \Delta G_2^o - \Delta G_1^o$$

Figure 2.7   Thermodynamic basis of the double decoupling method (taken with permission from ref. 73).

protein in solution. In both simulations the interactions of the ligand with its environment are decoupled.

We do not discuss here the various technical aspects of free energy simulations, but we note that sophisticated techniques are required to improve sampling and data analysis and to obtain meaningful estimations of binding free energies or their differences. Interested readers are referred to recent reviews.[73-75]

The quality of the potential energy function applied in the simulation is a crucial determinant of the accuracy of the free energy estimation. Most calculations use classical force fields. These give a reasonable description of the proteins owing to the limited variability of protein sequences, but they may be less appropriate for diverse ligands or cofactors. A particularly challenging aspect of force fields is the proper account of polar interactions. The evaluation of long-range electrostatic interactions is computationally demanding and their best approximations invoke either periodicity or a dielectric continuum beyond a certain cutoff distance. A proper description of polar interactions is problematic also at short interatomic separations. The oversimplified representation of molecular charge densities by atomic point charges and the neglect of polarization may affect the quality of the description adversely. Although improved models by distributed multipoles[76] or by effective multipoles[77] are well established, their use is hampered by the complexity of storing and evaluating the interactions of higher rank multipole moments. The application of polarizable force fields is rare and both their testing and implementation are scarce. We note that improved quality electrostatics and polarization can be treated on an equal footing since both require the inclusion of higher rank moments in the force fields.[78]

All methods described above aim at estimating the binding free energy. If we wish to calculate its enthalpy and entropy components we are faced with additional difficulties. In principal, the enthalpy should be easier to evaluate than the free energy owing to the smaller fluctuations of the ensemble averages of the former (see *e.g.* ref. 69). However, these fluctuations are still too high to obtain meaningful results with reasonable computational effort. For the same reason, the evaluation of an enthalpy difference as the difference of ensemble averages is highly inaccurate. Various methods have been proposed to calculate enthalpy or entropy differences as ensemble averages (rather than the difference of ensemble averages). These methods are typically based on formulas for free energy differences and exploit relationships between thermodynamic quantities.[78,79] Unfortunately, they are unable to achieve the accuracy of the best techniques for evaluating free energy differences (*vide supra*); they give results with reasonable accuracy for simple solute–solvent systems, but they are not appropriate for treating ligand–protein binding.

Binding free energy differences can be estimated based on the recognition that in TI calculations only the λ-dependent part of the Hamiltonian contributes to the free energy difference of a perturbation, while the λ-independent parts of enthalpy and entropy contributions exactly cancel each other.[80] As the λ-dependent part converges faster, it allows a good estimation of the enthalpy differences owing to the perturbation. This makes it possible not only to evaluate the free energy differences of ligand binding to DNA with different nucleotide sequences, but also to decompose the enthalpy changes into various interactions.[81]

2.4.2   Other Computational Methods

2.4.2.1   *Estimation of Free Energy*

Molecular simulation based methods give a theoretically well founded and potentially accurate description of ligand binding thermodynamics. On the other hand, primarily for practical reasons partly discussed above, they do not offer a general solution to calculate binding free energies and their components. This prompted the development of a plethora of other methods to calculate the binding free energy or its specific contributions. Some of them include simulation-based estimates of certain properties, but they invoke additional approximations with respect to the methods discussed in Section 2.4.1.

MM-PBSA[82] calculates the binding free energy as the difference between the free energies of the solvated complex and those of the solvated unbound components. The free energy is approximated with the following terms:

$$G = E_{MM} + G_{PBSA} - TS_{MM} \tag{2.22}$$

where E_{MM} is the molecular mechanical energy, G_{PBSA} is the solvation free energy and TS_{MM} is the solute entropy.

Several variants for the calculation of these terms have been proposed. E_{MM} can come from simple molecular mechanical minimization or from MD trajectories. In this latter case the energy of the unbound molecules can be obtained from simulations performed for the unbound molecules or from the simulation performed for the complex. G_{PBSA} is calculated with a numerical solution of the Poisson–Boltzmann equation and an estimate of the nonpolar free energy with a surface area term. TS_{MM} usually includes an estimate of the conformational entropy obtained by normal-mode analysis. MM-PBSA was found to be appropriate to improve virtual screening results when applied as a post-docking filter and also to prioritize design compounds. On the other hand, it is expected to correctly rank compounds with free energy differences of at least 3 kcal mol^{-1}, at best.[83]

The Linear Interaction Energy (LIE) method[84,85] estimates the standard binding free energy as the sum of an electrostatic (V^{el}) and a van der Waals (V^{vdw}) term:

$$G = \frac{1}{2}\left(\langle V^{el}\rangle_{bound} - \langle V^{el}\rangle_{free}\right) + \alpha\left(\langle V^{vdw}\rangle_{bound} - \langle V^{vdw}\rangle_{free}\right) \qquad (2.23)$$

where the ½ comes from the assumption of linear response and α is an adjustable parameter.

Again, several variants of the method have been proposed. They include the replacement of the ½ coefficient of the electrostatic term by an adjustable parameter,[86] the addition of a term proportional to the solvent-accessible surface area to account for cavity formation[87,88] and the replacement of the molecular mechanical electrostatic and van der Waals energy terms by quantum mechanical/molecular mechanical (QM/MM) interaction energy calculated for the time averaged structure.[89] The application of the LIE method requires the knowledge of some binding free energies to perform the calibration of the adjustable parameters. A related approach is the PDLD/S-LRA/β method[90] that calculates the electrostatic term with the protein dipoles–Langevin dipoles model[91] assuming a linear response; it is claimed to perform effectively without the need of system-specific parameterization. These linear response-based methods were shown to give reasonable results for certain series of ligands. In other cases, LIE estimations are subject to important errors and owing to the approximations involved an *a priori* assessment of the quality of the results is difficult if at all possible.

Scoring functions, designed for a fast ranking of ligand–protein complexes, also estimate binding free energies (see refs. 92 and 93 for recent reviews). In typical applications, scores for a large number of ligands complexed with the same protein are calculated and then a selection of top ranked ligands gives a set enriched with compounds showing reasonable binding affinity towards the protein. Various schemes are used to derive scoring functions and they largely differ in the way the various free energy components are approximated. As scoring functions are typically used for treating a large number of compounds (often in the range of 10^5–10^6), accuracy and rigor in the theoretical foundation are sacrificed for speed. As a consequence, the correlation between scores and binding free energies is poor, and the enthalpy and entropy components cannot be straightforwardly identified. On the other hand, with the improvement of methodology and with the advancement of available computer power the notion of scoring function starts to expand and to include more refined methods. In addition to those already discussed in this section, at least two others are worth briefly describing here. An assumption that the thermodynamics of ligand binding is dominated by the displacement of water molecules from the protein binding site led to a scoring function whose principal parameters are the interaction energy and excess entropy terms for the displaced water molecules.[94,95] The protein hydration sites are determined with short molecular dynamics simulations rather than derived from experimental structures. In contrast to the usual scoring functions, this method is designed to compare binding free energies of congeneric series. Reported results include binding free energy differences for a set of factor Xa inhibitors showing good correlation with experimental values[94] and for a set of kinase inhibitors explaining selectivity profiles.[96] Another scoring function was derived by using structural data and calorimetric measurements to separately fit enthalpic and entropic contributions.[97] Various models were generated and it was found that in spite of fitting to the enthalpy and entropy these components are poorly reproduced, while the description of the total free energy compares favorably with that of other scoring functions. The model is obtained with a least-square fit of 42 descriptors and thus it gives limited insight into the physics behind the scoring function.

2.4.2.2 *Estimation of Enthalpy*

Quantum mechanics (QM) offers a potentially highly accurate description of intermolecular interactions. Its advantages over molecular mechanics (MM) include that no parameterization is required and thus compounds with unusual structural motifs can be treated, and the accuracy of the description can be systematically increased. Unfortunately, the high computational demand represents a serious limitation to sampling the configurational space by QM. At the same time, a limitation of the QM evaluation of the energy of configurations generated by MM stems from the differences of the QM and MM potential surfaces.[98] An alternative approach, the approximation of the enthalpy of binding by semiempirical QM calculations for a single configuration, was also proposed. The X-ray geometries of ligand–protein complexes were reasonably reproduced by semiempirical energy-minimized structures.[99] Furthermore, the comparison of semiempirical and *ab initio* methods with partial accounts for electron correlation showed fair agreement between the calculated interaction energies for a series of complexes that simulate typical ligand–protein interactions.[99] The AM1 approximations showed superior results over PM3 parameterization. In another study, experimental ligand–protein binding enthalpies were compared to AM1 calculated values. The experimental binding enthalpies were reproduced for eight complexes with an rms error of about 2 kcal mol^{-1} for the best model. This model included water molecules added to the unbound compounds to replace H-bonds broken when compounds were removed from the complex.[100] Further improvement of the results was reported with a description of solvation by a combination of explicit water molecules and an implicit continuum solvation model.[101] These highly accurate results are somewhat unexpected, taking into account the approximations involved. In particular, the application of a semiempirical QM method for calculating intermolecular interaction energies and the evaluation of the energy at a single configuration for a simplified model may significantly affect accuracy. A deeper understanding of the reasons for the good agreement with experimental binding enthalpies and further examples are needed to increase the confidence in the performance of the method. A final remark concerning the application of QM methods is that a highly accurate description of selected factors does not necessarily result in higher quality thermodynamic quantities that emerge as the sum of several partially cancelling contributions.[102]

The thermodynamic characterization of ligand–protein binding using structural data with an empirical parameterization stems from the successful application of this type of approach for the description of protein folding (see *e.g.* ref. 103). A key parameter in predicting thermodynamic quantities, ΔG, ΔH, ΔS and ΔC_p, in protein folding is the change of solvent accessible surface areas (ΔASA) and its dissection into apolar (ΔASA_{apolar}) and polar parts (ΔASA_{polar}). These descriptors were then also used in applications to ligand–protein binding.[104] An empirical parameterization specific for ligand–protein binding enthalpy and applicable for a wider range of systems complements ΔASA_{apolar} and ΔASA_{polar} with a protein-specific constant interpreted as the enthalpy of conformational change that accompanies binding.[105] An equation for these structural parameters and binding enthalpy was set up for seven HIV-1 complexes. This equation with three fitted parameters reproduced the binding enthalpies for the seven complexes within 0.5 kcal mol^{-1} rmsd. Equations with the same coefficients for the polar and nonpolar surface area changes and with the refitting of the constant for six other proteins yielded similar quality results for 18 additional complexes. In spite of these encouraging results, three important limitations for the method are expected. Firstly, the correction for the protonation/deprotonation contribution to the experimental enthalpy is a prerequisite for a good agreement between calculated and measured values. Secondly, as the equation uses a conformational enthalpy contribution that is protein and not complex dependent, the same excess conformational enthalpy is assumed for the protein in each of its complexes and also for each ligand of a protein. Finally, a further feature of this approach is that it does not include an explicit term for the ligand–protein interaction, rather the contribution of this interaction to the enthalpy is implicit in the surface area terms. This, however, seems to be a simplification in light of the known sensitivity of the enthalpy to the geometry of the interacting partners.[106]

2.4.2.3 Estimation of Entropy

Various approximate methods have been proposed to estimate the entropy change upon ligand binding. They typically address certain components of the entropy, most often the configurational entropy change of the solute. These methods cannot be directly compared to experimental results as they do not provide us with measurable quantities. It is also important to realize that the usual decomposition of the entropy into translational, rotational and vibrational components as $S = S_{trans} + S_{rot} + S_{vib}$ is somewhat arbitrary. Similarly, the hard (bond length and angles) and soft (dihedral and external) coordinates may couple and it is an approximation to treat selected components separately.

Normal mode analysis estimates the entropy from an energy-minimized structure assuming harmonic potentials.[107–109] The value of the calculated $T\Delta S$ was found to vary with the selected minimized structure by 5 kcal mol^{-1} in unfavorable cases.[110] Another factor affecting the utility of the normal mode analysis is the validity of the harmonic approximation. In a relevant example, the majority of the normal modes in lysozyme were found to be harmonic, but it was also found that the few anharmonic modes were responsible for the overwhelming majority of the atomic fluctuations.[111]

The quasi-harmonic (QH) method calculates the configurational entropy assuming a multivariate Gaussian distribution for the Boltzmann probabilities and deriving the covariance matrix of the coordinates from computer simulations.[112,113] A systematic analysis of the QH method as applied to biochemical systems revealed shortcomings that include an overestimation of the entropy and slow convergence.[114] This latter finding confirmed previous results.[115,116]

The "mining minima" approach[117–119] is able to estimate configurational entropy and is exempt from assumptions of the previous methods. It identifies local minima of the potential surface, *i.e.* predominant low-energy conformations, and their contributions to the configurational integral is evaluated taking anharmonicity also into account. The computational intensive search is currently practical with implicit solvent models only.

The restraint release (RR) approach[120] and its modified version[121,122] evaluate the entropy as the free energy change associated with the release of an appropriately defined restraint. This was applied in an attempt to estimate the total binding entropy as the sum of three components, namely configurational, polar solvation and hydrophobic terms.[123] Different thermodynamic cycles and approximations were used to calculate the three contributions. The configurational entropy is similar to that of other studies while total entropies calculated for four systems overestimate experimental values.

2.4.3 Summary of Binding Free Energy Calculations

Significant efforts have been devoted to developing methods for the calculation of the thermodynamic quantities of ligand–protein binding. Among them, the free energy of binding has received the most attention owing to its direct relation to ligand affinity, whose improvement is a principal goal in the development of drug candidates. The most successful methods are based on atomic level simulations using molecular mechanics force fields. In favorable cases these methods can reproduce experimental binding free energies, or their differences, with an accuracy of 1–2 kcal mol^{-1} that roughly corresponds to one order of magnitude uncertainty of binding constants. The calculation of enthalpy and entropy with simulation-based methods faces serious convergence difficulties. On the other hand, the increased interest in these quantities and their larger sensitivity to structural variations may prompt further efforts to develop improved methods to calculate them. Various other approaches are able to give reasonable estimates of certain thermodynamic quantities in specific cases. Improvements are hindered by the dependence of binding thermodynamics on several factors, like polar and hydrophobic solvation, configurational and rigid body entropy, ligand–protein interactions, *etc.*, whose contributions cancel to a large extent. For this reason a better representation of selected aspects may not result in an overall improvement. Altogether, there is room for significant progress in evaluating binding related thermodynamic quantities at a level of accuracy required for drug design applications. Nevertheless, calculations are able to support decision making in hit generation and, to a lesser extent, also in hit-and-lead optimizations. Another impact of these calculations is their contribution to our understanding of the binding process. The interpretation of the various contributions is only possible with theoretical and computational analyses of the binding event. Calculations complement experimental data by providing details at an atomic level and they drive our understanding of ligand–protein binding.

2.5 Impact of Thermodynamics on Medicinal Chemistry Optimizations

2.5.1 Property Shifts and Medicinal Chemistry Optimizations

Since Oprea[123] revealed in 2001 that the properties of drug molecules are distinct from those of their original leads, it has been demonstrated that recent clinical candidates have higher lipophilicity and molecular weight relative to their corresponding leads and historic drugs.[124] Assuming that lead optimization programs were started from good quality leads, one can conclude that it is the lead optimization strategy responsible for the observed property shift. There is, however, another option that suboptimal clinical candidates are typically identified by optimizing leads with disadvantageous physico-chemical characteristics. Comparing the physico-chemical profile of recent leads to historic leads, we recently showed that the property shift can be traced back to the lead discovery phase. We found that recent leads are more lipophilic and have higher molecular weight than that of historic leads and even drugs.[125] Since, in the current drug discovery paradigm, leads are typically developed from hits of virtual and/or experimental screening, the physico-chemical properties of the screening collection have a major impact on lead quality. Physico-chemical profiles of screening compounds were first investigated by

Tegaue *et al.*[126] in 1999, concluding that polar and low molecular weight starting points were more easily converted into leads than lipophilic and higher molecular weight hits. They proposed that a suitable screening library should consist of compounds with a molecular weight range between 100 and 350 and that $c\log P = 1$–3. Hits from such lead-like libraries would provide wider chemistry space for optimization of potency, physico-chemical and ADME properties. Investigating the effect of molecular complexity on ligand binding, Hahn and co-workers[36] showed that large and complex molecules more readily form suboptimal or repulsive interactions upon binding to proteins. In addition, compounds with high molecular weight and lipophilicity have a greater chance of being promiscuous, but they typically have limited solubility and ADMET (absorption, distribution, metabolism, excretion, toxicity) problems. Analysis of more than 2000 compounds investigated in Cerep Bioprint assays indicated that the risk of promiscuity increases for compounds with $c\log P$ larger than 4.[124] Based on the *in vitro* ADME dataset consisting of 30 000 GSK compounds, Gleeson demonstrated that compounds with $c\log P$ less than 4 and MW less than 400 have better ADME profiles than more complex and more lipophilic compounds.[127] Investigating almost 250 compounds *in vivo* using tolerability and toxicology studies, Hughes *et al.*[128] concluded that the prevalence of serious toxicological outcomes increases for compounds with $c\log P$ larger than 3. These observations finally led to the generally accepted conclusion that less complex, polar, low molecular weight hits serve as better starting points for hit-to-lead optimization.

In our previous work[125,129] we demonstrated in several examples that the increase in $\log P$ and molecular weight during hit-to-lead optimization is independent of the nature of the library screened, on the detection technology applied and on the lead discovery strategy used. Consequently, it is more than likely that medicinal chemistry optimizations are the primary source of the property shift. Figure 2.8 shows typical medicinal chemistry optimizations including several hit-to-lead and historic lead-to-candidate processes in the MW–$\log P$ space. Comparing hit-to-lead optimizations to historic lead optimization, it is obvious that virtually all of the optimizations show similar ΔMW/$\Delta\log P$ gradients. This suggests that the strategy of medicinal chemistry optimization is similar in both historic and recent programs and does not differentiate between lead discovery and lead optimization. The second observation is that the length of the optimization pathway depends on the nature of the starting point. Interestingly, there is only a minor shift in both $\log P$ and MW for historic lead optimization that starts from pre-optimized leads. This pathway is similar to that of the HTS-based hit-to-lead process that converts reasonably active hits to leads. Starting from typically less active virtual screening hits and natural product starting points, the optimization path to the corresponding lead is longer. Finally, hit-to-lead optimization of the least-active fragment hits shows the longest pathway on the MW–$\log P$ space.

Based on the comparison of different medicinal chemistry optimizations starting from compounds with a wide range of potency (from drug leads to

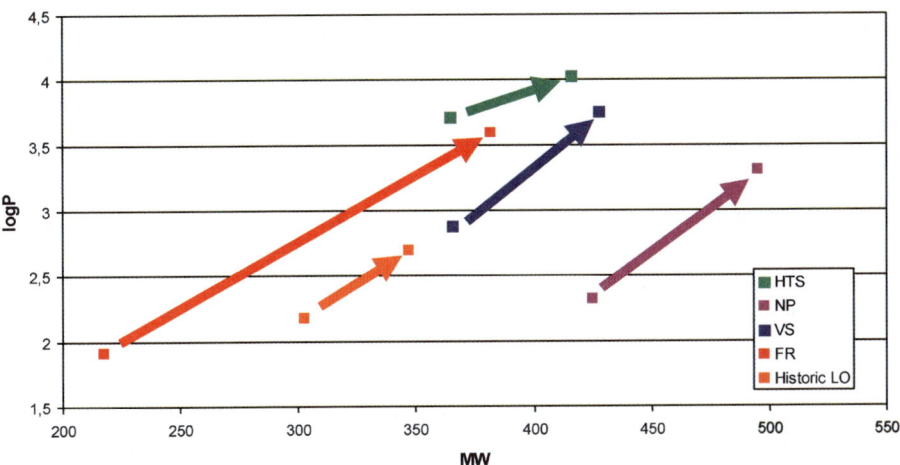

Figure 2.8 Averaged log*P*–MW paths for HTS-based, fragment-based (FR) and historical optimizations.

fragment hits), it seems that potency optimization is one of the key drives of the property shift.

Potency-driven optimization has a long tradition in drug discovery.[130] Emblematic drug hunters searched for chemical weapons efficacious against diseases in an approach that considered compounds with high *in vivo* potency more beneficial. In the modern era, increased potency might help increasing target specificity which could contribute to improved safety profile and higher therapeutic index. In addition, high potency allows the reduction of therapeutic doses that is beneficial from both safety and pharmaco-economical points of view. Since high potency seems to be a longstanding goal for drug discovery teams, the recently observed inflation in physico-chemical properties could not be rationalized solely by potency-driven optimization. The other factor contributing to this disadvantageous property shift is biological in nature. Recent clinical candidates are typically identified in mechanism-based discovery programs utilizing hierarchical screening cascades. These cascades are started by simple *in vitro* assays monitoring potency and selectivity on the target. More and more complex biological systems, including *in vivo* disease models, safety and toxicology studies, are typically located at the late phase tertiary cascade. Early phase *in vitro* screening allows high molecular weight, complex and lipophilic molecules to progress. At the end, compounds with suboptimal physico-chemical profiles are entered into late-stage testing that typically requires special formulations and solid state chemistry optimization. In contrast, the historic drug discovery paradigm was based on early phenotypic screening using *in vivo* models that filtered out compounds with suboptimal physico-chemical properties. Consequently, potency-driven optimization in conjunction with extensive *in vitro* profiling by early-phase screening cascades results in highly potent compounds with compromised physico-chemical and ADME properties.

Table 2.1 Potency change and property shift in medicinal chemistry optimizations.

Process	pXC50 change	MW change	logP change
Fragment hit-to-lead	2.76	172.8	1.22
HTS hit-to-lead	1.39	51.5	0.27
Historic lead optimization	1.7	42.0	0.5

The role of potency-driven optimization in the disadvantageous shift of physico-chemical properties can be supported by analysis of hit-to-lead[125,131] and lead-to-candidate[36,132] programs (Table 2.1), indicating that both molecular weight and lipophilicity correlate positively with potency. Optimization strategies focusing on potency only, however, represent a considerable risk for discovery programs, as demonstrated in Figure 2.9. Historic lead optimization was typically started from lead-like compounds that were optimized to historic drugs remaining well within the Lipinski zone. In contrast, present medicinal chemistry practice optimizes hits to leads located in the center of drug-like space. Consequently, further optimization with the same practice would shift the resulting candidate to the edge of the Lipinski zone, with suboptimal physico-chemical properties. Lead optimization programs starting from these leads are therefore faced with "reverse" optimization, *i.e.* they should reduce both molecular weight and log*P* to reach the average property of phase II compounds. Since most of the medicinal chemistry programs typically increase both parameters, this type of reverse optimization is challenging or even unrealistic within the typical timeframe of a discovery program.

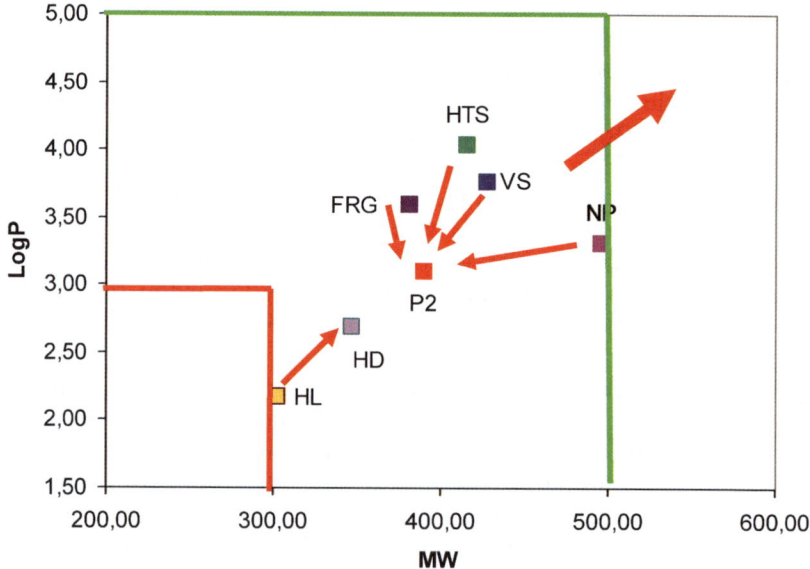

Figure 2.9 Reverse optimization from suboptimal leads.

Owing to the technical difficulties associated with reverse optimization, this situation is usually treated with some hope that the trade-off between potency and ADME properties, *i.e.* potency and physico-chemical properties, would be easier. Actually, there is a strong belief that high potency could compensate limited bioavailability. Unfortunately, limited bioavailability pushes the therapeutic doses higher and finally this would result in dose-limiting safety problems. Considering that the most significant part of clinical attrition is associated with lack of efficacy and safety issues, this approach would not be an ultimate solution to improve the performance of discovery teams. In fact, however, potency-driven medicinal chemistry optimization is basically responsible for the suboptimal combination of physico-chemical features (and disadvantageous ADME profile) that may affect further development adversely.

2.5.2 Optimization Strategies

Potency-driven medicinal chemistry optimizations can be categorized using the basic principles of binding thermodynamics. Equation (2.3) shows that efforts to increase the potency result in an improvement in the dissociation constant and consequently in the Gibbs free energy of binding. Since ΔG_{bind} can be decomposed into enthalpic and entropic terms (eqn 2.4), potency optimizations can be classified as enthalpy-driven and entropy-driven processes. Enthalpy–entropy compensation (see details in Section 2.2.2.2) prevents executing pure enthalpic and entropic optimizations. Enthalpy-driven optimizations can be identified as those where the gain in potency comes from improving the binding enthalpy that dominates over disfavored ΔS changes. Similarly, entropic optimizations are those that improve potency through increasing binding entropy that compensates for ΔH penalties.

Considering the multistep procedure of ligand binding in the framework of equilibrium thermodynamics, conformational rearrangements and solvation effects that occur at both the ligand and the protein side contribute to binding thermodynamics. Conformational rearrangements typically represents enthalpic penalties. Desolvation of the ligand reorganizes water clusters, thus resulting in significant entropic reward, while breaking hydrogen-bonding interactions between water molecules and the polar atoms of the ligand is associated with a disadvantageous change in binding enthalpy. Desolvation thermodynamics of protein binding sites is more complex, since it might be both enthalpic and entropic, depending on the binding interactions of the replaced waters. Formation of new interactions between the ligand and its binding site upon binding is associated with improvement in binding enthalpy. Ligand binding, however, limits the external rotational and translational freedom of the ligand and decreases the flexibility of the interacting partners that represents an entropic penalty. Ligand binding can be classified by the relative importance of enthalpic or entropic factors as enthalpy or entropy dominated. Comparing the binding thermodynamics of the starting ligand to that of the optimized compound, medicinal chemistry optimizations can be categorized as being basically enthalpy or entropy driven, depending on which component contributes more significantly to the affinity improvement.[106]

2.5.2.1   Enthalpic Optimization

Improvements in binding enthalpy typically require the formation of new contacts between the ligand and the protein. These beneficial interactions might include hydrogen bonds, salt bridges and van der Waals contacts. To obtain the maximal gain in enthalpy, one needs the interacting partners in optimal geometry. This is a challenging task even in structure-based optimization programs. On the other hand, these new interactions reduce the conformational flexibility of both the ligand and the protein, which represents an entropic penalty. In addition, heteroatoms typically used for the construction of new interaction sites disfavor the desolvation of the ligand, which results in an enthalpic penalty. These penalties could easily compensate the enthalpic reward of hard-to-form new interactions. Consequently, enthalpy optimizations are considered time consuming and challenging processes.

2.5.2.2   Entropic Optimization

Considering the entropy components of ligand binding, improvements in binding entropy could be realized by affecting conformational rearrangements and desolvation upon binding. Since desolvation of the ligand can be easily facilitated by increasing its lipophilicity, significant gain in desolvation entropy could be straightforwardly achieved in this way. More lipophilic compounds displace water molecules more easily from hydrophobic binding sites, which contributes further to entropic reward associated with increasing lipophilicity of the ligand. Gains in conformational entropy can be captured from approaches limiting the flexibility of the ligand. These are well-known tactics of medicinal chemistry (usually referred to as chain–ring strategies) that typically increase molecular weight and complexity with the simultaneous decrease of the entropic penalty. Entropic optimization therefore does not require specific interactions, but it could be realized by increasing the lipophilicity and complexity of the ligand. Consequently, increasing the potency could be more easily achieved in this way compared to enthalpic optimizations. In fact, entropy-driven optimization is a straightforward approach for medicinal chemistry teams working with strict timelines. Entropic optimization used routinely in most of the medicinal chemistry programs gives therefore a thermodynamic rationale for the undesirable shift in physico-chemical properties.

2.5.3   Thermodynamics-Guided Medicinal Chemistry Optimization

Thermodynamic characterization of ligand binding goes beyond the usual description of binding affinity and we argue that it could support medicinal chemistry decisions along the hit-to-drug-candidate path. From a thermo-dynamic point of view the optimization challenge is overriding the enthalpy–entropy compensation. Although enthalpy–entropy compensation is not a natural law, it represents a significant barrier for medicinal chemistry optimizations. Since this goal seems to be more easily achieved by improving entropic rather than enthalpic components, thermodynamically balanced optimization requires the measurement of the binding free energy components.

Evaluation of binding thermodynamics is obviously reasonable at decision points such as hit prioritization, lead selection and development candidate identification. Since most of the medicinal chemistry optimizations are entropic, the more enthalpic compounds identified at the milestone would be less affected by the disadvantageous shift in physico-chemical properties. Investigating the thermodynamics of low-potency enthalpic hits allows the identification of hot spots and also compounds that form the highest number of specific contacts with the target protein. At lead selection the enthalpy content of binding can be used for the evaluation of the optimization potential of different compound series. This comparative analysis might indicate whether an affinity gain with a favorable physico-chemical profile is realistic. Finally, a retrospective study suggests that thermodynamic signatures can be used to discriminate best-in-class compounds at the stage of candidate selection.[106]

In addition to thermodynamic characterization, ligand efficiency metrics are simple and useful tools governing medicinal chemistry optimizations. The original version of ligand efficiency (*LE*) as defined by Hopkins *et al.*[133] could be used to control the molecular weight and complexity, which are both scaled by the number of non-hydrogen atoms (N_{heavy}):

$$LE = \frac{G}{N_{\text{heavy}}} \tag{2.24}$$

Replacing the Gibbs free energy of binding by binding enthalpy, Ladbury and co-workers[134] suggested a new metric, enthalpic efficiency (*EE*), as a measure of binding enthalpy per non-hydrogen atoms:

$$EE = \frac{H}{N_{\text{heavy}}} \tag{2.25}$$

More recently, Reynolds and co-workers[32] demonstrated that *LE* shows significant size dependency. This observation prompted Nissink[34] to define the size-independent ligand efficiency (*SILE*) that can be used for the unbiased comparison of compounds with different sizes (see eqn 2.14). Investigating the size dependency of *EE*, we concluded that the unbiased evaluation of enthalpic efficiency needs *SIHE* (eqn 2.13). *SILE* and *SIHE* are useful for the selection of compounds with promising thermodynamic parameters in terms of both ΔG and ΔH. Monitoring *SILE* values could help maintaining or improving the binding efficiency along the optimization, while *SIHE* allows checking the contribution of more beneficial enthalpic factors. Very recently, Reynolds and Holloway[69] analyzed the size dependency of both enthalpic and entropic efficiencies. In line with our results,[31] they found similar trends for ligand efficiencies with respect to molecular size and concluded that these are primarily a consequence of enthalpic, not entropic, effects.

Entropic terms are considered differently, mostly *via* lipophilic ligand efficiency (*LLE*) introduced by Leeson and Sprigthorpe:[124]

$$LLE = pK_i - clogP \tag{2.26}$$

LLE helps to control lipophilic contribution to binding affinity optimizations. Since there is a correlation between apolar surface area and entropic factors,[68] *LLE* might be used for monitoring the relative contribution of entropic terms to the free energy of binding.

Up to now, there is only one single parameter that combines both molecular complexity and lipophilicity to a single metric: *LELP* (see eqn 2.27) depicts the price of ligand efficiency paid in log*P*, and consequently it can be used as a new metric of lipophilic efficiency:[126]

$$LELP = \frac{clogP}{LE} \tag{2.27}$$

Considering that *LE* reflects basically enthalpic contributions while log*P* is connected to entropic factors, *LELP* would be a unique metric that could help medicinal chemistry teams optimizing compounds towards thermodynamically balanced milestones. Interestingly, recently a Pfizer group reported[135] that *LELP*, unlike *LE* and *LLE*, was able to discriminate successful drugs and development candidates. More importantly, they demonstrated that candidates with poor *LELP* (*LELP* > 10) yielded only compounds that had been terminated.

In addition to the direct comparison of compounds (including similar or different chemotypes) delivered by medicinal chemistry programs, one can monitor changes in binding thermodynamics along optimizations. This kind of monitoring activity typically includes hit-to-lead and lead-to-candidate optimizations.[136] Although this is a more resource intensive approach, it can be effectively combined with conventional screening cascades. The evaluation of binding thermodynamics could be ideally inserted after primary screening that focuses only on promising compounds. Ideally, binding affinity could be increased with only a limited increase of molecular weight and lipophilicity. Based on the principles of binding thermodynamics, this can be achieved by the improvement of binding enthalpy forming optimized polar interactions. Since enthalpic interactions have significant positional restraints, increasing the polarity of the compounds does not necessarily have a direct correlation with binding enthalpy. On the other hand, however, polar groups are required for high binding enthalpy. Geometrical requirements of enthalpic interactions are most straightforwardly fulfilled in a structure-based optimization program. Even with available structural data in hand, the complexity of the binding event usually allows only the qualitative interpretation of binding thermodynamics. It should be noted, however, that in the lack of structural information, thermodynamic experiments can provide valuable information on the binding interactions, and thus give experimental feedback of the success of compound design.

Comparative analysis of high affinity and high enthalpy compounds revealed that high enthalpy compounds, although less potent, have typically lower molecular weight and lipophilicity and are less complex (Table 2.2).[31] This is in line with the size dependence of binding free energy and binding enthalpy presented in Figure 2.4: we found that in contrast to ΔG,[137] ΔH decreases with the increasing size of the ligand. Considering that, in the function of non-hydrogen atoms, ΔG and ΔH have opposite slopes, the two functions intersect at a molecular size represented by 20–30 heavy atoms. Based on this finding, we concluded that the opportunity for the enthalpic optimization of compounds larger than this threshold diminishes gradually with increasing size. It is interesting to see that the location of the intersection is around the typical size of lead compounds.[138] This suggests that there is a higher chance for successful enthalpic optimization in the lead discovery phase. Thermodynamics-guided hit selection and hit-to-lead optimization would provide viable leads with high binding enthalpy and might contribute to the success of subsequent lead optimization. High enthalpy leads have a higher chance to interact with protein hot spots. Since most proteins have a limited number of hot spots,[39,42] early phase identification of critical binding site elements would improve the outcome of the forthcoming lead optimization. The increasing separation of available ΔG and ΔH for larger compounds suggests that lead optimizations are typically entropy-driven. Since entropy-driven optimizations are associated with the undesired shift of physicochemical properties, classical lead optimization programs are characterized by the challenging trade-off between affinity and ADME related physicochemical properties. Consequently, the selection of high enthalpy leads is crucial for delivering good quality development candidates.

Table 2.2 Selected property means of large $(30 < HA < 50)$ high affinity $(pK_d > 10)$ and high enthalpy $(pK_H > 11)$ compounds.

	High affinity compounds	*High enthalpy compounds*
Heavy atoms	42.1	37.6
Molecular weight	602	542
*c*log*P*	3.3	0.2
H-bond acceptors	7.5	9.2
H-bond donors	3.7	3.7
Charged atoms	0.2	1.4
Apolar surface area (Å^2)	414	330

2.6   Case Studies

The case studies presented below are parts of drug discovery processes where the decision making was supported by thermodynamic considerations. In these examples the chemical starting point is optimized along alternative paths. Subsequent comparison of the optimized compounds reveals that although affinity is usually similar, thermodynamic signatures may be significantly different. In agreement with the discussion of the previous sections, compounds with a dominating enthalpic component were typically selected for further optimization. As compounds are obtained by the optimization of a common starting point, their thermodynamic signatures can be given relative to the predecessor. In this way, the alternative optimization paths can be characterized as enthalpy- or entropy-driven according to the change of these components by monitoring $\Delta\Delta H$ and $\Delta T\Delta S$ along the optimization paths. The $\Delta\Delta$ values allow tracking subtle changes in the thermodynamics of binding. Investigating the role of different substitution patterns on binding thermodynamics would be a typical example when changes in ΔG, ΔH and ΔS with respect to the unsubstituted core are meaningful to compare. We note, however, that the situation is different when compounds cannot be traced back to a common ancestor, as in the case of structurally unrelated hits and leads. In such a case, ΔH and $T\Delta S$ (in addition to ΔG) and also *SIHE* (in addition to *SILE* and *LELP*) are to be considered. Nevertheless, since here we are focusing on the optimization process and comparing optimization strategies applied to common starting points, we will use $\Delta\Delta H$ and $\Delta T\Delta S$.

2.6.1   Renin Inhibitors

Renin is an aspartic protease of the renin–angiotensin system that cleaves its natural substrate angiotensinogen to angiotensin I. Since the cleavage of angiotensinogen is the rate-determining step in the production of angiotensin II, renin inhibitors serve as promising therapy in hypertension. Diaminopyrimidine-type renin inhibitors were discovered by an HTS campaign and a subsequent parallel synthesis approach at Pfizer, identifying a 6.6 μM **Hit** (Figure 2.10A).[139] The X-ray structure of the **Hit**–renin complex identified five hydrogen bonds formed with the diaminopyrimidine part of the molecule. Interestingly, both the large S2 and the smaller hydrophobic S3 subpocket were unoccupied. Early optimization of the **Hit** involved tethering the molecule by the introduction of a tetrahydroisoquinoline (**Lead 1**) and a benzoxazinone (**Lead 2**) rings and extending them by a methoxypropyl side-chain toward the S3 subpocket.[140,141] X-Ray analysis of the renin–**Lead 1** complex showed that all hydrogen bonds stayed intact and, as designed, the methoxypropyl side-chain reached the S3 subpocket (Figure 2.10B). Although the X-ray structure of the renin–**Lead 2** complex is not publicly available, the structure of the 2,2-dimethylbenzoxazinone analog was solved. This structure revealed that all the hydrogen bonds identified for the **Hit** exist and that the S3

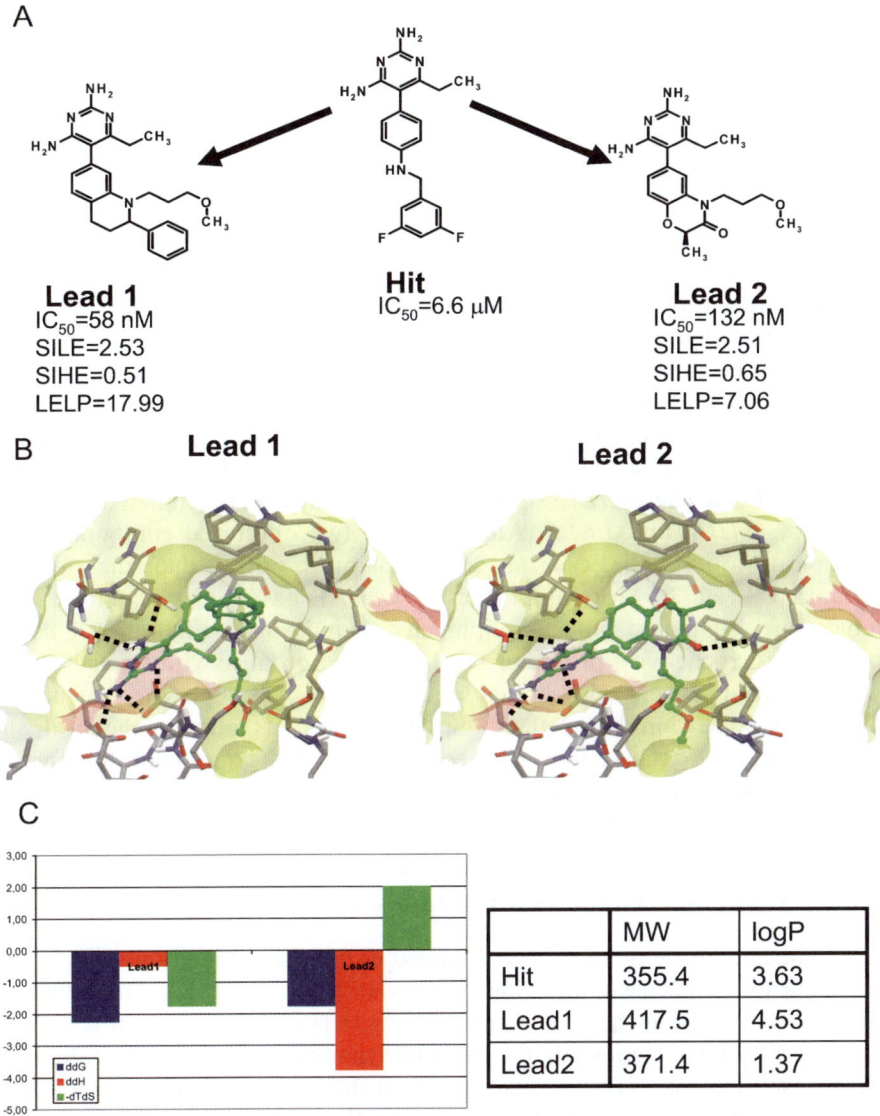

Figure 2.10 Thermodynamics-guided optimization of renin inhibitors. (A) The alternative optimization paths. (B) The binding mode of the optimized inhibitors. (C) The thermodynamic profile of the optimized inhibitors.

subpocket was filled by the methoxypropyl side-chain. In addition to the hydrogen bonds already present with the **Hit** and the filled S3 subpocket identified for **Lead 1**, the methyl group of **Lead 2** in position 2 formed beneficial van der Waals contacts and very probably the benzoxazinone group formed new polar contacts within the active site. Comparing the binding thermodynamics of **Lead 1** to that of the hit (Figure 2.10C), one can conclude that new hydrophobic van der Waals contacts were formed. These interactions resulted in a moderate gain in ΔH (less than 1 kcal mol^{-1}). Displacing ordered water molecules from the hydrophobic S3 subpocket, however, was entropically favored and compensated for the entropy loss associated with the decreased flexibility, resulting in an almost 2 kcal mol^{-1} gain in $T\Delta S$. Therefore, this optimization of the **Hit** can be classified as being entropy driven, resulting in **Lead 1** with reasonable ligand efficiency (*SILE*) and enthalpic efficiency (*SIHE*) but disadvantageous *LELP*. Evaluation of the binding thermodynamics for the **Lead 2** complex in relation to the Hit revealed that, in addition to the entropy gain caused by filling S3, significant enthalpic reward was realized (Figure 2.10C). This latter gain in enthalpy ($\Delta H \approx 4$ kcal mol^{-1}) can be rationalized by the favored van der Waals contacts of the 2-methyl group and the hypothesized new polar interactions formed with the benzoxazinone group. Owing to the significant enthalpy gain achieved, this optimization can be considered as being enthalpy driven and yielded **Lead 2** with ligand efficiency (*SILE*) similar to Lead 1, but with much better enthalpic efficiency (*SIHE*) and *LELP*. Analyzing the two optimized compounds, the entropically optimized **Lead 1** is somewhat more potent than the enthalpy optimized **Lead 2**. More importantly, **Lead 1** has a significantly higher molecular weight and higher lipophilicity than **Lead 2** (Figure 2.10C). The better physico-chemical profile of **Lead 2** is in accordance with the fact that it was delivered by enthalpy-driven optimization. Although the potency of **Lead 2** is somewhat lower than that of **Lead 1**, its physico-chemical and thermodynamic data support the selection of this compound for further optimization. In fact, optimized Pfizer compounds were reported around the benzoxazinone rather than the tetrahydroisoquinoline scaffold.[142]

2.6.2   Carbonic Anhydrase Inhibitors

Carbonic anhydrase is a well characterized enzyme that binds small, rigid and fragment-like benzenesulfonamides with relatively high affinity. Investigating the utility of binding thermodynamics characterization in fragment screening, Scott and Jones reported the thermodynamics-guided optimization of benzenesulfonamide (BSA) type fragment inhibitors (Figure 2.11A).[143,144] Binding of benzenesulfonamides is driven by the four hydrogen bonds formed with the sulfonamide moiety and also through the two hydrogen bonds to the zinc co-factor. Owing to these specific interactions, binding free energies have a favorable enthalpy component while the entropy contributes less to their binding thermodynamics. In this study the authors investigated the role of aromatic substituents on binding affinity. They analyzed the effect of halogen, methyl, amino and cyano groups at different positions of the benzene ring and compared binding free energies and their enthalpic and entropic components with respect to the unsubstituted BSA.

Comparing the thermodynamic signatures of BSA derivatives, anomalous behavior of F-substituted analogs was observed. The *ortho*-F derivative

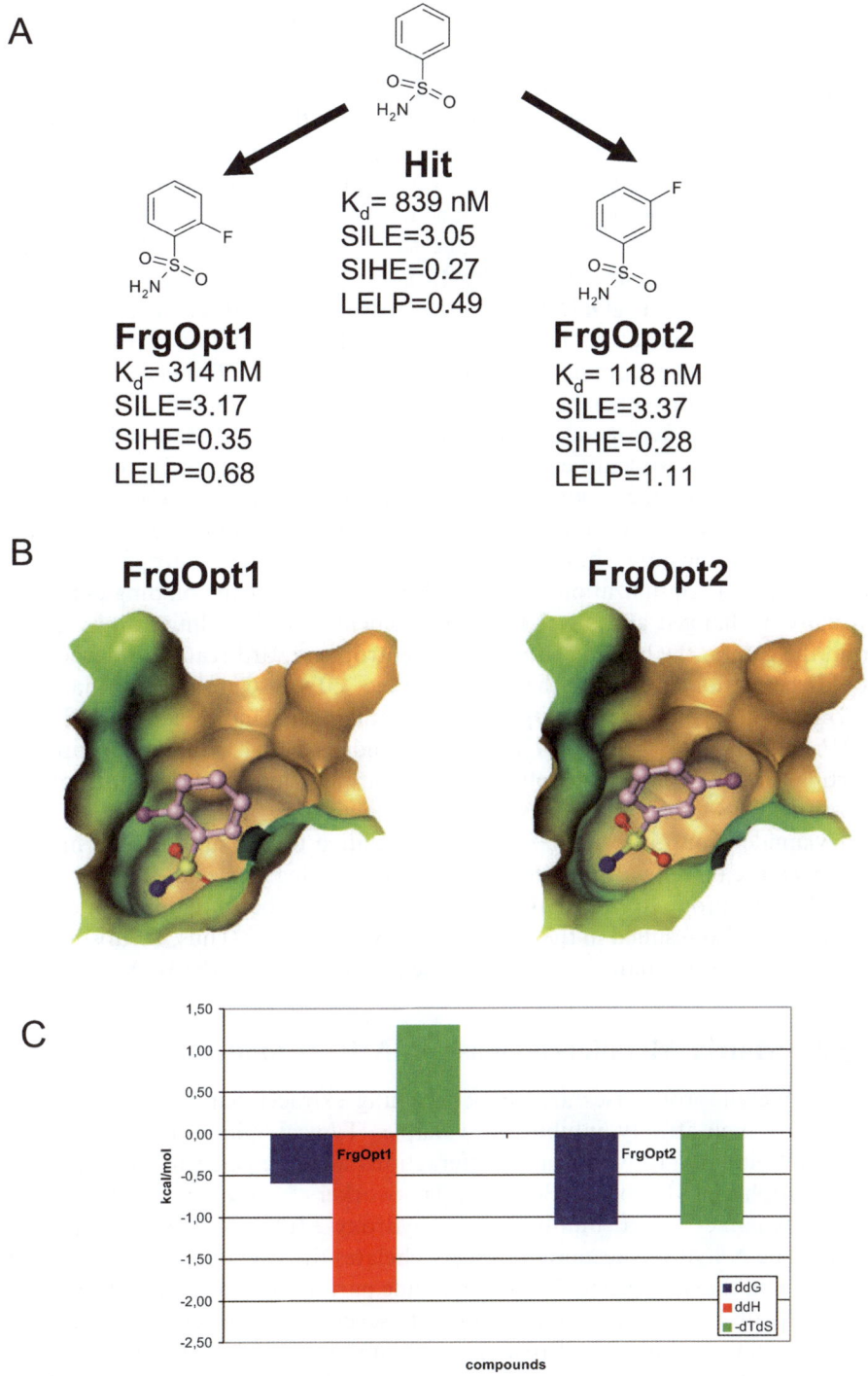

Figure 2.11 Thermodynamics-guided optimization of carbonic anhydrase inhibitors. (A) The alternative optimization paths. (B) The binding mode of the optimized inhibitors. (C) The thermodynamic profile of the optimized inhibitors.

(**FrgOpt1**) differs from all the other compounds in the series, as it has a large favorable enthalpic and a smaller, but still important, unfavorable entropic component. In contrast, the thermodynamic signature of the *meta*-F compound (**FrgOpt2**) shows a similar picture to that of the other analogs, having a smaller favorable entropic and an almost absent enthalpic component. Although the binding affinity and the ligand efficiency (*SILE*) of **FrgOpt1** was somewhat smaller than that of **FrgOpt2**, the better enthalpic efficiency (*SIHE*) and *LELP* values suggested FrgOpt1 should be followed up. Analysis of the X-ray structure of **FrgOpt1** revealed that, interestingly enough, the fluorine atom in **FrgOpt1** faces to the hydrophobic wall of the binding site (Figure 2.11B). Contrarily, the fluorine atom of **FrgOpt2** points in the opposite direction towards a dominantly hydrophobic surface of the enzyme that is preferred by all the other halogens placed at different positions. The unique orientation of the fluorine atom in the **FrgOpt1** complex was rationalized by a new polar contact identified with the main-chain amide NH of Thr200. Comparison of the thermodynamic profile of **FrgOPt1** and **FrgOpt2** is in line with the structural findings. Optimization of **FrgOpt1** improved the binding enthalpy significantly but was coupled with some entropic penalty (Figure 2.11C). Optimization of **FrgOpt2**, however, kept the binding enthalpy virtually unchanged and all of the improvement in the binding affinity came from the gain in the binding entropy. The entropic reward realized by **FrgOpt2** could be rationalized by the hydrophobic effect. Considering that the thermodynamic profile of **FrgOpt1** is more enthalpic (as indicated by the higher *SIHE* value), one can suggest this compound as being a more appropriate starting point for further optimization. Further optimization of fluorine-substituted BSA derivatives supported this conclusion. The introduction of a benzylamide group into the *para* position relative to the sulfonamide moiety improved the binding affinity in both cases. Interestingly, the addition of the *p*-benzylamide group to **FrgOp1** not only increased the enthalpy content of binding but also resulted in the highest affinity compound. Thus, F-substitution *ortho* to the sulfonamide group has privileged properties in the BSA series.

2.6.3   Matrix Metalloproteinase (MMP12) Inhibitors

Matrix metalloproteinases are zinc-containing extracellular proteases having numerous high-affinity inhibitors available. Typical inhibitors bind to the substrate binding groove, forming interactions with the zinc site and also with the hydrophobic S1′ pocket. Bertini and co-workers[145] investigated the binding thermodynamics of a number of *N*-hydroxy-2-(phenylsulfonamido)aceta-mides. The *N*-hydroxyacetamido moiety chelates the zinc ion while the phenyl group fits in the S1′ pocket. The sulfonamide part of the molecules is located in the substrate binding groove and contacts several polar side-chains. Starting from the unsubstituted hydroxamic sulfonamide ligand, the authors intro-duced a methoxy group into the *para* position of the phenyl ring (**Lead**) and the sulfonamide nitrogen atom was subsequently substituted by an isobutyl (**Opt1**) and a hydroxyethyl (**Opt2**) group (Figure 2.12A). Analysis of the corresponding X-ray structures revealed that the methoxyphenyl moiety fits well into the S1′ pocket (Figure 2.12B). Methoxy derivatives penetrate somewhat deeper into the S1′ pocket compared to the unsubstituted derivative. This suggests

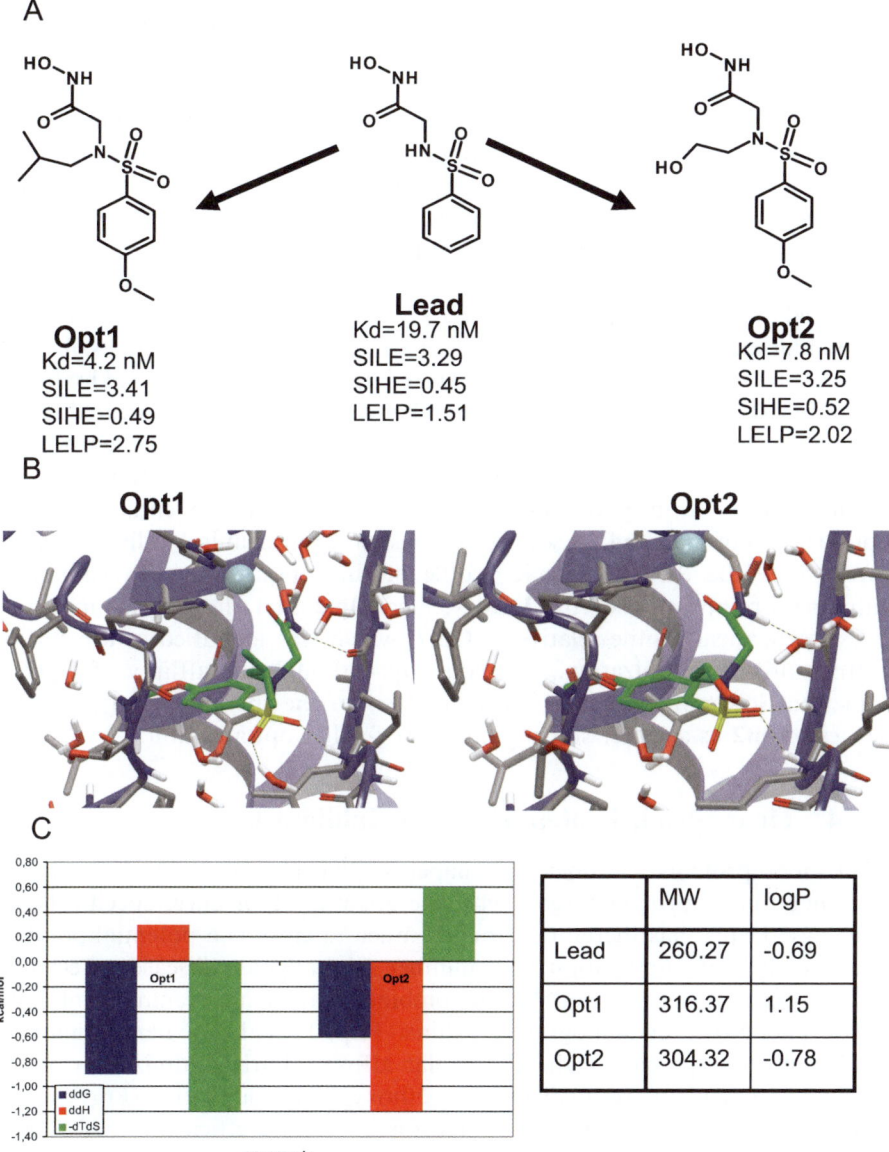

Figure 2.12 Thermodynamics-guided optimization of MMP12 inhibitors. (A) The alternative optimization paths. (B) The binding mode of the optimized inhibitors. (C) The thermodynamic profile of the optimized inhibitors.

that the methoxy group probably displaces water molecules from the hydrophobic S1' pocket, which would result in some gain in binding entropy. Although the methoxy group forms van der Waals contacts, it also disturbs the optimal hydrogen-bond interactions of the sulfonamide group and the coordination of the hydroxamic acid. Since these interactions are rather similar for both the **Lead** and the optimized compounds (**Opt1** and **Opt2**), the observed difference in their binding thermodynamics is basically related to the substituent of the sulfonamide nitrogen atom (Figure 2.12C). Introduction of the hydrophobic isobutyl group (**Opt1**) improved the ligand efficiency (*SILE*). The enthalpic efficiency (*SIHE*) remained virtually constant, but *LELP* became worse due to the increased lipophilicity. The thermodynamic profile of **Opt1** is in line with these findings. Improvement of the binding affinity has been achieved by entropy-driven optimization, as indicated by the beneficial change in binding entropy (Figure 2.12C). On the other hand, introducing the polar hydroxyethyl group kept the ligand efficiency (*SILE*) constant and improved the enthalpic efficiency (*SIHE*) to some extent. The observed reward in binding enthalpy could be attributed to van der Waals contacts formed between the ethyl spacer and Pro238. More importantly, the terminal hydroxyl group forms hydrogen bonds with a number of water molecules located at the entrance of the substrate binding site. The polar character of the substituent helps to control the lipophilicity and improves the *LELP* value. Improvement in the binding affinity of **Opt2** can be attributed to the enthalpy-driven optimization of the **Lead**, as seen from the significant gain in binding enthalpy revealed by the thermodynamic profile of this optimization. In summary, enthalpic optimization of the **Lead** yielded **Opt2** that showed an improved physico-chemical profile relative to **Opt1**, which was identified as a result of entropy-driven optimization. Consequently, although the affinity of **Opt2** is somewhat lower than that of **Opt1**, analysis of the thermodynamic profiles suggests **Opt2** as a better starting point for further optimization.

2.6.4   Heat Shock Protein (HSP90) Inhibitors

Heat shock proteins are molecular chaperons that play a role in the structure and function of protein targets with therapeutic importance, specifically in oncology. Hsp90 has therefore attracted much interest as a potential point of pharmacological intervention in a number of malignant diseases. Recently, researchers at Astex reported[146] that combined fragment screening using both NMR and X-ray crystallography against Hsp90 yielded a new aminopyrimidine chemotype with high micromolar affinity. Early optimization of the original fragment hit resulted in a 2-methoxyphenyl derivative (**Frg**) having improved low micromolar affinity, reasonable ligand efficiency (*SILE*) and acceptable *LELP* value. This compound was further optimized by introducing a second substituent at the 6-position of the 2-methoxyphenyl moiety. Improvement in the binding affinity was achieved by both the 6-methoxy (**FrgOpt1**) and the 6-chloro (**FrgOpt2**) substituents (Figure 2.13A). The

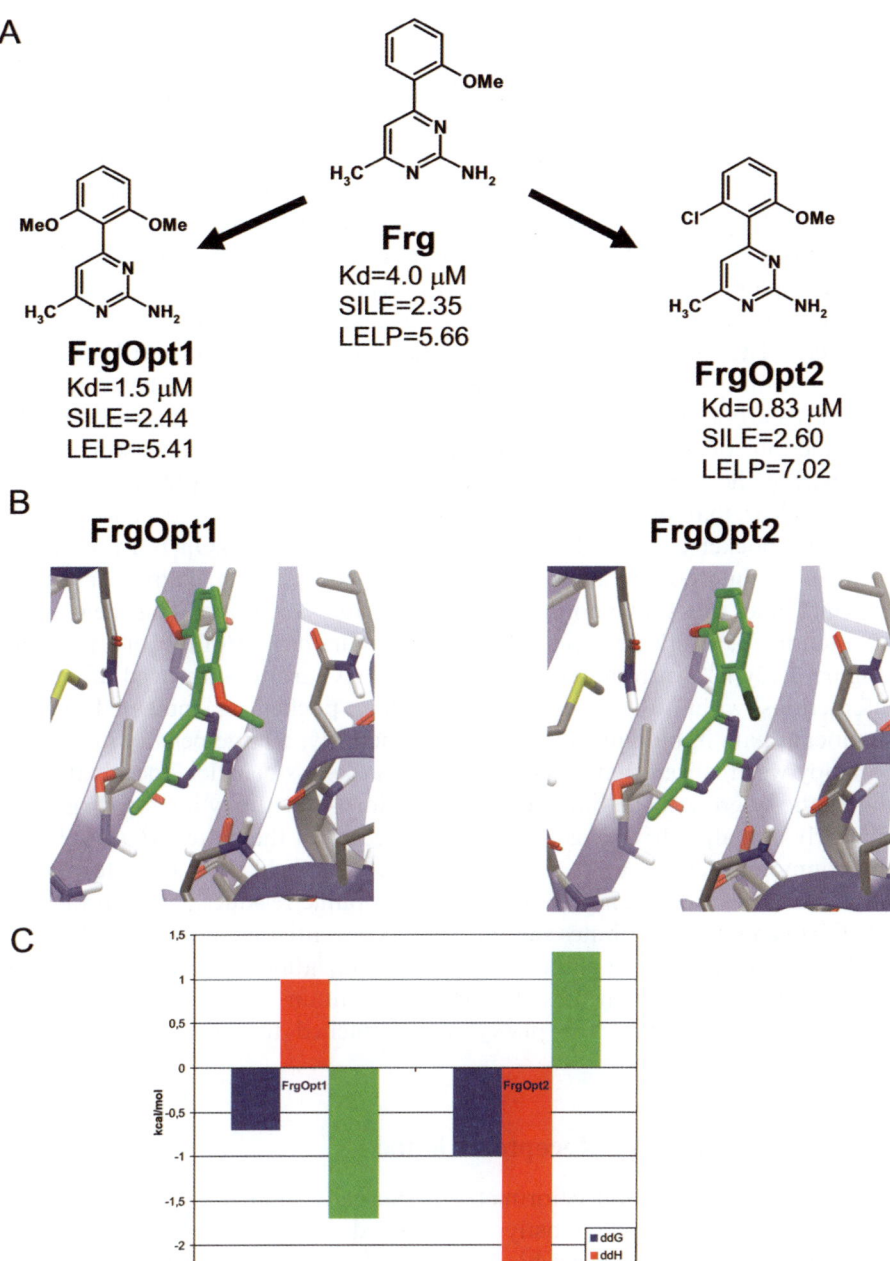

Figure 2.13 Thermodynamics-guided optimization of HSP90 inhibitors. (A) Alternative optimization paths. (B) The binding mode of the optimized inhibitors. (C) The thermodynamic profile of the optimized inhibitors.

dimethoxy derivative (**FrgOpt1**) showed 2.5-fold improvement in the binding affinity, while some marginal improvement in both the ligand efficiency (*SILE*) and *LELP* occurred. The affinity of the non-symmetrical 2-chloro-6-methoxy compound (**FrgOpt2**) had much improved affinity relative to the starting fragment (fivefold improvement), with increased ligand efficiency (*SILE*) and slightly worse *LELP*. Investigating the experimental binding modes of both optimized fragments, the authors identified several important differences. Interactions with the aminopyrimidine core are similar to that observed in the ADP complex of the enzyme. The 2-methoxyphenyl group fits in the proximal lipophilic pocket, suggesting that further lipophilic substitution would be beneficial. The conformation of the bicyclic system could also contribute to the binding affinity. It was hypothesized that the optimal geometry would be almost planar. Therefore the binding affinity could be further improved by the introduction of substituents stabilizing the protein-bound conformation. Comparing the X-ray structures of the **FrgOpt1** and **FrgOpt2** complexes, it became clear that the chlorine atom in **FrgOpt2** penetrates deeper into the lipophilic pocket compared to the methoxy group in **FrgOpt1**. In addition, there is an interesting favorable interaction between the chlorine atom of **FrgOpt2** and the surrounding backbone carbonyl groups. This new polar interaction is reflected in the thermodynamic profile of **FrgOpt2** (Figure 2.13C).[41] Introduction of the methoxy group improved lipophilic interactions within the hydrophobic binding pocket, as indicated by the significant gain in binding entropy. Simultaneously, a large penalty in binding enthalpy was realized that is probably due to the less beneficial desolvation of the methoxy group. The binding of the 2-chloro-6-methoxy analog (**FrgOpt2**) is clearly enthalpy driven, as this is also reflected in the better *SIHE* value of this compound relative to its alternative **FrgOpt1**. It is likely that the formation of beneficial van der Waals contacts with lipophilic residues and also the new Cl···C=O contact contributes significantly to the enthalpic rewards that makes **FrgOpt2** a better starting point for further optimization. In fact, the optimized derivatives reported from these laboratories contains 2-chloro rather than 2-methoxy substituents at the phenyl ring attached to the aminopyrimidine core.[146]

2.6.5   Adenosine A1 Receptor Ligands

Since isothermal titration calorimetry is basically limited to soluble targets, membrane receptors are mostly investigated by the classical van't Hoff analysis. In a typical scenario, binding affinities are measured in a radioligand displacement assay at several different temperatures that, with the assumption of $\Delta C_p \approx 0$, enables the calculation of thermodynamic parameters (see Section 2.3.2). This methodology was applied to adenosine A1 antagonists by Borea and co-workers.[147,148] Starting from theophylline, position 8 of the xanthine ring was substituted by lipophilic groups including cyclopentyl (**CPT**) and phenyl (**8-PT**) rings (Figure 2.14A). Theophylline is a non-selective adenosine

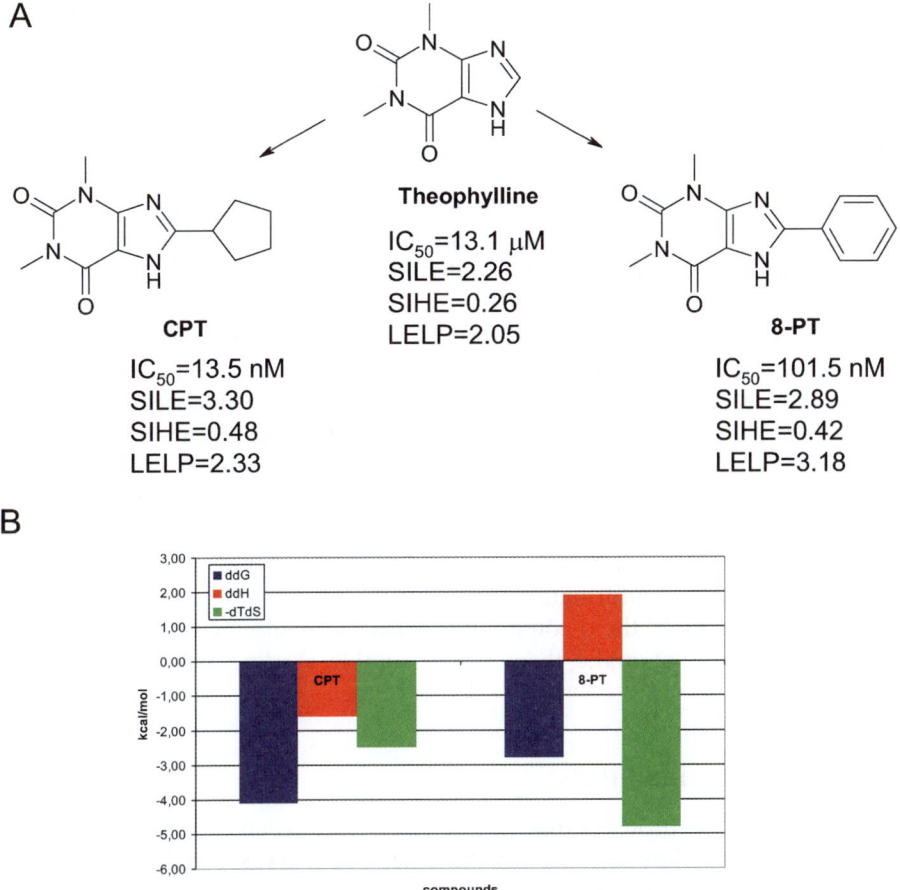

Figure 2.14 Thermodynamics-guided optimization of adenosine A1 ligands. (A) The alternative optimization paths. (B) The thermodynamic profile of the optimized inhibitors.

A1 and A2A antagonist that binds to A1 receptors with moderate affinity. Despite its micromolar IC_{50}, it shows reasonable ligand efficiency (*SILE*), moderate enthalpic efficiency (*SIHE*) and acceptable *LELP*. Introduction of the cyclopentyl substituent improved the binding affinity significantly. In fact, **CPT** has low nanomolar IC_{50} with much improved ligand efficiency and enthalpic efficiency, while there was only some marginal increase in *LELP*. This combination of efficiency indices suggests that, despite the lipophilic character of the cyclopentyl substituent, there is a significant improvement in the binding enthalpy. Introduction of the phenyl substituent (**8-PT**) was also beneficial in terms of binding affinity, ligand efficiency and enthalpic efficiency. It should be noted, however, that the efficiency profile of **8-PT** is much less promising than that of **CPT**, as indicated by smaller improvement in *SILE* and *SIHE* and more significant worsening of *LELP*. Based on this picture, one can conclude that the improved binding affinity of **8-PT** relative to theophylline is due to hydrophobic effects. Analysis of the thermodynamic profile supports these hypotheses (Figure 2.14B). Optimization of theophylline to **CPT** involves both enthalpic and entropic components. In contrast, the introduction of the 8-phenyl substituent resulted only in entropic gain that was associated with a significant enthalpic penalty. The lack of structural information makes the interpretation of binding thermodynamics very challenging. It is likely that most of the entropic reward is due to the facilitated desolvation of both **CPT** and **8-PT**. On the other hand, however, the cyclopentyl group of **CPT** might form some specific interactions with the receptor, probably *via* binding-site water molecules. In the case of the closely related adenosine A2A receptor, Sherman and co-workers[149] gave a reasonable interpretation for the experimental SAR on the basis of differential water displacement patterns. Therefore it can be hypothesized that the difference in the binding thermodynamics of **CPT** and **8-PT** can be rationalized by binding-site solvent thermodynamics. Increasing availability of GPCR structures would help in the interpretation of thermodynamic data. On the other hand, however, this example shows that van't Hoff thermodynamics, even with the lack of structural information, is useful in selecting more enthalpic compounds. As such, these compounds form more specific interactions with the binding site that make them more suitable starting points for further optimization.

2.6.6   Trypsin Inhibitors

Trypsin is a serine protease that can be considered as an interesting target of the blood coagulation cascade. On the other hand, trypsin and thrombin are both well-characterized proteins extensively used for the investigation of binding principles. Klebe and co-workers[150] reported a systematic study on a series of *N*-(2-naphthylsulfonyl)-L-3-amidinophenylalanine derivatives using protein crystallography and isothermal titration calorimetry. For demonstration purposes a piperidinyl carboxylic acid analog (**Lead**) was used as a starting point that was converted into the corresponding ester (**Opt1**) and an inner carboxamide (**Opt2**). The starting point shows moderate affinity but reasonable ligand efficiency (*SILE*) with acceptable enthalpic efficiency (*SIHE*) but suboptimal *LELP* (Figure 2.15A). Esterification of the carboxylic acid function yielded **Opt1** with much improved affinity, better *SILE* and *SIHE*, but only moderately improved *LELP*. The alternative path resulted in the carboxamide (**Opt2**) with reasonable affinity, promising *SILE* and *SIHE* and much improved *LELP*. All of the compounds bind to trypsin in a very similar way (Figure 2.15B). The characteristic benzamidino group forms interactions with an aspartate in the S1 pocket. The naphthylsulfonyl group fits in the S3/S4 pocket and shows only minor changes in its binding mode. Consequently, all changes in the binding affinity and thermodynamics might be attributed to the substituents of the piperidine ring. Comparing the binding modes of **Opt1**

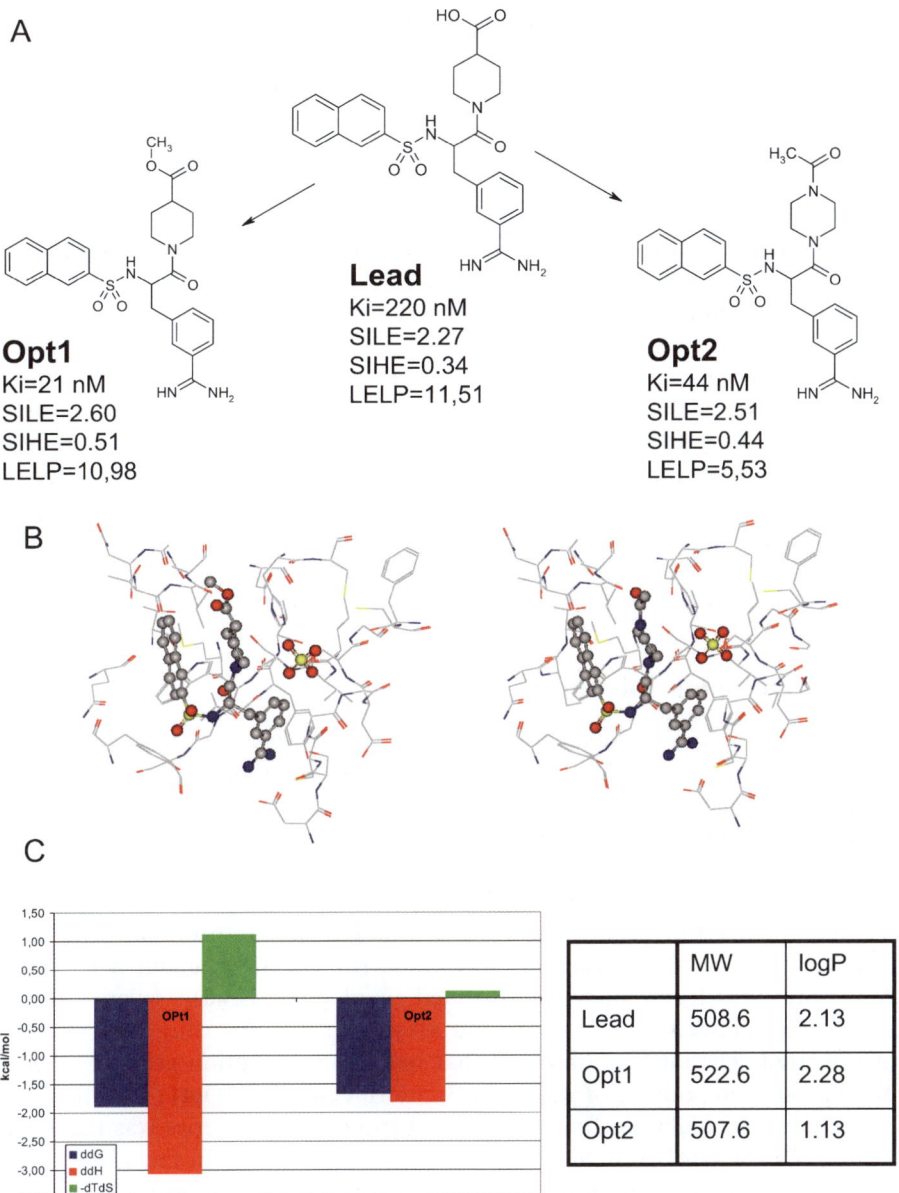

Figure 2.15 Thermodynamics guided optimization of trypsine inhibitors. (A) The alternative optimization paths. (B) The binding mode of the optimized inhibitors. (C) The thermodynamic profile of the optimized inhibitors.

and **Opt2**, it is important to note that the C-terminal carbonyl group forms a new bridged hydrogen bond to Ser96 through a water molecule. Although no structural information on the **Lead** is available, it is likely that it forms similar hydrogen bond. Since both the hydrogen bonding pattern and the buried surface area of the **Lead** remained unchanged, the achieved improvement in the binding affinity of **Opt1** and **Opt2** could only be rationalized on the basis of desolvation effects. In fact, the carboxylic group and its ester and amide derivatives are all pretty much exposed to the solvent. Considering that desolvation of an acid is more challenging than that of the ester and the amide, enthalpic rewards associated with the transformation of the free carboxylic acid group of the **Lead** seem to be logical. The desolvation enthalpy of the **Lead** could be readily reduced by esterification (**Opt1**) and formation of the inner amide (**Opt2**), since only one carbonyl oxygen is subjected to desolvation. This effect has been reflected in the thermodynamic profile of both optimization pathways (Figure 2.15C). Decreased flexibility of the inner amide relative to the carboxylic ester would be responsible for the entropic differences observed between the two derivatives of the **Lead**. Considering the character of the optimizations, one can conclude that both are enthalpy driven. Comparing efficiency metrics, **Opt1** is somewhat more promising than **Opt2**. Medicinal chemistry experience with metabolically labile esters as well as the significantly better physico-chemical profile and much improved *LELP* suggests **Opt2** as a compound for potential follow up.

2.6.7   Thrombin Inhibitors

Thrombin has been considered as a valuable cardiovascular target for a long time; however, most of the compounds utilize an amidino moiety mimicking the arginine head group of the natural substrate. The protonated nature of this group in physiological conditions limits their absorbtion significantly, thus making the development very challenging. Substituted benzylamines were found to be suitable replacements for the incriminated moiety. Klebe and co-workers[151] investigated a number of D-Phe-Pro-based thrombin inhibitors carrying substituted benzylamines at the C-terminal, S1 occupant part of the core. Starting from the unsubstituted analogue (**Lead**), several *meta*-substituted halogen (F, Cl, Br, I) and alkyl (Me, Et, iPr, tBu) analogs were investigated. The authors observed that halogen substitution generally yields more active compounds, of which the 3-chloro derivative (**Opt1**) was the most active one. The highest affinity compound from the alkyl series was the 3-Me analog (**Opt2**). **Opt1** showed almost 100-fold improvement in the affinity that was associated with significant improvements in ligand efficiency (*SILE*) and enthalpic efficiency (*SIHE*), while *LELP* was unchanged (Figure 2.16A). Although the improvements for **Opt2** were still remarkable in affinity (10-fold) and both *SILE* and *SIHE*, they are all less impressive than that found for **Opt1** and furthermore **Opt2** has suboptimal *LELP*. Investigating the binding modes it was concluded that all of the compounds bind very similar to thrombin. The phenyl ring of Phe occupies the S3 pocket and the central region of the ligand is involved in extensive hydrogen bonding networks with the enzyme. The phenyl groups of benzylamines are all located in the S1 pocket. Comparing the

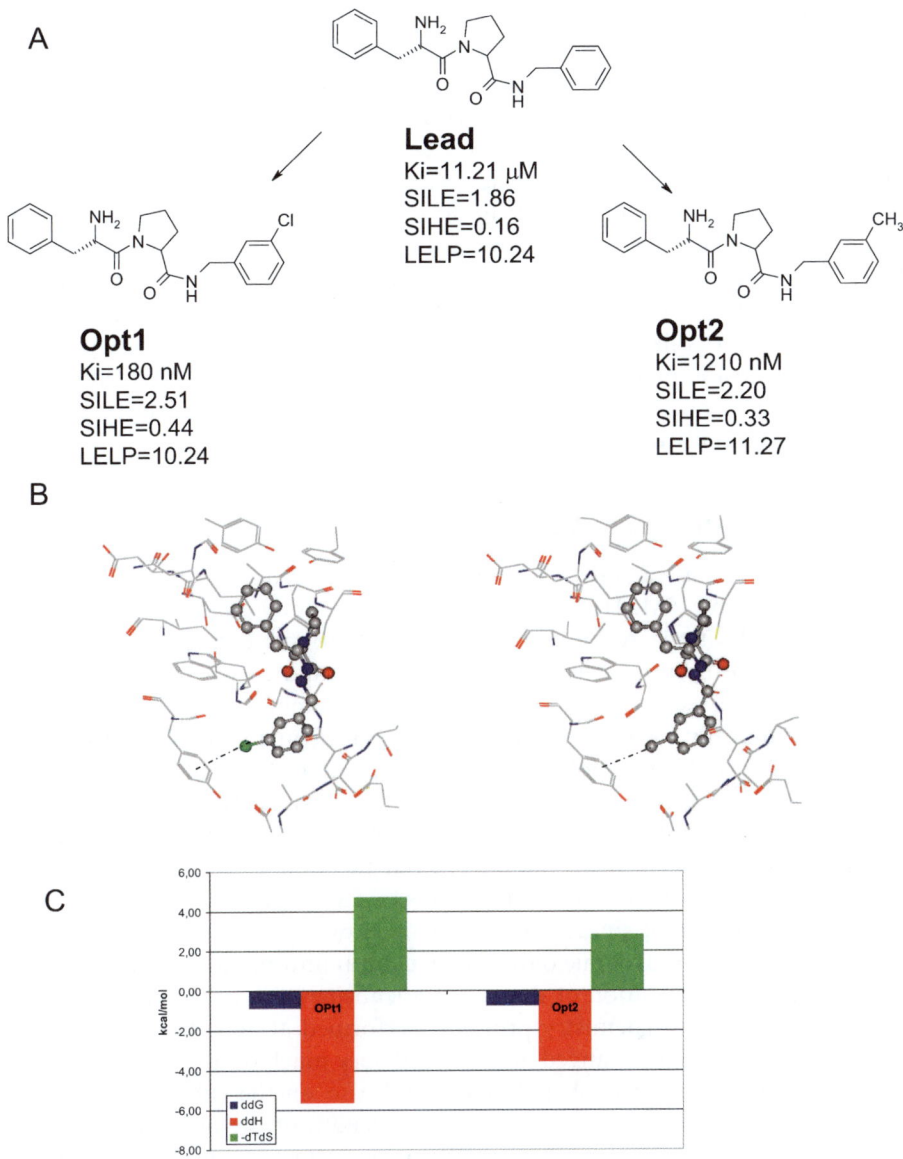

Figure 2.16 Thermodynamics-guided optimization of thrombin inhibitors. (A) The alternative optimization paths. (B) The binding mode of the optimized inhibitors. (C) The thermodynamic profile of the optimized inhibitors.

crystal structure of the apo form of the enzyme to that obtained by the unsubstituted benzylamine, it was found that the unsubstituted compound (**Lead**) displaces several water molecules except one in the close proximity of Phe227. The *meta* substituents could readily displace this water molecule that contributes to the improved affinity of **Opt1** and **Opt2** relative to the **Lead**. In addition, the chlorine atom of **Opt2** makes short contacts with two of the neighboring carbonyls (Phe227 and Trp215) and is located just above the aromatic plane of Tyr228 (Figure 2.16B). These contacts suggest that the chlorine atom of **Opt2** might form Cl···C=O and more generally Cl···π interactions with specific binding-site residues. Analyzing the thermodynamic profiles, one can conclude that both optimizations are enthalpy driven (Figure 2.16C). However, a higher reward in binding enthalpy was realized for **Opt1**. Since the desolvation energy of **Opt1** and **Opt2** are expected to be similar, and both compounds displace the incriminated water molecule from the S1 pocket, it is likely that the Cl···π type of polar interaction contributes significantly to the binding energetics of **Opt1**. Based on its better efficiency metrics and thermodynamic profile, **Opt1** would be an optimal starting point for further optimizations.

2.6.8   Adenosine A2A Antagonists

Adenosine A2A is a GPCR protein that was successfully crystallized recently[152] with ZM241385, a high-affinity antagonist developed in Parkinson indication. Although the binding thermodynamics of this compound have not been investigated so far, Baraldi and co-workers reported[153] an interesting series of compounds that closely resembles ZM241385 and was thermodynamically characterized by van't Hoff analysis. This series of pyrazolo[4,3-*e*]-1,2,4-triazolo[1,5-*c*]pyrimidines differs from the ZM series in that the 5-amino group of the 1,2,4-triazolo[2,3-*a*]-1,3,5-triazine ring system and the 4-position of the triazine are connected to a pyrazole ring with the simultaneous displacement of the nitrogen at position 4 to carbon. Owing to the structural similarity of these compounds to ZM241385, it is likely that their binding mode would be similar to that obtained by X-ray crystallography. The availability of thermodynamic data and structural information prompted us to investigate the optimization of this set of adenosine A2A antagonists in more detail. The medicinal chemistry program performed at Schering-Plough kept the heterocyclic core unchanged and focused very much on the linker between the core and the terminal phenyl group and also on the substitution of the latter. In this section we investigate how the length of the linker impacts the binding affinity and thermodynamics. To achieve this goal we started from a **Lead** that couples the terminal phenol with one CH_2 unit (Figure 2.17A). During the optimization the team obtained two compounds with two- and three-membered linkers, identified as **Opt1** and **Opt2**, respectively. The **Lead** compound showed reasonable affinity towards A2A receptors with high ligand (*SILE*) and enthalpic (*SIHE*) efficiencies and acceptable *LELP*. Inserting one CH_2 into the linker region resulted in 10-fold improvement in the affinity of **Opt1**, with exceptionally high ligand efficiency and almost identical *SIHE* and *LELP*. **Opt2** with the longest linker showed virtually the same potency and ligand efficiency as **Opt1**; in addition to the slightly increased *LELP*, the most

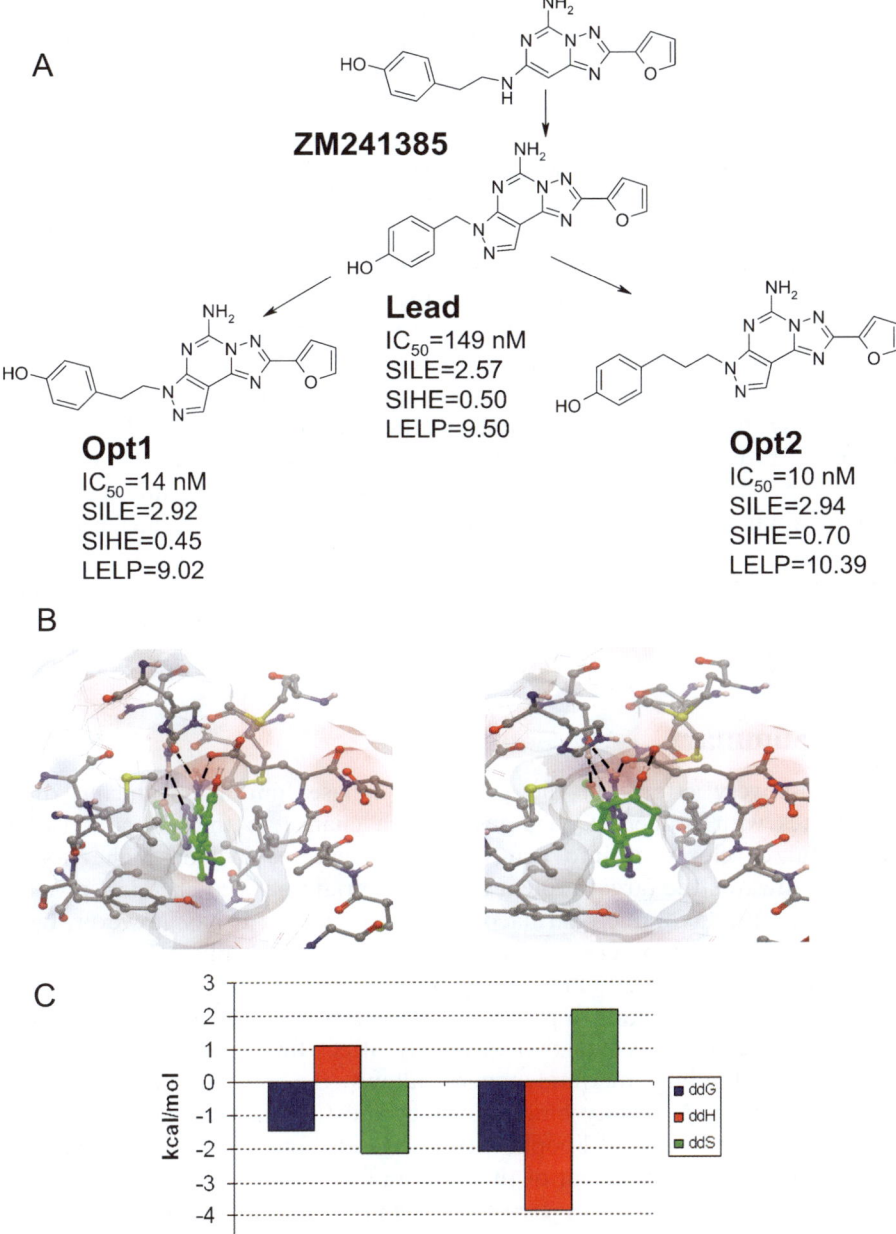

Figure 2.17 Thermodynamics-guided optimization of adenosine A2A antagonists. (A) The alternative optimization paths. (B) The binding mode of the optimized inhibitors. (C) The thermodynamic profile of the optimized inhibitors.

characteristic difference was the remarkable enthalpic efficacy as indicated by the high *SIHE*. Next, we docked both **Opt1** and **Opt2** into the binding pocket of the receptor using the crystallographic orientation of ZM241385 as constraint. Binding modes obtained in this way are depicted in Figure 2.17B. These calculations revealed that the heterocyclic core of **Opt1** and **Opt2** forms similar contacts as found for ZM241385 (Figure 2.17B). The heterocyclic core formed water-bridged H-bonds, including that to the free amino group and Asn253, and $\pi-\pi$ stacking interactions with Phe168. The linker orientated the terminal phenyl ring towards the extracellular space. In the case of **Opt1**, there was no specific contact found between the linker and the receptor that was identical with that found for ZM241385. On the other hand, however, elongation of the linker in **Opt2** helped the terminal phenol to reach Glu169 and form a new H-bond. The formation of this new specific contact impacts the binding thermodynamics of **Opt2** significantly (Figure 2.17C). Comparing the optimizations, it can be concluded that the insertion of one CH_2 unit was an entropy-driven process, while inserting two CH_2 groups into the linker region is an enthalpic approach. The former optimization increased the lipophilicity of the **Lead** that is behind its entropic character. In contrast, optimization to **Opt2** formed a new polar contact with the receptor that contributed significantly to the enthalpic gain responsible for the increased binding affinity.

2.7   Summary

Drug discovery efforts in the preclinical phase aim at producing drug candidates with properties that give a high chance to the compounds for successful development into drugs. Critical features include high enough potency towards the target protein and appropriate physico-chemical properties that ensure favorable pharmacokinetics and safety. It has, however, been recognized that high potency and appropriate physico-chemical properties are conflicting requirements in many drug discovery programs: a potency increase above a certain level tends to produce compounds with high molecular weight and lipophilicity, properties that adversely affect further development. Our analysis on binding thermodynamics allows the rationalization of this phenomenon. High-affinity binding can be most straightforwardly achieved by increasing the size and lipophilicity of ligands, and thereby increasing the favorable entropy component of the binding free energy. This entropy-driven ligand optimization realizes free-energy gain primarily by the release of solvating water molecules upon ligand–protein complex formation.

An alternative route to affinity increase is to introduce and optimize polar interactions between the ligand and protein and thereby increase the favorable enthalpy component of the binding free energy. This route is more challenging owing to the high sensitivity of the polar interactions to the geometrical arrangements of the interacting partners. It has also been shown that the potential of enthalpy-driven optimization is limited. Analyzing the dependence of affinity and binding enthalpy on ligand size, we pointed out that maximal achievable affinity increases and maximal favorable binding enthalpy decreases with increasing ligand size. An important consequence of these opposite trends is that enthalpic optimization of drug candidates is typically feasible up to the lead selection phase, but not in the lead optimization.

Binding thermodynamics offers a rationalization and hopefully alleviation of the observed conflict between affinity and physico-chemical property optimizations. As enthalpic optimization affects physico-chemical properties advantageously and is feasible at least up to the lead discovery phase, its application is recommended in the early stages of drug discovery projects. The experimental determination of thermodynamic signatures, most preferably with ITC or with the application of the van't Hoff analysis, makes it possible to apply thermodynamic considerations at decision points, like hit or lead selections. Hit-to-lead optimization supported by thermodynamic monitoring is an additional option and is expected to be increasingly available with the advance of experimental devices with higher throughput and lower protein consumption. Thermodynamic monitoring of optimization combined with structural studies facilitates the interpretation and, to a limited extent, the design of thermodynamic profiles.

Summarizing, binding thermodynamics contribute largely to the understanding of the hard-to-optimize relation between affinity and physico-chemical property. Thermodynamic considerations might reformulate present optimization practices to deliver better quality clinical candidates with improved developability.

Acknowledgements

The authors are grateful to Mike Hann (GSK) for comments on property inflation and Glyn Williams (Astex), Lyn Jones (Pfizer), Woody Scherman (Schrödinger), Ákos Tarcsay and Kinga Nyíri (Gedeon Richter) for discussing different case studies.

References

1. K. Sharp, *Protein Sci.*, 2001, **10**, 661.
2. A. Cornish-Bowden, *J. Biosci.*, 2002, **27**, 121.
3. E. B. Starikov and B. Nordén, *J. Phys. Chem. B*, 2007, **111**, 14431.
4. R. R. Krug, W. G. Hunter and R. A. Grieger, *J. Phys. Chem.*, 1976, **80**, 2335.
5. O. Exner, *Prog. Phys. Org. Chem.*, 1973, **10**, 411.
6. V. M. Krishnamurthy, G. K. Kaufman, A. R. Urbach, I. Gitlin, K. L. Gudiksen, D. B. Weibel and G. M. Whitesides, *Chem. Rev.*, 2008, **108**, 946.
7. L. Liu, C. Yang, T. W. Mu and Q. X. Guo, *Chin. Chem. Lett.*, 2001, **12**, 167.

8. D. H. Williams, E. Stephens, D. P. O'Brien and M. Zhou, *Angew. Chem., Int. Ed.*, 2004, **43**, 6596.

9. G. Graziano, *J. Phys. Soc. Jpn.*, 2000, **69**, 1566.

10. K. N. Houk, A. G. Leach, S. P. Kim and X. Zhang, *Angew. Chem., Int. Ed.*, 2003, **42**, 4872.

11. W. Blokzijl and J. B. F. N. Engberts, *Angew. Chem., Int. Ed.*, 1993, **32**, 1545.

12. M. Lucas, *J. Phys. Chem.*, 1976, **80**, 359.

13. B. Lee, *Biopolymers*, 1985, **24**, 813.

14. B. Lee, *Biophys. Chem.*, 1994, **51**, 271.

15. M. Kinoshita, *J. Chem. Phys.*, 2008, **128**, 024507.

16. T. Lazaridis, *J. Phys. Chem. B*, 2000, **104**, 4964.

17. T. Lazaridis, *Acc. Chem. Res.*, 2001, **34**, 931.

18. N. Muller, *Acc. Chem. Res.*, 1990, **23**, 23.

19. B. Lee and G. Graziano, *J. Am. Chem. Soc.*, 1996, **118**, 5163.

20. G. Graziano and B. Lee, *J. Phys. Chem. B*, 2005, **109**, 8103.

21. Y. S. Dijkaev and E. Ruckenstein, *J. Chem. Phys.*, 2010, **133**, 194105.

22. D. Chandler, *Nature*, 2005, **437**, 640.

23. E. Fisicaro, C. Compari and A. Braibanti, *Phys. Chem. Chem. Phys.*, 2004, **6**, 4156.

24. E. Fisicaro, C. Compari and A. Braibanti, *Biophys. Chem.*, 2010, **151**, 119.

25. A. Cooper, *Biophys. Chem.*, 2005, **115**, 89.

26. E. A. Meyer, R. K. Castellano and F. Diederich, *Angew. Chem., Int. Ed.*, 2003, **42**, 1210.

27. N. R. Syme, C. Dennis, S. E. V. Phillips and S. W. Homans, *ChemBioChem.*, 2007, **8**, 1509.

28. T. Liu, Y. Lin, X. Wen, R. N. Jorissen and M. K. Gilson, *Nucleic Acids Res.*, 2007, **35**, D198.

29. L. Li, J. J. Dantzer, J. Nowacki, B. J. O'Callaghan and S. O. Meroueh, *Chem. Biol. Drug Des.*, 2008, **71**, 529.

30. Scorpio database; http://www.biochem.ucl.ac.uk/scorpio/scorpio.html (accessed January 26, 2010).

31. G. G. Ferenczy and G. M. Keserű, *J. Chem. Inf. Model.*, 2010, **50**, 1536.

32. C. H. Reynolds, S. D. Bembenek and B. A. Tounge, *Bioorg. Med. Chem. Lett.*, 2007, **17**, 4258.

33. J. W. M. Nissink, *J. Chem. Inf. Model.*, 2009, **49**, 1617.

34. H. A. Carlson, R. D. Smith, N. A. Khazanov, P. D. Kirchhoff, J. B. Dunbar, Jr. and M. L. Benson, *J. Med. Chem.*, 2008, **51**, 6432.

35. M. Torheyden and G. Jansen, *Mol. Phys.*, 2006, **104**, 2101, Table 1.

36. M. M. Hann, A. R. Leach and G. Harper, *J. Chem. Inf. Comput. Sci.*, 2001, **41**, 856.

37. P. Selzer, H.-J. Roth, P. Ertl and A. Schuffenhauer, *Curr. Opin. Chem. Biol.*, 2005, **9**, 310.

38. A. Schuffenhauer, N. Brown, P. Selzer, P. Ertl and E. Jacoby, *J. Chem. Inf. Model.*, 2006, **46**, 525.

39. C. W. Murray and M. L. Verdonk, *J. Comput. Aided Mol. Des.*, 2002, **16**, 741.
40. G. Chessari and A. J. Woodhead, *Drug Discovery Today*, 2009, **14**, 668.
41. F. E. Torres, M. I. Recht, J. E. Coyle, R. H. Bruce and G. Williams, *Curr. Opin. Struct. Biol.*, 2010, **20**, 598.
42. P. A. Rejto and G. M. Verkhivker, *Proc. Natl. Acad. Sci. U. S. A.*, 1996, **93**, 8945.
43. S. M. Habermann and K. P. Murphy, *Protein Sci.*, 1996, **5**, 1229.
44. http://microcal.com/technology/itc.asp.
45. R. Perozzo, G. Folkers and L. Scapozza, *J. Recept. Signal. Transduct.*, 2004, **24**, 1.
46. D. J. Eatough, *Anal. Chem.*, 1970, **42**, 635.
47. Y.-L. Zhang and Z.-Y. Zhang, *Anal. Biochem.*, 1998, **261**, 139.
48. B. W. Sigurskjold, *Anal. Biochem.*, 2000, **277**, 260.
49. J. F. Brandts and L.-N. Lin, *Biochemistry*, 1990, **29**, 6927.
50. G. A. Holdgate and W. H. J. Ward, *Drug Discovery Today*, 2005, **10**, 1543.
51. B. M. Baker and K. P. Murphy, *Biophys. J.*, 1996, **71**, 2049.
52. A. Velazquez-Campoy, I. Luque, M. J. Todd, M. Milutinovich, Y. Kiso and E. Freire, *Protein Sci.*, 2000, **9**, 1801.
53. W. B. Peters, V. Frasca and R. K. Brown, *Comb. Chem. High Throughput Screening*, 2009, **12**, 772.
54. R. J. Falconer and B. M. Collins, *J. Mol. Recognit.*, 2011, **24**, 1.
55. J. Tellinghuisen, *Biophys. Chem.*, 2006, **120**, 114.
56. G. Maksay, *Prog. Biophys. Mol. Biol.*, 2011, **106**, 463.
57. H.-J. Wittmann, R. Seifert and A. Strasser, *Mol. Pharmacol.*, 2009, **76**, 25.
58. P. Gilli, V. Ferretti, G. Gilli and P.A. Borea, *J. Phys. Chem.*, 1994, **98**, 1515.
59. G. A. Holdgate, *Expert Opin. Drug Discovery*, 2007, **2**, 1103.
60. E. Edink, C. Jansen, R. Leurs and I. J. P. de Esch, *Drug Discovery Today: Technol.*, 2010, **7**, e189.
61. J. R. Horn, D. Russell, E. A. Lewis and K. P. Murphy, *Biochemistry*, 2001, **40**, 1774.
62. Y. S. N. Day, C. L. Baird, R. L. Rich and D. G. Myszka, *Protein Sci.*, 2002, **11**, 1017.
63. M. A. Wear and M. D. Walkinshaw, *Anal. Biochem.*, 2006, **359**, 285.
64. T. Liu, Y. Lin, X. Wen, R. N. Jorissen and M. K. Gilson, *Nucleic Acids Res.*, 2007, **35**, D198.
65. L. Li, J. J. Dantzer, J. Nowacki, B. J. O'Callaghan and S. O. Meroueh, *Chem. Biol. Drug Des.*, 2008, **71**, 529.
66. Scorpio database; http://www.biochem.ucl.ac.uk/scorpio/scorpio.html; accessed 20.04.2011.
67. T. S. G. Olsson, M. A. Williams, W. R. Pitt and J. E. Ladbury, *J. Mol. Biol.*, 2008, **384**, 1002.

68. C. H. Reynolds and M. K. Holloway, *ACS Med. Chem. Lett.*, 2011, **2**, 433.
69. L. D. Beveridge and F. M. Dicapua, *Annu. Rev. Biophys. Biophys. Chem.*, 1989, **18**, 431.
70. C. Jarzynski, *Phys. Rev. Lett.*, 1997, **22**, 1420.
71. C. Jarzynski, *Phys. Rev. E: Stat. Phys., Plasmas, Fluids, Relat. Interdiscip. Top.*, 1997, **56**, 5018.
72. M. K. Gilson, J. A. Given, B. L. Bush and J. A. McCammon, *Biophys. J.*, 1997, **72**, 1047.
73. A. Pohorille, C. Jarzynski and C. Chipot, *J. Phys. Chem. B*, 2010, **114**, 10235.
74. T. Steinbrecher and A. Labahn, *Curr. Med. Chem.*, 2010, **17**, 767.
75. J. Michel and J. W. Essex, *J. Comput. Aided Mol. Des.*, 2010, **24**, 639.
76. A. J. Stone and M. Alderton, *Mol. Phys.*, 1985, **56**, 1047.
77. G. G. Ferenczy, P. J. Winn and C. A. Reynolds, *J. Phys. Chem. A*, 1997, **101**, 5446.
78. N. Lu, D. A. Kofke and T. B. Woolf, *J. Phys. Chem. B*, 2003, **107**, 5598.
79. C. Peter, C. Oostenbrink, A. van Dorp and W. F. van Gunsteren, *J. Chem. Phys.*, 2003, **120**, 2652.
80. H. A. Yu and M. Karplus, *J. Chem. Phys.*, 1988, **89**, 2366.
81. J. Dolenc, S. Gerster and W. F. van Gunsteren, *J. Phys. Chem. B*, 2010, **114**, 11164.
82. P. A. Kollman, I. Massova, C. Reyes, B. Kuhn, S. Huo, L. Chong, M. Lee, T. Lee, Y. Duan, W. Wang, O. Donini, P. Cieplak, J. Srinivasan, D. A. Case and T.E. Cheatham III, *Acc. Chem. Res.*, 2000, **33**, 889.
83. B. Kuhn, P. Gerber, T. Schulz-Gasch and M. Stahl, *J. Med. Chem.*, 2005, **48**, 4040.
84. J. Åqvist, C. Medina and J.-E. Samuelsson, *Protein Eng.*, 1994, **7**, 385.
85. J. Åqvist, V. B. Luzhkov and B. O. Brandsdal, *Acc. Chem. Res.*, 2002, **35**, 358.
86. T. Hansson, J. Marelius and J. Åqvist, *J. Comput. Aided Mol. Des.*, 1998, **12**, 27.
87. H. A. Carlson and W. L. Jorgensen, *J. Phys. Chem.*, 1995, **99**, 10667.
88. D. K. Jones-Hertzog and W. L. Jorgensen, *J. Med. Chem.*, 1997, **40**, 1539.
89. A. Khandelwal, V. Lukacova, D. Comez, D. M. Kroll, S. Raha and S. Balaz, *J. Med. Chem.*, 2005, **48**, 5437.
90. N. Singh and A. Warshel, *Proteins*, 2010, **78**, 1705.
91. A. Warshel and M. Levitt, *J. Mol. Biol.*, 1976, **103**, 227.
92. R. Rajamani and A. C. Good, *Curr. Opin. Drug Discovery Dev.*, 2007, **10**, 308.
93. T. Cheng, X. Li, Y. Li, Z. Liu and R. Wang, *J. Chem. Inf. Model.*, 2009, **49**, 1079.
94. T. Young, R. Abel, B. Kim, B. J. Berne and R. A. Friesner, *Proc. Natl. Acad. Sci. U. S. A.*, 2007, **104**, 808.

95. R. Abel, T. Young, R. Farid, B. J. Berne and R. A. Friesner, *J. Am. Chem. Soc.*, 2008, **130**, 2817.
96. D. D. Robinson, W. Sherman and R. Farid, *ChemMedChem*, 2010, **5**, 618.
97. Y. T. Tang and G. R. Marshall, *J. Chem. Inf. Model.*, 2011, **51**, 214.
98. A. Weis, K. Katebzadeh, P. Söderhjelm, I. Nilsson and U. Ryde, *J. Med. Chem.*, 2006, **49**, 6596.
99. R. Villar, M. J. Gil, J. I. García and V. Martínez-Merino, *J. Comput. Chem.*, 2005, **26**, 1347.
100. E. Nikitina, V. Sulimov, V. Zayets and N. Zaitseva, *Int. J. Quantum Chem.*, 2004, **97**, 747.
101. E. Nikitina, V. Sulimov, F. Grigotiev, O. Kondakova and S. Luschekina, *Int. J. Quantum Chem.*, 2006, **106**, 1943.
102. P. Söderhjelm, J. Kongsted, S. Genheden and U. Ryde, *Interdiscip. Sci.: Comput. Life Sci.*, 2010, **2**, 21.
103. K. P. Murphy, V. Bhakuni, D. Xie and E. Freire, *J. Mol. Biol.*, 1992, **227**, 293.
104. B. M. Baker and K. M. Murphy, *Methods Enzymol.*, 1998, **295**, 294.
105. I. Luque and E. Freire, *Proteins*, 2002, **49**, 181.
106. E. Freire, *Drug Discovery Today*, 2008, **13**, 869.
107. N. Gō and H. A. Scheraga, *J. Chem. Phys.*, 1969, **51**, 4751.
108. N. Gō and H. A. Scheraga, *Macromolecules*, 1976, **9**, 535.
109. D. A. Case, *Curr. Opin, Struct. Biol.*, 1994, **4**, 285.
110. B. Kuhn and P. A. Kollman, *J. Med. Chem.*, 2000, **43**, 3786.
111. A. Kitao, S. Hayward and N. Gō, *Proteins*, 1998, **33**, 496.
112. M. Karplus and J. N. Kusshick, *Macromolecules*, 1981, **14**, 325.
113. M. M. Teeter and D. A. Case, *J. Phys. Chem.*, 1990, **94**, 8091.
114. C.-E. Chang, W. Chen and M. K. Gilson, *J. Chem. Theory Comput.*, 2005, **1**, 1017.
115. H. Gohlke and D. A. Case, *J. Comput. Chem.*, 2004, **25**, 238.
116. S.-T. D. Hsu, C. Peter, W. F. van Gunsteren and A. Bonvin, *Biophys. J.*, 2005, **88**, 15.
117. M. S. Head, J. A. Given and M. K. Gilson, *J. Phys. Chem. A*, 1997, **101**, 1609.
118. C.-E. Chang and M. K. Gilson, *J. Am. Chem. Soc.*, 2004, **126**, 13156.
119. C.-E. Chang, W. Chen and M. K. Gilson, *Proc. Natl. Acad. Sci. U. S. A.*, 2007, **104**, 1534.
120. J. Hermans and L. Wang, *J. Am. Chem. Soc.*, 1997, **119**, 2707.
121. N. Singh and A. Warshel, *J. Phys. Chem. B*, 2009, **113**, 7372.
122. N. Singh and A. Warshel, *Proteins*, 2010, **78**, 1724.
123. T. I. Oprea, A. M. Davis, S. J. Teague and P. D. Leeson, *J. Chem. Inf. Comput. Sci.*, 2001, **41**, 1308.
124. P. D. Leeson and B. Springthorpe, *Nat. Rev. Drug Discovery*, 2007, **6**, 881.
125. G. M. Keserű and G. M. Makara, *Nat. Rev. Drug Discovery*, 2009, **8**, 203.
126. S. J. Teague, A. M. Davis, P. D. Leeson and T. I. Oprea, *Angew. Chem., Int. Ed. Engl.*, 1999, **38**, 3743.

127. M. P. Gleeson, *J. Med. Chem.*, 2008, **51**, 817.

128. J. D. Hughes, J. Blagg, D. A. Price, S. Bailey, G. A. DeCrescenzo, R. V. Devraj, E. Ellsworth, Y. M. Fobian, M. E. Gibbs, R. W. Gilles, N. L. Greene, E. Huang, T. Krieger-Burke, J. Loesel, T. Wager, L. Whiteley and Y. Zhang, *Bioorg. Med. Chem. Lett.*, 2008, **18**, 4872.

129. G. M. Keserű and G.M. Makara, *Drug Discovery Today*, 2006, **11**, 741.

130. M. M. Hann, *Med. Chem. Commun.*, 2011, **2**, 349.

131. G. G. Ferenczy and G. M. Keserű, *Drug Discovery Today*, 2010, **15**, 919.

132. J. P. Overington, B. Al-Lazikani and A. L. Hopkins, *Nat. Rev. Drug Discovery*, 2006, **5**, 993.

133. A. L. Hopkins, C. R. Groom and A. Alex, *Drug Discovery Today*, 2004, **9**, 430.

134. J. E. Ladbury, G. Klebe and E. Freire, *Nat. Rev. Drug Discovery*, 2010, **9**, 23.

135. T. T. Wager, R. Y. Chandrasekaran, X. Hou, M. D. Troutman, P. R. Verhoest, A. Villalobos and Y. Will, *ACS Chem. Neurosci.*, 2010, **1**, 420.

136. E. Freire, *Chem. Biol. Drug Des.* 2009, **74**, 468.

137. I. D. Kuntz, K. Chen, K. A. Sharp and P. A. Kollman, *Proc. Natl. Acad. Sci. U. S. A.*, 1999, **96**, 9997.

138. T. I. Oprea, T. K. Allu, D. C. Fara, R. F. Rad, L. Ostopovici and C. G. Bologa, *J. Comput. Aided Mol. Des.*, 2007, **21**, 113.

139. D. Holswortha, M. Jalaiea, T. Belliottia, C. Caia, W. Collarda, S. Ferreiraa, N. A. Powella, M. Stiera, E. Zhanga, P. McConnella, I. Mochalkina, M. J. Ryana, J. Bryanta, T. Lib, A. Kasanib, R. Subedib, S. N. Maitib and J. J. Edmunds, *Bioorg. Med. Chem. Lett.*, 2007, **17**, 3575.

140. R. W. Sarver, J. Peevers, W. L. Cody, F. L. Ciske, J. Dyer, S. D. Emerson, J. C. Hagadorn, D. D. Holsworth, M. Jalaie, M. Kaufman, M. Mastronardi, P. McConnell, N. A. Powell, J. Quin III, C. A. Van Huis, E. Zhang and I. Mochalkin, *Anal. Biochem.*, 2007, **360**, 30.

141. N. A. Powell, F. L. Ciske, C. Cai, D. D. Holsworth, K. Mennen, C. A. Van Huis, M. Jalaie, J. Day, M. Mastronardi, P. McConnell, I. Mochalkin, E. Zhang, M.J. Ryan, J. Bryant, W. Collard, S. Ferreira, C. Gu, R. Collins and J. J. Edmunds, *Bioorg. Med. Chem.*, 2007, **15**, 5912.

142. Science Integrity; http://integrity.prous.com.

143. A. D. Scott, C. Phillips, A. Alex, M. Flocco, A. Bent, A. Randall, R. O'Brien, L. Damian and L. H. Jones, *ChemMedChem*, 2009, **4**, 1985.

144. L. H. Jones, presented at MEDI 456, 238th National Meeting and Exposition, August 16–20, 2009, Washington, DC.

145. I. Bertini, V. Calderone, M. Fragai, A. Giachetti, M. Loconte, C. Luchinat, M. Maletta, C. Nativi and K. J. Yeo, *J. Am. Chem. Soc.*, 2007, **129**, 2467.

146. C. W. Murray, M. G. Carr, O. Callaghan, G. Chessari, M. Congreve, S. Cowan, J. E. Coyle, R. Downham, E. Figueroa, M. Frederickson, B. Graham, R. McMenamin, M. A. O'Brien, S. Patel, T. R. Phillips, G.

Williams, A. J. Woodhead and A. J. Woolford, *J. Med. Chem.*, 2010, **53**, 5942.

147. P. A. Borea, K. Varani, L. Guerra, P. Gilli and G. Gilli, *Mol. Neuropharmacol.*, 1992, **2**, 273.
148. P. A. Borea, A. Dalpiaz, K. Varani, S. Gessi and G. Gilli, *Life Sci.*, 1996, **59**, 1373.
149. C. Higgs, T. Beuming and W. Sherman, *ACS Med. Chem. Lett.*, 2010, **1**, 160
150. F. Dullweber, M. T. Stubbs, D. Musil, J. Stürzebecher and G. Klebe, *J. Mol. Biol.*, 2001, **313**, 593.
151. B. Baum, M. Mohamed, M. Zayed, C. Gerlach, A. Heine, D. Hangauer and G. Klebe, *J. Mol. Biol.*, 2009, **390**, 56.
152. V. P. Jaakola, M. T. Griffith, M. A. Hanson, V. Cherezov, E. Y. Chien, J. R. Lane, A. P. Ijzerman and R. C. Stevens, *Science*, 2008, **322**, 1211.
153. P. G. Baraldi, B. Cacciari, G. Spalluto, M. Bergonzoni, S. Dionisotti, E. Ongini, K. Varani and P. A. Borea, *J. Med. Chem.*, 1998, **41**, 2126.

CHAPTER 3

Continuum Solvation in Biomolecular Systems

TRAIAN SULEA AND ENRICO O. PURISIMA*

Biotechnology Research Institute, National Research Council of Canada, 6100 Royalmount Avenue, Montreal, Quebec, Canada H4P 2R2
*E-mail: enrico.purisima@nrc.gc.ca

3.1 Solvation Models in Binding Affinity Prediction and Scoring Functions

Water plays a central role in virtually all cellular systems. This role is not limited to providing a passive embedding medium for proteins, nucleic acids, and other biomolecules. Water also takes an active role by contributing to the dynamics and the outcome of many biophysically interesting phenomena. The thermodynamic contribution arising from the change in hydration of the evolving molecular objects is particularly important in intermolecular recognition and association, folding and other conformational equilibria, reaction kinetics, and phase partitioning.

Intermolecular association in biological systems, *e.g.* protein–ligand binding, is typically accompanied by dehydration of the interacting surfaces and reorganization of the solvent water around the ensuing complex. Owing to the aqueous environment, seemingly favorable interactions between the protein and ligand (*e.g.*, hydrogen bonds, salt bridges, and even van der Waals interactions) come with the cost of losing similar interactions with the surrounding water molecules. The changes in solvation thermodynamics during complex formation are hence a crucial element of binding free

RSC Drug Discovery Series No. 23
Physico-Chemical and Computational Approaches to Drug Discovery
Edited by F. Javier Luque and Xavier Barril
© The Royal Society of Chemistry 2012
Published by the Royal Society of Chemistry, www.rsc.org

energies[1–4] and have to be accounted for with carefully calibrated solvation models. A number of solvation models presenting varying degrees of complexity and levels of theory have been developed and parametrized and will be reviewed in this chapter.

In the past two decades, the development of theoretical methods for predicting the strength of molecular association, *i.e.* the binding affinity, has been fuelled by a perceived benefit to drug discovery. Binding affinity prediction methods span several levels of theory, with a corresponding trade-off between prediction accuracy and computational demand.[3,4] On the one hand, there are the relatively slow but thermodynamically rigorous pathway approaches such as free energy perturbation (FEP) and thermodynamic integration (TI).[5,6] On the other hand, there is a large and ever-increasing number of faster approaches relying on binding affinity scoring functions that can be classified into three main categories: force-field-based, knowledge-based, and empirical. Regardless of their type, computational methods for predicting/scoring binding affinities depend on solvation models to account for changes in hydration free energy during complex formation.

Pathway-based free energy methods generally use an explicit solvent model. The linear interaction energy (LIE) approach, based on linear response approximation (LRA),[7–10] is an end-point approximation that also typically retains the explicit description of the solvent. More recent LIE variants replace the explicit solvent with continuum models to run the molecular dynamics (MD) simulations and/or calculate average interaction energies.[11–13] The LIE method yields predictions comparable to the FEP-like methods, but has the benefit of being more tractable and generally applicable.

An emergent group of end-point force-field-based scoring functions that represent a reasonable compromise between time, computational resources, and accuracy combine molecular mechanics (MM) force fields with a continuum treatment of solvation. Methods like MM-PB(GB)/SA[14–18] combine terms that describe electrostatic contributions to solvation, *e.g.* through the generalized Born (GB) or Poisson–Boltzmann (PB) continuum electrostatic solvation methods, with terms that describe the nonpolar contribution to solvation, that is, van der Waals interaction with the solvent and cavity formation based on a linear surface area (SA) approximation. The free and bound end-points are obtained from separate MD simulations of the complex, protein, and ligand in explicit solvent. The MD trajectories are then stripped off the explicit solvent, which is replaced by continuum solvent, to calculate average potential and solvation energies. The configurational entropy change calculated between the end points using the rigid-rotor/harmonic-oscillator approximation for selected trajectory snapshots is sometimes included in the predicted binding affinities. A common approximation that improves the convergence of energy averages relies on simulating only the MD trajectory for the solvated complex,[19–21] from which the protein and ligand free-state trajectories are rigidly extracted, with cancellation of internal energies and configurational entropies.

Solvated interaction energy (SIE)[22,23] is a similar end-point force-field-based scoring function that approximates the protein–ligand binding affinity by an interaction energy contribution and a desolvation free energy contribution. Electrostatic solvation effects are calculated with the boundary element solution to the Poisson equation, while nonpolar solvation is based on molecular SA. As in the single-trajectory MM-PB(GB)/SA approach, the free state is commonly obtained by rigid separation from an energy-minimized or MD-sampled complex. Calibration of several physical parameters, including the dielectric constant, Born radii, surface tension coefficient, and enthalpy–entropy compensation scaling factor, was based on a diverse dataset of protein–ligand complexes.[22] The SIE scoring function parametrized in this manner achieves a reasonable transferability across a wide variety of protein–ligand systems, consistently returning absolute binding affinities within the experimental range, as demonstrated by test cases published in the literature.[24] External testing of the standard SIE parametrization was also done in the CSAR-2010 scoring challenge (http://csardock.org/) consisting in a curated dataset of 343 protein–ligand complexes diverse with respect to ligands and targets, affording a binding affinity prediction rms error of 2.5 kcal mol^{-1}.[25]

Other force-field-based scoring functions implementing continuum solvation models in the spirit of MM-PB(GB)/SA and SIE methods have been reported.[14,26–29] There are also force-field-based scoring functions that implement atomic solvation parameters (ASP) usually scaled to the surface portion of the protein and ligand that is buried during complexation, along with other force-field and entropy-related terms.[30–34] Some of these ASP-based scoring functions can in fact be classified as empirical or regression-based scoring functions, which are QSAR-like models with coefficients for empirical energy terms obtained by fitting to binding affinities from a training set of complexes with known structures. Empirical scoring functions provide fast affinity ranking but have limited transferability outside the chemical space of their training sets. Many empirical scoring functions have been developed over the years, with solvation included indirectly *via* descriptors like buried nonpolar surface, inclusion of interstitial water in hydrogen bond parameters, polar and nonpolar contact surface, octanol/water partition coefficient, or hydrophobic contacts.[35–39] Finally, knowledge-based scoring functions, also known as statistical potentials or potentials of mean force (PMF), derive pairwise interaction potentials from the occurrence frequency of atom pairs from large structural databases of diverse protein–ligand complexes instead of fitting to experimental affinity data.[40–45] In theory, all effects occurring in binding, including desolvation, are implicitly convoluted within the PMFs of these functions.

With the advent of faster computers over the past decade, large-scale *in silico* docking–scoring (aka virtual screening) of small-molecule libraries has become appealing due to its speed and cost efficiency.[46] Unquestionably, the success (or failure) of virtual screening relies mostly on the quality of the underlying docking and scoring function(s). The challenge in virtual screening is exacerbated by the fact that, in order to be relevant in a drug discovery pipeline, accurate docking–scoring has to be achieved under the constraint of fast computing. To this end, a fast yet accurate solvation model is of paramount importance. This is afforded by continuum solvation models. In the next section we will summarize the ideas behind the more common continuum solvation models. In the section after, we will discuss the limitations of continuum solvation and present current trends in developing the next generation of solvation models that will retain the efficiency of the current continuum approximation, but will be able to capture aspects of the physics of hydration that are dependent on the discrete properties of water.

3.2 Continuum Solvation Models

Continuum solvation models typically replace the microscopic view of explicit-atom solvation models with a macroscopic view in which the solvent is described as a surrounding medium imbued with certain bulk properties such as a dielectric constant and surface tension coefficient. Standard physical theories are then applied, treating the solute as a macroscopic object embedded in the medium. Despite this drastic approximation, continuum solvation models work surprisingly well. Compared to explicit solvent simulations, continuum solvation models have a much lower computational cost, largely because the need for solvent relaxation is non-existent. In this section we will give a brief overview of continuum solvation models. More detail can be found in several reviews on continuum solvation models.[47–55] In the discussion that follows, when we refer to solvent we will invariably mean water, the most relevant environment for most biological systems.

3.2.1 Continuum Electrostatics

In a continuum electrostatics model, a hydrated solute molecule and its partial charges are treated as a charge distribution in a low-dielectric cavity that is in turn embedded in a high-dielectric medium representing water. At zero ionic strength, the Poisson equation describes the dependence of the electric potential on the solute charge distribution and on the geometry of the dielectric volumes:

$$\nabla \cdot \varepsilon(\mathbf{r}) \nabla \phi(\mathbf{r}) = -\rho(\mathbf{r}) \tag{3.1}$$

where $\varepsilon(\mathbf{r})$ is the spatially dependent dielectric constant, $\phi(\mathbf{r})$ is the electric potential and $\rho(\mathbf{r})$ is the charge distribution.

In the presence of mobile monovalent ions (non-zero ionic strength) the use of Debye–Hückel theory leads to the nonlinear Poisson–Boltzmann (PB) equation. This is usually simplified to a linearized form using a first-order approximation, $\sinh \phi(\mathbf{r}) \approx \phi(\mathbf{r})$, under the assumption that $\phi(\mathbf{r})$ is small:

$$\nabla \cdot \varepsilon(\mathbf{r}) \nabla \phi(\mathbf{r}) - \kappa^2 \phi(\mathbf{r}) = -\rho(\mathbf{r}) \tag{3.2}$$

In eqn (3.2), κ is the Debye–Hückel inverse screening length, which can be expressed as:

$$\kappa^2 = \frac{2I}{\varepsilon(\mathbf{r})kT} \tag{3.3}$$

where I is the ionic strength.

The Poisson equation can be solved analytically for simple geometric shapes such as a sphere. For example, we have the well-known Born equation for the free energy of hydration (gas to water) of a spherical ion:

$$G_h = -\frac{q^2}{2R}\left(1 - \frac{1}{\varepsilon_w}\right) \tag{3.4}$$

where R is the radius of the ion and ε_w is the dielectric constant of water.

In general, the Poisson and Poisson–Boltzmann equations must be solved numerically. The finite-difference method (FDM) solves the differential equations by discretizing the region of interest into grid points (typically a cubic grid). The generally off-grid solute partial charges are fractionally distributed among the nearby grid points and dielectric constants assigned to each grid according to the geometry of the dielectric boundary. The second derivatives of the potential at each grid point can be expressed in terms of the potentials at neighboring points. The coupled expressions for the potentials on the grid produce a linear system of equations that can be solved to yield the potential at each grid point. The potential at any other point of interest can be obtained through interpolation. Most continuum electrostatic models use a sharp dielectric boundary between the solute and solvent. A more realistic smooth transition from the solute to solvent dielectric using an exponential of a sum of atom-centered Gaussian functions has also been used.[56] Related to FDM is the finite-element method (FEM), which has the advantage of more easily handling variable-sized meshes, finer where needed and coarser elsewhere. A detailed review of finite-difference and finite-element solutions of the PB equation is available.[57] Continuum solvation models using FDM or FEM have become a standard component of many commercial and academic molecular modeling packages.

The boundary element method (BEM) is radically different from the finite-difference method. Instead of solving for the potential at grid points in a volume, BEM solves for the induced surface charge density (ISCD) at the dielectric boundary.[49,52,58] The surface defining the dielectric boundary is tessellated into (usually) triangular surface patches and the ISCD at each triangle is calculated. The ISCDs are obtained by solving a linear system of equations relating the ISCD at each surface patch to the normal component of the electric field from the solute charges. The matrix relating these two quantities reflects the details of the shape of the dielectric boundary. The interaction of the solute charges with the ISCD then captures the electrostatic interaction of the solute with the surrounding solvent. BEM has the advantage over FDM in reducing the dimensionality of the problem from 3D to 2D, significantly lowering the memory requirements for a given mesh/grid resolution. It also provides a more faithful representation of the molecular shape and solute charge distribution than FDM.

In quantum mechanical applications of BEM, the solute wave function rather than partial charges are used to compute the electric field. This resulting ISCD in turn perturbs the original wave function. The modified wave function is then used to generate a new ISCD and the process is iterated until convergence to a self-consistent reaction field (SCRF) is achieved.

Both FDM and BEM rigorously solve the Poisson (or Poisson–Boltzmann) equation. Although significantly faster than explicit water simulations, they are still much slower than a typical force-field energy calculation. A simpler and computationally less expensive continuum electrostatics model is provided by the Generalized Born (GB) model,[59] which tries to capture the essential features of the solution to the Poisson equation. The method is a generalization of the Born equation for a charged sphere (eqn 3.5):

$$G = \frac{1}{2}\left(1 - \frac{1}{\varepsilon_w}\right)\sum_{i,j}\frac{q_i q_j}{f_{GB}} \tag{3.5}$$

The most common form for f_{GB} is:

$$f_{GB}(r_{ij}) = \left[r_{ij}^2 + R_i R_j \exp\left(-r_{ij}^2/4R_i R_j\right)\right]^{1/2} \tag{3.6}$$

where R_i and R_j are the effective Born radii of atoms i and j, and r_{ij} is the interatomic distance.

When r_{ij} is large, f_{GB} reduces to r_{ij} and the contribution to the free energy is small. At intermediate distance, f_{GB} depends strongly on the effective Born radii. The effective Born radius of atom i is the radius of a sphere with the same partial charge as atom i such that its Born energy is identical to the reaction field energy of atom i in the solute molecule with all other partial charges zeroed. Much of the development work in the GB field has been devoted to rapid ways of obtaining the effective Born radii without resorting to a full-blown solution of the Poisson equation.[60–66] A variation to the solution has been to convert the volume integral into a surface integral using Green's theorem. This led to the Surface Generalized Born (SGB)[67–70] and Analytic Generalized Born (AGB) models.[71,72]

GB, in its common implementations, reduces the electrostatic solvation calculation to an essentially pairwise interaction, making it easy to incorporate in MD programs.[73] The pairwise nature of the method also facilitates decomposition of free energies into individual atomic contributions.[74] An extensive review of the GB model has been published.[75] One study comparing the performance of GB *versus* FDM and BEM finds that GB agrees with FD and BEM results to within 1% or better.[76]

An even simpler model with just two fitted parameters is the Sheffield solvation model:[77]

$$f(r_{ij}) = -\frac{1}{2}\left(1 - \frac{1}{\varepsilon_{\mathrm{w}}}\right)\sum_{i,j}\frac{q_i q_j}{\left(aR_i R_j + br_{ij}^2\right)^{1/2}} \tag{3.7}$$

where the R_i and R_j are just normal atomic radii, *e.g.* van der Waals radii from the force field used, and the parameters a and b are fitted to reproduce PB-derived solvation energies for a large set of molecules.

It should be noted that neither the GB nor the Sheffield model exhibits the proper asymptotic behavior for long-range interactions or for charges approaching a dielectric boundary.[77] However, both are useful approximations for estimating the overall electrostatic solvation energy of a molecule.

3.2.2   Nonpolar Contribution

Continuum electrostatics captures only one component of the solvation free energy of a molecule. Just as important is the nonpolar component which, in fact, can dominate in hydrophobic interactions. The nonpolar component comes from two sources: the free energy cost of forming the cavity in water that the solute molecule occupies, and the van der Waals interactions of the solute with the surrounding water molecules:

$$G_{\mathrm{np}} = G_{\mathrm{cav}} + G_{\mathrm{vdW}} \tag{3.8}$$

It is common to combine both of these into a single surface area (SA)-based term:

$$G_{\mathrm{np}} = \gamma \cdot SA + C \tag{3.9}$$

This is motivated by the good linear correlation between the transfer free energies of hydrocarbons and their solvent-accessible surface (SAS) area.[78,79] The coefficient, γ, is often referred to as a surface tension coefficient by analogy with the macroscopic surface tension property of liquids. One can also use the solvent-excluded surface (SES), also called the molecular surface (MS), which is the surface enclosing the volume that is excluded from a solvent probe sphere by the solute atoms.[80,81] The specific values of the coefficients in eqn (3.9) are derived empirically and their magnitudes depend on the choice of surface and atomic radii used. The coefficients are generally fitted on a training set of hydrocarbon molecules. GB and PB models supplemented by an SA term are generally referred to as GBSA and PBSA. How transferrable are the parameters to other molecule classes? It is reasonable to expect that the cavity component depends only on the size and shape of the cavity and not the identity of the atoms in the molecule. However, we expect the solute–solvent van der Waals interactions to be highly dependent on the nature of the solute atoms. In fact, the use of atom-type-dependent SA terms with PB and GB can improve agreement with experiment.[82] In the SGB/NP solvation model, both SA-dependent and independent terms are added for each atom with a heuristic to adjust for the degree of atom burial.[83] The use of such terms implicitly accounts for the variation in van der Waals interactions of solute atoms with water and at the same time provides extra parameters to correct for deficiencies in the electrostatic model (*e.g.*, lack of explicit hydrogen-binding terms). The SMx family[61,84–86] of solvation models applied in QM calculations also uses a more elaborate version of these SA terms that take into account the nature of the neighboring atoms.

Although the use of atom-type-dependent SA terms is an improvement over a single universal SA term, representing dispersion interactions in terms of SA has its limitations. In particular, buried atoms, which are completely ignored in SA-based models, can contribute significantly to the dispersion interaction.[50,87] An alternative to SA terms is to use a continuum van der Waals (CVDW) model.[88–91] In this model, one replaces the explicit water molecules with a uniform continuum density, much like in the continuum electrostatic model. Instead of calculating pairwise van der Waals interactions of solute atoms with explicit water centers, one calculates the integral of the interaction of the solute atoms with the continuum density over the entire volume from the solvent-accessible surface extending out to infinity, *i.e.* all possible locations of the water interaction center. For example, for an r^{-6} dispersion potential:

$$G_{\text{disp}} = -\rho_{\text{N}} \sum_i \int_{\Omega} \frac{B_{iw}}{|\mathbf{r} - \mathbf{r}_i|^6} \, d\Omega \tag{3.10}$$

where ρ_{N} is the solvent number density for bulk water (0.03333 Å^{-3} at 25 °C and 1 atm), B_{iw} is the Lennard-Jones force-field parameter for atom i and water, $\mathbf{r}_i$ is the position of solute atom i and Ω is the integration domain.

The volume integral can be converted into a surface integral,[88,90,91] which is much easier to compute:

$$G_{\text{disp}} = -\rho_{\text{N}} \sum_i B_{iw} \int_S \frac{(\mathbf{r} - \mathbf{r}_i) \cdot \mathbf{n}}{3|\mathbf{r} - \mathbf{r}_i|^6} \, dS \tag{3.11}$$

where the integral is over the SAS and $\mathbf{n}$ is the unit normal vector at each surface point.

The surface integral for the repulsive term can be obtained similarly, using the general formula to convert from volume to surface integrals for r^{-m} potentials ($m \neq 3$):[90,91]

$$\int_{\Omega} |\mathbf{r} - \mathbf{r}_i|^{-m} d\Omega = \int_S \frac{(\mathbf{r} - \mathbf{r}_i) \cdot \mathbf{n}}{(m - 3)|\mathbf{r} - \mathbf{r}_i|^m} \, dS \tag{3.12}$$

The surface integrals are evaluated numerically by tessellating the surface and using the discretized surface to carry out the summation. The solvent probe radius for generating the SAS is used as an adjustable parameter calibrated against reference calculations using explicit water simulations.[87,92] Continuum van der Waals models are superior to simple SA terms in improving the accuracy of continuum solvation models.[87,92–97]

The best functional form for the other nonpolar component, the cavity term, is not entirely clear. It is usually treated as being proportional to SA. In this approach, the choice of either the SAS or the SES has been a matter of taste and convenience. However, the use of the SES may have some advantages. Its associated surface tension coefficient is closer to macroscopic values.[98] It eliminates the need for curvature correction,[99,100] required when using the SAS. It reproduces the small energy barrier between the dimer contact minimum and separated methane molecules seen in explicit water simulations.[101] Alternatively, the cavity contribution can be expressed by a term proportional to the solvent-excluded volume (SEV), the volume enclosed by the SES, rather than the area or as a combination of both SA and volume terms. It has been argued that for small solutes a volume description is appropriate and for larger solutes a SA term works better, with the transition occurring at about a solute radius of 10 Å.[102,103] On the other hand, it has been suggested that the cross-over point may be specific to a purely spherical shape and is less clear for irregularly shaped molecules, and that a usable cavity term can be parameterized for either the SAS, SES, or SEV, with no evidence for a transition at the 10 Å range.[93]

A more elaborate model for cavitation is given by scaled particle theory (SPT), which expresses the free energy of a spherical cavity in terms of a cubic polynomial involving the sum of the cavity and solvent radii.[104,105] Extension to general molecular shapes involves summing over the cavitation energy of each (spherical) atom weighted by a factor proportional to the exposed SA of the atom.

3.3 Capturing Discrete Water in Continuum Solvation Models

One limitation that affects the accuracy of implicit solvation models is the continuum replacement of particulate structure in the first hydration shell around the solute. The local ordering of water in the first shell is different from isotropic bulk water and varies depending on solute polarity. Around a hydrophobic solute surface, interactions within the first hydration shell itself are favoured over interactions with the solute or with bulk solvent.[106] Around polar solute surfaces, water molecules interact strongly with the solute but orient differently around positively and negatively charged atoms, a phenomenon known as the charge asymmetry of water.[107] Local dewetting events can appear near the solute caused by particular molecular topologies in combination with the solute–solvent interactions.[108] Conversely, ion-pair interactions bridged by first-shell water molecules can become more stable than the contact ion pairing depending on the adjacent solute surface geometries.[109] The approximations made in current popular implicit models in describing the physics of hydrophobic and polar hydration limit their ability to describe such effects. Several recent simulations in explicit solvent are noteworthy as they uncover significant details in the water structure and energetics.

The microscopic water structure around a completely uncharged (zero partial charges) protein is considerable. Continuum models based on uniform distribution of solvent outside a characteristic surface are unable to address this property and its macroscopic and energetic consequences. As shown by MD in explicit solvent,[110] a positive potential of 13–24 kT e^{-1} can be calculated inside an uncharged biomolecule, even higher at cusps and other concave features of the surface. This arises primarily from a positive average charge density of 0.008 e $Å^{-3}$ at 1.0 Å from the uncharged protein surface, a consequence of a highly ordered first solvation shell that includes a fairly small population of water molecules with their dipoles pointing toward the protein, but which are potent for affecting the solute electrostatic potential.

Using MD in explicit water and Potential of Mean Force (PMF) free-energy calculations, it was shown that the interaction between positive and negative ions in water can be modulated by the presence and curvature of an adjacent hydrophobic surface.[109] Interestingly, the solvent-separated ion pair (SSIP), in which two ions are separated by one water molecule, can be made more stable than the contact ion pair (CIP) by the presence of a surface. Also, a surface-embedded positive ion attracting a negative ion from water is different from an embedded negative ion attracting a positive ion, with larger charge asymmetries for convex than concave ion-embedding surfaces. Neither the SSIP state nor the charge asymmetry are captured by PMF of ions with usual implicit solvent models, *e.g.* PBSA. Another explicit-solvent MD study interpreted the ion–ion PMFs by detailed solvent structure analysis at different stages of ion pairing in solution.[111] Significant non-homogeneity in water density, hydrogen bonding, and localized interaction energies in the back of the interacting cation and anion, and orientational restriction (caging), particularly for the shared bridging water molecules, was observed. Such non-homogeneities cannot be captured with an isotropic, bulk-water continuum solvent model.

Explicit-solvent MD simulations on a generic model for hydrophobic cavity–ligand interactions revealed a non-trivial solvent behavior of the receptor (concave), which is sensitive to the distance of the approaching (convex) ligand.[112] The ligand induces and controls an intermittent switching between dry and wet states of the hosting pocket, which determines the range and magnitude of the pocket–ligand attractions. As the ligand approaches, the pocket becomes completely dry even though it could easily accommodate the first-shell of hydration around the ligand. *A priori* defined solvent–solute interfaces, like SAS and SES typically used in continuum nonpolar solvation models, cannot account for such drying effects, which require explicit coupling of solute topography with solute–solvent interactions. Interestingly, dewetting in this hydrophobic model system was shown to be enthalpy driven rather than entropy driven as more commonly observed.[113] This study hence revealed an intriguing but legitimate possibility that the water confined in the cavity can have more entropy than bulk water and its release eliminates solvent fluctuations inside the pocket, while the enthalpy gain of dehydration results from the release of water molecules from the hydrophobic environment to the bulk water.

Extension of these cavity–ligand association studies to model systems of varying polarity provided an unprecedented, nanoscale picture of hydration thermodynamics.[114] A broad set of thermodynamic signatures is found depending on solute polarity, which captures patterns for charge asymmetry, enthalpy–entropy compensation, and localized water structure, ultimately showing that it is the water rather than the interacting solutes that drives receptor–ligand binding or rejection. The detailed picture obtained from these explicit-solvent simulations in models systems of cavity–ligand interactions help disentangle the thermodynamics of solvation in the more complex association of actual biomolecular systems.

The need to account for the first-shell water ordering effects prompted several recent developments within the realm of implicit solvation. The idea is to learn from the physics captured by explicit solvent models, which in principle can describe first hydration shell effects. These methods seek to maintain the computational efficiency characteristic of continuum models in order to allow portability and tractability for applications like virtual screening and hit optimization. Therefore, the next-generation solvation models are devised to mimic the accuracy and generality of explicit solvation models at the speed of current popular implicit solvation models. These characteristics distinguish the recent developments presented in this review from other discrete-water inspired improvements of implicit solvation models[115–117] and hybrid explicit–implicit approaches.[118–125]

3.3.1 Semi-Explicit Assembly (SEA) Continuum Model

The SEA continuum solvation model accounts for the particulate nature of water in the first shell about a solute by harvesting a wealth of details of the solvation response from MD simulations in an explicit solvent.[126,127] The underlying idea is to treat a solute molecule as a collection of atomic spheres of different types and sizes and precalculate the behavior of explicit water molecules around those component building-block spheres in isolation. Precalculations on spheres are the slow step of the method, but are carried out only once for each of the polar and nonpolar contributions to solvation free energy. The harvested information is then rapidly assembled into solvation free energies of actual solute molecules, by means of interpolation or fitted relationships between properties that can be readily calculated on actual molecules and the precalculated data on spheres. Important for model transferability is that SEA parametrization is based entirely on the physics of an explicit solvent, and therefore lacks free parameters from statistical fits to measured hydration free energies.

For the SEA nonpolar contribution to solvation, ΔG_{np}, data on spheres represent gas-to-water transfer free energies of uncharged spheres, based on TI calculations in vacuum and in explicit solvent. The obtained ΔG_{np} is the free energy of forming the solute cavity in water and includes the solute–water dispersion interactions. These transfer free energies are precalculated on spheres with varying parameters σ and ε of the Lennard-Jones (LJ) 12-6 potential, thus establishing a relationship between ΔG_{np} and any (σ, ε) combination. This is an essential component of the SEA method. The LJ dispersion-potential field is rapidly calculated around an actual solute molecule and then region-averaged near each exposed atom. Newly derived local (σ, ε) parameters assigned to exposed solute atoms can then be converted to nonpolar transfer free energies by linear interpolations from the precalculated $\Delta G_{np}(\sigma, \varepsilon)$ array and taking into account the fraction of surface exposure. This entire procedure, which also relies on an SAS positioned at the average location of the first-shell water as deduced from MD simulations on spheres, encodes information about the full solute structure and interactions into the solvent exposed regions of the molecule. Hence, the nonpolar SEA term more properly addresses dispersion interactions, accounts for effects of solute shape not only size, and solves non-additivity problems of traditional SA-based nonpolar solvation models (*e.g.*, γA). Owing to its detailed incorporation of dispersion interactions, nonpolar SEA calculations on hydrocarbons revealed hot spot regions that contribute very favorable solvation in water.

The SEA polar contribution to solvation is calculated from two terms, $\Delta G_{pol,surf}$ and $\Delta G_{pol,bulk}$, describing electrostatic interactions of the solute with the first-shell water molecules and with the more distant water molecules, respectively. In order to arrive at the first-shell component, $\Delta G_{pol,surf}$, several one-time precalculations were performed on charged spheres of various sizes using explicit-solvent MD simulations and varying the sphere's point-charge and LJ parameters. One descriptor that is obtained is the average dipole (and quadrupole) moment that captures the polar response of the first-shell water molecules. A simple functional relationship between the average first-shell water dipole moment *versus* the strength of the electrostatic field from the solute sphere is then established with a saturating function, *e.g.*:

$$f(x) = \frac{1}{c_0 + \exp\,(c_1 \cdot x)} + c_2 \tag{3.13}$$

where c_0, c_1, and c_2 are curve fitting coefficients.

Since water's dipole is asymmetric, there is asymmetry in the resulting function, with separate fits needed for the positive and negative spheres.

Around positive spheres, the limit of the sigmoidal curve coincides with the dipole moment of the explicit water model because water's average dipole points away from the solute surface along the normal vector. Around negative spheres, water points one hydrogen toward the solute center and only a partial projection of a water molecule's dipole is along this normal vector. This water-dipole–solute–electric-field association function (eqn 3.13) is key to rapidly assemble the free energy of a given solute molecule from data precomputed on spheres. Having calculated the electric field generated by all solute partial charges at any given solvation site, the appropriate dipole moment can be assigned and aligned with the electric field line at this solvation site. The $\Delta G_{pol,surf}$ is then obtained as a pairwise Coulombic sum of solute–solvent charge–dipole interactions and solvent–solvent dipole–dipole interactions over all solvation sites distributed on an SAS (constructed at the first-shell average position and density based on precalculations in spheres). Because positional and orientational distributions of surface water molecules are factored into the dipole representations, this pairwise sum approximates a free energy rather than simple potential energy.

The bulk component, $\Delta G_{pol,bulk,}$, is calculated with a continuum electrostatic model,[128] as the reaction field energy due to solute and first-shell water charges embedded in a spherical cavity, and added to the first-shell component. Final SEA hydration free energy is obtained from a Boltzmann-weighted sum over several first-shell solvation configurations.

A test of the non-polar component of SEA on a diverse set of 504 small molecules was found to correlate well with the nonpolar component from explicit water simulations,[129] to be superior to other advanced nonpolar solvation models that have relatively few free fitting parameters,[93,94] and be as fast to compute as γA methods that essentially do not correlate with explicit solvent calculations. Full SEA predictions on a diverse set of 504 small molecules, and on the 89 small molecules from the SAMPL1 and SAMPL2 challenges,[130,131] correlate well with the explicit-solvent FEP/TIP3P predictions.[129,132] Retrospective predictions on the SAMPL1 set (rmse = 4.1 kcal mol^{-1}) and SAMPL2 set (rmse = 1.8 kcal mol^{-1}) are comparable with the accuracy of the explicit-solvent FEP approach, slightly worse for SAMPL1 and better for SAMPL2.[126,132,133] It was noted that both SEA and FEP/TIP3P methods underestimate the transfer free energy of alcohols, amines, and sugars, and do not handle well molecules containing polyvalent sulfur and phosphorus, signalling deficiencies in the force field adopted for these tests. The SEA method with dipolar representation of first-shell water is able to capture the charge asymmetry but underestimates the associated free energy difference relative to explicit-solvent simulations. The more expensive quadrupolar water representation improves SEA predictions in terms of charge asymmetry, but not the agreement with hydration free energies. Thus, the SEA continuum method with the dipolar first-shell water appears to capture the necessary physics for general applications, achieving similar prediction accuracy and $\sim 10^{7}$-fold speedup relative to the explicit-solvent FEP method.[129] It is about 10-fold slower than the GB methods but significantly more accurate (rmse of 1.3 kcal mol^{-1} with SEA *vs.* 2.8 kcal mol^{-1} with the GB method[63] for the 504 small-molecule set).[129]

3.3.2 First Shell of Hydration (FiSH) Continuum Model

The FiSH continuum solvation model has been developed to mimic the accuracy of hydration calculations based on explicit-water MD simulations.[96] Specifically, it was calibrated on the explicit-solvent hydration model of the end-point LIE approach.[9,134,135] This was preferred over the more expensive pathway methods like FEP due to its term decomposition that is simpler and more compatible with the solvation contributions typically calculated with continuum models, and its similar accuracy in predicting hydration free energies.[136] The use of hydration data derived from explicit-solvent simulations instead of experiment allows independent calibration of electrostatic and nonpolar components and ensures transferability of the FiSH model, a philosophy also adopted for the development of the SEA continuum model.

The polar term of the FiSH continuum model has been adapted to capture the water's charge asymmetry and reproduce the polar component of the LIE explicit-solvent model. As mentioned earlier, water has an asymmetric response to positively and negatively charged solutes,[137–140] and implicit solvation models are completely symmetric with respect to charge inversion. This discrepancy in behavior has been explored in depth using charge-inverted pairs of neutral model systems.[107] A solution to this problem is to have Born radii that are automatically adjusted to reflect the local polarity, *i.e.* to use induced surface charge density (ISCD)-corrected Born radii.[141] In this approach, the ISCD is initially calculated using force-field radii for the atoms by solving the Poisson equation with a boundary element method (BEM).[142,143] Average ISCDs are calculated for each exposed atom. The final atomic Born radii are then calculated according to the sign and magnitude of the atom's ISCD in the initial run. In its simplest formulation, the ISCD-corrected Born radius of an atom is given by:[141]

$$r = \begin{cases} r_0 - c_+ \sigma & \text{if } \sigma \geq 0 \\ r_0 + c_- \sigma & \text{if } \sigma < 0 \end{cases} \tag{3.14}$$

where r_0 is the force-field van der Waals radius, σ is the average ISCD of the atom, and c_+ and c_- are coefficients fitted to reproduce explicit-solvent simulations[107] (c_+ and c_- are universal and independent of atom type).

The BEM reaction field energy is then recalculated using the adjusted Born radii. The correction is designed to capture the asymmetric response of water molecules, depending on the local electric field. Noteworthy, σ is a molecular property that depends on all atoms in the molecule and not just the atom that owns the surface patch. This approach has been shown to restore charge asymmetry in a purely continuum electrostatics manner.[141]

A further refinement implemented in FiSH is the nonlinear dependence of the Born radius on the ISCDs.[96] Clearly, the ISCD correction cannot increase/decrease indefinitely with increasing ISCD magnitude. The correction must level off and perhaps display a maximum/minimum. At moderately large negative ISCD, the Born radius is larger than the LJ radius because this reflects the orientation of the first-shell water molecule with the water hydrogen atoms pointing away from the surface. However, as σ becomes even more negative, the water molecule will be drawn closer to the solute molecular surface and the effective Born radius should decrease. For positive ISCD, increases in the value of σ are associated with a decrease in the Born radius as the water hydrogen is pulled closer to the solute, effectively decreasing the Born radius. As these decreases in σ become larger, a leveling off should occur since the van der Waals repulsion will start to become significant. Also, we expect the Born radius correction to depend upon the well-depth of the LJ potential. The rational function below is one functional form that exhibits the desired behavior:

$$r_i = r_{i,0} + \frac{A\sigma_i + B(\varepsilon_i/r_{i,0})}{D\sigma_i^2 + E\sigma_i + 1} \tag{3.15}$$

where A, B, D, and E are fitting parameters (applicable to all atom types), ε_i and $r_{i,0}$ are the force-field LJ well-depth and radius, respectively, and the fitting is carried out to reproduce LIE reaction field energies from MD simulations on a training set of solute molecules.[96]

The nonpolar component of the FiSH continuum model follows the term decomposition of the LIE explicit-solvent model and includes a term for solute–solvent van der Waals interactions and a term for cavity formation in water. Typical CVDW models assume a uniform solvent density.[93,95] However, radial distribution functions of the water oxygen around the solute show that the first shell can be far from being a homogeneous distribution relative to bulk water. The FiSH model addresses this by dividing the solvent volume into two regions.[96] The first region is the first solvation shell and is treated as if the centers of the water oxygens are constrained to lie exactly on the SAS. This makes the contribution of the first shell a two-dimensional integral approximated as a summation (eqn 3.16):

$$U_1^{\text{vdW}} = \rho_S \sum_{i=1}^{n\,\text{atoms}} \sum_{j=1}^{n\,\text{patches}} \left(\frac{A_{iw}}{r_{ij}^{12}} - \frac{B_{iw}}{r_{ij}^6} \right) SA_j \tag{3.16}$$

where ρ_s is the number density of water along the SAS, SA_j is the area of patch j, and A_{iw} and B_{iw} are the force-field LJ parameters for atom i and a TIP3P oxygen atom.

The second region starts at the SAS + 2.8 Å and a uniform solvent density is assumed from that point outward. The contribution of the second region is computed using a usual CVDW model for all solute atoms.[90,91,93] To obtain the total nonpolar component in the FiSH approach, a term proportional to the solute's molecular SA trained on pseudo-experimental free energies is employed to account for the cavity cost and added to the two-zone solute–solvent van der Waals interaction component. The surface tension of the cavity term and a constant are the only FiSH parameters fitted to experiment.

External tests on simple compounds and drug-like molecules yielded high correlations between all components of the FiSH continuum model and the corresponding components of the LIE explicit-solvent model, which carried over to total hydration free energies ($R^2 = 0.97$, slope 0.93–0.97 depending on the test set).[136] This prediction performance comes with a computing speed comparable to that of the underlying BEM for calculating reaction field energies. Therefore, the FiSH continuum model achieved its objective of being an efficient surrogate of an explicit-solvent hydration model. The performance of the FiSH implicit model in predicting experimental hydration free energies was very similar to that of the LIE explicit model, as measured retrospectively on the SAMPL1 challenge dataset (MUE of 2.2 kcal mol^{-1}; rmse of 2.7 kcal mol^{-1}) and prospectively in the SAMPL2 challenge (MUE of 1.7 kcal mol^{-1}; rmse 2.2 kcal mol^{-1}).[96,97,136]

Problematic chemical groups for the FiSH continuum model parallel the outliers of the SEA continuum model,[126] likely related to the underlying force field used for explicit-solvent simulations. The advances in handling charge asymmetry and van der Waals interactions, together with the almost complete elimination of free-fitting parameters, enhanced the performance and transferability of the FiSH model over its predecessor continuum electrostatics-dispersion solvation model,[95] on both SAMPL1 and SAMPL2 test sets.[96,97] It is expected that these improvements will carry over to solvated interaction energy calculations of binding affinity in biomolecular systems.

3.3.3 Analytical Generalized Born plus Nonpolar Implicit Solvent Model 2 (AGBNP2)

The AGBNP2 continuum model,[71] a modification of the original AGBNP model consisting of GB continuum electrostatics, SA-based cavity, and continuum van der Waals terms,[72] addresses first-shell effects by focusing on short-range hydrogen bonding, which is poorly described by continuum dielectric models. This new empirical component is based on a geometrical procedure that measures how well a solute atom can interact with hydration sites on the solute surface. A spherical water molecule is placed in a position that provides near-optimal interaction with a hydrogen-bonding donor or acceptor solute atom. The magnitude of the hydrogen bonding free energy correction, ΔG_{hb}, corresponding to each water sphere is a function of the predicted water occupancy of the location corresponding to the water sphere, w_s, which is the fraction of the volume of the water site sphere that is accessible to water molecules without causing steric clashes with solute atoms:

$$G_{hb} = \sum_s h_s\, S(w_s; w_a, w_b) \tag{3.17}$$

where h_s is the maximum correction energy which depends on the type of solute–water hydrogen bond and has to be optimized, and $S(w_s; w_a, w_b)$ is a polynomial switching function that smoothly interpolates from 0 to 1 between w_a and w_b (between 15% to 50% water occupancy in the study).

An analytical prescription to identify and measure the volume of hydration sites on the solute surface is a key ingredient of the model. Hydration sites that are deemed too small to contain a water molecule do not contribute to the solute hydration free energy. Conversely, hydration sites of sufficient size form favorable interactions with nearby polar groups. The hydrogen bonding correction function favors conformations of the solute in which polar groups are oriented so as to form hydrogen bonds with the surrounding water solvent, thereby achieving a more balanced equilibrium with respect to the competing intramolecular hydrogen bonds. This model thus incorporates the effects of both water granularity and nonlinear first-shell hydration effects.

It was shown that inclusion of the parametrized hydrogen bonding correction in the AGBNP2 model improves the agreement with experimental hydration free energies of small molecules. It is also expected that the final implementation of the AGBNP2 computer code will reach similar or better computing speed than the original AGBNP. In fact, AGBNP2 has been formulated for MD conformational sampling applications, which require potential models of low computational complexity and favorable scaling characteristics, and with analytical gradients. The results of MD simulations of a series of miniproteins show that the new model produces conformational ensembles in substantially better agreement with reference explicit-solvent ensembles than the original model with respect to both structural and energetics measures.[71]

The AGBNP2 implicit solvation has been recently employed with the novel binding energy distribution analysis method (BEDAM) for protein–ligand binding affinity prediction.[144] In BEDAM, the binding constant is computed by means of a weighted integral of the probability distribution of the binding energy obtained in the canonical ensemble in which the ligand is positioned in the binding site, but the receptor and the ligand interact only with the solvent continuum. Hence, the AGBNP2 model plays a central role in BEDAM, which correctly discriminated the known binders from the known nonbinders to the L99A and L99A/M102Q mutants of T4 lysozyme receptor, with computed standard binding free energies of the binders in reasonably good agreement with reported calorimetric measurements.

3.3.4 Three-Dimensional Reference Interaction Site Model (3D-RISM)

A different approach to obtain detailed information on the discrete structure and the thermodynamics of water around the solute is provided by the 3D-RISM,[145–147] which is conceptually very different from the classical mechanical approach underpinning MD. RISM is an integral equation theory solvation model based on the direct correlation function between interacting particles with indirect contributions propagating throughout the system.[148-150] Although RISM does not capture the solvent dynamics of the system, it yields the equilibrium statistical mechanical distribution of the solvent particles around the solute. The method was applied in modeling macromolecular hydration, successfully predicting localized water molecules in several protein structures.[118,119,151] A recent comparative study shows that the statistical mechanical model with the Kovalenko–Hirata closure (3D-RISM-KH) yields water density distributions that are very similar to those from classical MD simulations up to a 0.5 Å resolution, but for significantly reduced computational cost.[152] Also, 3D-RISM-KH produces smoother local water densities, since MD fluctuations give rise to noise in the average densities whereas 3D-RISM directly yields only the equilibrium distributions. Hence, 3D-RISM is a very promising application of integral equation theories of solutions to large biomolecules, alone or in combination with MD simulations by providing enhanced solvent sampling.[153,154]

3.3.5 Other Recent Models Capturing First-Shell Effects

The level-set Variational Implicit Solvent Model (VISM), developed by McCammon and co-workers, includes a dynamic solvent boundary whose optimal configuration is obtained by minimization of surface-dependent solvent free energy functional.[155,156] The method was shown to be able to predict the dewetting event and its thermodynamic effect in a hydrophobic receptor–ligand system.[112,157] Parametrization of the VISM based on the deeper understanding of cavity–ligand recognition and improved description of water thermodynamics will enable its applications for electrostatically interacting solutes.[114]

Setny and Zacharias described a Mean-Field Cellular Automata (MFCA) procedure using a body-centered cubic (bcc) grid to simulate the discrete distribution of explicit water molecules around a single-conformation representation of the solute.[158] Through an iterative procedure, they optimize the position and orientation of water molecules on the grid and are able to calculate both electrostatic and nonpolar solvation energies. The method is not a continuum model, but does not have all the degrees of freedom and associated computational cost of a full explicit-water MD simulation either. It has been tested on the SAMPL0 set[159] with encouraging results, using the SAMPL1 molecules as a training dataset.[130] Further developments should consider including bulk solvent electrostatic effects beyond the first shell of water-occupied grid points, and differential weighting of grid points corresponding to buried, highly localized, water molecules.

3.4 Concluding Remarks

Predicting binding affinities in aqueous solution requires carefully calibrated solvation models to account for changes in solvation thermodynamics upon binding. Traditional continuum solvation models replacing the discrete water structure with certain bulk properties such as a dielectric constant and surface tension coefficient work surprisingly well for their level of approximation. They are also suitable for large-scale applications, with computational cost savings arising from instantaneous solvent relaxation. Nonetheless, the accuracy of popular continuum models is limited if one looks at the detailed structure and energetics within the first-shell water around the solute. Charge asymmetry, local dewetting, non-homogeneity in water density, hydrogen bonding and localized interactions, bridging water caging and energetics, which are revealed by explicit-solvent simulations, are not captured by these continuum models. An emergent direction is to address these shortcomings of continuum modeling by mimicking the behavior of explicit-solvent models. Models like SEA, FiSH, and AGBNP2 restore in a purely implicit manner some physical aspects of the discrete solvent in the all-important first shell around the solute and are able to capture some of the effects arising from this particulate, nonbulk water structure. The improvements in these models also come without compromising the main advantage of the continuum approach: computational efficiency. Because of their physics and speed, the next-generation continuum solvation models hold much promise for improving large-scale applications for drug discovery and understanding of biomacro-molecular systems.

References

1. A. A. Rashin, *Prog. Biophys. Mol. Biol.*, 1993, **60**, 73.
2. B. Honig, K. Sharp and A.-S. Yang, *J. Phys. Chem.*, 1993, **97**, 1101.
3. M. K. Gilson and H. X. Zhou, *Annu. Rev. Biophys. Biomol. Struct.*, 2007, **36**, 21.
4. H. Gohlke and G. Klebe, *Angew. Chem., Int. Ed.*, 2002, **41**, 2644.
5. *Free Energy Calculations in Rational Drug Design*, ed. M. R. Reddy and M. D. Erion, Kluwer/Plenum, New York, 2001.
6. J. D. Chodera, D. L. Mobley, M. R. Shirts, R. W. Dixon, K. Branson and V. S. Pande, *Curr. Opin. Struct. Biol.*, 2011, **21**, 150.
7. F. S. Lee, Z.-T. Chu, M. B. Bolger and A. Warshel, *Protein Eng.*, 1992, **5**, 215.
8. J. Åqvist, V. B. Luzhkov and B. O. Brandsdal, *Acc. Chem. Res.*, 2002, **35**, 358.
9. J. Aqvist and J. Marelius, *Comb. Chem. High Throughput Screening*, 2001, **4**, 613.
10. J. Åqvist, C. Medina and J.-E. Samuelsson, *Protein Eng.*, 1994, **7**, 385.
11. J. Carlsson, M. Andér, M. Nervall and J. Åqvist, *J. Phys. Chem. B*, 2006, **110**, 12034.
12. R. Zhou, R. A. Friesner, A. Ghosh, R. C. Rizzo, W. L. Jorgensen and R. M. Levy, *J. Phys. Chem. B*, 2001, **105**, 10388.
13. Y. Su, E. Gallicchio, K. Das, E. Arnold and R. M. Levy, *J. Chem. Theory Comput.*, 2006, **3**, 256.

14. X. Zou, Y. Sun and I. D. Kuntz, *J. Am. Chem. Soc.*, 1999, **121**, 8033.
15. P. A. Kollman, I. Massova, C. Reyes, B. Kuhn, S. Huo, L. Chong, M. Lee, T. Lee,Y. Duan, W. Wang, O. Donini, P. Cieplak, J. Srinivasan, D. A. Case and T. E. Cheatham, *Acc. Chem. Res.*, 2000, **33**, 889.
16. B. Kuhn, P. Gerber, T. Schulz-Gasch and M. Stahl, *J. Med. Chem.*, 2005, **48**, 4040.
17. H. Gohlke and D. A. Case, *J. Comput. Chem.*, 2004, **25**, 238.
18. S. P. Brown and S. W. Muchmore, *J. Med. Chem.*, 2009, **52**, 3159.
19. I. Massova and P. A. Kollman, *J. Am. Chem. Soc.*, 1999, **121**, 8133.
20. M. S. Lee and M. A. Olson, *Biophys. J.*, 2006, **90**, 864.
21. S. P. Brown and S. W. Muchmore, *J. Chem. Inf. Model.*, 2006, **46**, 999.
22. M. Naïm, S. Bhat, K. N. Rankin, S. Dennis, S. F. Chowdhury, I. Siddiqi, P. Drabik, T. Sulea, C. Bayly, A. Jakalian and E. O. Purisima, *J. Chem. Inf. Model.*, 2007, **47**, 122.
23. Q. Cui, T. Sulea, J. D. Schrag, C. Munger, M.-N. Hung, M. Naïm, M. Cygler and E. O. Purisima, *J. Mol. Biol.*, 2008, **379**, 787.
24. T. Sulea and E. O. Purisima. In *Computational Drug Discovery and Design: Methods in Molecular Biology*, **Vol 819**, ed. R. Baron, Humana Press, New York, 2012.
25. T. Sulea, Q. Cui and E. O. Purisima, *J. Chem. Inf. Model.*, 2011, **51**, 2066.
26. N. Froloff, A. Windemuth and B. Honig, *Protein Sci.*, 1997, **6**, 1293.
27. F. Polticelli, P. Ascenzi, M. Bolognesi and B. Honig, *Protein Sci.*, 1999, **8**, 2621.
28. T. Zhang and D. E. Koshland, *Protein Sci.*, 1996, **5**, 348.
29. B. Shoichet, A. R. Leach and I. D. Kuntz, *Proteins*, 1999, **34**, 4.
30. J. Novotny, R. E. Bruccoleri and F. A. Saul, *Biochemistry*, 1989, **28**, 4735.
31. S. Krystek, T. Stouch and J. Novotny, *J. Mol. Biol.*, 1993, **234**, 661.
32. D. H. Williams, J. P. L. Cox, A. J. Doig, M. Gardner, U. Gerhard, P. T. Kaye, A. R. Lal, I. A. Nicholls, C. J. Salter and R. C. Mitchell, *J. Am. Chem. Soc.*, 1991, **113**, 7020.
33. S. Vajda, Z. Weng, R. Rosenfeld and C. DeLisi, *Biochemistry*, 1994, **33**, 13977.
34. Z. Weng, S. Vajda and C. Delisi, *Protein Sci.*, 1996, **5**, 614.
35. H. J. Böhm, *J. Comput. Aided Mol. Des.*, 1994, **8**, 243.
36. R. D. Head, M. L. Smythe, T. I. Oprea, C. L. Waller, S. M. Green and G. R. Marshall, *J. Am. Chem. Soc.*, 1996, **118**, 3959.
37. J. S. Tokarski and A. J. Hopfinger, *J. Chem. Inf. Comput. Sci.*, 1997, **37**, 792.
38. M. D. Eldridge, C. W. Murray, T. R. Auton, G. V. Paolini and R. P. Mee, *J. Comput. Aided Mol. Des.*, 1997, **11**, 425.
39. R. Wang, L. Lai and S. Wang, *J. Comput. Aided Mol. Des.*, 2002, **16**, 11.
40. H. F. G. Velec, H. Gohlke and G. Klebe, *J. Med. Chem.*, 2005, **48**, 6296.
41. H. Gohlke and G. Klebe, *Curr. Opin. Struct. Biol.*, 2001, **11**, 231.
42. S.-H. Huang and X. Zou, *Annu. Rep. Comput. Chem.*, 2010, **6**, 281.
43. I. Muegge and Y. C. Martin, *J. Med. Chem.*, 1999, **42**, 791.

44. S.-Y. Huang and X. Zou, *J. Chem. Inf. Model.*, 2010, **50**, 262.
45. A. V. Ishchenko and E. I. Shakhnovich, *J. Med. Chem.*, 2002, **45**, 2770.
46. C. McInnes, *Curr. Opin. Chem. Biol.*, 2007, **11**, 494.
47. C. J. Cramer and D. G. Truhlar, *Chem. Rev.*, 1999, **99**, 2161.
48. C. J. Cramer and D. G. Truhlar, *Rev. Comput. Chem.*, 1995, **6**, 1.
49. J. Tomasi, *Theor. Chem. Acc.*, 2004, **112**, 184.
50. F. J. Luque, C. Curutchet, J. Muñoz-Muriedas, A. Bidon-Chanal, I. Soteras, A. Morreale, J. L. Gepi and M. Orozco, *Phys. Chem. Chem. Phys.*, 2003, **5**, 3827.
51. J. Tomasi, B. Mennucci and R. Cammi, *Chem. Rev.*, 2005, **105**, 2999.
52. J. Tomasi and M. Persico, *Chem. Rev.*, 1994, **94**, 2027.
53. M. Feig, and C. L. Brooks, III, *Curr. Opin. Struct. Biol.*, 2004, **14**, 217.
54. J. Chen, C. L. Brooks, III and J. Khandogin, *Curr. Opin. Struct. Biol.*, 2008, **18**, 140.
55. B. Roux and T. Simonson, *Biophys. Chem.*, 1999, **78**, 1.
56. J. A. Grant, B. T. Pickup and A. Nicholls, *J. Comput. Chem.*, 2001, **22**, 608.
57. N. A. Baker, *Methods Enzymol.*, 2004, **383**, 94.
58. S. Miertus, E. Scrocco and J. Tomasi, *Chem. Phys.*, 1981, **55**, 117.
59. W. C. Still, A. Tempczyk, R. C. Hawley and T. Hendrickson, *J. Am. Chem. Soc.*, 1990, **112**, 6127.
60. G. D. Hawkins, C. J. Cramer and D. G. Truhlar, *Chem. Phys. Lett.*, 1995, **246**, 122.
61. G. D. Hawkins, C. J. Cramer and D. G. Truhlar, *J. Phys. Chem.*, 1996, **100**, 19824.
62. M. Scarsi, J. Apostolakis and A. Caflisch, *J. Phys. Chem. A*, 1997, **101**, 8098.
63. A. Onufriev, D. Bashford and D. A. Case, *J. Phys. Chem. B*, 2000, **104**, 3712.
64. M. S. Lee, M. Feig, F. R. Salsbury, Jr. and C. L. Brooks, III, *J. Comput. Chem.*, 2003, **24**, 1348.
65. M. S. Lee, F. R. Salsbury, Jr. and C. L. Brooks, III, *J. Chem. Phys.*, 2002, **116**, 10606.
66. A. Onufriev, D. Bashford and D. A. Case, *Proteins: Struct., Funct., Bioinf.*, 2004, **55**, 383.
67. A. Ghosh, C. S. Rapp and R. A. Friesner, *J. Phys. Chem. B*, 1998, **102**, 10983.
68. Z. Yu, M. P. Jacobson and R. A. Friesner, *J. Comput. Chem.*, 2006, **27**, 72.
69. L. Y. Zhang, E. Gallicchio, R. A. Friesner and R. M. Levy, *J. Comput. Chem.*, 2001, **22**, 591.
70. J. Mongan, C. Simmerling, J. A. McCammon, D. A. Case and A. Onufriev, *J. Chem. Theory Comput.*, 2007, **3**, 156.
71. E. Gallicchio, K. Paris and R. M. Levy, *J. Chem. Theory Comput.*, 2009, **5**, 2544.

72. E. Gallicchio and R. M. Levy, *J. Comput. Chem.*, 2004, **25**, 479.
73. A. Onufriev, *Annu. Rep. Comput. Chem.*, 2008, **4**, 125.
74. H. Gohlke, C. Kiel and D. A. Case, *J. Mol. Biol.*, 2003, **330**, 891.
75. D. Bashford and D. A. Case, *Annu. Rev. Phys. Chem.*, 2003, **51**, 129.
76. M. Feig, A. Onufriev, M. S. Lee, W. Im, D. A. Case and C. L. Brooks, III, *J. Comput. Chem.*, 2004, **25**, 265.
77. J. A. Grant, B. T. Pickup, M. J. Sykes, C. A. Kitchen and A. Nicholls, *Chem. Phys. Lett.*, 2007, **441**, 163.
78. R. B. Hermann, *J. Phys. Chem.*, 1972, **76**, 2754.
79. J. A. Reynolds, D. B. Gilbert and C. Tanford, *Proc. Natl. Acad. Sci. U. S. A.*, 1974, **71**, 2925.
80. F. M. Richards, *Annu. Rev. Biophys. Bioeng.*, 1977, **6**, 151.
81. J. Greer and B. L. Bush, *Proc. Natl. Acad. Sci. U. S. A.*, 1978, **75**, 303.
82. R. C. Rizzo, T. Aynechi, D. A. Case and I. D. Kuntz, *J. Chem. Theory Comput.*, 2006, **2**, 128.
83. E. Gallicchio, L. Y. Zhang and R. M. Levy, *J. Comput. Chem.*, 2002, **23**, 517.
84. C. C. Chambers, G. D. Hawkins, C. J. Cramer and D. G. Truhlar, *J. Phys. Chem.*, 1996, **100**, 16385.
85. C. J. Cramer and D. G. Truhlar, *Acc. Chem. Res.*, 2008, **41**, 760.
86. C. J. Cramer and D. G. Truhlar, *Science*, 1992, **256**, 213.
87. R. M. Levy, L. Y. Zhang, E. Gallicchio and A. K. Felts, *J. Am. Chem. Soc.*, 2003, **125**, 9523.
88. M.-J. Huron and P. Claverie, *J. Phys. Chem.*, 1972, **76**, 2123.
89. F. M. Floris, A. Tani and J. Tomasi, *Chem. Phys.*, 1993, **169**, 11.
90. F. M. Floris, J. Tomasi and J. L. Pascual-Ahuir, *J. Comput. Chem.*, 1991, **12**, 784.
91. F. M. Floris and J. Tomasi, *J. Comput. Chem.*, 1989, **10**, 616.
92. M. Zacharias, *J. Phys. Chem. A*, 2003, **107**, 3000.
93. C. Tan, Y. H. Tan and R. Luo, *J. Phys. Chem. B*, 2007, **111**, 12263.
94. J. A. Wagoner and N. A. Baker, *Proc. Natl. Acad. Sci. U. S. A.*, 2006, **103**, 8331.
95. T. Sulea, D. Wanapun, S. Dennis and E. O. Purisima, *J. Phys. Chem. B*, 2009, **113**, 4511.
96. C. R. Corbeil, T. Sulea and E. O. Purisima, *J. Chem. Theory Comput.*, 2010, **6**, 1622.
97. E. O. Purisima, R. C. Corbeil and T. Sulea, *J. Comput. Aided Mol. Des.*, 2010, **24**, 373.
98. I. Tuñón, E. Silla and J. L. Pascual-Ahuir, *Protein Eng*, 1992, **5**, 715.
99. A. Nicholls, K. A. Sharp and B. Honig, *Proteins*, 1991, **11**, 281.
100. K. A. Sharp, A. Nicholls and R. F. Fine, *Science*, 1991, **252**, 106.
101. J. Pitarch, V. Moliner, J. L. Pascual-Ahuir, E. Silla and I. Tuñón, *J. Phys. Chem.*, 1996, **100**, 9955.
102. K. Lum, D. Chandler and J. D. Weeks, *J. Phys. Chem. B*, 1999, **103**, 4570.
103. D. Chandler, *Nature*, 2005, **437**, 640.

104. R. A. Pierotti, *J. Phys. Chem.*, 1965, **69**, 281.

105. R. A. Pierotti, *Chem. Rev.*, 1976, **76**, 717.

106. T. M. Raschke and M. Levitt, *Proc. Natl. Acad. Sci. U. S. A.*, 2005, **102**, 6777.

107. D. L. Mobley, A. E. Barber, C. J. Fennell and K. A. Dill, *J. Phys. Chem. B*, 2008, **112**, 2405.

108. B. J. Berne, J. D. Weeks and R. Zhou, *Annu. Rev. Phys. Chem.*, 2009, **60**, 85.

109. I. Chorny, K. A. Dill and M. P. Jacobson, *J. Phys. Chem. B*, 2005, **109**, 24056.

110. D. S. Cerutti, N. A. Baker and J. A. McCammon, *J. Chem. Phys.*, 2007, **127**, 155101.

111. C. J. Fennell, A. Bizjak, V. Vlachy and K. A. Dill, *J. Phys. Chem. B*, 2009, **113**, 6782.

112. L.-T. Cheng, Z. Wang, P. Setny, J. Dzubiella, B. Li and J. A. McCammon, *J. Chem. Phys.*, 2009, **131**, 144102.

113. P. Setny, R. Baron and J. A. McCammon, *J. Chem. Theory Comput.*, 2010, **6**, 2866.

114. R. Baron, P. Setny and J. A. McCammon, *J. Am. Chem. Soc.*, 2010, **132**, 12091.

115. A. Papazyan and A. Warshel, *J. Phys. Chem. B*, 1998, **102**, 5348.

116. J. Florián and A. Warshel, *J. Phys. Chem. B*, 1997, **101**, 5583.

117. J. Florián and A. Warshel, *J. Phys. Chem. B*, 1999, **103**, 10282.

118. T. Imai, R. Hiraoka, A. Kovalenko and F. Hirata, *Proteins: Struct., Funct., Bioinf.*, 2007, **66**, 804.

119. N. Yoshida, T. Imai, S. Phongphanphanee, A. Kovalenko and F. Hirata, *J. Phys. Chem. B*, 2009, **113**, 873.

120. G. King, and A. Warshel, *J. Chem. Phys.*, 1989, **91**, 3647.

121. D. Beglov and B. Roux, *J. Chem. Phys.*, 1994, **100**, 9050.

122. W. Im, S. Berneche and B. Roux, *J. Chem. Phys.*, 2001, **114**, 2924.

123. V. Lounnas, S. K. Lüdemann and R. C. Wade, *Biophys. Chem.*, 1999, **78**, 157.

124. M. S. Lee, F. R. Salsbury, Jr. and M. A. Olson, *J. Comput. Chem.*, 2004, **25**, 1967.

125. D. Shivakumar, Y. Deng and B. Roux, *J. Chem. Theory Comput.*, 2009, **5**, 919.

126. C. J. Fennell, C. W. Kehoe and K. A. Dill, *Proc. Natl. Acad. Sci. U. S. A.*, 2011, **108**, 3234.

127. C. J. Fennell, C. Kehoe and K. A. Dill, *J. Am. Chem. Soc.*, 2009, **132**, 234.

128. L. Onsager, *J. Am. Chem. Soc.*, 1936, **58**, 1486.

129. D. L. Mobley, C. I. Bayly, M. D. Cooper, M. R. Shirts and K. A. Dill, *J. Chem. Theory Comput.*, 2009, **5**, 350.

130. J. P. Guthrie, *J. Phys. Chem. B*, 2009, **113**, 4501.

131. M. Geballe, A. Skillman, A. Nicholls, J. P. Guthrie and P. J. Taylor, *J. Comput. Aided Mol. Des.*, 2010, **24**, 259.

132. D. L. Mobley, C. I. Bayly, M. D. Cooper and K. A. Dill, *J. Phys. Chem. B*, 2009, **113**, 4533.

133. P. Klimovich and D. Mobley, *J. Comput. Aided Mol. Des.*, 2010, **24**, 307.

134. H. A. Carlson and W. L. Jorgensen, *J. Phys. Chem.*, 1995, **99**, 10667.

135. M. Almlof, J. Carlsson and J. Aqvist, *J. Chem. Theory Comput.*, 2007, **3**, 2162.

136. T. Sulea, C. R. Corbeil and E. O. Purisima, *J. Chem. Theory Comput.*, 2010, **6**, 1608.

137. W. M. Latimer, K. S. Pitzer and C. M. Slansky, *J. Chem. Phys.*, 1939, **7**, 108.

138. A. A. Rashin and B. Honig, *J. Phys. Chem.*, 1985, **89**, 5588.

139. B. Roux, H.-A. Yu and M. Karplus, *J. Phys. Chem.*, 1990, **94**, 4683.

140. C. S. Babu and C. Lim, *J. Phys. Chem. B*, 1999, **103**, 7958.

141. E. O. Purisima and T. Sulea, *J. Phys. Chem. B*, 2009, **113**, 8206.

142. E. O. Purisima, *J. Comput. Chem.*, 1998, **19**, 1494.

143. E. O. Purisima and S. H. Nilar, *J. Comput. Chem.*, 1995, **16**, 681.

144. E. Gallicchio, M. Lapelosa and R. M. Levy, *J. Chem. Theory Comput.*, 2010, **6**, 2961.

145. Q. Du, D. Beglov and B. Roux, *J. Phys. Chem. B*, 2000, **104**, 796.

146. D. Beglov and B. Roux, *J. Chem. Phys.*, 1995, **103**, 360.

147. T. Imai, A. Kovalenko and F. Hirata, *Chem. Phys. Lett.*, 2004, **395**, 1.

148. D. Chandler and H. C. Andersen, *J. Chem. Phys.*, 1972, **57**, 1930.

149. F. Hirata, B. M. Pettitt and P. J. Rossky, *J. Chem. Phys.*, 1982, **77**, 509.

150. S. M. Kast, *ChemPhysChem*, 2004, **5**, 449.

151. T. Imai, A. Kovalenko and F. Hirata, *Mol. Simul.*, 2006, **32**, 817

152. M. C. Stumpe, N. Blinov, D. Wishart, A. Kovalenko and V. S. Pande, *J. Phys. Chem. B*, 2010, **115**, 319.

153. T. Luchko, S. Gusarov, D. R. Roe, C. Simmerling, D. A. Case, J. Tuszynski and A. Kovalenko, *J. Chem. Theory Comput.*, 2010, **6**, 607.

154. N. Blinov, L. Dorosh, D. Wishart and A. Kovalenko, *Biophys. J.*, 2010, **98**, 282.

155. J. Dzubiella, J. M. J. Swanson and J. A. McCammon, *J. Chem. Phys.*, 2006, **124**, 084905.

156. L.-T. Cheng, J. Dzubiella, J. A. McCammon and B. Li, *J. Chem. Phys.*, 2007, **127**, 084503.

157. P. Setny, Z. Wang, L. T. Cheng, B. Li, J. A. McCammon and J. Dzubiella, *Phys. Rev. Lett.*, 2009, **103**, 187801.

158. P. Setny and M. Zacharias, *J. Phys. Chem. B*, 2010, **114**, 8667.

159. A. Nicholls, D. L. Mobley, J. P. Guthrie, J. D. Chodera, C. I. Bayly, M. D. Cooper and V. S. Pande, *J. Med. Chem.*, 2008, **51**, 769.

CHAPTER 4

Bioavailability Prediction at Early Drug Discovery Stages: In Vitro *Assays and Simple Physico-Chemical Rules*

JORDI MUNOZ-MURIEDAS

GlaxoSmithKline, Medicines Research Centre, Gunnels Wood Road, Stevenage SG1 2NY, UK
E-mail: jordi.4.munoz-muriedas@gsk.com

4.1 Bioavailability

Oral administration is the most widely used method for delivering drugs into the systemic circulation.[1] Unless the characteristics of the drug and the conditions to be faced under drug administration make it not viable, drug design projects will optimize the properties of their compounds in order to produce an orally administered drug. This is one of the less invasive routes for the patient and a relatively economic way of administration. However, the delivery of an oral dose to treat a systemic condition is complicated by the presence of physiological barriers. The orally administered drug must undergo several processes, each one subject to an extent of variability, before reaching systemic circulation: the drug must dissolve in the fluid at the absorption site, cross the membrane lining the gastrointestinal tract and be exposed to metabolism in the gut, gastrointestinal epithelial cells and liver. The onset and extent of therapeutic response will be linked to the amount of drug available at

RSC Drug Discovery Series No. 23
Physico-Chemical and Computational Approaches to Drug Discovery
Edited by F. Javier Luque and Xavier Barril
© The Royal Society of Chemistry 2012
Published by the Royal Society of Chemistry, www.rsc.org

systemic level. Oral pharmacokinetics (PK) describe these processes by measuring key parameters like the peak concentration in plasma after oral administration (C_{max}), the time needed to achieve peak concentration (T_{max}) and the area under the concentration *vs.* time curve (AUC).

Bioavailability (F) is defined as the fraction of an administered dose which reaches either its site of action or a biological fluid from which the drug has access to its site of action (normally it is measured in systemic circulation). Its value (when expressed as percentage) ranges from 0% (no availability) to 100% (total availability). Intravenous administrations are totally available and are used as reference to measure the bioavailability of other routes of administration. In the case of oral bioavailability, it is mathematically defined as:

$$F(\%) = \frac{AUC_{po} \cdot D_{iv}}{AUC_{iv} \cdot D_{po}} \cdot 100 \tag{4.1}$$

where D_{po} and D_{iv} are the doses of the oral and intravenous administrations, and AUC_{po} and AUC_{iv} stand for the area under the curves of the oral and intravenous administrations, respectively.

We can also express bioavailability as the product of the fraction absorbed in the gastrointestinal tract (F_a) times the fraction that escapes metabolism in the liver and gut wall (F_m):

$$F = Fa \times Fm \tag{4.2}$$

Low bioavailability results in erratic assessment of key PK parameters. Small variations in any of the factors determining bioavailability can translate into dramatic differences in the levels of the drug in systemic circulation. This may cause lack of response or the appearance of toxic effects. Poor bioavailability increases the risk of high variability of response among patients, especially if the compound is metabolized by enzymes with different expression levels in the population, and can also have implications in modulating the risk of drug–drug interactions. In addition, high bioavailability reduces the amount of drug to be administered and the costs associated with the therapeutic treatment of the disease.

Many drugs fail in clinical phases due to poor PK profiles. The adoption of new high-throughput assays to measure physico-chemical properties like solubility or permeability and a closer interaction between PK scientists and medicinal chemists may have helped to reduce the attrition rate due to unsatisfactory PK. However, the rate of attrition is still significant.[2] For these reasons, it is very important to have a good estimation of the bioavailability before reaching clinical phases, where a failure means a great loss of time and money. In a drug design project, oral bioavailability measurements are usually carried out in rats for selected compounds that achieved satisfactory results in previous steps in the drug design project decision cascade. The criteria may vary from one project to another, but typically they include factors such as low intrinsic clearance, good solubility, good *in vitro* permeability and the absence of cytochrome P450 inhibition. It is common to consider as low bioavailability values those below 30% and as very low bioavailability values those below 10%.

Rat is one of the most used animals in preclinical oral absorption studies. High correlations between drug fractions absorbed in rats and humans have been shown in several studies.[3,4] However, those studies only measured the fraction absorbed in the intestine and not the fraction that actually reached systemic circulation. Sometimes, the terms "fraction absorbed" and "bioavailability" are used interchangeably, but this is only true if the drug does not experience any metabolization on its way to systemic circulation. Cao *et al.*[5] reported a good correlation for the expression levels of transporters in the small intestine, and Chiou *et al.*[6] stated that intestinal mucosa membranes are comparable across mammal species. Taken together, both findings provide an explanation for the similar profiles of absorption in human and rat. Nevertheless, Cao *et al.* also found big differences in the expression of drug metabolizing enzymes between human and rat, which indicates possible differences in drug metabolism into rat and the difficulty in translating bioavailability measures from rat to human.[7] To assess this issue, bioavailability of the more advanced compounds in a drug design project (potential candidates) can also be measured in other species.

Accurate predictions of bioavailability are still one of the most challenging problems a drug design project must face. The gastrointestinal absorption component of bioavailability is affected by many physico-chemical and physiological properties, and also by factors related to the dosage form. However, the work by Amidon *et al.* revealed that the two main components controlling gastrointestinal absorption are permeability and solubility/dissolution of the drug at the absorption site.[8] On the top of that, a prediction of the amount of drug lost after absorption due to first-pass metabolism in the liver is also necessary.

This chapter is intended to provide a review about the prediction of three main components of bioavailability: permeability, solubility and first-pass metabolism from the point of view of an early phase drug design project. Accordingly, emphasis is put on the use of high-throughput assays and the definition of simple rules based on physico-chemical properties to foresee the amount of drug able to reach systemic circulation.

4.2 Internal Dataset Used in This Chapter

An internal set of 6429 compounds from past drug design projects with measured bioavailability in rat has been used to illustrate and test the predictive power of the computational and *in vitro* approaches described in this chapter. Bioavailability was measured by comparing the area under the curve of plasmatic levels *vs.* time after 1 hour intravenous infusion at 1 mg kg^{-1} and oral administration at 3 mg kg^{-1}. The bioavailability median is 22.9%, whereas the values for the first and third quartiles amount to 7.5% and 50.3%, respectively. Bioavailability was classified as high ($> 60\%$), medium (30–60%), low (10–30%) and very low ($< 10\%$). The pie chart in Figure 4.1 shows the distribution of compounds in each category.

The following data were also determined for subsets of the compounds in the dataset by searching GSK internal databases: rat microsome intrinsic clearance ($n = 2791$), octanol/water log D at pH 7.4 ($n = 1653$), artificial membrane permeability ($n = 3249$) and MDCK-2 permeability in the presence of a P-glycoprotein inhibitor ($n = 564$). For the comparison between octanol/ water log D and artificial membrane permeability (see Section 4.3.2), a set of 22564 compounds from drug design projects was used.

4.3 Permeability

Permeability is a key component to ensure optimal bioavailability. In order to reach systemic circulation, a drug needs to be absorbed in the intestine, a process that involves permeation through a membrane formed by the gastrointestinal epithelial cells. Different mechanisms of membrane permeation are possible, such as passive diffusion, endocytosis, active transport and paracellular permeation. Since passive diffusion is the most common mechanism and it is also the most influenced by physico-chemical properties, this chapter will focus on it. For recent surveys on other mechanisms the reader is addressed to the reviews by Sugano *et al.*[9] and Dobson and Kell.[10]

Passive diffusion is a process driven by a concentration gradient, where molecules move from a higher concentration area, *i.e.* the gastric lumen in the case of intestinal absorption, to a lower concentration area, *i.e.* the capillary blood vessels. In its journey, the molecule will cross the apical lipid bilayer

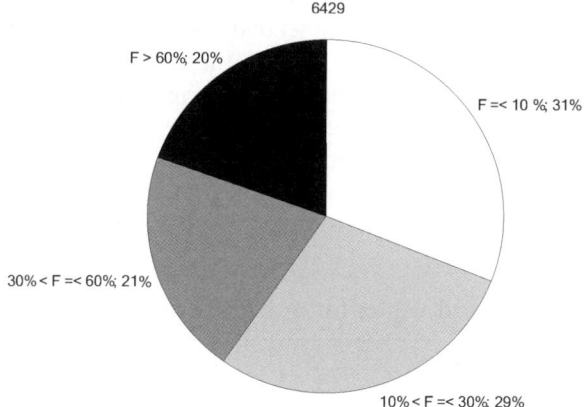

Figure 4.1 Distribution of bioavailability values in the internal dataset used for this chapter.

membrane of the enterocyte, then pass through the cytoplasm and exit the cell through the basolateral membrane.

4.3.1 Permeability *In Vitro* Assays

There are several *in vitro* assays that measure the capacity of a molecule to permeate through a membrane. In these assays, a buffered solution containing the test compound is placed on one side of the membrane (donor compartment) and buffer with no test compound is placed on the other side (receptor compartment). The compound diffuses through the membrane from the donor to the receptor, and the permeation rate is calculated by measuring the concentrations in each compartment. The permeation rate will be determined by a combination of the physicochemical properties of the test compound, the environment in which it is placed and the nature of the membrane it must cross. The most popular *in vitro* assays include lower throughput cell-based assays such as Caco-2 (human colon carcinoma cell line)[11,12] or MDCK-2 (Madin–Darby Canine Kidney cell line),[13] and higher throughput assays using artificial membranes such as PAMPA (Parallel Artificial Membrane Permeation Assay)[14,15] or black lipid membranes.[16]

Results from artificial membranes, MDCK-2 and Caco-2 assays correlate well for compounds absorbed by passive diffusion. Data may vary for actively absorbed or efflux compounds.[17,18] Compared to artificial membranes, cell-based assays represent a step further into modeling the intestinal epithelium. MDCK-2 cell permeability assays are usually preferred to Caco-2 in lead optimization stages due to culture times. MDCK-2 can provide reliable results after 3 days of culture, in contrast to the 21 days required for Caco-2. In addition, MDCK-2 has also the advantage of a reduced expression of transporters, which makes it suitable to study passive diffusion.[19] Both assays provide a good prediction of gastrointestinal absorption for highly permeable compounds, but are of limited applicability for compounds with low or moderate permeability.[17] The recommended cut-offs for permeability can be found in Table 4.1, while Figure 4.2 shows the correlation between available data from MDCK-2 and artificial membrane permeability measures within the internal dataset.

It can be seen from Figure 4.2 how the artificial membrane assay tends to classify more compounds as highly permeable; however, this classification does not necessarily translate into a high bioavailability. Figures 4.3 and 4.4 show

Table 4.1 Recommended values for *in vitro* permeability classification.

Permeability class	MDCK-2 (nm s⁻¹)	Artificial membrane (nm s⁻¹)
High	200	100
Medium	50–200	10–100
Low	<50	10

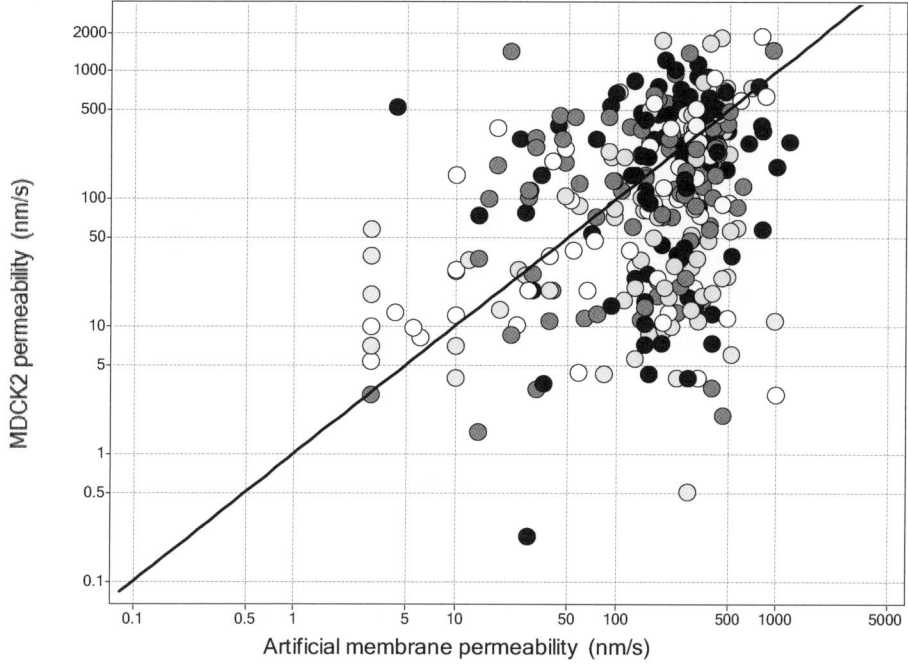

Figure 4.2 MDCK-2 *vs.* artificial membrane permeability. Data points coloured by bioavailability (*F*) class (white: $F \leq 10\%$; light grey: *F* between 10 and 30%; dark grey: *F* between 30 and 60%; black: $F > 60\%$).

the distribution of bioavailability (*F*) classes found in the internal dataset for each permeability bin defined by the cut-offs in Table 4.1.

If we use the permeability measurements to predict systemic exposure, the artificial membrane permeability shows high specificity (95%), but very low sensitivity (20%) for low bioavailability compounds ($F \leq 10\%$). This means that the assay can be trusted as a predictor of low systemic exposure, but, when used as a filter, will allow the progression of many compounds with low bioavailability to the next step. In the case of compounds with high bioavailability ($F \geq 30\%$), the assay is very unspecific (37%) but very selective (82%), which means that predictions for high systemic exposure are not reliable. However, when used as a filter, the assay will progress almost all the high bioavailability compounds to the next step.

Regarding the MDCK-2 assay, its specificity and sensitivity for low bioavailability are 37% and 82%, respectively. This makes it less suited as a filter for low bioavailability compounds, since many compounds with good bioavailability will be classified as low bioavailability and potentially discarded. In the case of high bioavailability compounds, the MDCK-2 assay shows good specificity (68%) and moderate sensitivity (48%).

The results from these *in vitro* assays can be used sequentially as part of the list of parameters in the decision cascade used to prioritize compounds for

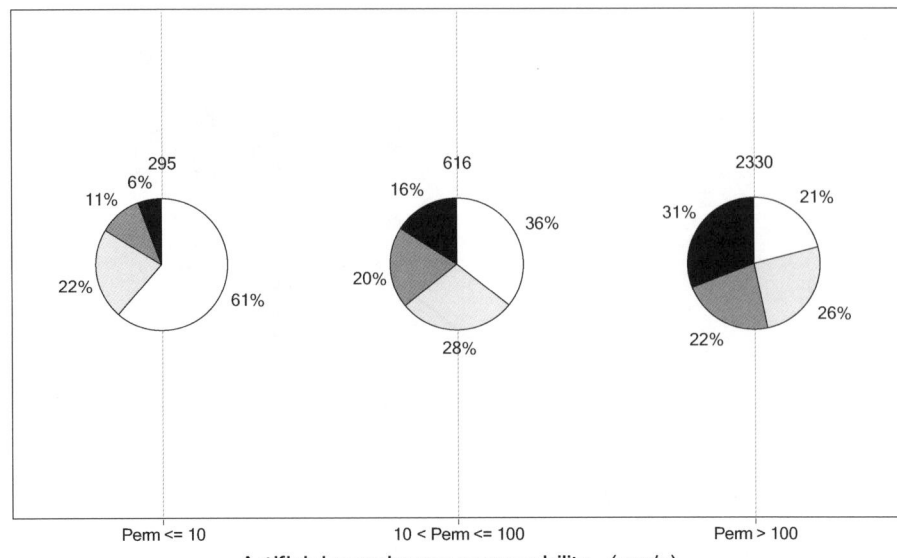

Artificial membrane permeability   (nm/s)

Figure 4.3   Distribution of bioavailability values per artificial membrane permeability class. Portions in pie charts coloured by bioavailability (F) class
(white: $F \leq 10\%$; light grey: F between 10 and 30%; dark grey: F between
30 and 60%; black: $F > 60\%$).

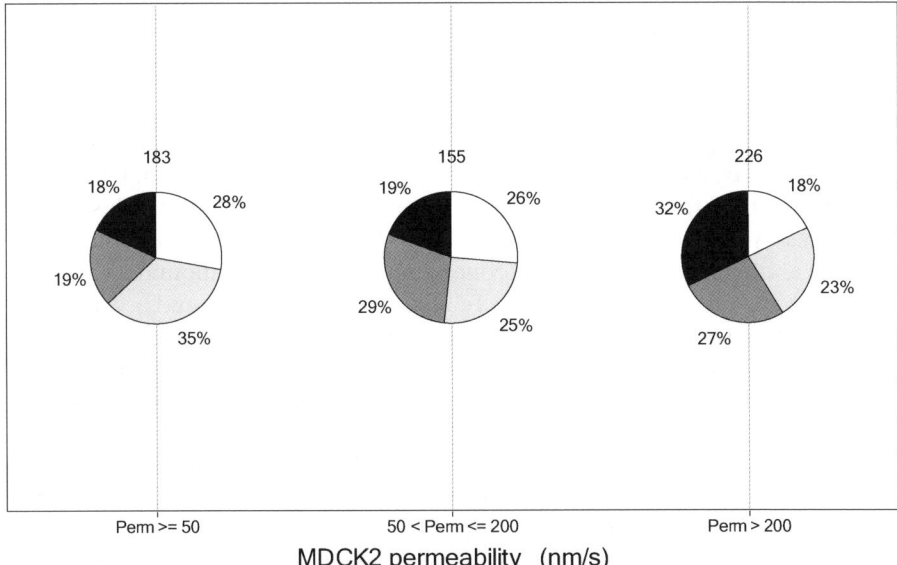

MDCK2 permeability   (nm/s)

Figure 4.4   Distribution of bioavailability values per MDCK-2 permeability class.
Portions in pie charts coloured by bioavailability (F) class (white: $F \leq$
10%; light grey: F between 10 and 30%; dark grey: F between 30 and 60%;
black: $F > 60\%$).

further progression in a drug design project. Given its high specificity for low bioavailability, the artificial membrane permeability is a good assay to be used in very early stages of a project as a filter to discard compounds that are very likely to produce low systemic exposure values. Meanwhile, the MDCK-2 assay will be able to help in the selection of compounds with more chances of high bioavailability among compounds selected after the artificial membrane assay.

4.3.2 Permeability: Simple Rules

In a recent publication presenting a set of rules of thumb generated for a number of key ADMET (absorption, distribution, metabolism, excretion, toxicity) assays run within GSK,[20] Gleeson found that acidic compounds are the least permeable, which can be attributed to their electrostatic properties, followed by zwitterionic species, bases and neutral compounds (in that order). Only neutral molecules show a nonlinear dependency with hydrophobicity, while for ionized species the hydrophobicity seems to lead to an increase in permeability. Figure 4.5 shows a comparison between artificial membrane permeability data and the octanol/water distribution coefficient (log D) at pH 7.4. The plot reveals a parabolic relationship, which can be explained by the fact that very hydrophobic compounds may fuse with the lipidic membrane instead of crossing it. The optimal log D value for the maximum permeability appeared to be around 2.5.

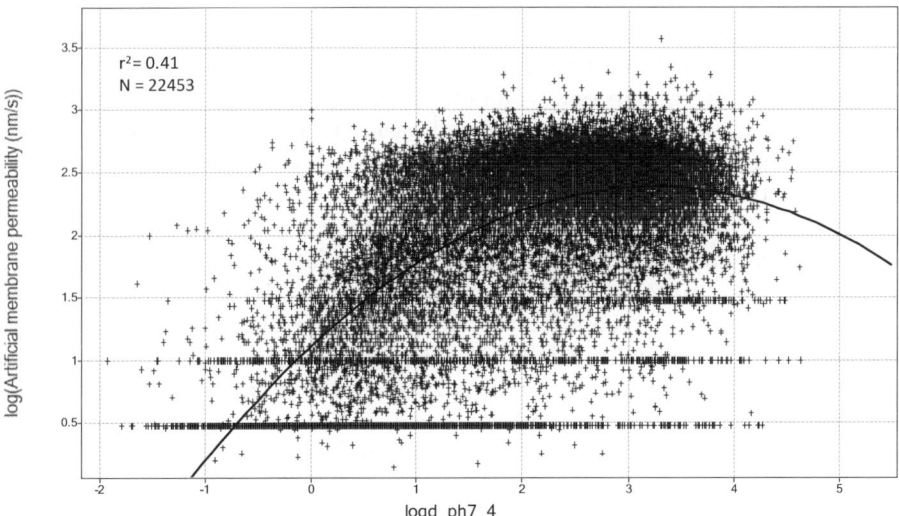

Figure 4.5 Representation of the artificial membrane permeability *vs.* octanol/water distribution coefficient at pH 7.4. The result of fitting to a quadratic polynomial function is shown.

In some cases, the use of log D is preferred to the use of the octanol/water partition coefficient (log P). While log P only takes into account the concentration of neutral species in the organic and aqueous phases, log D will vary according to the ability of the molecule to ionize and is often considered a measure of the true lipophilicity of the molecule at a given pH; log P is sometimes referred to as the intrinsic lipophilicity. On the other hand, log D assays are more adequate for large scale throughput assays than log P assays.

The Lipinski Rule of 5, also known as "Lipinski rules",[21] predicts that poor absorption or permeation is more likely when there are more than 5 hydrogen bond donors, 10 hydrogen bond acceptors, the molecular weight is greater than 500 or log P (expressed in the original paper as calculated C log P) is greater than 5. The rules were derived by analyzing the properties of compounds in the United States Adopted Names (USAN) Directory that had survived phase I clinical trials and had entered phase II. The move to phase II was taken as an indicator that the compound had the right properties to ensure a good absorption, and that the compounds with the poorer properties should be among those that failed in phase I. Later, Veber *et al.* stated that the commonly applied molecular weight cutoff of 500 does not itself significantly to separate compounds with poor oral bioavailability in rat from those with acceptable values in a set of 1000 internal GSK compounds and suggested a set of additional rules, the "Veber rules",[22] for predicting the chances of molecules for good bioavailability in rat. In particular, two additional criteria were proposed: (i) the molecule should have 10 or fewer rotatable bonds and (ii) the polar surface area should be equal to or less than 140 Å^2 (or 12 or fewer hydrogen-bond donors and acceptors).

Figures 4.6 and 4.7 show the performance of the Lipinski rules in predicting bioavailability and how metabolism affects this prediction. In general, oral drugs have no more than one violation of the aforementioned criteria. When no distinctions are made regarding metabolism, it can be seen from Figure 4.6 that similar profiles for the distribution of bioavailability values can be found in the subsets of compounds that violate none, one or two of the Lipinski rules. When three or more rules are violated, the number of compounds with very low bioavailability increases dramatically. It must be pointed out that Lipinski rules take into account only the component in bioavailability due to permeation; therefore, the effect of metabolism on bioavailability is out of the domain of applicability of the rules. On the other hand, Figure 4.7 shows that there is no significant correlation between the number of violated rules and the distribution of bioavailability values for compounds with a high hepatic extraction rate ($E > 70\%$). For compounds with low hepatic extraction rate ($E < 30\%$), the subset of compounds that violate less than two rules shows the best bioavailability profiles, with around 40% of them with bioavailabilities greater than 60%. This number significantly decreases to 15% when two or more rules are violated. Compounds that are substrates for biological transporters are also exceptions to the rules.

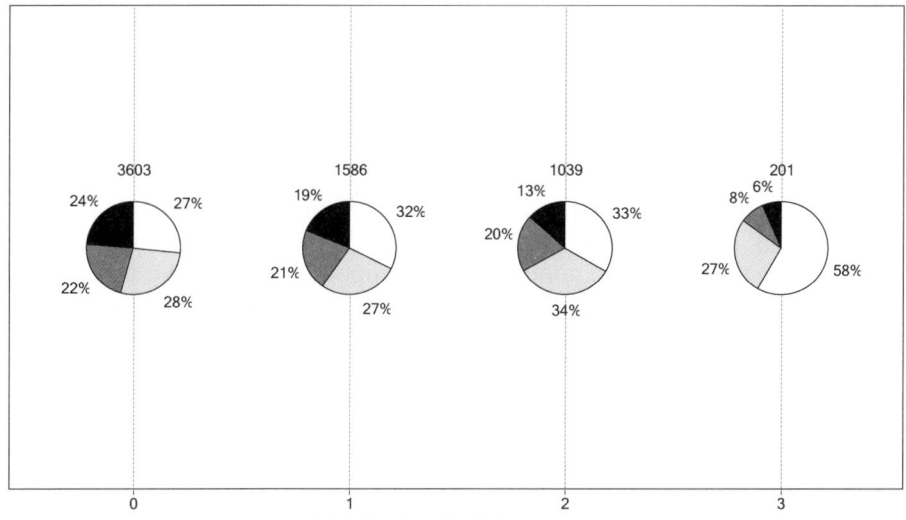

Figure 4.6 Distribution of bioavailability values (*F*) per number of violations of the Lipinski rule. Portions in pie charts coloured by bioavailability class (white: *F* ≤ 10%; light grey: *F* between 10 and 30%; dark grey: *F* between 30 and 60%; black: *F* > 60%).

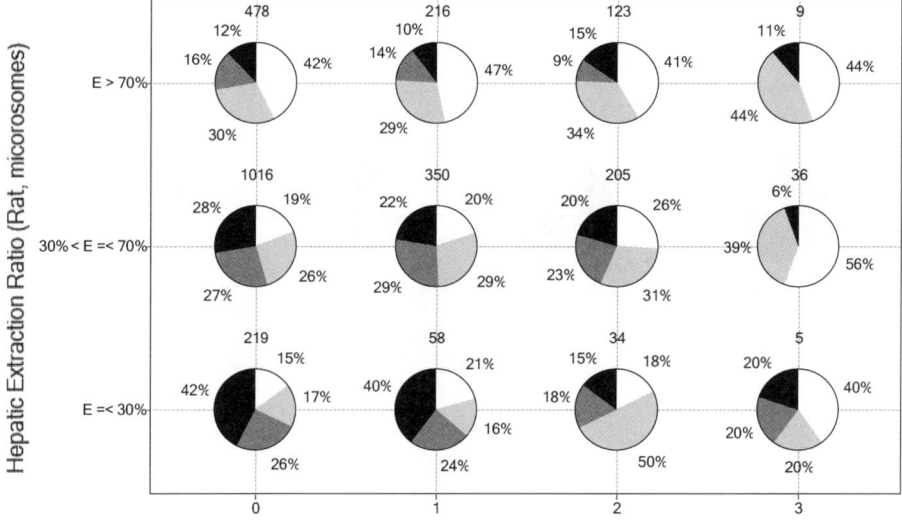

Figure 4.7 Distribution of bioavailability values (*F*) per number of violations of the Lipinski rule and hepatic extraction (*E*) class. Portions in pie charts coloured by bioavailability (*F*) class (white: *F* ≤ 10%; light grey: *F* between 10 and 30%; dark grey: *F* between 30 and 60%; black: *F* > 60%).

As expected, the Lipinski rules correlate well with the results from *in vitro* permeability assays. Figure 4.8 shows the correlation of the rules with the artificial membrane permeability results.

While the Lipinski and Veber rules define a "drug-like" set of properties, the term "lead-like" applies to molecules identified from high-throughput screening campaigns. Leads are expected to increase in complexity during their optimization towards drugs and a set of rules,[23,24] known as the "Rule of 3" or "Astex rules",[25] were suggested as guidelines in the selection of leads. According to the rule of 3, a good lead-like compound should have a molecular weight lower than 300, a number of hydrogen bond donors equal to or less than 3, a number of hydrogen bond acceptors equal to or less than 3, and a calculated log *P* equal to or less than 3. In addition, the rule also suggests a number of rotatable bonds equal to or less than 3.

Based on the knowledge that lipophilicity and size are important properties in determining absorption by passive diffusion, Hill *et al.* published a very simple qualitative model that can be used in conjunction with the "Lipinski Rule of 5" to gain insight into variations in the gastrointestinal absorption of project compounds.[26] The model, based on human gastrointestinal absorption data, is a partial least-squares discriminant analysis (PLS-DA) and classifies compounds on the basis of values for calculated log *D* at pH 7.4[27] and the calculated molar refractivity (*CMR*) of compounds.[28] Positive values for the

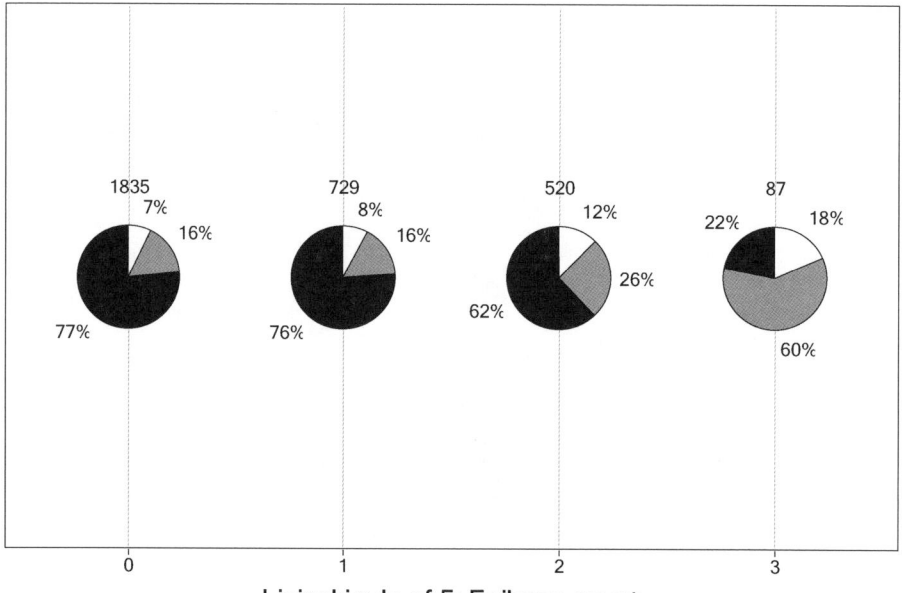

Lipinski rule of 5: Failures count

Figure 4.8   Distribution of *in vitro* permeability classes (artificial membrane) per number of violations of Lipinski rules. Portions in pie charts coloured by permeability (*P*) class (white: $P \leq 10$ nm s^{-1}; grey: *P* between 10 and 100 nm s^{-1}; black: $P > 100$ nm s^{-1}).

linear discrimination function (*ldf – distance*; see eqn 4.3) predict absorption to be greater than 33%. Figure 4.9 shows the performance of this model in predicting *in vitro* permeability within the internal dataset.

$$ldf - distance = 2.56 + 0.41 \log D\,(\text{pH7.4}) - 0.23 CMR \tag{4.3}$$

As for Lipinski rules, the log *D/CMR* model does not take into account either metabolism or transport effects. Therefore it cannot be used directly to predict bioavailability. Figure 4.10 shows the distribution of the bioavailability data on the log *D/CMR* plot for different hepatic extraction rates. As can be seen from the plot, the model lacks predictability for high bioavailability in compounds with high hepatic extraction rates.

4.4 Solubility

Solubility is a desirable property for a drug. Ligands interact with their receptors as single entities, not as solid crystals, and the same applies to absorption. Hence, the administered solid dosage form must dissolve. Only the amount of drug in solution will be available for permeation across the gastrointestinal membrane and absorption into systemic circulation. Suboptimal solubility is likely to be an issue in programs trying to optimize oral

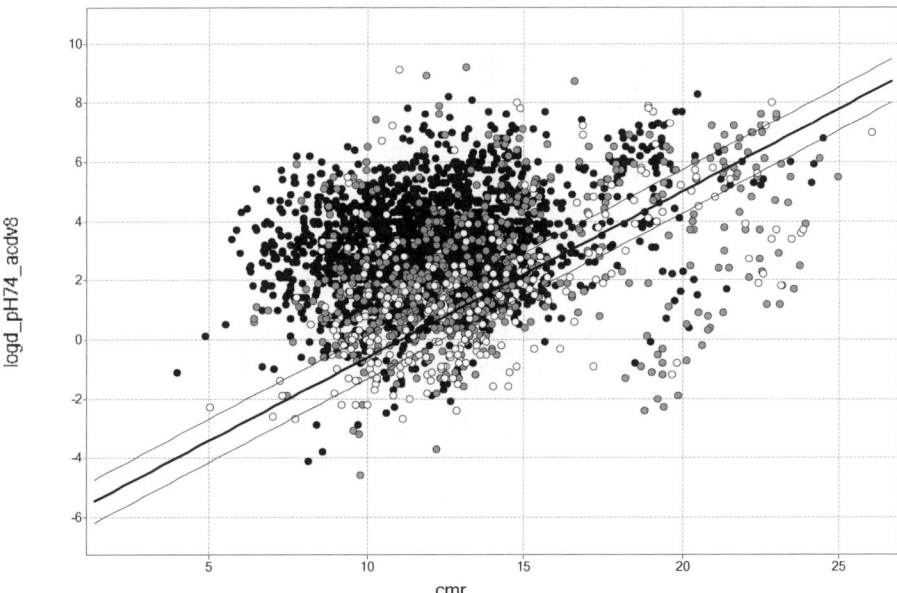

Figure 4.9 Predicted octanol/water log *D* at pH 7.4 *vs.* *CMR*. The line on the plot represents the linear discrimination function for permeation. Data points coloured by artificial membrane permeability (*P*) class (white: $P \leq 10$ nm s^{-1}; grey: *P* between 10 and 100 nm s^{-1}; black: $P > 100$ nm s^{-1}).

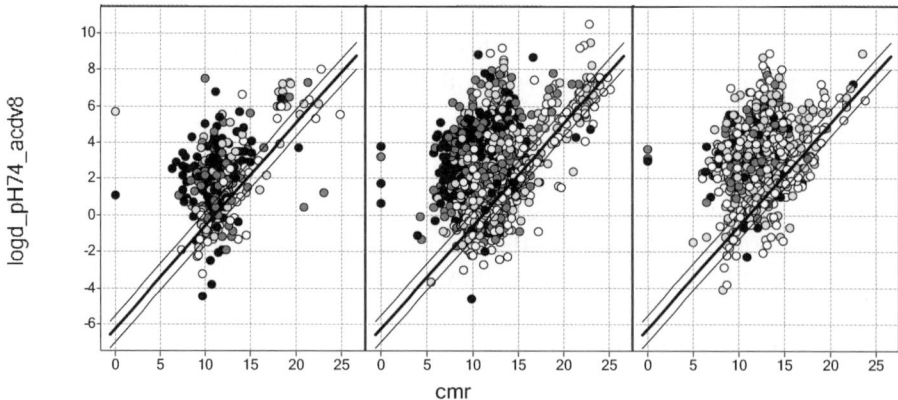

Figure 4.10 Predicted octanol/water log *D* at pH 7.4 *vs. CMR*. The line on the plot represents the linear discrimination function for permeation. Compounds coloured by bioavailability class (*F*). Trellis on hepatic extraction (*E*) class (left: $E \leq 30\%$; middle: *E* between 30 and 60%; right: $E > 60\%$). Data points coloured by bioavailability (*F*) class (white: $F \leq 10\%$; light grey: *F* between 10 and 30%; dark grey: *F* between 30 and 60%; black: $F > 60\%$).

bioavailability, since low solubility may lead to erratic and incomplete absorption.

4.4.1 Solubility *In Vitro* Assays

It is important to assess the solubility of compounds of interest before developing them any further; for that purpose, high-throughput solubility screens have been incorporated in early stages of drug design projects. High-throughput solubility screens normally measure kinetic solubility, where first the solid compound is dissolved in a solubility enhancer solvent like dimethyl sulfoxide (DMSO), and then an aliquot of the solution is added to an aqueous buffer. For example, in the particular case of the GSK in-house kinetic solubility assay, 5 μL of 10 mM DMSO stock solution is diluted to 100 μL using a pH 7.4 phosphate-buffered saline solution. The mixture is then equilibrated for 1 hour at room temperature and filtered through Millipore Multiscreen HTS-PCF filter plates. The eluent is quantified by chemiluminescent nitrogen detection (CLND). This assay can measure solubilities within the range of 1 to 500 μM. Guidelines for solubility recommended to the discovery project teams in early stages can be found in Table 4.2.

Nevertheless, the minimum solubility required for a compound will depend on its permeability and its dose. To this end, it is convenient to determine the maximum absorbable dose (*MAD*),[29] which is defined as the minimum amount of drug that can be absorbed at a certain dose. The *MAD* can be estimated from the solubility of the compound and its permeability through the intestinal barrier using expressions such as that shown in eqn (4.4):

Table 4.2 Recommended values for kinetic solubility (5% DMSO) classification.

Solubility class	μg mL⁻¹	μM
High	>100	200
Medium	10–100	20–200
Low	<10	20

$$MAD = S \times K_a \times SIWV \times SITT \tag{4.4}$$

where S is the solubility (mg mL^{-1}) at pH 6.5, K_a is the intestinal absorption rate (min^{-1}), $SIWV$ is the small intestine water volume ($\sim$250 mL) and $SITT$ is the small intestine transit time ($\sim$270 min).

The values from kinetic solubility measurements are very useful in drug discovery since they reproduce the conditions of many biological assays and will identify compounds that do not have a good kinetic solubility even in the presence of a solubility enhancer. Those compounds will be flagged as compounds with potential permeability problems. However, kinetic solubility experiments are not good predictors of bioavailability, especially for compounds where intestinal absorption is limited by the dissolution rate. Figure 4.11 shows the proportions of bioavailability (F) classes found in the internal dataset for each kinetic solubility bin defined by the cut-offs in

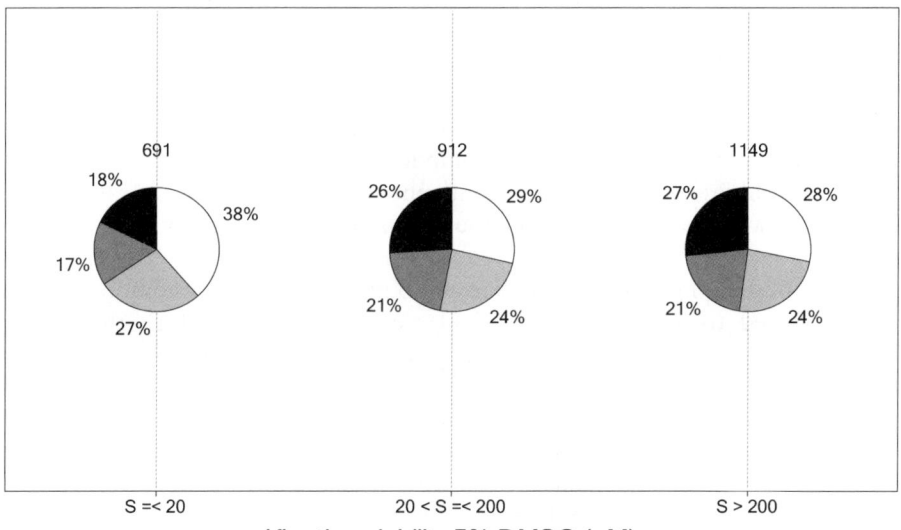

Figure 4.11 Distribution of bioavailability values per kinetic solubility (S) class. Portions in pie charts coloured by bioavailability (F) class (white: $F \le$ 10%; light grey: F between 10 and 30%; dark grey: F between 30 and 60%; black: $F > 60\%$).

Table 4.2. It can be seen how the proportion of compounds with low bioavailability is slightly higher in compounds with low kinetic solubility, but no predictions can be made for compounds with high or medium kinetic solubility.

At this point, thermodynamic solubility measurements are needed. In these experiments the aqueous buffer, water or other relevant solvents, or conditions simulating biological fluids, is added directly to the solid compounds and equilibrated for a longer period of time, normally 24–72 hours. The amount of dissolved compound in the filtered supernatant is then measured by high performance liquid chromatography (HPLC).

Thermodynamic assays are low throughput but there are *in silico* models that can also be used in the early stages. These models can calculate solubility profiles at different pH values and have been normally fitted using experimental values from solubility in water, although some models now are also predicting solubility in DMSO.[30] They can be used as a reasonable tool for ranking compounds, but the use of the raw numbers can be dangerous without a previous validation. For a review on the performance of several of the available solubility predictors, the reader is referred to the work of Hewitt *et al.*[31]

4.4.2 Solubility: Simple Rules

Solubility at a given pH can be described as a function of pK_a, log P and melting point. While the pK_a will determine the fraction ionized and more soluble of the compound, the octanol/water partition coefficient (log P) will take into account the intrinsic hydrophobicity. The melting point involves the intermolecular interactions between the molecules of the compound in the crystal lattice of its solid state. The stronger these interactions are, the higher the melting point and the lower the solubility. Models or rules not taking into account variables accounting for the influence of the properties of a compound in its solid state will only predict the component of solubility restricted by solvation.[32]

Depending on the restricting step in solubility, low solubility compounds are often referred to as "brick dust" (*i.e.*, stable crystals in which the strong intermolecular bonds within the crystal restrict the solubility) or "greasy balls" (*i.e.*, highly lipophilic compounds with low solvation in water).[33] Poorly soluble drugs that have reached the market normally belong to the second group, probably because their solubility can be largely improved by incorporating excipients such as disintegrants, solubility enhancers, wetting agents, cyclodextrins and lipids in the formulation.[33]

In Gleeson's rules of thumb publication,[20] it is stated that the solubility decreases as the molecular weight and log P increase, and that neutral compounds with a low log P can be highly soluble. Regarding ionization, it is also stated that zwitterionic molecules are more soluble than acids, followed by bases. On the other hand, the Yalkowski equation (eqn 4.5) takes into account hydrophobicity (represented by log P) and the crystal lattice energy represented by the melting point (MP):[34,35]

$$\log S = 0.5 - 0.01\,(MP - 25) - \log P \tag{4.5}$$

Equation (4.5) works well for uncharged molecules. Its main drawback is that melting points are not commonly measured in the early stages of a drug design project owing to their cost and low throughput. Equation (4.5) can also be corrected to take ionization into account by replacing log P with log D at a given pH (*i.e.*, 7.4):

$$\log S\,(\text{pH}7.4) = 0.5 - 0.01\,(MP - 25) - \log D\,(\text{pH}7.4) \tag{4.6}$$

The Abraham extended solubility equation (eqn 4.7),[36] based on the Abraham descriptors,[37] offers a distinct physicochemical approach to solubility:

$$\log S = 0.518 - 1.004 R_2 + 0.771 \pi_2^H + 2.168 \sum \alpha_2^H + 4.238 \sum \beta_2^H$$
$$- 3.362 \sum \alpha_2^H \sum \beta_2^H - 3.987 V_X \tag{4.7}$$

where R_2 is an excess molar refraction (in units of cm^3 mol^{-1}/10), π_2^H is the dipolarity/polarizability, $\Sigma\alpha_2^H$ is the summation of hydrogen-bond acidity, $\Sigma\beta_2^H$ is the summation of hydrogen-bond basicity and V_X is the McGowan characteristic volume (in units of cm^3 mol^{-1}/100).[38]

According to eqn (4.7), the two main properties that lead to an increase in insolubility are the hydrogen-bond acidity ($\Sigma\alpha_2^H$) and particularly the hydrogen-bond basicity ($\Sigma\beta_2^H$), which describe the capacity to establish hydrogen-bond interactions. However, if the compound is itself a hydrogen-bond acid and a hydrogen-bond base, then intermolecular hydrogen-bond interactions will lead to an increase in melting point and to a decrease in solubility. This effect is described by the cross term $\Sigma\alpha_2^H\Sigma\beta_2^H$ with a negative coefficient in eqn (4.7). The other polar term in the equation is π_2^H, with a positive but small contribution to solubility. It may be due to the fact that dipole solute/solute interactions counteract to some extent the dipole solute/water interactions that would eventually lead to an increase in solubility. R_2, describing the ability of a compound to interact with surrounding σ and π electrons, has a negative effect on solubility, which means that this type of electronic effect is stronger in solute/solute than in solute/solvent interactions. Lastly, V_X, with a large negative coefficient, is directly related to the size of the compounds, and hence with the energy needed to disrupt the solvent structure in order to create a cavity to accommodate the solute.

It has also been reported that the aromaticity of the compound can play a major role in the prediction of solubility. Lamanna *et al.* published a simple regression tree model based on molecular weight and the proportion of aromatic atoms in the compound. Both properties have a negative effect on solubility.[39] In a similar approach, the solubility forecast index (*SFI*),[40] is defined as:

$$SFI = \log D\,(\mathrm{pH7.4}) + N_{ar} \tag{4.8}$$

where N_{ar} is the number of aromatic rings.

There is a reasonable chance of having good solubility for *SFI* values below 5. The average *SFI* for oral drugs is 2.4. The distribution of kinetic solubility values within the internal dataset for each *SFI* value can be seen in Figure 4.12.

4.5 First-Pass Metabolism

On its way to systemic circulation, the amount of drug absorbed in the intestine will be collected by the portal vein and carried into the liver, from where the drug will access systemic circulation *via* the cava vein. In the liver, the drug will be exposed to drug metabolizing enzymes present in the hepatocytes and, depending on its properties, can be metabolized to a certain degree. The drug can also be eliminated *via* bile, which flows from the liver to the duodenum *via* the bile duct and gall bladder. As a result of its pass through the liver, a percentage of the oral dose will be lost before reaching systemic

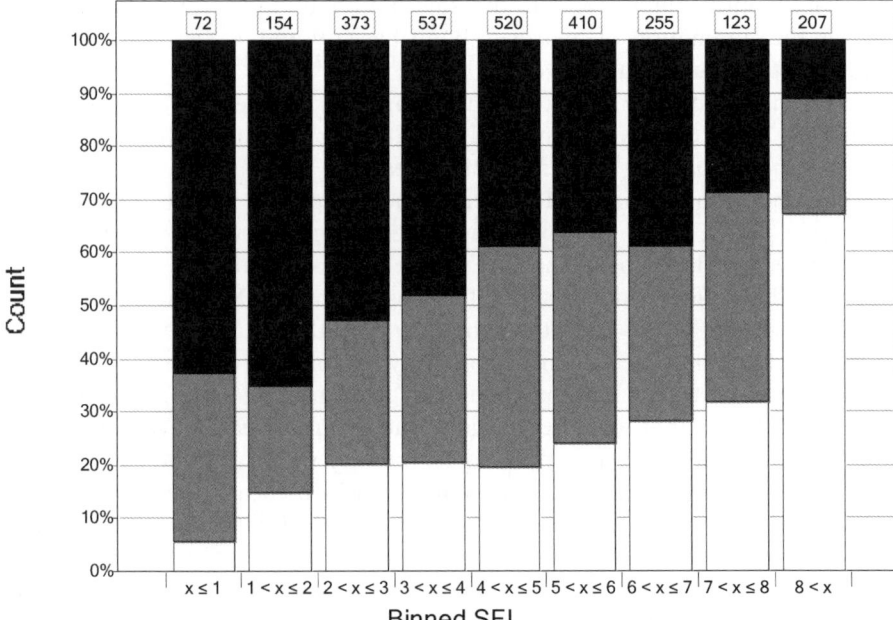

Figure 4.12 Representation of calculated solubility forecast indices. Proportions in columns coloured by measured kinetic solubility (*S*) class (white: $S \leq 50\ \mu\mathrm{M}$; grey: *S* between 50 and 200 $\mu\mathrm{M}$; black: $S > 200\ \mu\mathrm{M}$).

circulation. This is the so called first-pass effect and will reduce the bioavailability of a drug. It is worth mentioning that drugs are also exposed to metabolism in the enterocytes, especially metabolism by cytochrome 3A4 systems.

The hepatic clearance is defined as the volume of blood cleared of drug by the liver per unit of time. The amount of blood cleared by the liver will be limited by the blood flow accessing it. The relationship between these properties can be expressed as in eqn (4.9):

$$Cl_h = \Phi \times (C_{in} - C_{out})/C_{in} = \Phi \times E \tag{4.9}$$

where Cl_h is the hepatic clearance, Φ is the blood flow at the liver, C_{in} and C_{out} denote the amounts of drug accessing and leaving the liver per unit of time, and the quotient $(C_{in} - C_{out})/C_{in}$ is known as the extraction ratio (E) and defines the fraction of drug eliminated from portal blood during the absorption phase.

4.5.1 Intrinsic Clearance *In Vitro* Assays

It is possible to foresee the extent of the first-pass effect by means of assays that will measure the intrinsic metabolic capacity of *in vitro* systems mimicking the liver. We define as intrinsic clearance the maximum clearing capacity in the absence of any other confounder such as blood flow, protein binding or saturation. Metabolization of a drug is a saturable process and can be described by means of a Michaelis–Menten equation, where Cl_h can be expressed as:

$$Cl_h = \frac{V_{max}}{K_m + C_{u,liver}} \tag{4.10}$$

where $C_{u,liver}$ is the unbound fraction of drug accessing the liver.

The total amount of drug metabolized in the liver per unit of time (V) can be expressed as:

$$V = Cl_h \times C_{u,liver} \tag{4.11}$$

At concentrations much lower than K_m, the liver will reach its maximum clearance capacity and the hepatic intrinsic clearance can be expressed as:

$$Cl_{int} = \frac{V_{max}}{K_m} \tag{4.12}$$

The results obtained from an intrinsic clearance assay will depend on the *in vitro* system used to represent the liver. Each enzyme involved in the metabolic process will have its individual V_{max} and K_m, but it would be too difficult to measure all of them individually. Instead, the assays measure the intrinsic clearance capacity of the *in vitro* system used as a whole. The most common *in vitro* systems are (i) hepatic microsomes, which mainly contain cytochromes and phase I drug metabolizing enzymes, (ii) S9 fractions, which contain phase I and phase II enzymes and (iii) hepatocytes, which contain all the hepatic drug metabolizing enzymes and are also useful to measure hepatocyte permeability. All of these systems are available from different species, the most commonly used being human, rat, mouse, monkey and dog. Assays based on microsomes are the higher throughput and less expensive. They are also the most used in lead optimization stages.

There are different physiological models that allow the extrapolation of intrinsic clearance measurements into hepatic clearance values. One of the most commonly used is the "well-stirred" model (eqn 4.13), which assumes that the liver is a well-stirred compartment and that the concentration of unbound drug in the emergent blood is in equilibrium with the unbound drug in the liver. This equilibrium is assumed to be achieved instantaneously.

$$Cl_h = \Phi \times E = \frac{\Phi \times Cl_{int} \times f_u}{\Phi + Cl_{int} \times f_u} \tag{4.13}$$

where f_u is the unbound drug fraction.

The extraction rates obtained from intrinsic clearance assays allow the prediction of the amount of drug lost due to the first hepatic pass and the expected maximum bioavailability of the drug (assuming the absence of problems due to low permeability and solubility). Extraction ratios below 0.3 are considered low and desirable, while ratios above 0.7 are considered high. Figure 4.13 shows the proportions of bioavailability (F) classes found in the internal dataset for each extraction ratio class.

4.5.2   Intrinsic Clearance: Simple Rules

Predictions of clearance based merely on physicochemical descriptors are complex. There is a trend for metabolism to increase with log *P*. The hydrophobicity increases the permeability of compounds and so their penetration into hepatocytes and access to drug metabolizing enzymes. Reactivity and affinity to a drug metabolizing enzyme also play a key role: while log *P* predicts how likely is a compound to approach the metabolizing systems, affinity and reactivity predict how likely is the compound to be metabolized once it is there. Chemical groups in compounds that tend to be metabolized include benzylic or allylic positions, electron-rich aromatic rings, N-methyl or O-methyl units and the hydroxyl group. Strategies to reduce metabolization are normally focused on deactivation by addition of electron-withdrawing groups or addition of steric hindrance at the site of metabolization.

Several families of enzymes are responsible for the metabolization of drugs; the most prominent among them is the cytochrome P450 family. Drugs might

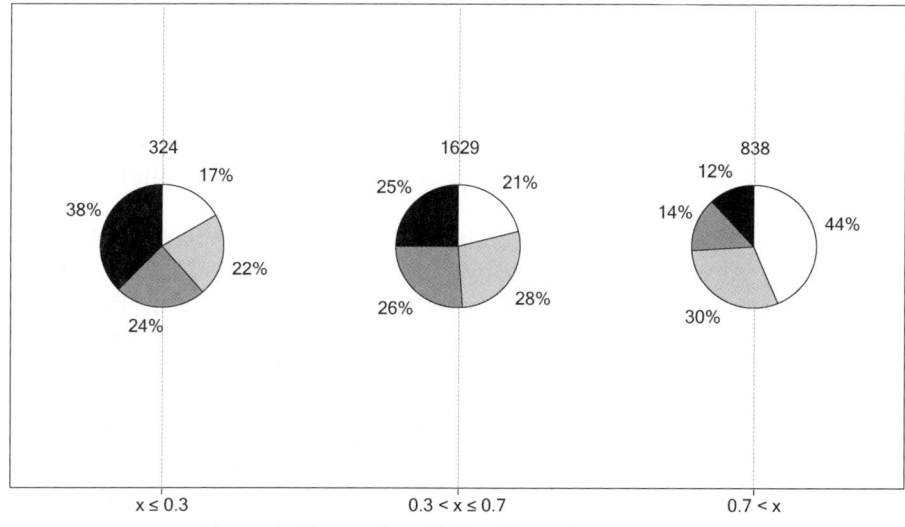

Hepatic Extraction Ratio (Rat micorosomes)

Figure 4.13 Distribution of bioavailability values per hepatic extraction ratio (E) class. Portions in pie charts coloured by bioavailability (F) class (white: $F \leq 10\%$; light grey: F between 10 and 30%; dark grey: F between 30 and 60%; black: $F > 60\%$).

be metabolized by more than one enzyme, but different enzymes may present differences in the characteristics of the drugs they metabolize. The identification of these characteristics provides useful information to design molecules with a lower clearance. In the case of the cytochrome isoforms most relevant in drug design, the following trends have been observed:[41] 1A2, planar polyaromatic amines and amides; 2C19, amides or weak bases with two hydrogen bonds; 2C9, weak acids with a hydrogen bond acceptor; 2D6, basic compounds with an ionizable nitrogen at 4–7 Å from the site of metabolization; and 3A4, relatively large molecules of diverse structure.

It is also useful to combine experimental data from intrinsic clearance assays with *in silico* predictions for metabolization. Predictions from models help to generate ideas on what is driving metabolization in compounds with high clearance and identify potential sites of metabolism. Two of the most popular pieces of software available for metabolism prediction are Metasite,[42,43] with models for interactions with cytochrome P450 based on GRID-derived molecular interaction fields,[44] and Meteor,[45] based on rules generated from an expert knowledge database.

4.6 The Whole Picture: Putting the Pieces Together

Data from the *in vitro* assays mentioned in this chapter can be combined to provide useful tools to predict bioavailability. The Biopharmaceutical

Classification System (BCS; see Figure 4.14) categorizes drugs into one of four so-called biopharmaceutical classes according to their water solubility and membrane permeability.[8]

Class I: high permeability/high solubility. Drugs in this class are well absorbed. If bioavailability is low it may be due to metabolism. The limiting step to drug absorption is dissolution or gastric emptying if dissolution is very rapid.

Class II: high permeability/low solubility. Drug dissolution *in vivo* will be the limiting step. Absorption will occur over an extended period of time and much more of the intestine will be exposed to the drug. Consequently, the dissolution profile must be determined for several time points and at several physiological pH values. Dissolution media reproducing *in vivo* conditions should also be used.

Class III: low permeability/high solubility. Permeability is the rate-controlling step. Rate and extent of drug absorption may be highly variable for this class of drug. If dissolution is quick, variability can be attributed to gastrointestinal transit and contents and membrane permeability.

Class IV: low permeability/low solubility. Drugs in this class are expected to have low bioavailability.

In 2004, Wu and Benet suggested a modified version of the BCS to take into account metabolism and the effect of transporters.[46] This classification is based on the authors' claim that high permeability (BCS Class I and Class II compounds) also enhances penetration into hepatocytes and access to drug metabolizing enzymes, while for low permeability compounds the major routes of elimination must be renal and/or biliar. Regarding the effect of transporters, this should be minimal for compounds with high solubility (BCS Class I and Class III compounds), which can achieve high concentrations able to saturate any transporter in the gastrointestinal membrane. However, for those high solubility compounds with low permeability (BCS Class III), an uptake transporter will be necessary to ensure absorption. In the case of high permeable/low solubility compounds (BCS Class II), the inability to saturate efflux transporters will affect the rate of absorption. Finally, for low

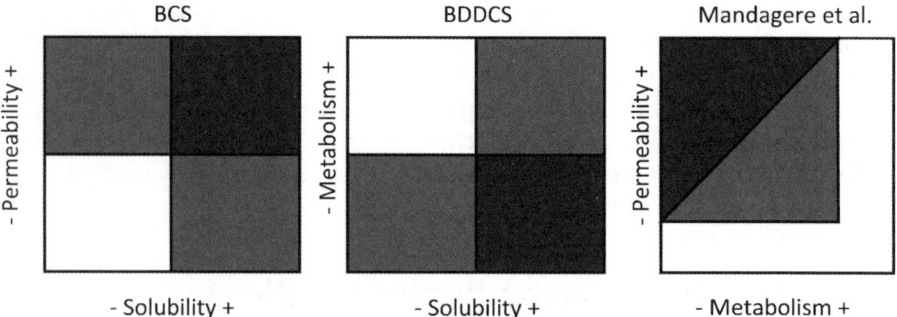

Figure 4.14   BCS, BDDCS and Mandagere classifications. Areas coloured by expected bioavailability from high (darker) to low (white).

permeability/low solubility compounds the effect of both absorptive and efflux transporters could be important. Given the high correlation found between permeability and metabolization, the authors suggested the use of a new classification, the Biopharmaceutics Drug Disposition Classification System (BDDCS; see Figure 4.14), where the permeability classification is replaced with metabolism criteria. Thus, a compound would be classified in one of four categories: (i) Class 1, high solubility/extensive metabolism; (ii) Class 2, low solubility/extensive metabolism; (iii) Class 3, high solubility/poor metabolism; and (iv) Class 4, low solubility/poor metabolism.

In a similar trend, Mandagere *et al.* also proposed a graphical method to predict oral bioavailability by plotting *in vitro* permeability *vs. in vitro* metabolic stability (see Figure 4.14).[47] According to its coordinates in the plot, the method predicts the compound's bioavailability as high (compounds with high permeability and low metabolization), low (low permeability and/or high metabolization) and medium (combinations of intermediate values).

Following the same philosophy, Figure 4.15 combines the three main components of bioavailability discussed in this chapter. The plot contains the values for intrinsic clearance (extrapolated to hepatic extraction ratio), kinetic solubility and artificial membrane permeability available for the internal

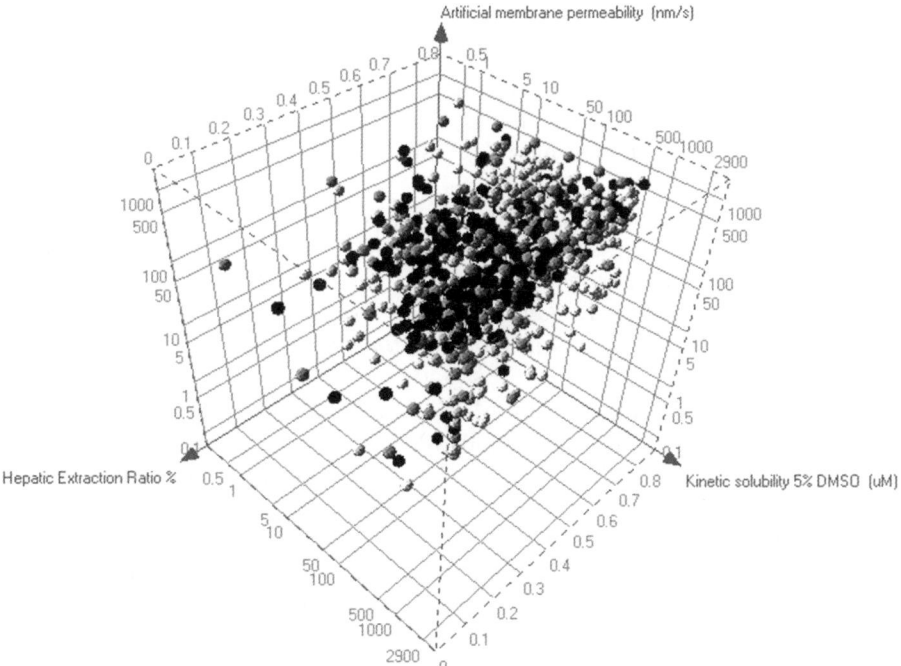

Figure 4.15 Representation of hepatic extraction clearance *vs.* kinetic solubility *vs.* artificial membrane. Data points coloured by bioavailability (*F*) class (white: *F* ≤ 10%; light grey: *F* between 10 and 30%; dark grey: *F* between 30 and 60%; black: *F* > 60%).

bioavailability dataset. The three-dimensional plot shows how the areas where the *in vitro* assays predict high permeability, high solubility and low metabolism provide the higher chances for proper *in vivo* oral bioavailability.

The usefulness of *in vitro* techniques has been acknowledged by regulatory agencies. The BCS classification system was designed to replace the use of *in vivo* assays in bioequivalence studies for immediate release oral drugs with *in vitro* permeability and solubility/dissolution tests. Its adoption by regulatory agencies like the US Food and Drug Administration (FDA), the European Medicines Agency (EMA) and the World Health Organization (WHO) reveals the key role that *in vitro* assays play in oral drug development.

References

1. H. Rosen and T. Abribat, *Nat. Rev. Drug Discovery*, 2005, **4**, 381.
2. I. Kola and J. Landis, *Nat. Rev. Drug Discovery*, 2004, **3**, 711.
3. W. L. Chiou and A. Barve, *Pharm. Res.*, 1998, **15**, 1792.
4. Y. H. Zhao, M. H. Abraham, J. Le, A. Hersey, C. N. Luscombe, G. Beck, B. Sherborne and I. Cooper, *Eur. J. Med. Chem.*, 2003, **38**, 233.
5. X. Cao, S. T. Gibbs, L. Fang, H. A. Miller, C. P. Landowski, H. C. Shin, H. Lennernas, Y. Zhong, G. L. Amidon, L. X. Yu and D. Sun, *Pharm. Res.*, 2006, **23**, 1675.
6. W. L. Chiou and P. W. Buehler, *Pharm. Res.*, 2002, **19**, 868.
7. X. Cao, L. X. Yu, C. Barbaciru, C. P. Landowski, H. C. Shin, S. Gibbs, H. A. Miller, G. L. Amidon and D. Sun, *Mol. Pharmaceutics*, 2005, **2**, 329.
8. G. L. Amidon, H. Lennernas, V. P. Shah, and J. R. Crison, *Pharm. Res.*, 1995, **12**, 413.
9. K. Sugano, M. Kansy, P. Artursson, A. Avdeef, S. Bendels, L. Di, G. F. Ecker, B. Faller, H. Fischer, G. Gerebtzoff, H. Lennernaes and F. Senner, *Nat. Rev. Drug Discovery*, 2010, **9**, 597.
10. P. D. Dobson and D. B. Kell, *Nat. Rev. Drug Discovery*, 2008, **7**, 205.
11. P. Artursson and J. Karlsson, *Biochem. Biophys. Res. Commun.*, 1991, **175**, 880.
12. S. Yamashita, T. Furubayashi, M. Kataoka, T. Sakane, H. Sezaki, and H. Tokuda, *Eur. J. Pharm. Sci.*, 2000, **10**, 195.
13. J. D. Irvine, L. Takahashi, K. Lockhart, J. Cheong, J. W. Tolan, H. E. Selick and J. R. Grove, *J. Pharm. Sci.*, 1999, **88**, 28.
14. M. Kansy, F. Senner and K. Gubernator, *J. Med. Chem.*, 1998, **41**, 1007.
15. M. Bermejo, A. Avdeef, A. Ruiz, R. Nalda, J. A. Ruell, O. Tsinman, I. Gonzalez, C. Fernandez, G. Sanchez, T. M. Garrigues and V. Merino, *Eur. J. Pharm. Sci.*, 2004, **21**, 429.
16. M. Thompson, R. B. Lennox and R. A. McClelland, *Anal. Chem.*, 1982, **54**, 76.
17. U. Fagerholm, *J. Pharm. Pharmacol.*, 2007, **59**, 905.
18. D. A. Volpe, *J. Pharm. Sci.*, 2008, **97**, 712.
19. H. van de Waterbeemd, *Basic Clin. Pharmacol. Toxicol.*, 2005, **96**, 162.

20. M. P. Gleeson, *J. Med. Chem.*, 2008, **51**, 817.
21. C. A. Lipinski, F. Lombardo, B. W. Dominy and P. J. Feeney, *Adv. Drug Delivery Rev.*, 1997, **23**, 3.
22. D. F. Veber, S. R. Johnson, H. Y. Cheng, B. R. Smith, K. W. Ward and K. D. Kopple, *J. Med. Chem.*, 2002, **45**, 2615.
23. M. M. Hann, A. R. Leach and G. Harper, *J. Chem. Inf. Comput. Sci.*, 2001, **41**, 856.
24. T. I. Oprea, A. M. Davis, S. J. Teague and P. D. Leeson, *J. Chem. Inf. Comput. Sci.*, 2001, **41**, 1308.
25. M. Congreve, R. Carr, C. Murray and H. Jhoti, *Drug Discovery Today*, 2003, **8**, 876.
26. A. P. Hill, R. M. Hyde, A. D. Robertson, P. M. Woolard, R. C. Glen and G. R. Martin, *Headache*, 1994, **34**, 308.
27. Advanced Chemistry Development, Inc., 110 Yonge Street, 14th Floor, Toronto, Ontario, M5C 1T4, Canada; http://www.acdlabs.com.
28. Daylight Chemical Information Systems, Inc., 120 Vantis, Suite 550, Aliso Viejo, CA 92656, USA; http://www.daylight.com.
29. K. C. Johnson and A. C. Swindell, *Pharm. Res.*, 1996, **13**, 1795.
30. P. Japertas, P. Maas and A. Petrauskas, http://www.acdlabs.com/download/publ/2003/ddt03_dmso_poster.pdf.
31. M. Hewitt, M. T. Cronin, S. J. Enoch, J. C. Madden, D. W. Roberts and J. C. Dearden, *J. Chem. Inf. Model.*, 2009, **49**, 2572.
32. C. A. Bergstrom, *Basic Clin. Pharmacol. Toxicol.*, 2005, **96**, 156.
33. C. A. Bergstrom, C. M. Wassvik, K. Johansson and I. Hubatsch, *J. Med. Chem.*, 2007, **50**, 5858.
34. S. H. Yalkowsky and S. C. Valvani, *J. Pharm. Sci.*, 1980, **69**, 912.
35. N. Jain and S. H. Yalkowsky, *J. Pharm. Sci.*, 2001, **90**, 234.
36. M. H. Abraham and J. Le, *J. Pharm. Sci.*, 1999, **88**, 868.
37. M. H. Abraham, *Chem. Soc. Rev.*, 1993, **22**, 73.
38. M. Abraham and J. McGowan, *Chromatographia*, 1987, **23**, 243.
39. C. Lamanna, M. Bellini, A. Padova, G. Westerberg and L. Maccari, *J. Med. Chem.*, 2008, **51**, 2891.
40. A. P. Hill and R. J. Young, *Drug Discovery Today*, 2010, **15**, 648.
41. D. F. Lewis, *Pharmacogenomics.*, 2004, **5**, 305.
42. I. Zamora, L. Afzelius and G. Cruciani, *J. Med. Chem.*, 2003, **46**, 2313.
43. Molecular Discovery, Via Stoppani 38, 06135 Ponte San Giovanni, Perugia, Italy; http://moldiscovery.com.
44. P. J. Goodford, *J. Med. Chem.*, 1985, **28**, 849.
45. Lhasa Ltd., 22–23 Blenheim Terrace, Woodhouse Lane, Leeds LS2 9HD, UK; http://lhasalimited.org.
46. C. Y. Wu and L. Z. Benet, *Pharm. Res.*, 2005, **22**, 11.
47. A. K. Mandagere, T. N. Thompson and K. K. Hwang, *J. Med. Chem.*, 2002, **45**, 304.

CHAPTER 5

Molecular Descriptors for Database Mining. Translating Empirical Chemistry into Mathematics: Tools for QSAR and In Silico Screening Based on the Hydrophobicity of Small Molecules

FRANCESCA SPYRAKIS*[a,b], PIETRO COZZINI[a,b] AND GLEN E. KELLOGG[c]

[a] Department of General and Inorganic Chemistry, University of Parma, Parma, Italy; [b] INBB, Biostructures and Biosystems National Institute, Rome, Italy; [c] Department of Medicinal Chemistry and Institute for Structural Biology & Drug Discovery, Virginia Commonwealth University, Richmond, Virginia, 23298-0540, USA
*E-mail: francesca.spyrakis@unipr.it

5.1 Introduction

Even in the era of GPU (graphics processing unit) and of inexpensive multi-core CPU (central processing unit) processors, the problem of discovering new drugs *in silico*, with a reasonable, *i.e.*, small number of false negatives and,

RSC Drug Discovery Series No. 23
Physico-Chemical and Computational Approaches to Drug Discovery
Edited by F. Javier Luque and Xavier Barril
© The Royal Society of Chemistry 2012
Published by the Royal Society of Chemistry, www.rsc.org

most of all, few false positives, coupled with the utopian desire of avoiding long and expensive experimental trials, still represents one of the most demanding challenges of computational chemists.[1] Many computational strategies, starting with simple docking and progressing to extensive molecular dynamics simulations, found their basis in the primary assumption that the three-dimensional structure of a molecule and its biological activity are intimately related. Nearly from the beginning, two different approaches to discovery, *i.e.*, ligand-based and structure-based design, have tried to exploit this structure–activity assumption with different viewpoints. While ligand-based methods look for molecules that are as similar as possible (with somewhat limited structural extrapolation) to known active molecules, structure-based strategies look for molecules that are ideally complementary to known active or binding sites. First, ligand-based methods are broadly considered in terms of their dimensionality, with 1D or 2D methods being considered separately from 3D methods. The former are described as "classical" (*vide infra*) Quantitative Structure–Activity Relationship (QSAR) methods, where whole molecule descriptors that are independent of molecular conformation are exploited to construct models of activity as functions of structure. Three-dimensional ligand-based methods do incorporate molecular conformation and can be further categorized into pharmacophore-, shape- and molecular fields-based approaches.

Pharmacophore-based methods, such as CATALYST[2] and UNITY,[3] rely on the three-dimensional identification of the key interaction points that would stabilize the formation of a complex between a set of the ligands and the protein target, in particular identifying hydrophobic groups, hydrogen-bond donors and acceptors and positively and negatively charged centers. Shape-based strategies, such as ROCS,[4] exploit the spatial/shape information provided by the ensemble of atoms in ligands, looking for an overall shape similarity rather than for a pharmacophore similarity, *i.e.*, comparing atomic radii instead of atom types. Finally, molecular fields-based methods such as CoMFA,[5] CoMSIA[6] and GOLPE based on GRID,[7,8] compare molecules by placing them in a Cartesian system using different alignment rules and calculating the similarity between them through the superimposition of steric, electrostatic or hydrophobic potential fields, but also can include fields encoding hydrogen-bond donors and acceptors.

Structure-based methods obviously have the significant benefit of knowing the specifics of the target site when designing a ligand. Many reviews have been written on structure-based design.[9–16] However, a sizeable number of targets are not structurally known, and many of these are therapeutically important. Thus, ligand-based methods continue to be used and developed. This review will focus on the complementary roles that ligand-based and structure-based methods can play in the "virtual" screening of databases to identify and optimize compounds with biological activity.

While all computational methods have advantages and disadvantages, which are often dependent on the set of molecules being studied, none alone has been proven capable of discovering a novel lead compound, optimizing it for activity, selectivity and toxicology, and otherwise developing the candidate into a drug. In fact, consensus approaches,[17,18] applying the combination of different tools and taking advantage of their complementary strengths (and offsetting their respective weaknesses), might represent the winning strategy.[19] One glaring weakness in most computational methodology is a poor or nonexistent evaluation of entropic effects provided by both the ligands and their binding sites.[20–22] Secondly, better parameterization of the dynamic nature of molecules, which would enable identification of new conformations and binding modes,[23] is also highly desirous.[24]

In this review, we reflect on QSAR, which is, in our view, the historical foundation of much of computational chemistry as it applies to biological molecules and particularly their interactions with small molecules leading to biological effects. Our specific focus will be on the rich value of log $P_{o/w}$ (the partition coefficient for solvent partitioning between octan-1-ol and water), an empirical descriptor that encodes the thermodynamics of biomolecular interactions.

5.2   The History of QSAR

Among the numerous strategies utilized for designing and discovering new drugs, QSAR methods have always occupied a representative place since the landmark 1964 paper of Hansch and Fujita.[25] In their seminal work, these authors observed for the first time that, by combining the information contained in electrostatic and hydrophobic "descriptors", it was possible to rationalize the effects of a series of substituents on the biological activity of different benzoic acid derivatives. This was neatly tied together with a simple equation or QSAR that represented activity as a function of descriptors that are in reality a stand-in for structure. The vision that compounds with similar structure could have a similar activity, and be rationally and predictably modified, was perhaps the basis for the development of many brilliant and successful computational approaches and applications over the next five decades. As was pointed out by Martin, in the QSAR acronym are combined both the originality and the physical chemical basis of this extremely simple, but also sometimes enigmatic, approach.[26] Thus, "Q" and "R" distinguish these strategies from other computational methods by describing how structural changes relate to corresponding changes in properties like biological activity, while "S" designates the structures of the small ligands exerting a biological activity "A" on the selected target.

From the almost primitive analysis described by Hansch and Fujita, employing a restricted number of simple and easily interpretable physico-chemical descriptors and relatively commonplace statistical descriptions, the QSAR world has moved towards more complex multi-parametric and nonlinear approaches.[27] In the last 20 years the number and diversity of descriptors has significantly increased in accord with the increasing size and chemical diversity of data sets, while mathematics and statistical procedures have become more and more complicated with machine learning, genetic algorithms and other cutting edge statistical methodologies. Thus, QSAR technology and science has become more akin to the fields of data mining, knowledge discovery and database management, and away from its origins in physical organic chemistry.[28,29]

Different QSAR approaches may be differentiated with respect to the target property values, the chosen descriptors and/or the optimization algorithm used to relate descriptors and target properties, *i.e.*, continuous properties like IC_{50} or binding constants, categorical unrelated properties such as the class of metabolic stability, and categorical unrelated properties like the pharmacological classes of the compounds. The last two categories require classification modeling approaches rather than the simpler multi-linear regression type modeling.[27] In any case, as defined by the OECD (Organization for Economic Co-operation and Development), valid QSAR models to be used in regulatory assessment of chemical safety should be characterized by: (i) a defined endpoint, (ii) an unambiguous algorithm, (iii) a defined domain of applicability, (iv) appropriate measures of goodness-of-fit, robustness and predictivity and (v) a mechanistic interpretation.[30] Thus, even if recent models have been accurately improved by adding numerous and new descriptors and by developing better and better algorithms, there is still substantial room for further advancement of the field *via*, for instance, improving data mining algorithms and model validation strategies.[27] We believe that the rapid and recent growth of experimental publicly available SAR data reported in PubChem,[31] NCI,[32] WOMBAT[33] and BINDING DB,[34] and other databases, has opened a novel era for QSAR where the problem of developing robust QSAR models is becoming more challenging. It is now clear that no universal QSAR approach is able to produce the best models for all data sets,[27] and that even the method(s) (in addition to the selected set of variables and the best resulting model) must be optimized for each data set.

Recently described QSAR approaches have been much more applied to data mining, combinatorial chemistry/high throughput and virtual screening analyses[35] for the prediction of potential new lead compounds, rather than towards an increasingly obsolete evaluative approach that is largely suitable for traditional one compound at a time lead optimization (Figure 5.1).

5.3 The Evolving Dimensionality of QSAR Methods

The dimensionality of QSAR has evolved from its earliest incarnations to the present day.[27,36] First-generation QSAR techniques, also known as 1D-QSAR, were typically based on the correlation of the biological activity with a single global molecular property as, for example, the solubility, the pK_a or log $P_{o/w}$ value. Then, the inclusion of chemical graph descriptors proposed by Hansch, Fujita, Free and Wilson led to the development of 2D-QSAR, providing the means to also consider the contributions of bi-dimensional molecular

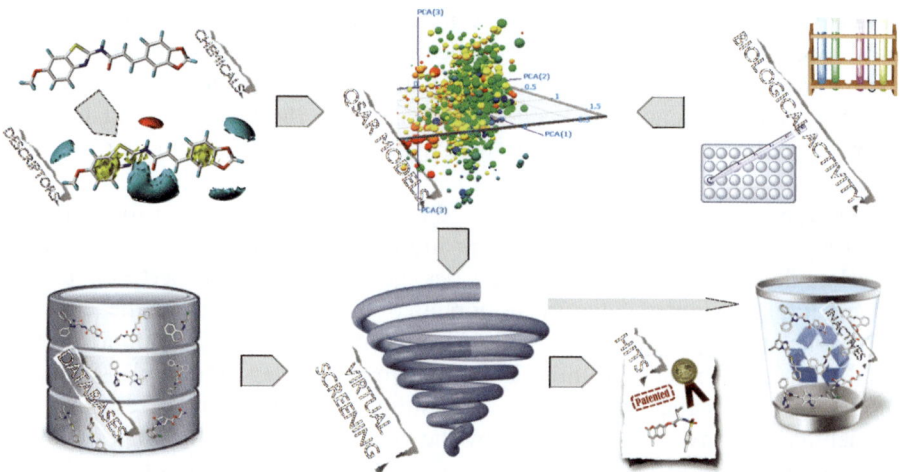

Figure 5.1   Scheme for the generation of QSAR models and application to virtual
screening analysis.

connectivity to biological activity.[37–39] Later, a number of purely non-empirical
1D and 2D descriptors based on graph theory were developed and described by
Kier and Hall[40] and others.[41,42] In 1988, Cramer *et al.* introduced Comparative
Molecular-Field Analysis (CoMFA).[5] This innovative and pioneering approach,
soon recognized as 3D-QSAR, now provided three-dimensional pharmacophoric "images" of compounds by placing them on a surface or a grid, where at each
grid point the interactions of selected chemical probes with the ligand are
energetically evaluated. The new statistical method of partial least squares
(PLS)[43,44] was employed to abstract the QSAR. CoMSIA,[6] GOLPE,[7,8] HASL[45]
and other 3D methods are largely similar in approach, differing in how grid
point values are calculated and in how the models are generated.

The major obstacle in applying 3D-QSAR is related to the prerequisite of
identifying the correct bioactive conformation for each ligand and then to
perform a proper alignment of all ligands in the set. This is necessary as the
underlying assumption behind 3D-QSAR is that the set of ligands is bound in
their "receptor" in exactly these superimposed conformations. Thus, in the
absence of a target structure, to be used in docking simulations with the aim of
identifying the correct ligand orientation(s), this superposition task is not at all
obvious, with a high risk of generating useless and unsuccessful models.
Alternatively, this problem may be tackled by considering each molecule in
many different conformations, orientations, by enumerating tautomers,
stereoisomers and protonation states. In 1998, Klein and Hopfinger[46]
proposed the 4D-QSAR approach, where a number of different ligand
configurations are considered in a single simulation.[5,47–49] As reported by Lill
in a recent review, 4D-QSAR models can be thought of as an extension of 3D-
QSAR to address uncertainties during the alignment process and can thus be
adopted to solve problems involving receptors known to accommodate ligands

in different binding poses, such as cytochrome P450.[36] Later, the desire to account for induced-fit effects[23,50,51] led to the development of 5D-QSAR.[52]

The field of high dimensionality QSAR is clearly still under development. These novel QSAR adaptations can take into account receptor adjustments induced by steric, electrostatic, hydrogen bond or lipophilic potentials. The ligands in a data set are truly not monotonic and may bind in different sub-pockets or in different forms or modes of the binding site and, thus, may experience different interaction fields of the protein.[53] Moreover, as recently reported by Polanski,[54] the inclusion of multiple solvation scenarios, and real receptor or target-based receptor model data, might necessitate the application of 6D-[52] or 7D-QSAR.[27] Nevertheless, as with any statistical tool, increasing the number of fitted parameters carries a significant risk of over-fitting: each added parameter should be carefully handled and evaluated with respect to its physical interpretation as well as statistical relevance. With a large enough parameter set, any quantity or relation can be modeled, but the predictive power of such models can be quite low.

5.4 Descriptors and Molecular Properties

In the beginning, QSARs were little more than simple mathematical models relating the biological activity of a limited series of molecules to their electronic and steric attributes and relative hydrophobicities.[25,55] Even though over the last five decades the quantity and complexity of available descriptors has increased almost exponentially, the aforementioned three molecular attributes still represent the most important and fundamental properties to which ligand activities are most often related.[56]

5.4.1 Electronic Descriptors

In its original formulation, the well-known Hammett σ constant gave a measure of the electronic effect of a particular substituent, measuring the electronic charge distribution of the benzene ring where the substituent has been virtually placed. To calculate this value, Hammett determined the equilibrium constants for the ionization in water of a series of substituted benzoic acids:[57–60]

$$\sigma_X = \log K_X - \log K_H \tag{5.1}$$

In particular, positive σ values characterize electron-withdrawing substituents, while electron-donating groups present negative values.

Subsequently, the ρ reaction constant, defining the susceptibility of a reaction to substituent effects, was introduced into the Hammett equation:

$$\log \frac{K_X}{K_H} = \rho \times \sigma \tag{5.2}$$

In the case of positive ρ values, reactions are sustained by the effects of electron withdrawal, while negative ρ values suggest quite the opposite, *i.e.*, the help of electron-donating effects. However, only substituents in *meta* and *para* ring positions follow this trend. Owing to proximity and steric effects, *ortho* substitutions are differently parameterized to consider the significant effects that *ortho* substituents might have on both enthalpy and entropy.[61,62]

As reported by Selassie, the Hammett constants have been used in 7000/8500 equations in the physical organic chemistry database and in about 1600/8000 in the biology database. These dual databases of QSAR equations relate, respectively, physico-chemical and bio-activities to structural parameters belonging to the CQSAR package[63] and were screened to confirm the effect of electronic properties on the chemical and biological reactivities of molecules.[56]

Nevertheless, the Hammett equation presents a number of limitations that should be taken into account. First, primary σ values are directly obtained from experimental analyses performed on substituted benzoic acids, but mathematical models derive the secondary values. Thus, their accuracy is dependent on both the accuracy of the experimental measurements and on the reliability of the mathematical model. Second, the proximal effects of substituents normally neglected should be considered since they tend to distort electronic contributions. Moreover, changes in the mechanism or transition state, causing discontinuities in Hammett plots, and changes in solvent, leading to dissimilarities in reaction mechanism, should also be taken into account for a more reliable representation of chemical-physical properties for molecular substituents.[56]

More recently, *ab initio* and semiempirical methods have been used to calculate quantum chemical descriptors such as net atomic charges, HOMO and LUMO energies or frontier orbital electron densities. The approach proposed by Sullivan and co-workers, combining frontier orbital theory and topological parameters, demonstrated the advantages of including quantum chemical descriptors, as it was shown that it was possible to explain intra- and intermolecular interactions with descriptors easily derivable from molecular theoretical structure.[64]

5.4.2 Steric Descriptors

The steric effect was first parameterized by Taft in 1956,[65] through the introduction of the Taft constant:

$$E_S = \log(k_X/k_H) \tag{5.3}$$

with k_X and k_H being, respectively, the hydrolysis rates of the XCH_2COOR and CH_3COOR esters.

E_S has been demonstrated to be intimately correlated with van der Waals radii.[66] Molar refraction (MR), first described by Tute,[67] incorporated a polarizability component directly related to London dispersion forces. Unfortunately, the MR parameter does not contain any shape information and it has been mostly used in biological QSARs that are dominated by intermolecular interactions. In order to better represent 3D molecular space, Verloop and co-workers proposed the STERIMOL descriptor made up of five different factors: *L*, *B1*, *B2*, *B3* and *B4*, representing the length and the width of a particular substituent.[68] Extensive studies demonstrated the usefulness of just three of them: length and the minimum and maximum widths.[69] In cellular systems and in distribution and transport studies essentially based on diffusion, the molecular weight (MW) and log MW have been extensively used as steric descriptors.[66,67] Of particular note, this parameter has been commonly adopted in QSAR studies investigating cross resistance of different drugs in multidrug-resistant cell lines.[56] Nevertheless, despite the fundamental role played by molecular shape and size in regulating biological interaction and subsequent drug effects, steric effects are among the most challenging to parameterize.

5.4.3 Hydrophobic Descriptors

Between the 19th and the 20th centuries, Meyer[70] and Overton[71] were the first to recognize the importance of hydrophobic interactions in drug activity, while Barclay and Butler[72] showed a linear relationship between heat of hydration and entropy of hydration. In particular, the observation that the partition coefficient between olive oil and water was related to the activity of small organic molecules, and in particular to their narcotic potency, was first reported in 1899 by Meyer.[70] Ferguson extended this concept by thermodynamically relating the hydrophobic properties of a series of depressants to their depressant action.[73] Nevertheless, it was not until 1962 that Hansch published his critical study, in which both electronic and hydrophobic descriptors were applied in a multiparameter approach, yielding a QSAR for a series of plant growth regulation factors.[74] This first introduced and ascertained the fundamental role played by hydrophobicity in regulating biological interactions in many different systems: protein–ligand complexes, assembly of lipids in biomembranes, protein folding processes, detergency, coagulation, *etc.*[75–77] Most importantly, this work provided the basis for the later development of much more reliable and successful QSAR models.

The standard QSAR hydrophobic parameter for a specific substituent is π, which, according to the following equation, can be defined as the difference between the log *P* of the substituted molecule and that of the parent molecule:

$$\pi_X = \log P_{R-X} - \log P_{R-H} \tag{5.4}$$

Analogous to electronic descriptors, hydrophobic descriptors were first determined for aromatic substituents, without taking into account any correction factor. These were later introduced to consider unsaturation, branching and ring fusion effects. The manual approach adopted for calculating hydrophobic parameters developed by Hansch and Leo was based on an extensive series of fragment constants that were supported, in a constructionist way, by a number of correction factors for bonds and proximity effects. The manual calculation was soon superseded by CLOGP, the first automated system for estimating log P values for small molecules.[78,79] Later, many other programs based on substructure (atom or fragment) or on whole molecule algorithms were developed for rapidly and facilely calculating this fundamental hydrophobic parameter.[80] Whole molecule-based approaches exploit molecular or spatial properties for log P calculation,[81,82] while sub-structure-based methods sum the properties of fragments or atoms.[83–85] A completely different approach was recently proposed by Kamlet and co-workers,[86] who reported the use of solvatochromic parameters, which were later extended by Abraham for refinement of molecular descriptors and to characterize hydrophobicity scales.[87] The next two sections of this review explore the hydrophobic effect and estimation of log P, respectively, in more detail, with particular emphasis on the significant information content encoded in log P.

5.5   Connectivity and Related Descriptors

Other types of descriptors are somewhat difficult to characterize or actually encode multiple effects. These descriptors can be used to take into account specific properties of specific molecule sets, as, for example, electronic descriptors like E_{LUMO} that describes the energy of the lowest unoccupied molecular orbital, I_{α}, indicating the presence of an aceanthrylene ring, or I_1, related to the number of fused rings,[56] or the family of molecular connectivity indices first proposed by Randic[42] and then extensively developed by Kier and Hall.[88] These latter indices were shown to well model many different physico-chemical properties such as boiling point, molar refraction, polarizability and partition coefficient.[40] Kier and Hall also introduced the atomistic E-State, encoding for both molecular electronic and steric properties, and later developed the HE-State and atom-type E-State descriptors.[39] These electro-topological parameters have been extensively used in diverse studies;[89,90] however, as pointed out above, complex descriptors must be physically interpretable and used with care due to the risk of obtaining unreliable or chance correlations.[91]

5.6   Information Encoded in Empirical Hydrophobic Descriptors

Much as the ancient Roman god, Janus Bifrons, who is usually depicted with two opposite faces, the properties and behavior of water produce many contradictions and seemingly inconsistent interpretations. For example, explanations of the hydrophobic effect from the perspective of water can be confusing. Consider the following two statements: (1) water and oil do not mix (see Figure 5.2); (2) water generally attracts *itself* more strongly than other molecules.[92] Substances that do not mix with water are termed as "hydrophobic" (water fearing), but the reason that water and oil do not mix is related to water's ability to self-organize by forming up to four hydrogen bonds per water molecule. This mutual attraction between water molecules causes the observed segregation of oil from water and results in an apparent oil–oil attraction.[93]

In the early 1950s, the *hydrophobic interaction* was identified as an important factor in protein stability by Kauzmann, who first coined the term.[94] Amino acids are divided into those that are water-like (polar) and those that are oil-like. This notion began to explain why proteins with specific sequences fold in specific 3D structures. We now know that hydrophobic interactions are a major driving force in biological processes; however, these events are "concerted", as stated by Dill,[93,95] and cannot be represented or computationally modeled as the sum of specific "sub-events".

As noted above, QSAR analyses have often employed hydrophobic descriptors. It may even be possible to say that the early successes of QSAR were made possible by these descriptors, as the hydrophobic effect is ubiquitous and a major driving force in biomolecular structure and association. Here, we will describe some of the thermodynamic basis for this

Figure 5.2 Hydrophobicity in Nature. The simple principle that "oil and water don't mix" is applicable on many scales, from this leaf and drop of water to the association of molecules in biological media.

and explore the wide range of information that is encoded within hydrophobic descriptors.

5.6.1 Nano-Scale Icebergs and Entropy

The formation of an iceberg of waters surrounding nonpolar molecules as defined by Frank and Evans[96] is generally accepted and validated experimentally by crystallography.[97] The variation of pK_a, molecular volume and the change of expected behavior of protein functional groups has been explained by Klotz[98] using this concept. The formation of a larger cage structure around the protein "solute" produces a smaller surface area than the combined area of isolated, *i.e.*, small molecule, solutes; this maximizes the amount of free water and thus the entropy of the system. In simple terms, the hydrophobic effect can be described as follows: a hydrophobic molecule disrupts the structure of bulk water, yielding an entropy increase due to the rearrangement of water molecules surrounding the solute and the concomitant stabilization of nonpolar entities. Thus, hydrophobic "bonds" are formed and are endothermic, *i.e.*, their force decreases with increasing temperature, with a maximum at around 60 °C. Interestingly, hydrogen bonds become weaker when the temperature increases.

Many examinations of molecular recognition and hydrophobic interactions have been reported. The early proposal by Frank and Evans of an iceberg model in which water forms frozen patches or microscopic icebergs around hydrophobic solute molecules, with an associated significant entropic loss, suggested that the extent of the iceberg formation increases in accord with the size of the solute.[96] This model was then further developed thermodynamically by Nemethy and Scheraga, who described the hydrogen-bonded ice-like clusters of water molecules as being in equilibrium with a non-hydrogen-bonded liquid, thus quantitating the different number of hydrogen bonds formed in the clusters.[99] While Kauzmann first described the existence of van der Waals interactions between nonpolar portions of molecules immersed in water solutions to interpret protein denaturation, only the later explanation provided by Haymet, Silverstein and Dill[100,101] equated hydrophobicity and the hydrophobic effect as the consequence of entropy gain by desolvation processes occurring at each hydrophobic–hydrophobic interface. Nevertheless, as recently pointed out by Blokzijl and Engberts,[102] despite the continuing efforts for a better understanding of the hydrophobic effect, traditional models employing conventional molecular mechanics have been generally demonstrated to be deeply inadequate, and the driving forces guiding the hydration of apolar molecules and the interactions between them are still not well understood.

5.6.2 Hydrophobic Fields

Do hydrophobic fields exist? From the American Heritage Dictionary: "a field (as used in physics) is defined as a region of space characterized by a physical property, such as gravitational or electromagnetic force or fluid pressure, having a determinable value at every point in the region".[103] Thus, since hydrophobic residues, functional groups or atoms, by themselves, do not exert a force upon one another that drives their interaction (in the way that opposite electrostatic charges are attracted), there really is not a hydrophobic field. In this regard, hydrophobic "moments" may be useful parameters, but they should not be viewed as a force that is defined by analogy to electrostatic fields. Similarly, the language used to define a "lipophilicity force field" could also be misleading. However, if we use the dictionary definition of a field by representing the grid region of space to be characterized by a physical property, e.g., the free energy of solvent transfer (log *P*) of the atoms involved, one may be able to produce a free energy field rather than a force field. There are numerous technical issues with such a representation, e.g., how can the molecular property of log *P* be treated as atomistic or how can a free energy have distance dependence as a force does? Perhaps the most compelling reason to consider the representation of the hydrophobic effect as a field is that it provides a useful and pragmatic construct to understand, explain and (possibly) predict biomolecular interactions.

5.6.3 Additional Free Energy Information

In a recent *Journal of Medicinal Chemistry Perspective*, Bissantz *et al.*[104] described how "*an understanding of typical noncovalent protein–ligand interactions can arise from a synergistic use of crystal structure database searching, structures and association properties of biomolecular and synthetic receptor–ligand systems and various computational models*". To understand, interpret and therefore model biological interactions, we have to go back to the Gibbs fundamental equation of free energy, $\Delta G = \Delta H - T\Delta S$. Actually, we have to consider, in addition to entropy and enthalpy, effects arising from flexibility and cooperativity, solvation and desolvation. The combination of hydrophobic interactions and hydrogen bonds (ionic-reinforced and other) are the major forces that drive protein–ligand, protein–protein and protein–DNA associations.

Another parameter that influences the free energy of hydration is the shape of the molecule or its cavity.[105,106] Leo and Hansch related the nature of the solute and the molecular volume (calculated as CPK volume) with hydrophobicity.[107] Thereafter, using the partition coefficient for octan-1-ol/water (log $P_{o/w}$) as the primary measure of hydrophobicity became widely accepted.

5.6.4 Absorption, Distribution, Metabolism and Excretion (ADME)

Lipophilicity/hydrophobicity probably should not be discussed without citing Lipinski's "Rule of Five". This is a rule of thumb to evaluate a lead compound or to screen a set of compounds with a certain pharmacological or biological activity and predict its likelihood of being an orally active drug in humans. The rule was formulated in 1997 by Lipinski, based on the observation that most drugs are relatively small and lipophilic molecules.[108] It describes molecular properties important for drug pharmacokinetics in the human body, including their absorption, distribution, metabolism and excretion (ADME). However, the rule does not predict if a compound is pharmacologically active.

Thus, the Rule of Five (so named because the cutoffs for each of the four parameters are all close to 5 or a multiple of 5) evaluates the drug-likeness of a compound by enumerating simple molecular properties that ensure oral bioavailability:

- molecular weight of less than 500 Daltons
- limited lipophilicity (*i.e.*, log $P_{o/w} < 5$)
- maximum of 5 H-bond donors (expressed as the sum of OHs and NHs)
- maximum of 10 H-bond acceptors (expressed as the sum of Os and Ns)

While these rules are not inviolable, they represent a set of guidelines that many researchers apply in drug discovery, particularly when screening large databases with *in silico* methods.

5.7 Estimation of log *P*

The 1971 paper of Leo and Hansch[96] became a milestone for understanding hydrophobicity and its importance in drug design; it provided the history and theory of partition phenomena with (most importantly) a comprehensive set of log *P* values. A recent complete review of hydrophobic interactions can be found in Meyer *et al.*[109]

Because of the burgeoning importance of hydrophobicity and log *P* for QSAR, drug discovery and, later, protein folding, several diverse methods to computationally estimate log *P* values were created.[80,110–112] We can roughly divide the algorithmic approaches into two major categories: those based on substructures and those based on the whole molecule.

The first category includes both fragment-based methods and atom-based methods. The second includes molecular lipophilicity potential (MLP) and related approaches, topology descriptions and molecular property descriptions. Famous examples of the fragment-based algorithms are Rekker's method,[113] the CLOGP method that originated in Hansch's laboratory[78,79,114–116] and the ACD/log *P* method.[117] The first fragment-based method was that of Rekker, who deconvolved experimental log *P* values into partial values for molecular fragments. The underlying assumption is that the total log *P* for the molecule can be calculated as a sum of all these contributions. In addition, a number of correction factors were derived to estimate and incorporate molecular environment effects. The main advantages of substructure approaches are their highly interpretable results and the ability to train the calculation algorithm with relatively few experimental data. In contrast, limits arise when the compound of interest contains fragments that are "missing" from the training set or when fragments are "arbitrary", *i.e.*, non-coincident with recognizable organic functional groups.

Atom-based methods are an extension of fragment-based methods where the assumption is that the total log P of a molecule is the sum of contributions from each individual atom. Well-known examples are the XLOGP[118–120] and Ghose-Crippen[121–124] methods. The major criticism for this approach is that sometimes, because of a perhaps overly holistic view of the molecule, the molecule is more than the sum of the parts into which it has been divided. Similarly, the "chemistry" of the molecule is obscured when it is considered as just a collection of atoms.

Two basic ideas are behind the molecular methods. The first is to find correlation between selected molecular properties and experimental log P values. These methods are influenced less by specific fragments, but sometimes more data are needed to accurately model log P, and the contributing chemical descriptors can often be physically or chemically uninterpretable or even meaningless. The second is to use quantum mechanical calculations to study the interactions between solute and solvents. Thus, the relative solubilities of the solute in different solvents, such as water and octan-1-ol, can be estimated. The early work of Roger and Cammarata,[125,126] that of Hopfinger and Battershell,[127] Klopman[128] and, most recently, Bravi and Wikel[129] by adapting neural network and artificial intelligence methods, have implemented many tools to estimate log P over the last two decades. Despite these developments, the empirical methods are most commonly used in applications.

5.8 HINT: an Empirical log *P*-Based Paradigm for Drug Discovery

Starting with the ideas of Leo and Rekker, Abraham and Leo extended the concept of log P *via* a fragmental approach to calculate amino acid and side-chain partition coefficients.[130] This work coincided with a growing interest in protein secondary structure prediction based on patterns of hydrophobic and polar side-chains, which led to the development of different hydrophobicity scales, as recently classified by Biswas and co-workers.[76,131–133] The Abraham and Leo paper concluded with the suggestion that extending fragment parameters to atomic entities could be valuable to evaluate docking and protein folding phenomena.[130] The implementation of these concepts led to the design of the "natural" force field, HINT (Hydropathic INTeractions), by Kellogg and Abraham.[84,134,135] HINT is a noncovalent force field derived from experimental log P data, *i.e.*, from partition coefficients based on the Hansch and Leo log P estimation method.

The HINT log P approach is based on the assumption that the octan-1-ol/water environment can represent or model the biological system in that it contains both hydrophobic and hydrophilic media. Just as log P is "constructed" from fragmental contributions, it can be also be considered as the sum of individual lipophilic propensities for each atom that we term hydrophobic atom constants (a_i), such that $\log P = \sum a_i$. As described above, there are numerous methods available for the calculation of a_i. HINT uses an adaptation of the Hansch and Leo CLOGP method.[114–116]

In practice, HINT calculates scores for each atom against all other atoms:

$$b_{ij} = a_i S_i a_j S_j T_{ij} R_{ij} + r_{ij} \tag{5.5}$$

where a_i is the hydrophobic atom constant for the i-th atom, *i.e.*, the specific contribution of the i-th atom to the total $\log P$ value of the molecule, S_i is the solvent accessible surface area for the i-th atom, T_{ij} is a logic variable that assumes the values +1 or −1 depending on the polar properties of the interacting atoms, R_{ij} is the exponential term $-|r|$ (r is the distance between atoms i and j) and r_{ij} is the Lennard-Jones van der Waals term.

Thus, the total HINT score is:

$$H_{\text{total}} = \sum \sum b_{ij} = \sum \sum a_i S_i a_j S_j T_{ij} R_{ij} + r_{ij} \tag{5.6}$$

Since $\log P$ is a measure of an equilibrium, we can write $\Delta G = -2.303\, RT \log P$ or $\log P = -\Delta G/(2.303\, RT)$. Now, $\log P = \sum a_i$ and because the HINT score is proportional to a functional $f(a_i)$, the HINT score can be thought of as representing the free energy of a molecular association.

As $\log P$ encodes both enthalpic and entropic contribution to ΔG of a biomolecular association by measuring an equilibrium, HINT is able to evaluate the strength of all nonbonding interactions, including hydrogen bonding, Coulombic interaction and hydrophobic effects. Thus, the most important application of HINT is in the evaluation of intermolecular interactions, as these calculations produce an interaction score that has been demonstrated to correlate well with the free energy of interactions.[136–139] For reference, based on numerous investigations we estimate that, on average, for protein–ligand associations, 515 HINT score units correspond to 1 kcal mol^{-1} free energy of binding.[134] However, in modeling and drug discovery, absolute scores are not as important as the differences in score values for analogous systems. Similarly, the difference of Gibbs free energy between two states ($\Delta\Delta G$) is a very important thermodynamic parameter in modeling biological systems as it is the basis for differences in measured binding (and efficacy). With this in mind, the error in prediction across a very diverse set of protein–ligand complexes was around ± 2.6 kcal mol^{-1}, but can decrease to less than ± 1.0 kcal mol^{-1} for sets of similar ligands complexed with the same protein.

Because biological interactions are not (usually) a simple interaction between two bodies but actually n-body interactions involving water molecules or other cofactors, HINT has tools to consider the important waters that can establish hydrogen bridges between protein and ligand, between protein and protein or between protein and DNA. Amadasi *et al.* developed a robust method to calculate the relevance of binding site waters and the contribution of bridging waters to the overall free energy of binding.[138,140,141] In addition, to build more realistic models for docking, *etc.*, we also need to take into account nonstandard ionization states on residues and/or ligand functional groups by properly placing protons in modeling our system. Fornabaio *et al.*[137] described the HINT Computational Titration approach to optimize proton placement.

The evaluation of both enthalpic and entropic contributions to the free energy of binding is the most relevant advantage of HINT with respect to other scoring functions, whether they are built-in or separated from docking packages. Generally, scoring functions mainly consider enthalpic effects. This difference was evaluated in a comparison of four scoring functions, FlexX, GOLD, Autodock and HINT.[139] That said, no extant scoring function, including HINT, is accurate over *all* data sets. It is a much better practice to calibrate and validate the scoring function for the particular data set of interest!

For further discussion about the relationship between hydrophobicity and drug discovery, see Sarkar and Kellogg.[142] As already noted, the HINT score is most often used in evaluating docked ligand poses in both drug discovery settings (virtual screening) and in drug design/optimization settings, where the goal is fine-tuning the ligand's functional groups with respect to features of the pocket. It is interesting to note that HINT, docking and scoring, virtual screening and many other tools in the arsenal of the molecular modeler are really just generating QSARs, *i.e.*, equations that relate the structure of a compound to its activity or binding.

5.9 Treating Hydrophobicity as a 3D Property

The successes of 1D and 2D QSAR over the past five decades are even more stunning when one reconsiders the sketchiness of the "structure" information contributing to the model. Clearly the advent of 3D+ QSAR was a revolutionary change in computationally driven drug discovery and design if only because "structure" became a 3D metric, just like the molecules themselves. The earliest programs for 3D QSAR focused only on steric and electrostatic fields because, as discussed above, hydrophobics are more difficult and perhaps less rigorous to implement as a field.

5.9.1 The HINT 3D Maps

Perhaps undaunted only by naivety of the challenges in creating hydrophobic fields (*vide supra*), we developed a suite of map types to represent hydropathy: a property map, an inverse (complement) map and an interaction map. The first is directly analogous to mapping an electrostatic field in that a set of grid points that act as probe atoms is superimposed over the molecule(s) of interest and each point's response to the molecule is recorded. This actually creates two maps: one that differentiates (by sign) hydrophobic from polar regions and another that further defines the polar regions into acidic (hydrogen bond donor) and basic (hydrogen bond acceptor). The inverse map, which is most useful for mapping the hydropathic properties of a cavity,[143] is calculated by projecting the properties of neighboring atoms into space such that hydrogen-bond donors beget acceptor grid points, acceptors beget donor grid points and hydrophobic atoms yield hydrophobic grid points. Lastly, the HINT interaction map is an interesting construct wherein the strengths and types of interactions, *i.e.*, favorable polar (H-bond, coulombic), unfavorable polar, favorable hydrophobic and unfavorable hydrophobic (interacting with a polar group, *e.g.* desolvation), are mapped to the regions of space where these interactions occur. See Figure 5.3 for example maps of these three types.

The HINT 3D field set was shown to provide another map type for the CoMFA[5] implementation of 3D QSAR with a fair degree of success.[85]

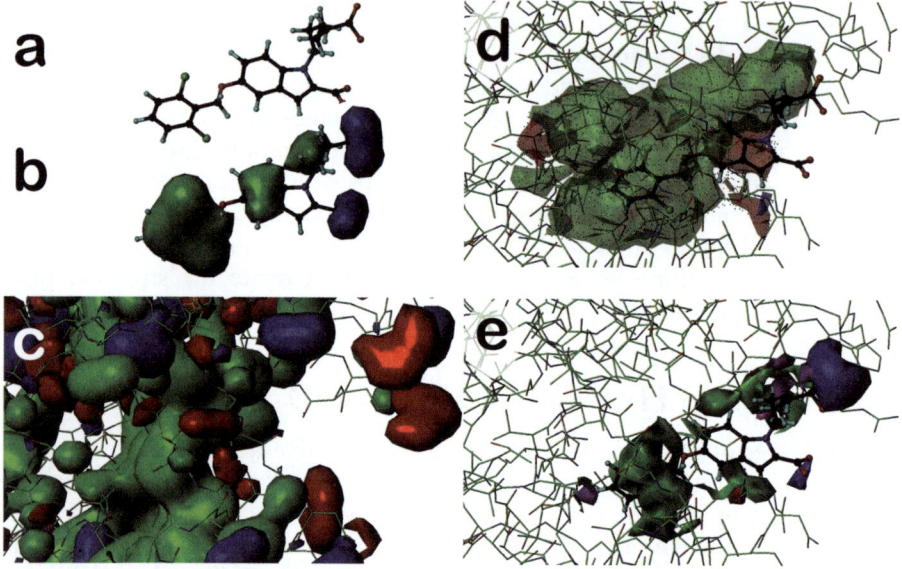

Figure 5.3   HINT maps for 1MZS. X-Ray co-crystal structure of a bacterial FabH condensing enzyme (*E. coli* β-ketoacyl-ACP-synthase III) and a small molecule inhibitor (dichlorobenzyloxyindolecarboxylic acid). (a) The inhibitor (black: carbon; cyan: hydrogen; red: oxygen; green: chlorine; blue: nitrogen). (b) HINT hydropathic map for inhibitor (green: hydrophobic; blue: polar Lewis base). (c) HINT hydropathic map for FabH protein in binding site region (green: hydrophobic; blue: polar Lewis base; red: polar Lewis acid). (d) HINT complement map for binding site (projection of site residues into cavity) with superposition of ligand (same colors). Note that the cavity is mostly hydrophobic, but does have a deeply buried region (far left) that would favor a ligand acidic group. (e) HINT interaction maps showing character [blue: favorable polar (hydrogen bond or acid/base); red: unfavorable polar (acid/acid, base/base); green: hydrophobic/hydrophobic; violet: hydrophobic/polar], loci and strength of interactions between the protein and ligand. Note that the ligand's carboxylate group, which was not in the pocket, forms a strong hydrogen bond with the protein at the pocket entrance.

However, the information encoded within a HINT hydropathic map partially correlates with both the steric and electrostatic fields; the former because, while the hydrophobic effect is a complex higher order phenomenon, the direct atom–atom London forces illustrate obvious attractions between hydrophobic atoms; and the latter because (absent sign) the HINT atom constants track with partial charge.[134] It was shown that, just as for conventional 1D and 2D QSAR, "variable" selection is required to optimize the field set applied to each data set.[144] Not surprisingly, molecular sets with significant hydrophobic character and presumed hydrophobic-driven binding are more likely to produce better models when including the hydrophobic field. The CoMSIA 3D QSAR approach of Klebe *et al.*[6] also provides a comprehensive field set including hydrophobicity.

5.10 Virtual Screening and Modeling with an Empirical Force Field

We created the HINT force field to provide an alternative, and totally empirical, approach to modeling and understanding interactions in biological media. The core information comes from an experiment that, in a sense, measures relative free energy of association between small molecules and two solvents that were intended to be prototypical of two key biological environments: water and lipid.[145] In total, the HINT force field is effective at predicting associations, *etc.*, but it is a simple model based on fairly imprecise log $P_{o/w}$ measurements and empirically determined distance relationships. Clearly, electrostatics are better handled with methods based on Poisson–Boltzmann[146] (*e.g.*, DelPhi[147] or Zap[148]) using charges derived by quantum methods and a better and far more detailed understanding of association energetics is available through free energy perturbation or QM/MM studies, but these sorts of tools are mostly inaccessible for virtual screening and database mining on a large scale due to their computational complexity and cost.

The hydropathic description of a molecule, particularly in 3D, is pragmatically useful because hydropathy encodes enthalpy and entropy, and represents steric, electrostatic and hydrophobic interactions, as well as solvation and desolvation. It is also a very fast calculation, which is an equally important attribute for database mining and virtual screening. We continue to develop tools based on this paradigm that expand its capabilities in terms of *de novo* identifying water molecules likely to be important, pre-calculating optimum ionization states (and the resulting manifold of iso-energetic representations), and accounting for flexibility in both the target and putative ligands.

References

1. R. Clark, *Curr. Top. Med. Chem.*, 2009, **9**, 791.
2. P. W. Sprague, *Perspect. Drug Discovery Des.*, 1995, **3**, 1.

3. R. W. Homer, J. Swanson, R. J. Jilek, T. Hurst and R. D. Clark, *J. Chem. Inf. Model.*, 2008, **48**, 2294.

4. T. S. Rush III, J. A. Grant, L. Mosyak and A. Nicholls, *J. Med. Chem.*, 2005, **48**, 1489.

5. R. D. Cramer III, D. E. Patterson and J. D. Bunce, *J. Am. Chem. Soc.*, 1988, **110**, 5959.

6. G. Klebe, U. Abraham and T. Mietzner, *J. Med. Chem.*, 1994, **37**, 4130.

7. M. Baroni, G. Costantino, G. Cruciani, D. Riganelli, R. Valig and S. Clementi, *QSAR Comb. Sci.*, 1993, **12**, 9.

8. M. Baroni, G. Cruciani, S. Sciabola, F. Perruccio and J. S. Mason, *J. Chem. Inf. Model.*, 2007, **47**, 279.

9. A. C. Anderson, *Chem. Biol.*, 2003, **10**, 787.

10. A. D. Andricopulo, L. B. Salum and D. J. Abraham, *Curr. Top. Med. Chem.*, 2009, **9**, 771.

11. P. M. Dean, D. G. Lloyd and N. P. Todorov, *Curr. Opin. Drug Discovery Dev.*, 2004, **7**, 347.

12. R. T. Kroemer, *Curr. Protein Pept. Sci.*, 2007, **8**, 312.

13. J. H. Lin, *Curr. Top. Med. Chem.*, 2011, **11**, 171.

14. C. Sotriffer and G. Klebe, *Farmaco*, 2002, **57**, 243.

15. M. van Dongen, J. Weigelt, J. Uppenberg, J. Schultz and M. Wikstrom, *Drug Discovery Today*, 2002, **7**, 471.

16. B. Waszkowycz, *Curr. Opin. Drug Discovery Dev.*, 2002, **5**, 407.

17. P. S. Charifson, J. J. Corkery, M. A. Murcko and W. P. Walters, *J. Med. Chem.*, 1999, **42**, 5100.

18. R. D. Clark, A. Strizhev, J. M. Leonard, J. F. Blake and J. B. Matthew, *J. Mol. Graph. Model.*, 2002, **20**, 281.

19. R. D. Clark, *Curr. Top. Med. Chem.*, 2009, **9**, 791.

20. J. D. Faraldo-Gomez and B. Roux, *J. Comput. Chem.*, 2007, **28**, 1634.

21. N. Singh and A. Warshel, *J. Phys. Chem. B*, 2009, **113**, 7372.

22. N. Singh and A. Warshel, *Proteins*, 2010, **78**, 1724.

23. F. Spyrakis, A. Bidon-Chanal, X. Barril and F. J. Luque, *Curr. Top. Med. Chem.*, 2011, **11**, 192.

24. P. Cozzini, G. E. Kellogg, F. Spyrakis, D. J. Abraham, G. Costantino, A. Emerson, F. Fanelli, H. Gohlke, L. A. Kuhn, G. M. Morris, M. Orozco, T. A. Pertinhez, M. Rizz and C. A. Sotriffer, *J. Med. Chem.*, 2008, **51**, 6237.

25. C. Hansch and T. Fujita, *J. Am. Chem. Soc.*, 1964, **86**, 1616.

26. G. Greco, E. Novellino and Y. Martin, in *Approaches to the Three-Dimensional Quantitative Structure–Activity Relationships*, ed. K. B. Lipkowitz and D. B. Boyd, Wiley, New York, 1997, p. 183.

27. A. Tropsha, in *Recent Advances in Development, Validation, and Exploitation of QSAR Models*, ed. D. J. Abraham and D. P. Rotella, Wiley, New York, 2010, p. 505.

28. A. Tropsha, in *Recent Trends in Quantitative Structure–Activity Relationships*, ed. D. J. Abraham, Wiley, New York, 2003, p. 49.

29. A. Tropsha, in *Predictive QSAR (Quantitative Structure–Activity Relationships) Modeling*, ed. Y. C. Martin, Elsevier, Amsterdam, 2006, p. 113.
30. OECD (Q)SAR project; http://www.oecd.org/document/23/ 0,3343,en_2649_34379_33957015_1_1_1_1,00.html.
31. PubChem; http://pubchem.ncbi.nlm.nih.gov/.
32. NCI; http://dtp.nci.nih.gov/docs/dtp_search.htpl.
33. WOMBAT; http://sunsetmolecular.com/.
34. BINDING DB; http://www.bindingdb.org/bind/index.jsp.
35. A. Tropsha, in *Application of Predictive QSAR Models to Database Mining*, ed. T. Oprea, Wiley-VCH, Weinheim, 2005, p. 437.
36. M. A. Lill, *Drug Discovery Today*, 2007, **12**, 1013.
37. S. M. J. Free and J. W. Wilson, *J. Med. Chem.*, 1964, **7**, 395.
38. L. B. Kier, L. H. Hall, W. J. Murray and M. Randic, *J. Pharm. Sci.*, 1975, **64**, 1971.
39. L. B. Kier and L. H. Hall, *Molecular Structure Description: The Electrotopological State*, Academic Press, New York, 1999.
40. L. B. Kier and L. H. Hall, *Molecular Connectivity in Chemistry and Drug Research,* Academic Press, New York, 1976.
41. H. Hosoya, *Bull. Chem. Soc. Jpn.*, 1971, **44**, 2332.
42. M. Randic, *J. Am. Chem. Soc.*, 1975, **97**, 6609.
43. *Handbook of Partial Least Squares*, ed. E. Esposito Vinzi, W. W. Chin, J. Henseler and H. Wang, Springer, Berlin, 2010.
44. M. Tenenhaus, V. E. Vinzia, Y.-M. Chatelinc and C. Laurob, *Comput. Stat. Data Anal.*, 2005, **48**, 159.
45. A. M. Doweyko, *J. Med. Chem.*, 1988, **31**, 1396.
46. C. D. Klein and A. J. Hopfinger, *Pharm. Res.*, 1998, **15**, 303.
47. A. J. Hopfinger, S. Wang, J. S. Tokarski, B. Jin, M. Albuquerque, P. J. Madhav and C. Duraiswami, *J. Am. Chem. Soc.*, 1997, **119**, 10509.
48. V. Lukacova and S. Balaz, *J. Chem. Inf. Comput. Sci.*, 2003, **43**, 2093.
49. A. Vedani, H. Briem, M. Dobler, H. Dollinger and D. R. McMasters, *J. Med. Chem.*, 2000, **43**, 4416.
50. H. A. Carlson, *Curr. Opin. Chem. Biol.*, 2002, **6**, 447.
51. S. J. Teague, *Nat. Rev. Drug Discovery*, 2003, **2**, 527.
52. A. Vedani, M. Dobler and M. A. Lill, *J. Med. Chem.*, 2005, **48**, 3700.
53. M. A. Lill, A. Vedani and M. Dobler, *J. Med. Chem.*, 2004, **47**, 6174.
54. J. Polanski, *Curr. Med. Chem.*, 2009, **16**, 3243.
55. C. Hansch, *Acc. Chem. Res.*, 1969, **2**, 232.
56. C. D. Selassie, in *History of Quantitative Structure–Activity Relationships*, ed. D. J. Abraham and D. P. Rotella, Wiley, New York, 2003, p. 1.
57. G. N. Burckhardt, W. G. K. Ford and E. Singleton, *J. Chem. Soc.*, 1936, 17.
58. L. P. Hammett, *Chem. Rev.*, 1935, **17**, 125.
59. L. Hammett, *J. Chem. Educ.*, 1966, **43**, 464.

60. L. Hammett, *Physical Organic Chemistry,* 2nd edn., McGraw-Hill, New York, 1970.
61. M. Charton, *Prog. Phys. Org. Chem.*, 1971, **8**, 235.
62. T. Fujita and T. Nishioka, *Prog. Phys. Org. Chem.*, 1976, **12**, 49.
63. http://www.biobyte.com/bb/prod/cqsarad.html.
64. J. J. Sullivan, A. D. Jones and K. K. Tanji, *J. Chem. Inf. Comput. Sci.*, 2000, **40**, 1113.
65. R. W. Taft, in *Steric Effects in Organic Chemistry*, ed. M. Newman, Wiley, New York, 1956, p. 556.
66. M. Charton, in *Steric Effects in Drug Design*, ed. M. Charton and I. Motoc, Springer, Berlin, 1983, p. 57.
67. M. S. Tute, in *Quantitative Drug Design*, ed. C. Ramsden, Pergamon, New York, 1990, p. 18.
68. A. Verloop, W. Hoogenstraaten and J. Tipker, in *Drug Design*, ed. E. Ariens, Academic Press, New York, 1976, p. 165.
69. A. Verloop, *The STERIMOL Approach to Drug Design*, Dekker, New York, 1987.
70. H. Meyer, *Arch. Exp. Pathol. Pharmakol.*, 1899, **42**, 109.
71. E. Overton, *Studien Uber die Narkose,* Fisher, Jena, , Germany, 1901.
72. I. M. Barclay and J. A. V. Butler, *Trans. Faraday Soc.*, 1938, **34**, 1445.
73. J. H. Ferguson, *Proc. R. Soc. London, Ser. B*, 1939, **127**, 387.
74. C. Hansch, P. P. Maloney, T. Fujita and R. M. Muir, *Nature*, 1962, **194**, 178.
75. J. N. Israelachvili, *J. Phys. Chem.*, 1992, **96**, 520.
76. G. D. Rose, A. R. Geselowitz, G. J. Lesser, R. H. Lee and M. H. Zehfus, *Science*, 1985, **229**, 834.
77. H. J. Schneider, *Angew. Chem., Int. Ed. Engl.*, 1991, **30**, 1417.
78. A. Leo, *Chem. Rev.*, 1993, **93**, 1281.
79. A. J. Leo and D. Hoekman, *Perspect. Drug Discovery Des.*, 2000, **18**, 19.
80. R. Mannhold and H. van de Waterbeemd, *J. Comput. Aided Mol. Des.*, 2001, **15**, 337.
81. A. K. Ghose and G. M. Crippen, *J. Med. Chem.*, 1985, **28**, 333.
82. T. Suzuki and Y. Kudo, *J. Comput. Aided Mol. Des.*, 1990, **4**, 155.
83. J. Devillers, D. Domine, C. Guillon and W. Karcher, *J. Pharm. Sci.*, 1998, **87**, 1086.
84. G. E. Kellogg and D. J. Abraham, *J. Mol. Graph.*, 1992, **10**, 212, 226.
85. G. E. Kellogg, S. F. Semus and D. J. Abraham, *J. Comput. Aided Mol. Des.*, 1991, **5**, 545.
86. M. J. Kamlet, P. W. Carr, R. W. Taft and M. H. Abraham, *J. Am. Chem. Soc.*, 1981, **103**, 6062.
87. J. A. Platts, M. H. Abraham, D. Butina and A. Hersey, *J. Chem. Inf. Comput. Sci.*, 2000, **40**, 71.
88. L. H. Hall and L. B. Kier, *J. Pharm. Sci.*, 1977, **66**, 642.
89. J. D. Gough and L. H. Hall, *J. Chem. Inf. Comput. Sci.*, 1999, **39**, 356.

90. V. E. Heinzen, V. Cechinel Filho and R. A. Yunes, *Farmaco*, 1999, **54**, 125.
91. H. Kubinyi, *Quant. Struct. Act. Relat.*, 1995, **14**, 149.
92. D. Chandler, *Nature*, 2002, **417**, 491.
93. L. R. Pratt, *Annu. Rev. Phys. Chem.*, 2002, **53**, 409.
94. W. Kauzmann, *Adv. Protein Chem.*, 1959, **14**, 1.
95. K. A. Dill, T. M. Truskett, V. Vlachy and B. Hribar-Lee, *Annu. Rev. Biophys. Biomol. Struct.*, 2005, **34**, 173.
96. H. S. Frank and M. W. Evans, *J. Chem. Phys.*, 1945, **13**, 507.
97. M. M. Teeter, *Proc. Natl. Acad. Sci. U. S. A.*, 1984, **81**, 6014.
98. D. W. Darnall and I. M. Klotz, *Arch. Biochem. Biophys.*, 1975, **166**, 651.
99. G. Nemethy and H. A. Scheraga, *J. Chem. Phys.*, 1962, **36**, 3382.
100. A. D. J. Haymet, K. A. T. Silverstein and K. A. Dill, *Faraday Discuss.*, 1996, **103**, 117.
101. K. A. T. Silverstein and K. A. Dill, *J. Chem. Phys.*, 2001, **114**, 6303.
102. W. Blokzijl and J. Engberts, *Angew. Chem., Int. Ed. Engl.*, 2003, **32**, 1545.
103. *The American Heritage Dictionary*, Houghton Mifflin, Boston, 1991.
104. C. Bissantz, B. Kuhn and M. Stahl, *J. Med. Chem.*, 2010, **53**, 5061.
105. R. B. Hermann, *J. Phys. Chem.*, 1972, **76**, 2754.
106. R. B. Hermann, *J. Phys. Chem.*, 1975, **79**, 163.
107. A. Leo, C. Hansch and P. Y. Jow, *J. Med. Chem.*, 1976, **19**, 611.
108. C. A. Lipinski, F. Lombardo, B. W. Dominy and P. J. Feeney, *Adv. Drug Delivery Rev.*, 2001, **46**, 3.
109. E. E. Meyer, K. J. Rosenberg and J. Israelachvili, *Proc. Natl. Acad. Sci. U. S. A.*, 2006, **103**, 15739.
110. P. Buchwald and N. Bodor, *Curr. Med. Chem.*, 1998, **5**, 353.
111. R. Mannhold and A. Petrauskas, *QSAR Comb. Sci.*, 2003, **22**, 466.
112. R. Mannhold, G. I. Poda, C. Ostermann and I. V. Tetko, *J. Pharm. Sci.*, 2009, **98**, 861.
113. G. G. Nys and R. F. Rekker, *Chem. Ther.*, 1974, **9**, 361.
114. A. Leo, P. Y. Jow, C. Silipo and C. Hansch, *J. Med. Chem.*, 1975, **18**, 865.
115. A. J. Leo, *J. Pharm. Sci.*, 1987, **76**, 166.
116. A. J. Leo, *Methods Enzymol.*, 1991, **202**, 544.
117. A. Petrauskas and E. A. Kolovanov, *Perspect. Drug Discovery Des.*, 2000, **19**, 99.
118. T. Cheng, Y. Zhao, X. Li, F. Lin, Y. Xu, X. Zhang, Y. Li and R. Wang, *J. Chem. Inf. Model.*, 2007, **47**, 2140.
119. R. X. Wang, Y. Fu and L. H. Lai, *J. Chem. Inf. Comput. Sci.*, 1997, **37**, 615.
120. R. Wang, Y. Gao and L. Lai, *Perspect. Drug Discovery Des.*, 2000, **19**, 47.
121. A. K. Ghose and G. M. Crippen, *J. Comput. Chem.*, 1986, **7**, 565.
122. A. K. Ghose and G. M. Crippen, *J. Chem. Inf. Comput. Sci.*, 1987, **27**, 21.
123. A. K. Ghose, A. Pritchett and G. M. Crippen, *J. Comput. Chem.*, 1988, **9**, 80.

124. A. K. Ghose, V. N. Vishwanadhan and J. J. Wendoloski, *J. Phys. Chem. A*, 1998, **102**, 3762.
125. K. S. Rogers and A. Cammarata, *Biochim. Biophys. Acta*, 1969, **193**, 22.
126. K. S. Rogers and A. Cammarata, *J. Med. Chem.*, 1969, **12**, 692.
127. A. J. Hopfinger and R. D. Battershell, *J. Med. Chem.*, 1976, **19**, 569.
128. G. Klopman and L. D. Iroff, *J. Comput. Chem.*, 1980, **2**, 157.
129. G. Bravi and J. H. Wikel, *Quant. Struct.-Act. Relat.*, 2000, **19**, 39.
130. D. J. Abraham and A. J. Leo, *Proteins*, 1987, **2**, 130.
131. J. Janin, *Nature*, 1979, **277**, 491.
132. J. Kyte and R. F. Doolittle, *J. Mol. Biol.*, 1982, **157**, 105.
133. R. Wolfenden, L. Andersson, P. M. Cullis and C. C. Southgate, *Biochemistry*, 1981, **20**, 849.
134. G. E. Kellogg and D. J. Abraham, *Eur. J. Med. Chem.*, 2000, **35**, 651.
135. G. E. Kellogg, J. C. Burnett and D. J. Abraham, *J. Comput. Aided Mol. Des.*, 2001, **15**, 381.
136. P. Cozzini, M. Fornabaio, A. Marabotti, D. J. Abraham, G. E. Kellogg and A. Mozzarelli, *J. Med. Chem.*, 2002, **45**, 2469.
137. M. Fornabaio, P. Cozzini, A. Mozzarelli, D. J. Abraham and G. E. Kellogg, *J. Med. Chem.*, 2003, **46**, 4487.
138. M. Fornabaio, F. Spyrakis, A. Mozzarelli, P. Cozzini, D. J. Abraham and G. E. Kellogg, *J. Med. Chem.*, 2004, **47**, 4507.
139. F. Spyrakis, A. Amadasi, M. Fornabaio, D. J. Abraham, A. Mozzarelli, G. E. Kellogg and P. Cozzini, *Eur. J. Med. Chem.*, 2007, **42**, 921.
140. A. Amadasi, F. Spyrakis, P. Cozzini, D. J. Abraham, G. E. Kellogg and A. Mozzarelli, *J. Mol. Biol.*, 2006, **358**, 289.
141. A. Amadasi, J. A. Surface, F. Spyrakis, P. Cozzini, A. Mozzarelli and G. E. Kellogg, *J. Med. Chem.*, 2008, **51**, 1063.
142. A. Sarkar and G. E. Kellogg, *Curr. Top. Med. Chem.*, 2010, **10**, 67.
143. A. Tripathi, J. A. Surface and G. E. Kellogg, *Methods Mol. Biol.*, 2011, **716**, 39.
144. G. E. Kellogg, *Med. Chem. Res.*, 1997, **7**, 417.
145. R. P. Schwarzenbach, P. M. Gschwend and D. M. Imboden, *Environmental Organic Chemistry*, 2nd edn., Wiley-Interscience, New York, 2002.
146. M. K. Gilson, A. Rashin, R. Fine and B. Honig, *J. Mol. Biol.*, 1985, **184**, 503.
147. http://www.ifm.liu.se/compchem/msi/doc/life/insight2K/delphi/delphiTOC.html.
148. http://www.eyesopen.com/zap-tk.

CHAPTER 6

Pharmacophore Models in Drug Design

VALERIE J. GILLET

Information School, University of Sheffield, Regent Court, 211 Portobello Street, Sheffield S1 4DP, UK
E-mail: v.gillet@sheffield.ac.uk

6.1 Introduction

A pharmacophore is an abstract concept introduced by Kier in the late 1960s[1] that is used to describe the steric arrangement of functional groups that enable a small molecule to bind to a receptor. The IUPAC definition is as follows: "A pharmacophore is the ensemble of steric and electronic features that is necessary to ensure the optimal supramolecular interactions with a specific biological target structure and to trigger (or to block) its biological response." Thus, a pharmacophore does not represent a real molecule; it describes the essential features for binding.

A pharmacophore is most often used when active compounds have been identified but the three-dimensional (3D) structure of the receptor site is unknown. Three-dimensional conformations of the active compounds are superimposed such that features of the molecules that could form the same interactions with a receptor are overlaid. The pharmacophore is then assumed to be the spatial arrangement of the common superimposed features. Figure 6.1 illustrates a basic pharmacophore derived from a series of 5HT2c antagonists. The process of determining a pharmacophore is known as pharmacophore elucidation. Despite considerable efforts over the last three decades, pharmacophore elucidation remains as a challenging problem. For a

RSC Drug Discovery Series No. 23
Physico-Chemical and Computational Approaches to Drug Discovery
Edited by F. Javier Luque and Xavier Barril
© The Royal Society of Chemistry 2012
Published by the Royal Society of Chemistry, www.rsc.org

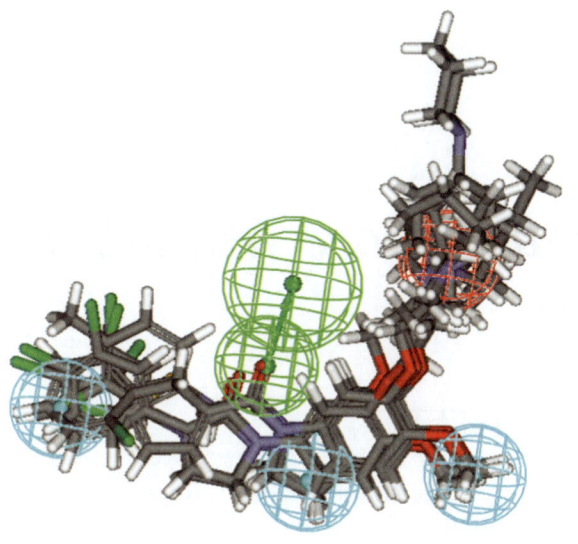

Figure 6.1 An illustration of the basic pharmacophore concept. The 5HT2c antagonists can be aligned to generate the 3D pharmacophore model shown with the following colour coding: Hydrogen bond acceptor (green), positive ionizable (red), hydrophobic (cyan). (Reprinted with permission of the American Chemical Society from A. R. Leach *et al.*, *J. Med. Chem.*, 2010, **53**, 539).

typical dataset there are usually many conformations of the ligands to be considered, together with many possible mappings between features, and the absence of the receptor is such that the problem is underdetermined.[2]

The last few years have seen a growing interest in using a protein–ligand complex to define a pharmacophore in what has become known as structure-based or receptor-based pharmacophore elucidation, to distinguish it from the more traditional ligand-based pharmacophore methods. Although these methods allow the interacting features of a single bound ligand to be identified, these features cannot be said to be those that are essential for binding since another ligand may bind in a different way.

Pharmacophore concepts are also embodied in a range of two-dimensional descriptors which describe the topological arrangement of potential binding features, usually in the form of a binary vector.[3,4] Such descriptors are commonly used in applications such as similarity searching and diversity analysis. These 2D pharmacophore descriptors will not be discussed further here.

This chapter provides an overview of 3D pharmacophore elucidation methods, focusing on the main issues involved both in program design and also in practical use of the techniques. The main applications of pharmacophores are discussed, including their use in searching for other compounds that may also exhibit the pharmacophore in a virtual screening setting, and the development of 3D quantitative structure–activity relationship (QSAR) models. Finally, several examples of the prospective application of pharmacophore methods to the discovery of novel bioactive compounds are described. Pharmacophore methods are currently of considerable interest to the computer-aided drug discovery community, as indicated by the numerous reviews that have appeared in the recent literature (see, for example, refs. 2, 5–8).

6.2 Ligand-Based Pharmacophore Elucidation

Given a set of active molecules that are assumed to bind to a receptor in a similar way, pharmacophore elucidation involves identification of pharmacophoric features within the ligands followed by alignment of the ligands such that features of the same type in different ligands are overlaid. The alignment step requires consideration of the conformational flexibility of the ligands while searching for mappings between the features. Typically there are many alignments that could be generated and so most methods also include a scoring function in order to prioritize alignments as pharmacophore hypotheses. The major components required in order to generate pharmacophore hypotheses are described below, illustrated by the first widely used programs. Some of the more recent pharmacophore elucidation programs are then mentioned briefly.

6.2.1 Feature Representation

The feature types generally include hydrogen bond donors, hydrogen bond acceptors, aromatic and hydrophobic features. Some methods also include charged feature types, although others consider these to be adequately represented as strong donors and acceptors. Often features are defined by functional groups expressed as SMARTS patterns and typically these are user-definable. The features are mapped to one or more points (which may also include associated vectors) that are used for fitting during the alignment, *e.g.* typically a least-squares fitting procedure is applied to a mapping identified between the points. Donor and acceptor features can be represented by points placed on the heavy atoms or at the presumed positions of the protein atom (also known as site points) or both. Inclusion of site points allows the generation of alignments in which the corresponding ligand atoms that could form a hydrogen bond to the same protein atom are in different locations (see Figure 6.2).

The handling of hydrophobic features varies in different programs. Several methods are based on the definitions introduced by Greene *et al.* and used in the CATALYST program:[9] a hydrophobicity value is calculated for each atom; neighbouring atoms with large values are clustered; and a hydrophobic point is used to represent each cluster. Some methods place points on all atoms that are not donors or acceptors; these can be considered as steric points that represent the shape of a molecule rather than hydrophobic points.[10] Others make a distinction between steric and hydrophobic points.[11] Hydrophobic and aromatic rings are usually represented by points placed at the centres of the rings.

Important issues to consider at the feature definition stage are tautomerism and protonation states, as these will have a direct impact on the features identified.[12] Thus, careful preparation of the molecules is required, both at the pharmacophore elucidation stage and in any subsequent database searching.

The different approaches taken to defining features and mapping them to points can lead to significant differences when using pharmacophore methods. For example, Spitzer *et al.* compared pharmacophore queries built using CATALYST, PHASE and MOE in virtual screening.[13] Despite attempting to

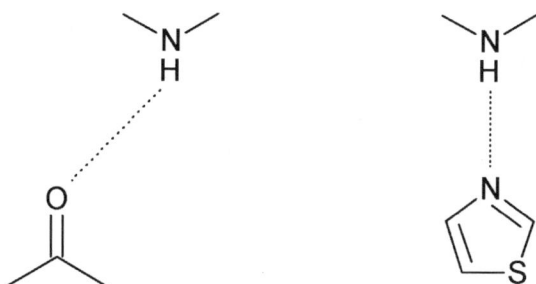

Figure 6.2 Ligand atoms can interact with the same protein atom from different positions in the binding site. Here the carbonyl oxygen and the thiazole nitrogen both form a hydrogen bond with the same protein NH group but with different geometries.

represent the same interaction types in each method, the authors found that considerable differences were seen in the hits retrieved.

Recent detailed reviews of feature definitions are provided by Wolber *et al.*[14] and by Leach *et al.*[2]

6.2.2 Alignment

The first computational approach to pharmacophore elucidation is the active analogue approach developed by Marshall *et al.*[15] This is a semi-automated method in that the features to be overlaid are assigned manually for each ligand. The computational step consists of finding a superposition of the ligands such that the specified features are overlaid. The process involves a constrained search of the conformation space of the ligands. The ligands are ordered on flexibility, with the most rigid ligand used to begin the search. A systematic search is made of the conformational space of this ligand and the distance ranges between the features are recorded. The distance ranges are then used to constrain the conformational search of the second ligand and a new set of (reduced) distance ranges is determined. This process continues with the distances becoming more and more restricted as the more flexible ligands are considered.

An early example of the application of this approach was to deduce the pharmacophore for inhibiting the angiotensin converting enzyme (ACE), which is involved in regulating blood pressure.[16] Typical ACE inhibitors contain a zinc-binding group such as a thiol or a carboxylic acid, a terminal carboxylic acid group that interacts with a basic residue in the protein, and an amido carbonyl group that acts as a hydrogen bond acceptor. The constrained systematic search derives distance ranges between the features that are accessible to all of the ACE inhibitors in low-energy conformations. A pharmacophore resulting from this analysis is shown in Figure 6.3; one of the pharmacophoric features is an extended point that corresponds to the presumed location of the zinc atom in the protein.

The first fully automated pharmacophore elucidation programs appeared in the mid-1990s, with the most well-known examples being the commercially available programs DISCO, CATALYST and GASP. These programs identify features automatically and explore potential mappings of the features simultaneously with conformations. This leads to an enormous increase in the search space compared to the active analogue approach and also a large increase in the number of plausible solutions that can be generated. In a typical application there are several features in each ligand that could potentially form interactions with a receptor, there are many different ways in which they can be mapped considering all ligands, and there are many different conformations accessible to each ligand.

As with most methods that involve 3D representations of molecules, there are two different approaches taken to handling conformational flexibility. Many programs, including DISCO and CATALYST, are based on the

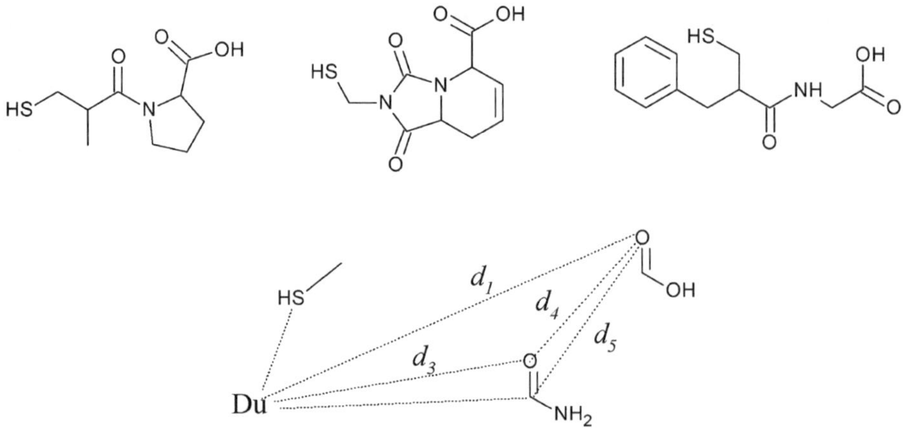

Figure 6.3 ACE inhibitors together with the features and the five distances used to define the 3D pharmacophore discovered by the constrained systematic search method. Du indicates the presumed location of the zinc atom in the enzyme (the extended or site point).

ensemble approach in which the conformers of the ligands are computed upfront and each conformer of each ligand is considered in turn when generating superpositions. In the GASP program, conformation is explored on-the-fly and a genetic algorithm (GA) is used to search conformational space of the ligands at the same time as exploring different alignments.

The alignment step in DISCO is based on clique detection, which is a common method for generating molecular alignments.[17] Each molecule is considered as a 3D graph consisting of its feature points only. Two graph representations are compared by first constructing what is known as a correspondence graph, which consists of pairs of matching nodes (*i.e.* representing features of the same type), one from each molecular graph. Edges are inserted into the correspondence graph if the inter-feature distances between nodes in the original graph are equal (within a tolerance). A clique in the correspondence graph is a completely connected subgraph and represents a mapping between features in the original graphs. Usually several different cliques exist, each representing a different mapping of the features. In DISCO, the molecule with fewest conformations is chosen as the reference molecule and each conformer of each other molecule is aligned to it in turn using clique detection to identify mappings to the reference molecule. DISCO requires that all ligands map to all pharmacophore points; however, it does allow one or more ligands to be omitted from the pharmacophore.

The HipHop module of CATALYST is based on identifying small configurations of features that are common to all ligands and then attempts to enlarge them.[18] The procedure starts by identifying all configurations (sets of points in 3D space) of two features in each ligand subject to some constraints. For example, they should be surface accessible and the distance between them should be greater than some user-defined distance. A search is then made over all conformations of all ligands and configurations that are not common to at least one conformation of each ligand are rejected. The configurations in the reference molecule are then enumerated to include all three feature configurations and again those that are not common across all ligands are rejected, and so on. CATALYST aims to identify all possible pharmacophore hypotheses, but this is subject to various parameters on conformation sampling, inter-point distance tolerances, *etc.* CATALYST allows pharmacophore hypotheses to be built in which not all features are required to be present in all ligands, *i.e.* partial pharmacophore points are permitted, and also allows one or more ligands to be omitted from the pharmacophore.

In contrast to DISCO and CATALYST, GASP takes a single conformer of each ligand as input and then explores conformational space simultaneously with alignment using a genetic algorithm.[19] The molecule with fewest pharmacophoric features is chosen as the base molecule. The GA then explores the torsional space of all the molecules together with mapping of each molecule onto the base molecule. The chromosomes are scored using a weighted fitness function which takes account of the conformational energy of the ligands, the goodness-of-fit of the features in each ligand to the pharmacophore points, and the total overlap volume of the aligned ligands. The weights on the features are user-definable. GASP finds full pharmacophore points only, *i.e.* the features must be present in all molecules in the set. In practice this has meant that datasets have had to be split into smaller subsets, each consisting of two or three ligands.

6.2.3   Scoring Functions

The large number of ways in which a typical set of ligands can be overlaid is such that a scoring function is required in order to prioritize them. The ensemble approaches tend to produce a large number of potential solutions as output, which are then scored, whereas in optimization methods such as the GA the scoring function is used to drive the optimisation algorithm towards solutions with optimal scores. Scoring functions are typically composed of a number of components which can include: a measure of how well the ligands fit the pharmacophore points; the conformational strain energy of the ligands; and the total volume of the overlay, for which smaller is better. Some methods also include a measure of the selectivity of the pharmacophore on the basis that a pharmacophore that occurs rarely within a set of random molecules is more likely to be indicative of the true requirements for activity than one that is very common.[18,20] The different scoring function components are often combined using a weighted-sum fitness function, which implies that the top scoring solution is most likely to be the correct solution. Some recent pharmacophore elucidation methods use Pareto ranking as a way of overcoming the need to assign relative weights to the different components of a scoring function.[21–23] This is in recognition of the underdetermined nature of ligand-based pharmacophore elucidation, where the absence of the receptor structure is such that it is unlikely that a scoring function would be able to find the "correct" solution as the top scoring solution.

6.2.4 Recent Methods

CATALYST, DISCO and GASP represent the first generation of fully automated pharmacophore elucidation programs. Although these programs were used widely, a comparative study of their performance over a number of datasets carried out in 2002 revealed that the state-of-the-art at that time was disappointing.[24] Since then, many new methods have been developed, some of which have superseded the earlier commercial programs and which have been designed to address specific limitations of the earlier methods.[10,20–23,25–37] Of these, current programs provided by the main software vendors include GALAHAD,[21] PHASE,[20] CATALYST[18] and the pharmacophore program within MOE.[26]

6.2.5 Evaluation of Pharmacophore Methods

A common way of evaluating pharmacophore programs is on their ability to reproduce known ligand binding modes. Patel *et al.* used this approach in their comparison of DISCO, CATALYST and GASP.[24] Five test sets were constructed, each consisting of a diverse set of ligands bound to the same protein. The "true" pharmacophore was constructed by superimposing the binding sites of the proteins to give an overlay of the ligands with the pharmacophore features determined following visual inspection. As well as evaluating the performance of the programs, the study concluded that multiple criteria should be used to judge the success, or otherwise, of the methods. Ideally all of the ligands should be overlaid exactly as seen in the crystal structures. In practice, however, this ideal is rarely seen, although useful hypotheses can be generated that are close to the ideal but which are lacking in one or more aspect. For example, the features may not overlay exactly to the true pharmacophore but they may be within an acceptable tolerance (however this is defined); it may be that some but not all of the ligands are superimposed correctly; the correct features may be identified, which can be useful for determining the relationship between structure and activity, but for the wrong conformations, which is not useful for database search, *etc.* Following its introduction, the Patel dataset was subsequently used by several groups to demonstrate the improved performance of new methods.[20,21] More recently, as more protein structures have become available, and as the need to careful curate such data sets has become more apparent, more extensive and higher quality datasets have been developed.[2,35]

Pharmacophore hypotheses can also be evaluated in retrospective virtual screening experiments performed on a database of known actives and inactives (or decoys, *i.e.* compounds assumed to be inactive). The compounds in the database can be partitioned into those that match the pharmacophore with performance assessed on how close this is to the active/inactive partitioning. Alternatively, the compounds can be scored and ranked on their fit to the pharmacophore with the ranking assessed using area under the curve (AUC) or enrichment factors. Another factor is the extent to which the pharmacophore queries are able to retrieve diverse chemotypes.[38] These issues are discussed further below.

6.2.6 Datasets for Ligand-Based Pharmacophore Elucidation

The choice of input molecules is a crucial factor in determining the likely success of identifying the true pharmacophore. Ideally the compounds should be diverse in structure so that common features are likely to be essential for binding rather than incidental, *e.g.* present due to the ligands sharing a common chemotype. At least one of the ligands should be relatively rigid. If all of the compounds are flexible, it may still be possible to find the correct mapping of the ligands; however, there is no guarantee that the correct spatial arrangement of the features will be found, making the pharmacophore of limited use in database searching. The molecules should also all bind at the same receptor site. Some methods allow some of the molecules to be excluded from a pharmacophore hypothesis, which can be beneficial if the ligands represent different binding modes.[18] Other methods have been developed that are capable of identifying more than one binding mode in a single set of input molecules.[28]

6.2.7 Validation of Pharmacophore Hypotheses

Given the underdetermined nature of ligand-based pharmacophore elucidation, due to the absence of the receptor, it is unrealistic to expect any method to produce a single correct solution. Ideally then, several plausible pharmacophore hypotheses should be considered and validated prior to their use. This could be done by leaving out some of the known active compounds to see if they would be predicted as actives by any of the hypotheses, or by carrying out some retrospective virtual screening to determine the hit rate on known actives before using the pharmacophore prospectively. As with many computational methods used in drug design, pharmacophore elucidation methods should be used in an iterative design, test and refine process. For example, newly discovered hits following the application of a pharmacophore model could be included in a second round of pharmacophore elucidation, with the aim of refining the initial hypothesis.

6.3 Application of Pharmacophores in Virtual Screening

The main use of a pharmacophore is in virtual screening, that is, to search a database of compounds to retrieve those that also contain the pharmacophore and which are then candidates for testing. A significant advantage that these approaches have over other ligand-based virtual screening methods, such as 2D similarity searching, is the diversity of the hits they retrieve. The more abstract nature of a pharmacophore, in which the focus is on generalized features (donors, acceptors, *etc.*) and inter-feature distances rather than on the atomic skeleton of structures, gives the potential of retrieving compounds that belong to different chemical series, *i.e.* "scaffold-hops".[38] Scaffold hopping is of interest for a number of reasons: there may be problems with an active series due to difficult synthesis; there may be off-target effects that need to be avoided; the known actives may have poor pharmacokinetic properties; or it may be necessary to change series in order to move into new intellectual property space.

Database searching usually involves a fast screening step prior to carrying out a full geometric search.[39,40] The screening can be based on simple feature counts where compounds in the database that contain fewer features of a given type than are present in the query are eliminated. Distance-based pharmaco-phore fingerprints can also be used which encode inter-feature distances of up to four features into a binary representation.[3,39] The pharmacophore key of the query can be quickly compared with the key of each database molecule using bit operations, and molecules that cannot contain the query can be eliminated. For those compounds that pass the screening step, their conformations must be examined to see if they can display the features in the correct geometric arrangement, within a defined tolerance, to match the pharmacophore.

As for pharmacophore elucidation, there are generally two ways of handling conformational flexibility: one is to generate ensembles of conformers and to use rigid searching; the other is to vary conformation during the search. The UNITY 3D search is based on a single input conformation and uses the directed tweak method[41] to vary the torsion angles in a database structure at search time to attempt to map it to the query. This approach can result in a large numbers of hits, depending on the strain energy that is permitted in the conformers. Several programs use ensembles of conformations where each conformation of each molecule is taken in turn and rigid mapping used to attempt to match it to the pharmacophore. An issue to consider here is the conformational sampling that is used to build the ensemble of conformers, since there is a trade-off between the size of the resulting database and, therefore, the time taken to carry out the search and the number of hits returned. Often an energy threshold will be applied to limit the conformers to acceptable strain energies. Another approach is to ensure a diverse set of conformers is generated, *e.g.* the poling approach used in CATALYST aims to identify a diverse set of conformers within a given energy threshold.[40] The CATALYST best method uses a combination of approaches where a rigid match is carried out initially using ensembles of conformers, and conformers that are close to the query are tweaked to see if they can be made to fit within a predefined energy threshold. Effective conformational sampling with respect to bioactive conformations is a much discussed topic in the literature, and a recent review of different conformer generation programs is provided by Schwab.[40]

Many of the issues encountered during pharmacophore elucidation are also issues in pharmacophore searching. Thus, among others, consideration should be given to: the definitions of the features used to characterize the molecules, such as hydrogen bond donors and acceptors; the tolerances on the distances that determine when a pharmacophore is found; and the resolution at which conformers are sampled. It is also important to ensure that tautomerism and protonation are handled appropriately. From a practical perspective, similar settings should be used for both elucidation and search, otherwise it could be that molecules used to build the pharmacophore would not be found as hits.

Pharmacophore-based virtual screening can produce large numbers of false positives, that is, compounds that exhibit the pharmacophore but that prove to be inactive. This can be due to the limited amount of information that is encoded in the query, especially if it is restricted to the pharmacophoric features. The number of false positives can be reduced by adding an excluded or an included volume to the query. Excluded volumes attempt to model the receptor site and a database compound which overlaps with the excluded volume would be excluded from the hit list. Included volumes are typically generated from the ligands used to build the query and a database molecule should be fully contained within the included volume as well as matching the pharmacophore features in order to be considered a hit. The number of false positives can also be restricted by reducing the distance tolerances that determine when a pharmacophore is deemed to match. While these techniques can be effective in reducing the number of false positives, they also run the risk of increasing the number of false negatives, *i.e.* of missing compounds that could be active. Pharmacophore-based virtual screening can also have high false negative rates due to a query not accounting for alternative binding modes. This can be a significant factor in receptor-based pharmacophores, as described below.[42]

6.4   Receptor-Based Pharmacophores

Pharmacophores for database searching can also be derived from a protein binding site either with or without a bound ligand. When the X-ray structure of a protein is available but the binding modes of potential ligands are unknown, a putative binding site can be analyzed to generate a map of potential interaction sites or hotspots that can be used as pharmacophore points. In the structure-based pharmacophore method (SBP),[43] interaction sites are identified using the rule-based program LUDI and converted into CATALYST pharmacophoric features. In contrast to the rule-based approach, the GRID program is an energy-based method for characterizing the binding sites of proteins. Various probes (atoms or functional groups) can be used to calculate the interaction energies between the probe and protein atoms at the vertices of a grid placed within the binding site. The most favourable positions for the different probes can then be used to define pharmacophores for search.[44,45]

The main issue with receptor-based pharmacophores is that a large number of interaction sites can be identified, many of which are not essential for binding and so methods are required to prioritize them, *e.g.* in order to select subsets for use in database search. This can be difficult to achieve manually and the recent HS-Pharm method was developed to automate the selection of hotspots within a binding site. It uses machine-learning techniques to discriminate between interacting and non-interacting protein atoms in order to identify hotspots from which to form pharmacophoric features.[46]

Recently, there has been a growing interest in deriving a pharmacophore from a protein–ligand complex, as in the LigandScout program.[47] This approach allows very precise queries to be defined: the features are based on known interactions; directional information can be included with features such as hydrogen bond donors and acceptors; and the search query can be augmented by an exclusion zone based on the protein cavity as an additional constraint in database searching.[39] A receptor-based pharmacophore query should be effective in reducing the number of false positives returned in a database search. However, different ligands in the same active series often make different interactions with the protein; hence a single receptor-based pharmacophore query would typically have a high false negative rate, that is, it would miss compounds that could bind *via* different interactions. One way of overcoming this limitation is to search using multiple pharmacophore queries and combine the hit lists or to base the pharmacophore query on interactions that are common in multiple protein–ligand complexes.[48]

Inclusion of such detailed information of the binding site raises the question of why not use other structure-based design methods such as docking? The value of receptor-based pharmacophore screening compared to docking lies in its speed and the simplicity of the approach. Database searching is a significantly faster process than protein–ligand docking and hence receptor-based pharmacophores can be used to filter compounds prior to a more costly docking procedure.[42,49] The relative simplicity of a pharmacophore, whereby complex intermolecular interactions are represented by the geometric arrangement of a small number of features, can also be beneficial. Tolerances on the feature distances introduce a degree of fuzziness to a pharmacophore that circumvents the issue of protein flexibility, to some extent, something that is very difficult to model in protein–ligand docking. Furthermore, although a receptor-based pharmacophore search can provide a ranking of compounds, this is usually based simply on a measure of how well a compound matches the pharmacophore; they do not attempt to estimate binding affinities which current docking scoring functions attempt but with limited success. Thus, receptor-based pharmacophore virtual screening can be used as an alternative to protein–ligand docking and as one of a number of filtering techniques in a hierarchical screening protocol, either to select compounds for the more computationally demanding docking or to filter the output from a docking run.

6.5 Pharmacophores and 3D QSAR

Some pharmacophore elucidation methods aim to develop a quantitative prediction of activity in the form of a 3D QSAR model, *e.g.* the HypoGen module of CATALYST and PHASE. These methods take both active and inactive molecules as input and attempt to find hypotheses that are common among the active compounds but uncommon in the inactives. Furthermore, the activity of each compound is estimated based on its fit to the pharmacophore, and the hypotheses are ranked on the extent to which the estimated activity values correlate with the experimental data. The models can then be used to predict the activities of previously unseen compounds. HypoGen constructs hypotheses in three steps: a constructive step, in which pharmacophore hypotheses that are present in most of the actives are identified; a subtractive step, in which hypotheses that are present in at least half of the inactives are eliminated; and an optimization step, in which small perturbations are made in an attempt to improve the scores of the remaining hypotheses. In PHASE, once an alignment has been generated, the molecules are surrounded by a grid and partial least-squares regression is used to correlate the locations of atoms or features with activity.

When pharmacophore generation is linked to the building of a 3D QSAR model, it is possible to evaluate the pharmacophores using statistical methods. Often this has been based on internal predictivities; however, it is also possible to apply the more rigorous test of prediction on an external test set. This is the approach taken by Evans *et al.*, who compared PHASE and CATALYST on eight series of compounds with known activities against different targets.[50] In each case, the compounds were divided into training and test sets. The compounds in the training sets were used to generate pharmacophore hypotheses, which were then tested on their abilities to predict the activities of compounds in the test sets. PHASE was found to outperform CATALYST on these datasets. However, even for PHASE, acceptable models were found for only four of the data sets, and it was often the case that the models with highest predictivities were not those scored highest by the programs. Moreover, in some cases the programs were found to be highly sensitive to the chosen parameters, an observation that was also reported in an earlier study, and which can make it very difficult to reproduce models reported in the literature.[51]

3D QSAR presents challenges even beyond traditional pharmacophore elucidation methods and, although having the ability to incorporate inactive molecules into pharmacophore modelling is very appealing, such molecules should be chosen with great care. Although it may be reasonable to assume that a set of active molecules bind in similar ways to a receptor and therefore share common features, inactivity can be due to a variety of different reasons.

It may be that some of the inactives contain the essential features of the pharmacophore but are inactive due to factors such as excessive entropy loss on binding or poor membrane permeability.[52] Inclusion of such inactives may have a detrimental effect and prevent the true pharmacophore from scoring highly. Also, as with all QSAR methods, it may be possible to find patterns in the data that have no physical meaning and are therefore of little or no use in prediction. Therefore, it is very important that any models produced using 3D QSAR methods are validated.

6.6 Example Applications of Pharmacophores

This section describes some successful prospective applications of pharmaco-phore methods reported in the literature. Further examples are provided by Markt *et al.*,[53] Matter and Sotriffer,[54] Hein *et al.*[42] and by Martin.[8]

6.6.1 Ligand-Based Pharmacophores

Marriott *et al.* used pharmacophore methods to identify novel molecules active against the muscarinic M_3 receptor.[55] The pharmacophore was derived using the DISCO program applied to three diverse compounds that had previously been reported in the literature. Five pharmacophore hypotheses, each containing at least four features, were examined visually and two were selected for the database search. Both pharmacophores contained a positively charged amine, a hydrogen bond acceptor and two hydrogen bond donor sites. The database searches returned 172 molecules, which were then tested experimentally; three of these showed significant activity (see Figure 6.4), with one being particularly potent and also amenable to rapid chemical synthesis.

Markt *et al.* used a ligand-based pharmacophore method to identify compounds active against the cannabinoid CB_2 receptor, which belongs to the class of G-protein coupled receptors (GPCRs) for which few X-ray structures exist.[56] Several ligand-based pharmacophore hypotheses were generated using the HipHop module of CATALYST applied to five known CB_2 agonists. Volume constraints were added to the hypotheses based on the input ligands. The hypotheses were validated using retrospective virtual screening on a dataset consisting of literature CB_2 ligands mixed with decoy compounds selected from the Derwent World Drug Index. The two best performing hypotheses were then used for prospective screening of over 900 000 commercially available compounds. The search resulted in approximately 30 000 hits, which were filtered based on physico-chemical properties. A diverse subset of 600 compounds was selected, visually inspected and 14 were chosen for experimental testing. Of the 14 compounds tested, seven were biologically active. The three most potent compounds are shown in Figure 6.5 aligned to the pharmacophore model.

Wang *et al.* developed a cannabinoid CB_1 receptor pharmacophore model based on a set of eight known active compounds that were extracted from the

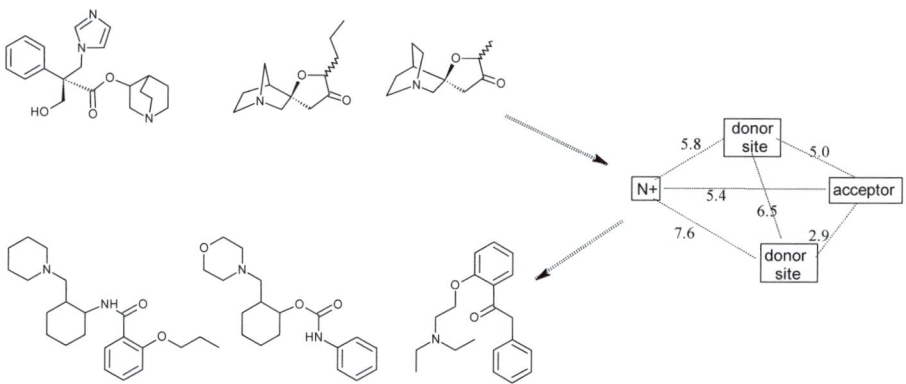

Figure 6.4 Identification of muscarinic M₃ receptor antagonists using 3D pharmacophore mapping and database searching.

literature.[57] The HipHop module of CATALYST was used and 10 pharmacophore hypotheses were generated. These were inspected visually by superimposing them onto the original compounds and one was chosen for database searching. A search was then carried out on a corporate database of half a million compounds. The number of compounds retrieved (~22 000) exceeded the screening capacity (420), and so the hits were first filtered on availability and simple physico-chemical properties. A Bayesian model was then used to prioritize the remaining compounds and a diverse set was selected for testing. Five compounds were found that showed good affinity for the receptor.

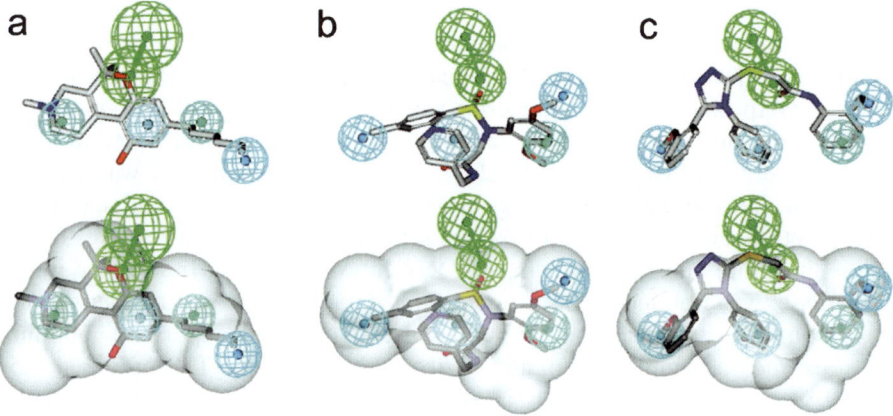

Figure 6.5 Hydrogen bond acceptor features are shown as a pair of green spheres, indicating the position of the heavy atom and the site point. The blue spheres represent hydrophobic features. The grey spheres indicate a volume constraint incorporated in the query. (Reprinted with permission of the American Chemical Society from P. Markt *et al.*, *J. Med. Chem.*, 2009, **52**, 369).

Evers and Klabunde used pharmacophore-based screening as a component of a hierarchical virtual screening protocol for which they had derived a homology model.[58] The Aventis corporate database was first filtered using simple physico-chemical properties; then pharmacophore searching was carried out using CATALYST based on models generated using HipHop; this left 22 950 compounds, which were then docked into the homology model using GOLD. The top scoring compounds were clustered using 2D fingerprints and a diverse subset was chosen for testing. Of these, 37 exhibited binding affinity below 10 μM. One conclusion of the study was that homology models can be used successfully in docking studies. However, Evers *et al.* then carried out a direct comparison of various ligand-based virtual screening methods with docking based on homology models of four biogenic amine binding GPCRs in a retrospective study.[59] The ligand-based methods included pharmacophore generation and search using the CATALYST software alongside other established ligand-based virtual screening methods. The active compounds used to build the pharmacophore models were extracted from the literature. Two pharmacophore models were built for each target, each based on a different class of compounds, and a shape constraint was included in the search query based on a representative molecule. The database used for search (and for the docking protocol) consisted of diverse subsets of actives for each of the targets identified in the MDDR database, together with inactive compounds chosen at random. The pharmacophores were found to outperform the docking in terms of the enrichment of active compounds.

6.6.2   Receptor-Based Pharmacophores

Schuster *et al.* describe the application of the receptor-based pharmacophore program LigandScout to the design of 17β-hydroxysteroid dehydrogense 1 inhibitors.[60] Two protein–ligand complexes from the Protein DataBank were selected (1equ and 1i5r). LigandScout was used to identify all protein–ligand interactions for each complex. The 1i5r model was refined in order that the features were compatible with CATALYST since this program was used to carry out the database searching. Both models were then validated using retrospective screening applied to a dataset consisting of known actives found in the literature and decoys from the Derwent World Drug Index. The initial 1i5r model was not successful in identifying hits; hence it was further refined based on the literature compounds by removing some hydrogen bonding features and merging some of the hydrophobic features. The modified 1i5r model was selected for virtual screening of a commercial database and the National Cancer Institute compound collection. Approximately 1500 hits were found and the highest scoring of these were docked into the protein binding site. A total of 14 compounds were then selected for testing based on fit to the pharmacophore, docking score and compound availability. Of these, four were found to be potent, two of which were steroidal but two were non-steroidal and therefore represent scaffold-hops.

Receptor-based pharmacophores have also been used in hierarchical screening protocols where the main computational tool is protein–ligand docking. For example, Engel *et al.* built a receptor-based pharmacophore model from a homology model of the thyrotropin releasing hormone receptor TRH-R1 using the GRID program. The pharmacophore was used to search a database of one million compounds using UNITY.[61] Relaxed tolerances were applied to the intra-feature distances in order to deal with protein flexibility and the imprecision of the homology model, and the search parameters were set to return 10% of the database as hits. A diverse subset of the hits was then docked into five different conformations of the homology model to mimic protein flexibility. A diverse set of the docking hits was then selected for experimental testing, with five antagonists found.

6.7 Conclusions

3D Pharmacophore methods have been an established technique in computer-aided drug design for many years. Traditionally their main use has been as a virtual screening technique when some active compounds are known, *e.g.* compounds identified in the literature or following in-house testing, and when the 3D structure of the receptor is unavailable. The pharmacophore can then be used to search a database to identify compounds that may also exhibit the pharmacophore and which are therefore good candidates for testing. In practice, the number of false positive can be large due to the imprecise nature of a pharmacophore, so that the number of hits returned from a pharmacophore search may exceed screening capacity. Thus, they have often been used as a component of a hierarchical screening protocol. Receptor-based pharmacophores have also become popular in recent years both as pre- and post-filters for docking runs and also as a complementary approach to docking. The issues here are somewhat different, with a greater risk of false negatives unless multiple protein–ligand complexes are taken into account to represent a variety of binding interaction patterns. As well as being used for virtual screening, pharmacophore elucidation can also be a useful tool to enable medicinal chemists to understand the key features required for activity during lead optimization. The large number of applications of pharmacophores in the literature demonstrates the success of the technique in drug discovery to date, and it is likely that pharmacophore methods will continue to play an important role in drug discovery in the future.

References

1. J. H. Van Drie, *Drug Discovery Today: Technol.*, 2010, **7**, e255.
2. A. R. Leach, V. J. Gillet, R. A. Lewis and R. Taylor, *J. Med. Chem.*, 2010, **53**, 539.
3. S. D. Pickett, J. S. Mason and I. M. McLay, *J. Chem. Inf. Comput. Sci.*, 1996, **36**, 1214.

4. G. Schneider, W. Neidhart, T. Giller and G. Schmid, *Angew. Chem., Int. Ed.*, 1999, **38**, 2894.

5. S.-Y. Yang, *Drug Discovery Today*, 2010, **15**, 444.

6. *Pharmacophores and Pharmacophore Searches*, ed. T. Langer and R. D. Hoffmann, Wiley-VCH, Weinheim, 2006.

7. Y. C. Martin, in *Comprehensive Medicinal Chemistry II*, ed. J. B. Taylor, D. J. Triggle and J. S. Mason, Elsevier, Amsterdam, 2007, vol. 4, p. 119.

8. Y. C. Martin, in *Comprehensive Medicinal Chemistry II*, ed. J. B. Taylor, D. J. Triggle and J. S. Mason, Elsevier, Amsterdam, 2007, vol. 4, p. 515.

9. J. Greene, S. Kahn, H. Savoj, P. Sprague and S. Teig, *J. Chem. Inf. Comput. Sci.*, 1994, **34**, 1297.

10. N. P. Todorov, I. L. Alberts, I. J. P. de Esch and P. M. Dean, *J. Chem. Inf. Model.*, 2007, **47**, 1007.

11. T. J. Cheeseright, M. D. Mackey, J. L. Melville and J. G. Vinter, *J. Chem. Inf. Model.*, 2008, **48**, 2108.

12. F. Oellien, J. Cramer, C. Beyer, W. D. Ihlenfeldt and P. M. Selzer, *J. Chem. Inf. Model.*, 2006, **46**, 2342.

13. G. M. Spitzer, M. Heiss, M. Mangold, P. Markt, J. Kirchmair, G. Wolber and K. R. Liedl, *J. Chem. Inf. Model.*, 2010, **50**, 1241.

14. G. Wolber, T. Seidel, F. Bendix and T. Langer, *Drug Discovery Today*, 2008, **13**, 23.

15. G. R. Marshall, C. D. Barry, H. E. Bosshard, R. A. Dammkoehler and D. A. Dunn, in *Computer-Assisted Drug Design*, ed. E. C. Olson and R. E. Christoffersen, American Chemical Society, Columbus, , OH, 1979, p. 205.

16. R. A. Dammkoehler, S. F. Karasek, E. F. B. Shands and G. R. Marshall, *J. Comput. Aided Mol. Des.*, 1989, **3**, 3.

17. Y. C. Martin, M. G. Bures, E. A. Danaher, J. Delazzer, I. Lico and P. A. Pavlik, *J. Comput. Aided Mol. Des.*, 1993, **7**, 83.

18. D. Barnum, J. Greene, A. Smellie and P. Sprague, *J. Chem. Inf. Comput. Sci.*, 1996, **36**, 563.

19. G. Jones, P. Willett and R. C. Glen, *J. Comput. Aided Mol. Des.*, 1995, **9**, 532.

20. S. L. Dixon, A. M. Smondyrev, E. H. Knoll, S. N. Rao, D. E. Shaw and R. A. Friesner, *J. Comput. Aided Mol. Des.*, 2006, **20**, 647–671.

21. N. J. Richmond, C. A. Abrams, P. R. N. Wolohan, E. Abrahamian, P. Willett and R. D. Clark, *J. Comput. Aided Mol. Des.*, 2006, **20**, 567.

22. S. J. Cottrell, V. J. Gillet, R. Taylor and D. J. Wilton, *J. Comput. Aided Mol. Des.*, 2004, **18**, 665.

23. S. J. Cottrell, V. J. Gillet and R. Taylor, *J. Comput. Aided Mol. Des.*, 2006, **20**, 735.

24. Y. Patel, V. J. Gillet, G. Bravi and A. R. Leach, *J. Comput. Aided Mol. Des.*, 2002, **16**, 653.

25. E. J. Gardiner, D. A. Cosgrove, R. Taylor and V. J. Gillet, *J. Chem. Inf. Model.*, 2009, **49**, 2761.

26. P. Labute, C. Williams, M. Feher, E. Sourial and J. M. Schmidt, *J. Med. Chem.*, 2001, **44**, 1483.
27. S. J. Cho and Y. X. Sun, *J. Chem. Inf. Model.*, 2006, **46**, 298.
28. J. Feng, A. Sanil and S. S. Young, *J. Chem. Inf. Model.*, 2006, **46**, 1352.
29. G. Wolber, A. A. Dornhofer and T. Langer, *J. Comput. Aided Mol. Des.*, 2006, **20**, 773.
30. F. Q. Zhu and D. K. Agrafiotis, *J. Chem. Inf. Model.*, 2007, **47**, 1619.
31. J. Marialke, R. Korner, S. Tietze and J. Apostolakis, *J. Chem. Inf. Model.*, 2007, **47**, 591.
32. A. V. Anghelescu, R. K. DeLisle, J. F. Lowrie, A. E. Klon, X. M. Xie and D. J. Diller, *J. Chem. Inf. Model.*, 2008, **48**, 1041.
33. D. Schneidman-Duhovny, O. Dror, Y. Inbar, R. Nussinov and H. J. Wolfson, *Nucleic Acids Res.*, 2008, **36**, W223.
34. J. Taminau, G. Thijs and H. De Winter, *J. Mol. Graphics Model.*, 2008, **27**, 161.
35. G. Jones, *J. Chem. Inf. Model.*, 2010, **50**, 2001.
36. O. Korb, P. Monecke, G. Hessler, T. Stutzle and T. E. Exner, *J. Chem. Inf. Model.*, **50**, 1669.
37. N. J. Richmond, P. Willett and R. D. Clark, *J. Mol. Graphics Model.*, 2004, **23**, 199.
38. G. Hessler and K.-H. Baringhaus, *Drug Discovery Today: Technol.*, 2010, **7**, e263.
39. T. Seidel, G. Ibis, F. Bendix and G. Wolber, *Drug Discovery Today: Technol.*, 2010, **7**, e221.
40. C. H. Schwab, *Drug Discovery Today: Technol.*, 2010, **7**, e245.
41. T. Hurst, *J. Chem. Inf. Comp. Sci.*, 1994, **34**, 190.
42. M. Hein, D. Zilian and C. A. Sotriffer, *Drug Discovery Today: Technol.*, 2010, **7**, e229.
43. P. D. Kirchhoff, R. Brown, S. Kahn, M. Waldman and C. M. Venkatachalam, *J. Comput. Chem.*, 2001, **22**, 993.
44. M. M. Ahlström, M. Ridderström, K. Luthman and I. Zamora, *J. Chem. Inf. Model.*, 2005, **45**, 1313.
45. C. Tintori, V. Corradi, M. Magnani, F. Manetti and M. Botta, *J. Chem. Inf. Model.*, 2008, **48**, 2166.
46. C. Barillari, G. Marcou and D. Rognan, *J. Chem. Inf. Model.*, 2008, **48**, 1396.
47. G. Wolber and T. Langer, *J. Chem. Inf. Model.*, 2005, **45**, 160.
48. J. Zou, H.-Z. Xie, S.-Y. Yang, J.-J. Chen, J.-X. Ren and Y.-Q. Wei, *J. Mol. Graphics Model.*, 2008, **27**, 430.
49. T. Langer and G. Wolber, *Drug Discovery Today: Technol.*, 2004, **1**, 203.
50. D. A. Evans, T. N. Doman, D. A. Thorner and M. J. Bodkin, *J. Chem. Inf. Model.*, 2007, **47**, 1248.
51. R. Kristam, V. J. Gillet, R. A. Lewis and D. Thorner, *J. Chem. Inf. Model.*, 2005, **45**, 461.

52. M. D. Hall, N. K. Salam, J. L. Hellawell, H. M. Fales, C. B. Kensler, J. A. Ludwig, G. Szakács, D. E. Hibbs and M. M. Gottesman, *J. Med. Chem.*, 2009, **52**, 3191.

53. P. Markt, D. Schuster and T. Langer, in *Virtual Screening. Principles, Challenges, and Practical Guidelines*, ed. C. A. Sotriffer, Wiley-VCH, Weinheim, 2011, p. 115.

54. H. Matter and C. A. Sotriffer, in *Virtual Screening. Principles, Challenges, and Practical Guidelines*, ed. C. A. Sotriffer, Wiley-VCH, Weinheim, 2011, p. 319.

55. D. P. Marriott, I. G. Dougall, P. Meghani, Y. J. Liu and D. R. Flower, *J. Med. Chem.*, 1999, **42**, 3210.

56. P. Markt, C. Feldmann, J. M. Rollinger, S. Raduner, D. Schuster, J. Kirchmair, S. Distinto, G. M. Spitzer, G. Wolber, C. Laggner, K.-H. Altmann, T. Langer and J. R. Gertsch, *J. Med. Chem.*, 2009, **52**, 369.

57. H. Wang, R. A. Duffy, G. C. Boykow, S. Chackalamannil and V. S. Madison, *J. Med. Chem.*, 2008, **51**, 2439.

58. A. Evers and T. Klabunde, *J. Med. Chem.*, 2005, **48**, 1088.

59. A. Evers, G. Hessler, H. Matter and T. Klabunde, *J. Med. Chem.*, 2005, **48**, 5448.

60. D. Schuster, L. G. Nashev, J. Kirchmair, C. Laggner, G. Wolber, T. Langer and A. Odermatt, *J. Med. Chem.*, 2008, **51**, 4188.

61. S. Engel, A. P. Skoumbourdis, J. Childress, S. Neumann, J. R. Deschamps, C. J. Thomas, A.-O. Colson, S. Costanzi and M. C. Gershengorn, *J. Am. Chem. Soc.*, 2008, **130**, 5115.

CHAPTER 7

Docking and Virtual Screening

GARRETT M. MORRIS

InhibOx Ltd., Oxford Centre For Innovation, New Road, Oxford OX1 1BY,
UK
E-mail: garrett.morris@inhibox.com

7.1 Introduction

In the last decade, the resoundingly successful Human Genome Project[1] has
uncovered a growing number of new therapeutic targets for drug discovery,
while new initiatives[2] and advances in high-throughput X-ray crystallography
and nuclear magnetic resonance spectroscopy have helped to elucidate a
growing number of three-dimensional structures of proteins and protein–
ligand complexes. This growth in newly discovered targets has prompted an
increased interest in structure-based design to help guide the synthesis of
biologically active molecules that could inhibit them.

Molecular docking and virtual screening have developed during the last
three decades at the fertile multi-disciplinary interface between chemistry,
biology and computer science. Despite unrealistic expectations early on,
docking and virtual screening have emerged as relatively successful and widely
used computational approaches in structure-based drug discovery. They have
been greatly facilitated by the dramatic growth in the availability, computational power and storage capacity of computers, and the free access to and
growing size of online small-molecule[3,4] and macromolecular[5-7] structure
databases.

The goal of automated molecular docking can be stated simply: to predict
both structurally the most likely binding modes, and energetically the binding
affinity, for a small molecule and a macromolecule. In virtual screening,

RSC Drug Discovery Series No. 23
Physico-Chemical and Computational Approaches to Drug Discovery
Edited by F. Javier Luque and Xavier Barril
Published by the Royal Society of Chemistry, www.rsc.org

sometimes referred to as *in silico* high-throughput screening (HTS), the goal is to rank active molecules above inactive molecules. Molecular docking is usually performed between a small organic molecule and a target macro-molecule, typically a protein, and usually an enzyme involved in a therapeutically relevant biochemical pathway. This is often referred to as ligand–protein docking, but there is a large and related field of protein–protein docking. This chapter focuses on ligand–protein docking and uses the more generic term "target" to refer to the protein, DNA or RNA macromolecule to which the small molecule or "ligand" is being docked, although terms such as "receptor" and "macromolecule" are also commonly used.

Molecular docking has a wide variety of uses and applications, including structure–activity studies, hit identification, lead optimization, providing binding hypotheses to facilitate predictions for mutagenesis studies, assisting X-ray crystallography in the fitting of substrates and inhibitors to electron density, chemical mechanism studies, and combinatorial library design. Docking has been applied in both molecular biology to identify novel chemical probes, and in drug discovery to identify hits that can be optimized into lead molecules and candidate drugs.

7.2   Molecular Docking Methods

This section focuses on the theory of molecular docking and provides an overview comparing and contrasting the methodologies of some of the most cited[8] and widely used docking tools, namely AutoDock,[9–14] DOCK,[15–17] FlexX,[18,19] GOLD[20–22] and ICM.[23,24] There are a number of excellent reviews of molecular docking methods[8,25,26] and a large number of publications comparing the performance of a variety of molecular docking tools,[27–38] often for high-throughput docking.

Since comparing docking methods is difficult,[28] and because there is evidence that some docking methods do better with certain classes of target than others,[26] it is advisable to try several docking methods to determine which method works best for the target of interest. The process of taking a known crystal structure of a complex of the target of interest, separating the ligand, and then docking the ligand back into the co-crystallized-form of the target is known as "re-docking." While it is important to compare the ability of a given docking method to re-dock a variety of ligands to the target of interest, it is arguably more important to consider how a docking method performs when cross-docking. Cross-docking involves taking a set of protein–ligand complexes and docking each ligand to each target; not only should the correct binding mode be recovered for a ligand with its cognate protein, but the scores of the ligands should correspond with experimentally observed values. In many ways, cross-docking is a better way of assessing the accuracy of a docking method than re-docking.[13,39–41] Success in binding mode prediction is often measured in terms of the root mean square deviation (RMSD) of the Cartesian coordinates of the symmetrically equivalent atoms of the ligand in the docked and experimentally observed conformations; a docking is generally regarded as successful if the RMSD is less than the somewhat arbitrary threshold of 2 Å. There are alternative (and arguably better) measures of success, such as whether the correct ligand–target interactions are recovered,[36] but these evaluation methods are harder to automate.

Docking methods fall into various categories, depending on how they treat the flexibility of the ligand and the receptor. Rigid docking methods such as DOCK[17,42] require either a single conformation or a multiconformer library for each ligand. AutoDock was the first docking method to treat the ligand as flexible during the docking,[9] and AutoDock 4 introduced side-chain flexibility in the receptor.[13,14,43] Other methods that consider backbone flexibility in the target will be discussed below. In order to explore these various possible binding modes for the ligand–protein system, the docking method requires a search algorithm, which is discussed in the next section.

7.2.1 Search Methods

Docking involves finding the most favorable binding mode(s) of a ligand to the target of interest. The binding mode or "state" of a ligand with respect to the receptor can be uniquely defined by its state variables. Each of these state variables describes one degree of freedom in a multidimensional search space, and the bounds of these degrees of freedom describe the extent of the search. In flexible docking, each independent molecule has a set of state variables that consist of its position: x-, y- and z-translations; its orientation, which can be described by Euler angles, axis-angle, or a quaternion; and, if the molecule is flexible, a set of torsion angles for each rotatable bond. Rigid body docking, where a single rigid conformation of the ligand is docked, is faster than allowing the ligand to change its conformation during the docking, partly because the search space is much smaller and partly because the software does not need to generate the new positions of the new conformation. However, if in a rigid body docking the chosen conformation of the ligand is not similar to the observed one, then there will be a lower probability of finding a complementary fit. In general, the more rotatable bonds in a ligand, the more difficult and time consuming the docking will tend to be. This is because the size of the search space increases exponentially with the number of rotatable bonds.

In addition to a search method to explore these state variables, all docking methods require a scoring function to quantify the binding mode. Some docking methods use two scoring functions, one during docking to rank candidate binding modes and another more accurate one after docking to re-score the predicted binding mode and estimate the binding affinity of the protein–ligand complex; such retrospective scoring, however, cannot affect the efficiency and accuracy of the primary scoring function.[44]

Scoring functions are discussed in greater detail elsewhere, but it is worth summarizing them here. They fall into one of four classes: empirical, force field-based, knowledge-based[25,26] or machine learning-based. The application of machine learning methods is noteworthy and represents a new class of scoring functions for evaluating docking results. By training a set of artificial neural networks on a large set of protein–ligand complexes with a wide range of known binding affinities, NNScore[45] was developed to predict whether a docked ligand was a strong binder or not and was shown to perform very well, being able to recover true actives very early in its ranked hit lists.

Search methods fall into two major categories: systematic and stochastic. Systematic search methods sample the search space at regular intervals, and are deterministic: the result of a systematic search is the same regardless of how many times the search is repeated, although the quality of the solution depends on the frequency or granularity of sampling of the search space. Owing to the combinatorial explosion in the size of the search space that occurs with increasing numbers of degrees of freedom, systematic methods are only really suitable for rigid body docking problems, although a coarser granularity can be used with larger search problems. Stochastic search methods, however, iteratively make random changes to the state variables to seek better values of the scoring function until a user-defined termination criterion is met, so the outcome of a stochastic search varies.[8]

Systematic search methods are commonly used in rigid protein–rigid protein docking, where there are only six degrees of freedom, in programs such as DOT,[46] GRAMM[47,48] and ZDOCK.[49] Stochastic search methods are more suitable for higher dimensional problems, such as flexible ligand–protein docking. Examples of stochastic search methods include Monte Carlo simulated annealing[9] and genetic algorithms.[12,20,21]

Search methods can also be classified by how broadly they explore a search space, being either local or global. Local search methods tend to find the nearest or "local" minimum energy to the starting configuration, whereas global methods search the whole state space for the best or "global" minimum energy. Hybrid global–local search methods have also been developed, where an iteration in the global search algorithm is followed by a local search. These hybrid methods have been shown to perform even better than global methods alone, being more efficient and able to find lower energies than either global or local search methods alone.[12]

In AutoDock 4, for example, there is the choice of two local search methods (Solis and Wets[50] and Pattern Search[51]), two global search methods [Monte Carlo (MC) simulated annealing (SA)[52] and the genetic algorithm (GA)],[53–55] and two hybrid global–local search methods [the Lamarckian GA (LGA), where the GA can be hybridized with either the Solis and Wets or Pattern Search local search methods].[12] AutoDock Vina[56] uses a simplified scoring function that can be differentiated analytically, and is thus able to be used with a local search method that takes advantage of gradient information, helping to accelerate the docking.

MCSA algorithms[42] begin with a random initial state (although it is possible to specify user-defined binding mode for a ligand) and proceed to generate a new binding mode by randomly modifying the state variables. If the new state has a lower energy, E, or better value of the scoring function, then this move is automatically accepted; if the energy is higher than the energy of the last state, however, it is accepted probabilistically according to the Metropolis criterion, $\exp(-\Delta E/k_\mathrm{B}T)$, which is dependent on the annealing temperature, T (k_B is the Boltzmann constant). Initially, the temperature is high and moves with large increases in energy tend to be accepted, but as the search proceeds the temperature is gradually lowered and the search becomes more local. MCSA typically terminates after a user-defined number of moves.

Genetic algorithms use a population of individuals; in docking, each individual represents a candidate binding mode. Each individual has an associated chromosome consisting of a sequence of "genes" that correspond to the state variables describing the candidate docking. A search can begin with a random initial population, or it can be "seeded" with an individual with user-defined genes. The GA proceeds by performing a series of genetic operations on the population, followed by selection of the best individuals to go through into the next generation. Fitness in this context is the scoring function value for a candidate docking. The genetic operations include some form of crossover between two parent chromosomes, typically a two-point crossover, where the sequence of genes of two parents are split into three sub-sequences of genes, and the second sub-sequence of genes is exchanged to generate two new offspring. It is crossover that allows the GA to make large jumps in the search space; without crossover, the search becomes an evolutionary algorithm (EA). Another common genetic operator is mutation, where a gene in a single individual is chosen at random and modified by some random amount. A GA terminates upon reaching some maximum number of generations, a maximum number of scoring function evaluations, or convergence of the population of docking solutions on a single binding mode.

Other biologically inspired search algorithms have been used in docking, such as the particle swarm optimization (PSO) algorithm in SODOCK[57] and pso@autodock,[58] and ant colony optimization (ACO) in PLANTS.[59] These "biological" approaches tend to use a population of candidate docking solutions, and have been shown to explore larger search spaces quite efficiently.

Of the other widely cited docking algorithms, DOCK uses an isomorphous subgraph matching method[60] for the rigid-body orientating component of its search to match chemical features between the ligand and the negative image of the binding site. FlexX matches ligand features with complementary interaction sites, building up the complete ligand by adding fragments of the molecule in a stepwise fashion. GOLD's global search method is a GA. The search method of ICM combines a biased MC procedure and a local energy minimization. Glide, another widely used docking method, uses a hybrid systematic–stochastic approach, by initially performing a rough systematic search to reduce the search space, followed by torsionally flexible energy optimization of the ligand, with the best binding modes being further refined using a Monte Carlo search.[33]

When setting up a docking, it is necessary to define the search space, which will depend on how much is known about the target: if there is no previous information regarding the location of the binding site, then the allowed search space should encompass the entire surface of the receptor. This is known as "blind docking," and has been shown to be successful with AutoDock.[61–63]

If the docking tool cannot feasibly explore a search space large enough to encompass the whole target, either because of limitations of the program or the time it would take, then probable sites such as cavities large enough to contain the ligand(s) should be investigated separately. These potential binding sites can be identified automatically using a number of binding pocket identification methods, such as AutoLigand.[64] Alternatively, the third-party tool BDT[65] can be used to set up staggered grid boxes for AutoGrid covering the entire protein; each grid box can then be explored with independent docking runs on a computational cluster. The second possibility is that there is either *a priori* information about the binding site, such as structures of complexes of ligands with known binding modes, or biochemical knowledge of active site residues and/or mutagenesis data; in such cases the search space can be centered on these key residues and reduced in size to focus on the region of interest, thus simplifying the search problem. It should be noted that some docking packages support constraints, *e.g.* GOLD[22,66] can include distance constraints to ensure a distance between a specified ligand atom and a protein atom (or between two ligand or protein atoms), which can be constrained to lie between minimum and maximum distance bounds.

7.2.2   Target Validation and Preparation

One of the most important aspects of docking and virtual screening is the selection, validation and preparation of the target structure. Ideally, the target structure should be experimentally determined, usually by either X-ray crystallography or nuclear magnetic resonance. Docking has been performed successfully against homology models,[67–70] although the reliability of the docking results depends heavily on the quality and bias of the homology model. The quality of an X-ray structure can be gauged from the resolution of the structure: usually a value less than 2.5 Å is preferred.

When validating a docking method using crystal structures of ligand–protein complexes, it is important to evaluate the quality of the ligand coordinates, in particular its bond lengths, bond angles and torsion angles; this can be done using Mogul,[71,72] a crystallographic analysis tool that exploits the structural data of the CSD.[3] Where possible, the coordinates of the atoms in the ligand assigned by the crystallographer should be compared with the original experimentally observed electron density. Sometimes these do not correspond, perhaps because the ligand contains an unusual value for a torsion angle; the Uppsala Electron Density Server[73] is a useful source of such electron density data. The Astex Diverse Set is an example of a small curated set of high-quality X-ray crystal structures of ligand–protein complexes.[74] ReLiBase+[5,75] can be used to assess the quality of a receptor–ligand complex, and provides specific information about another important aspect of ligand–protein docking discussed below, namely the structure of water in the complex; it can indicate "dubious" water molecules (sometimes a crystallographer mistakes an isoelectronic metal cation for a water molecule) and provides descriptors for individual water molecules and water-mediated protein–ligand contacts. ReLiBase+ can also indicate whether a ligand's binding mode is influenced by crystal packing artifacts.

In X-ray crystal structures, each atomic position has a corresponding temperature factor or *B*-value; this value indicates the atom's degree of structural mobility. Atomic positions that may be suspect are those with *B*-values higher than their surroundings. Incomplete side chains also occur sometimes, since the crystallographer did not have unambiguous electron density to assign a definite position. Furthermore, in some crystal structures, alternate locations of given side-chain atoms may be observed. In such cases, both alternative positions must be considered in subsequent docking calculations, particularly if these occur near a putative binding site.

In some cases, the biologically relevant form of the target structure, the biological unit, is a multimer, which means that the appropriate symmetry-related molecules must also be included in the coordinates of the target structure. Online databases such as the PDB[7] and Binding MOAD[6] provide target structures as a biological assembly suitable for docking studies.

Target selection involves gathering all available structures of the target, ideally with bound ligands, from publicly available sources such as the Protein Data Bank,[7,76] ReLiBase[5,75] and Binding MOAD.[6,77,78] It is preferable to use only high-resolution structures where available, ideally better than 2.5 Å. The structures should also be verified to contain the necessary cofactors, if any are required for biological activity. Structures that are incomplete or missing side chains should also be disregarded, or repaired if possible. If there is more than one target structure, they should ideally be overlayed by superimposing key residues in the binding site or region of interest using a least-squares superimposition method. SwissPdbViewer[79,80] is a freely available tool that offers several automated superimposition options under its "Fit" menu, such as "Magic Fit" and "Fit molecules (from selection)", and can also automatically reconstruct missing atoms in incomplete side chains. The extent of the conformational variability of the target should be determined. If the structures are very similar, then a representative should be selected. This is an important step and can greatly affect the success of a virtual screening experiment. A representative structure or "leader" for a 90% homology family of structures is already pre-calculated and available from Binding MOAD.[6,77,78] If there is a large degree of structural variation, and the docking method or computational resources allow, it is preferable to dock to several diverse conformations of the receptor, to decrease the chance of identifying false negatives.

For many docking methods, hydrogen atoms should be added to the target at the desired pH. Under physiological conditions at pH 7.2, the following residues would be expected to have ionized side chains: arginine, lysine, aspartic acid and glutamic acid. AutoDockTools (ADT) can be used to add hydrogen atoms; all hydrogens are required for the initial Gasteiger partial charge calculation, but the non-polar hydrogens will be merged later on. Each histidine side chain can be either neutral or positively charged at physiological pH. If it is neutral, either the delta or the epsilon nitrogen can be protonated. AutoDockTools offers a tool to help set the desired protonation state of each histidine side chain. Which protonation state it adopts will depend on its environment in the target.

The atomic assignments of imidazole rings in histidine and amido groups in asparagine and glutamine side chains can be ambiguous, since electron density is rarely if ever clear enough to distinguish carbon, nitrogen and oxygen atoms, and hydrogen atoms are never seen in X-ray crystallography. Tools such as REDUCE and its web interface, MOLPROBITY[81,82] and MOE's Protonate3D[83] can evaluate 180° flips of these groups to optimize the hydrogen-bond network and add hydrogen atoms appropriately. In addition, MOE's Protonate3D can optimize the placement of hydrogens on water molecules and ligands, and can also consider alternate protonation and tautomeric states of the ligand and cofactors.[83] With AutoDock, all water molecules except conserved water molecules should be removed from the structure; the ligand should also be removed, if a crystal structure of a complex is being used. If using the "wet ligand" technique[84] for treating bridging waters in AutoDock, all waters in the target should be removed. Consolv[85,86] can be useful in deciding which water molecules to keep in the target structure, since it can be used to predict which waters on a protein are likely to be conserved or displaced in other, independently solved crystallographic structures of the same protein.

Docking methods that do not use a force field, such as FlexX and GOLD, do not require partial charges to be assigned to the atoms in the ligand and receptor molecules. AutoDock 4 and UCSF DOCK, on the other hand, use scoring functions derived from the AMBER force field and, therefore, require partial atomic charges. The AutoDock 3 scoring function was calibrated using Kollman united-atom partial charges on the macromolecule and Gasteiger PEOE charges on the ligand, but in AutoDock 4, where the moving atoms during a docking could include flexible side chains in the protein, it uses Gasteiger PEOE charges for both ligand and macromolecule. Other AutoDock users have investigated the use of alternative methods of calculating partial charges on the ligand, *e.g.* Evans and Neidle concluded that the best charges to be used in AutoDock 3 for virtual screening of DNA minor groove binders came from calculations using AMSOL[87,88] with the AM1-CM2 Hamiltonian for non-polar organic solvents.[31]

There is evidence that the use of more accurate partial charges can help to improve the prediction of whether a ligand will bind or not, as demonstrated in various model binding sites in artificial mutants of T4 lysozyme.[89] More accurate treatment of partial charges can also help to improve binding mode prediction: if the ligand is allowed to polarize the target macromolecule by using quantum mechanical calculations to model both the ligand and nearby protein atoms in a candidate docking result, using these charges in subsequent dockings tends to improve binding mode prediction.[90]

Partial charges should be calculated or assigned if required by the docking tool (see Table 7.1 below). Some tools may use a dictionary of amino acid partial charges to simply assign the charges based on atom types. If there are any cofactors in the target structure, it will be necessary to compute the appropriate partial charges for these too. When using AutoDock 4 and AutoDock Vina, it is necessary to "merge" non-polar hydrogens, because they use a united-atom representation (see Table 7.1); merging involves adding the partial charge of the hydrogen atom to its parent non-polar carbon atom and then deleting the hydrogen atom. AutoDockTools (ADT) can be used both to calculate the partial charges and to merge the non-polar hydrogens on the target.

To accelerate the scoring calculation, some docking methods adopt a method first introduced by Goodford[91] to pre-calculate interaction energies for a range of atom types and functional groups with the target and then store these as grid maps. These can then be used with a trilinear interpolation method to calculate interaction energies between the atoms of the ligand and the target. A set of grid maps for a given receptor can be reused for parallel docking of a library of ligands, also saving time. For AutoDock 4, one grid map needs to be computed for each atom type in the ligand or set of ligands being docked, in addition to electrostatic potential and desolvation grid maps. These AutoDock 4 maps are pre-calculated by AutoGrid and stored separately, but the use of external grid maps also gives AutoDock 4 the

Table 7.1 Ligand input requirements for some of the most commonly cited docking software.

Docking tool	Auxiliary tools	File format	Hydrogens	Partial charges
AutoDock 4.2	AutoGrid, AutoDockTools, BDT	PDBQT	United Atom	Gasteiger PEOE[a]
AutoDock Vina	AutoDockTools	PDBQT	United Atom	None
DOCK 6	Chimera, Grid, Docktools, Nchemgrids, Sphgen, ANTECHAMBER	mol2	Explicit or United Atom	AM1-BCC, Gasteiger
FlexX 2	FlexV	mol2, SD	United Atom	Formal charges
GOLD 5	Hermes, GoldMine	mol2	Explicit	None
ICM 3.4	ICM-Pro, ICM-VLS	mol2, SD	Explicit	MMFF, ICM

[a]Alternative partial charge calculations have been used, e.g. AMSOL[87] with the AM1-CM2 Hamiltonian.[31]

versatility to be able to use maps computed by other methods and scoring functions, such as DrugScore[92] and AFMoC(obj).[93]

7.2.3   Target Flexibility

Many docking tools do not allow the target to be flexible, although this is a very important aspect of molecular recognition.[94,95] A target macromolecule may adopt different conformations in the unbound and bound states and may adopt different conformations with different ligands; examples of classes of proteins with varying degrees of structural change upon ligand binding are surveyed in Gunasekaran and Nussinov.[96] To tackle this and to explore the possibility of identifying allosteric binding sites, molecular dynamics has found an increasing number of applications in conjunction with molecular docking. These range from preparing the target before docking to accounting for receptor flexibility, solvent effects and induced fit, to calculating binding free energies and ranking docked ligands.[97] The so-called Relaxed Complex Scheme developed in the laboratory of McCammon[98] generates snapshots from molecular dynamics simulations[99–101] of the *apo* form of the target, and then applies a docking method such as AutoDock to dock the ligand of interest; the technique effectively takes into account induced fit and has been applied to develop novel inhibitors of HIV integrase.[102] Both FlexE[103] and GOLD[20–22,66,74,104,105] can also perform ensemble docking, using a super-imposed set of different conformations of the target.

Flexibility of the receptor can also be taken into account in AutoDock by combining the pre-calculated grid maps,[106] and this approach can also address another important aspect of ligand–protein docking, namely the presence or absence of water molecules in the active site. Much more accurate methods for treating the presence or absence of water molecules exist, such as inhomogeneous fluid solvation theory[107] and JAWS,[108] but these are time consuming and do not lend themselves to high-throughput docking and virtual screening.

7.2.4   Ligand Selection and Preparation

The type of ligands chosen for docking or virtual screening will depend on the goal. For lead discovery, crude filters such as molecular weight, polar surface area, solubility, total charge, commercial availability and price-per-compound can be applied to reduce the number of molecules to be docked. For lead optimization, filters such as similarity thresholds, pharmacophores, synthetic accessibility and absorption, distribution, metabolism, excretion and toxicology (ADME-Tox) properties are additionally applied. For focused lead optimization, and to inform optimal library design and to prioritize medicinal chemistry efforts, a custom virtual library is often constructed for docking of analogues that are related to the lead compound(s) and based on organic synthetic reactions available to the medicinal chemists.[109] For drug discovery,

Lipinski's so-called "Rule of Five"[110] provides guidelines to help reduce a set of ligands to a drug-like set of potentially more bioavailable molecules, although lead-like rules have also been devised to improve the success rate in lead discovery.[111] It should be stressed that these are rules of thumb and there are examples of drugs that violate these rules, such as HIV protease inhibitors which have molecular weights greater than 500.

No matter which docking method is selected, the user needs to prepare the appropriate input files. These will depend on the docking method used and, in particular, on the molecular representation used in that method. To assist the user in setting up a docking and in performing post-docking analysis, many docking programs include auxiliary tools, scripts and graphical user interfaces (GUI); Table 7.1 summarizes some of these.

AutoDock uses a United Atom model for the ligand and receptor, in which only polar hydrogens are present. It also requires partial atomic charges to be assigned to the ligand. The AutoDock scoring functions were calibrated using Gasteiger charges[112] on the ligand; thus, to use the scoring functions correctly, the ligand must be assigned Gasteiger partial charges. (It should be noted that alternative charge calculation methods for ligands have been successfully used in AutoDock.[31])

Most docking tools treat ligands as conformationally flexible during the docking, with the exception of flexible rings which are treated as a rigid fragment. The latest version of GOLD uses ring templates with conformations extracted from the approximately half a million X-ray crystal structures of small organic molecules in the Cambridge Structural Database.[3] Other methods compute the conformations of flexible ring systems during the docking, as can be done using FlexX.[113] By using appropriate parameters, it is also possible to model flexible rings in AutoDock.[114] Another strategy to explore ring flexibility is to perform conformational analysis on any ring-containing ligands prior to docking, which has been used extensively to study carbohydrate binding.[115–117]

In some cases, the small molecule may adopt more than one tautomeric or protomeric state, which can dramatically affect its physico-chemical properties and could make the difference between correctly and incorrectly predicting its binding mode. Software such as Epik[118–120] help to predict the pK_a and the most likely tautomeric state(s) of a ligand at a given pH.

Most docking tools require a 3D structure for each ligand, including explicit hydrogens. Depending on the source of the ligands (available molecules, molecules that have yet to be synthesized or vendor libraries), the steps required to process the molecules will vary. The following steps exemplify how to obtain these structures and how to process them for use in AutoDock.

ZINC[4,121] is one of the largest freely accessible collections of commercially available compounds; it claims it has 13 million purchasable compounds in its database. ZINC is a useful resource for molecular docking since it provides 3D structures of the small molecules in SYBYL MOL2 format. Subsets of compounds can be created by composing a query that specifies filtering criteria

for both molecular properties and two-dimensional (2D) chemical substructures. Services such as MolPort[122] and eMolecules[123] currently offer
approximately 9 million and 8 million commercially available compounds,
respectively, for searching in their online databases.

Ligands in the form of SMILES strings[124] can be converted into full 3D
atomic coordinates, including hydrogens, using tools such as CORINA,[125]
MOE[126] or ZINC.[4] Ligands in 2D SD format[127] can be converted into full 3D
atomic coordinates using CORINA,[125,128] MOE[126] or Ghemical.[129–131]
Ghemical can be used to sketch the ligand in 3D and then perform energy
minimization, molecular dynamics or a conformational search to identify low-
energy conformations. PRODRG[132–134] can take PDB format, MDL MOL files
or even ASCII text drawings of the molecule, instead of SD format. PRODRG is
available as a standalone executable or as a web service, where the user can
sketch the molecule in 2D and then convert the molecule into 3D (PRODRG is
also convenient for AutoDock 3, because it outputs PDBQ format).

It is important that the protonation, tautomeric and stereoisomeric forms of
the ligand be correct, otherwise subsequent calculations will be highly suspect.
The enumeration of all possible ligand tautomers can be achieved with such
programs as QUACPAC[135] (Open Eye), TAUTOMER (Molecular Networks)
and Schrödinger's LigPrep and Epik.[118,120]

When preparing ligands for AutoDock 4, either the GUI or command line
versions of AutoDockTools (ADT) can be used to set up the necessary input
files. The first step for AutoDock is to calculate Gasteiger partial charges[112]
and assign AutoDock atom types to each atom in the ligand. It is possible
manually to define the "root" of the torsion tree and choose interactively
which bonds should be rotatable during a docking using ADT, or the
command line script prepare_ligand4.py can be used to set automatically all
rotatable bonds to be rotatable. In ADT, the "Detect Root" option
automatically examines all the rotatable bonds in the ligand and chooses the
atom that is nearest to the center of the torsion tree. The "Choose Torsions..."
option in ADT displays all rotatable bonds as green or magenta, indicating
that they are active or inactive, respectively. Clicking on these bonds toggles
whether they are active or not. AutoDock 4 requires the ligand to be in
PDBQT format, which is very similar to PDB format but also includes the
partial atomic charge and the AutoDock atom type for each atom (ADT can
be used to "Save as PDBQT...").

7.2.5 Evaluating Docking Results

Regardless of the ligand–protein docking tool used, docking results should be
evaluated by considering its ability to reproduce known crystal structures and,
for unknown complexes, the shape and chemical complementarity between the
ligand and nearby atoms in the protein. Are all possible hydrogen bond donors
and acceptors in the ligand satisfied? Are the charged groups in the ligand
interacting with oppositely charged side chains in the receptor, or are they

buried in hydrophobic pockets? Are hydrophobic groups in the ligand buried in hydrophobic pockets in the receptor? An excellent review of molecular recognition in biological systems can be found in Bissantz *et al.*[136] When the structure of the ligand–protein complex is known, the parameters chosen for the docking can be judged by the docking tool's ability to reproduce the binding mode of a ligand to protein. The criterion usually used is the all-atom RMSD between the docked position and the crystallographically observed binding position of the ligand; success is typically regarded as having an RMSD from the crystallographic position of less than 2 Å.

When using stochastic methods, it is recommended that the experiment be repeated at least 10 times with different initial conditions, but a better idea of the conformational preferences can be obtained by repeating 50–100 times. The similarity of the predicted binding modes can be assessed by computing a matrix of pairwise RMSD values, sorting the docking results by energy or scoring function, and clustering docked conformations according to an RMSD threshold, typically 2 Å. If all of the dockings fall into one cluster, this indicates that the search parameters were sufficient for each docking to converge on the lowest value of the scoring function. If there is no clustering at all, then the dockings should be repeated but with increased sampling, either increasing the number of iterations per search or, if the method is population based, increasing the population size.

If the scoring function were perfect, the docked conformation with the lowest energy or best score would always correspond to the crystallographically observed binding mode, assuming that there are no bad contacts in the crystal structure. This is not always the case, and sometimes a slightly higher energy (slightly worse scoring) binding mode is observed more often than the lowest energy binding mode. Furthermore, almost all current docking methods tend to find the binding mode with the lowest possible interaction energy or best score for a given ligand, but this score does not necessarily indicate whether the ligand will even bind to the target.

In the last decade, there has been a growing realization that fragment-based drug design is a more effective way of exploring chemical space[137–139] than performing high-throughput screening using combinatorial libraries.[140] Recent work has shown that docking followed by rescoring using the implicit solvation model MM-GBSA (Molecular Mechanics/Generalized Born Surface Area) can improve the correlation between predicted binding scores and experimental data;[141] this approach has also proven to help improve the prediction of the binding modes of fragments.[142]

There has been growing interest in developing methods to distinguish binders from non-binders. One of the earliest reports that used docking successfully to discriminate binders from non-binders[143] proposed a simple, non-physical metric that summed the mean binding energy for all of the conformations in the binding mode cluster, and the total number of conformationally distinct clusters found out of 100 dockings. The more clusters and the weaker the mean energy, the less likely the ligand was to bind.

By building on statistical mechanical foundations, new methods have emerged to quantify this clustering phenomenon to estimate the contributions of translational and rotational entropy to binding affinity, by approximating the configurational entropy using the sizes of the clusters.[144–146]

After a high-throughput docking or virtual screening run, the results are gathered together and sorted, and the top scoring hits are selected for subsequent synthesis or purchase followed by biological assay. Many scoring functions, however, are additive and tend to give ligands with more atoms a better score. In order to normalize for this, the heavy atom ligand efficiency[147] can be readily calculated, by computing the ratio of the predicted binding affinity to the number of heavy atoms in the ligand; this can give a second set of top hits, and the union of these two sets is then used. A variety of methods for combining hit lists from docking, and from virtual screening, are discussed in the next section.

7.3 Virtual Screening

Virtual screening is the process whereby a large number of small molecules or "ligands" are evaluated computationally for their similarity to known biologically active molecules, or their likelihood to bind to a target.[148] Virtual screening is based on the similar property principle, which states that molecules with similar structures have similar properties.[149] Virtual screening can be performed using 2D descriptors,[150] 3D (shape-based) descriptors, experimental or calculated physico-chemical properties, and cheminformatic fingerprints of the ligands, and can be very useful in itself in finding hits and leads through library enrichment.[151] Since virtual screening using pre-calculated descriptors tends to be faster than molecular docking, it is often used as a computational filter to reduce the size and "enrich" the library of small molecules before molecular docking.

Ligand-based virtual screening uses only information about biologically active small molecules and no information about the structure of the target; indeed, the target may not be known. It has been reported that ligand-based virtual screening does as well as, if not better than, high-throughput docking; see, for example, Hawkins *et al.*[152] However, molecular docking has the advantage that when the structure of the target is known, it helps to provide 3D structural hypotheses of how a ligand may interact with its target, which is invaluable for lead optimization. Molecular docking is often used as the final stage after virtual screening. Virtual screening has been reported to have better rates of ligand discovery than *in vitro* high-throughput screening, which has thus prompted heightened interest in its use in hit identification.[153] Machine learning methods have also been used to great effect in virtual screening,[154,155] being trained on sets of active and inactive small molecules and then applied to large databases of small molecules to identify quickly potential hits.

The effectiveness of virtual screening has been assessed in a variety of ways,[156] but one of the most common metrics used is the enrichment rate, ER, or enrichment factor, usually quoted at some percentage of the ranked hit list of molecules that has been screened, typically 1%, 5% or 10%. In docking, the ligands are ranked by decreasing scoring function value, and in virtual screening by decreasing similarity. ER measures the ability to separate a small subset of active compounds from the remaining inactive molecules. If the screening method is perfect, then all the known active molecules will be ranked at the top of the list; if it is not, then the active molecules will be distributed randomly throughout the list. An alternative metric looks at the area under the curve (AUC) of a receiver operating characteristic (ROC) curve.[157] Here the true positive rate is plotted against the true negative rate as the library is traversed in rank order. An ideal method would have a ROC curve with an AUC of almost unity and would ascend as rapidly as possible near the *y*-axis, as all the true positive actives are ranked first, and then flatten out immediately as the remaining inactive molecules are encountered.

Metrics such as the AUC of a ROC curve may give a false impression of the effectiveness for a given high-throughput docking method or virtual screening approach, because the most important part of the ranking is near the top of the ranked list, where "early enrichment" is desirable. These shortcomings led to the development of a superior metric called the Boltzmann-enhanced discrimination of receiver operating characteristic (BEDROC),[158] which combines the discrimination power of the robust initial enhancement (RIE)[159] metric and the statistical significance of ROC curves.

Often, the best results in virtual screening can be found by using several different virtual screening methods run on the same active ligand.[160] The hit lists can be combined using data fusion techniques,[161] which have tended to give better enrichment rates than the individual methods alone; data fusion is also the same principle behind consensus scoring in docking methods. Data fusion rules define how the rank of a given compound in the hit lists from the various virtual screening techniques should be combined; with the SUM rule, a compound's ranks are simply added together to give its new rank position in the new fused hit list.[162] Alternative data fusion rules have been investigated, such as taking the maximum rank, or MAX rule, which chooses the maximum rank among the various hit lists, or the MIN rule, which uses the minimum rank.[163–165] More recently, a probabilistic framework was described that combines the results of both structure-based and ligand-based screening using a cumulative belief data fusion rule, along with probability assignment curves (PAC), empirically derived curves that relate a given metric to the probability of a compound being active.[166] This novel approach provides a quantitative probability that a given molecule will be active and increases the diversity of recovered actives.

Improved enrichment can also be achieved even when there is only one active molecule known. By identifying top scoring hits by virtual screening using just the initial active, and then repeating virtual screening on the most similar molecules found in the first screening, a more enriched set of ligands can be identified; this is called turbo similarity searching.[167]

There is some debate about the value of using 3D descriptors in virtual screening *versus* using 2D descriptors. The former is based on the shape of the molecule, while the latter is based on the chemical structure and functional groups present in the molecule. Molecular recognition occurs between three-dimensional molecules, so it would be reasonable to expect 3D methods to perform better than 2D ones. A comparison of the ability of 2D and 3D descriptors to distinguish between active and inactive molecules using different clustering methods[168,169] showed that 2D MACCS fingerprints were the most effective. A related study looking at the effectiveness of various descriptors in accurately predicting the property of a structure, given known values of other structures, showed a similar trend.[170] This apparent strength of 2D descriptors may be due to the biased nature of the validation sets used in these comparisons: the actives often consist of chemically related analogues synthesized as part of a single lead optimization effort (so-called analogue bias[171]). A second problem with many validation sets is artificial enrichment, caused by actives being too dissimilar from the decoys.[172] Furthermore, such decoys are often putative inactives: experimental data about experimentally confirmed inactive molecules is not published as often as for active molecules.

On the 3D side, poor handling of conformational flexibility may contribute to the observed difference in performance between 2D and 3D descriptor-based virtual screening methods. One great advantage of using shape and 3D descriptors, however, is that such methods are more likely to "scaffold hop" and find novel chemotypes that would be otherwise undetectable if using just 2D chemical substructure-based features. Thus the best approach is to use both 2D and 3D methods and combine the results using data fusion methods.

Clearly, the validation of docking and virtual screening methods requires carefully curated datasets of active and inactive molecules. One widely used set, the Directory of Useful Decoys, or DUD,[173] consists of 40 targets, each target having a set of known active compounds, and a larger set of "decoys" chosen to have similar physical properties, such as molecular weight, calculated log *P*, *etc.* A refined version of the DUD datasets was introduced to address the problem of analogue bias in the active sets, also adding large diverse datasets collated using WOMBAT.[171] A more recent benchmarking resource, the Maximum Unbiased Validation (MUV) datasets,[174,175] was constructed from PubChem bioactivity data[176] so as to remove any bias apparent in simple low-dimensional molecular properties, and to ensure the quality of the activity data, by removing compounds with dubious bioactivities and so-called "frequent hitters". MUV has 17 target datasets, each consisting of 30 active compounds and 15 000 corresponding decoy compounds. MUV has the advantage that the decoys have been experimentally verified as inactive.

In order to screen databases with tens of millions of compounds with potentially billions of pre-calculated low-energy conformations, it is necessary to use very efficient virtual screening algorithms. One such algorithm that was recently introduced, Ultrafast Shape Recognition, or USR,[177] derives part of its speed from the fact that it is non-superpositional: it does not require an optimal superposition of the two structures in order to calculate the similarity of their shapes. USR uses a set of four centroids defined with respect to a small molecule's atom positions, and then computes the distances from each centroid to all the atoms in the molecule. For each of these four distance distributions, the first three statistical moments are calculated, giving a vector descriptor for the ligand's shape that consists of 12 real valued numbers. The similarity between two molecules is then simply the inverse of one plus the Manhattan distance between the two molecule's USR descriptors; the Manhattan distance is the sum of the least absolute differences of the corresponding moments.

USR has proven to be a successful approach in prospective virtual screening against arylamine *N*-acetyltransferases.[178] One limitation of the USR approach, however, is that it does not recognize the difference between two chiral molecules. This shortcoming was addressed with the introduction of the Chiral Shape Recognition, or CSR,[179] method, which differed from the original USR method only in the placement of the four centroids, thus retaining the original speed of the method. In ElectroShape,[180] a further improvement in molecular recognition was obtained by adding a fourth dimension of partial charge to each atom's three Cartesian coordinates and introducing a fifth centroid placed strategically in this fourth "electrostatic" dimension. When validated against the DUD, an almost doubling in average enrichment factors at 1% was obtained, increasing from 7.4 for USR to 13.3 for ElectroShape using MMFF94 modified partial charges, as implemented in MOE.[126] It should also be noted that the original partial charges supplied with release 2 of DUD differed in the calculation methods used for the active and corresponding decoy sets, making the enrichment factors initially found by ElectroShape unreasonably high; a new version of the DUD sets with re-calculated and consistent partial charges is now being distributed by InhibOx[181] and being mirrored by ZINC.[182]

7.4 Conclusion

Despite the claims of many comparative studies, there is still no docking method or virtual screening approach that is reliably able to identify active molecules or "binders". Docking methods have particular strengths and weaknesses and tend to be suitable for particular classes of targets; no method is universal. There has been a lot of progress in the last decade, but there remain several challenges, including the accurate treatment of solvation and waters, metals, covalently bound molecules, target flexibility and perhaps one of the most challenging aspects of *in silico* screening: the prediction of the binding affinity. This requires a proper treatment of both the enthalpic and the entropic components of binding, which is necessarily computationally expensive, and diametrically opposed to the requirement that virtual screening and high-throughput docking be fast enough to explore libraries that nowadays can reach hundreds of millions of compounds. Significant progress has been made in the fields of molecular docking and virtual screening, although the quest for the best ways to validate, combine and improve these methods continues.

Acknowledgments

The author of this manuscript would like to thank Prof. Arthur J. Olson, Dr David S. Goodsell, Dr Ruth Huey, Dr Rita Lim, Dr Paul Finn, Dr John Liebeschuetz and Dr Woody Sherman for their helpful discussions.

References

1. F. S. Collins, M. Morgan and A. Patrinos, *Science*, 2003, **300**, 286.
2. J. Weigelt, *Exp. Cell Res.*, 2010, **316**, 1332.
3. F. H. Allen, *Acta Crystallogr., Sect. B: Struct. Sci.*, 2002, **58**, 380.
4. J. J. Irwin and B. K. Shoichet, *J. Chem. Inf. Model.*, 2005, **45**, 177.
5. M. Hendlich, *Acta Crystallogr., Sect. D: Biol. Crystallogr.*, 1998, **54**, 1178.
6. L. Hu, M. L. Benson, R. D. Smith, M. G. Lerner and H. A. Carlson, *Proteins*, 2005, **60**, 333.
7. H. M. Berman, J. Westbrook, Z. Feng, G. Gilliland, T. N. Bhat, H. Weissig, I. N. Shindyalov and P. E. Bourne, *Nucleic Acids Res.*, 2000, **28**, 235.
8. S. F. Sousa, P. A. Fernandes and M. J. Ramos, *Proteins*, 2006, **65**, 15.
9. D. S. Goodsell and A. J. Olson, *Proteins*, 1990, **8**, 195.
10. R. Huey, G. M. Morris, A. J. Olson and D. S. Goodsell, *J. Comput. Chem.*, 2007, **28**, 1145.
11. G. M. Morris, D. S. Goodsell, R. Huey and A. J. Olson, *J. Comput. Aided Mol. Des.*, 1996, **10**, 293.
12. G. M. Morris, D. S. Goodsell, R. S. Halliday, R. Huey, W. E. Hart, R. K. Belew and A. J. Olson, *J. Comput. Chem.*, 1998, **19**, 1639.
13. G. M. Morris, R. Huey, W. Lindstrom, M. F. Sanner, R. K. Belew, D. S. Goodsell and A. J. Olson, *J. Comput. Chem.*, 2009, **30**, 2785.
14. http://autodock.scripps.edu (last accessed 2011).
15. I. D. Kuntz, J. M. Blaney, S. J. Oatley, R. Langridge and T. E. Ferrin, *J. Mol. Biol.*, 1982, **161**, 269.
16. T. J. A. Ewing and I. D. Kuntz, *J. Comput. Chem.*, 1997, **18**, 1175.
17. http://dock.compbio.ucsf.edu/ (last accessed 2011).
18. M. Rarey, B. Kramer, T. Lengauer and G. Klebe, *J. Mol. Biol.*, 1996, **261**, 470.
19. http://www.biosolveit.de/FlexX (last accessed 2011).
20. G. Jones, P. Willett and R. C. Glen, *J. Mol. Biol.*, 1995, **245**, 43.
21. G. Jones, P. Willett, R. C. Glen, A. R. Leach and R. Taylor, *J. Mol. Biol.*, 1997, **267**, 727.
22. http://www.ccdc.cam.ac.uk/products/life_sciences/gold/ (last accessed 2011).

23. R. A. Abagyan, M. M. Totrov and D. N. Kuznetsov, *J. Comput. Chem.*, 1994, **15**, 488.
24. http://www.molsoft.com/docking.html (last accessed 2011).
25. R. D. Taylor, P. J. Jewsbury and J. W. Essex, *J. Comput. Aided Mol. Des.*, 2002, **16**, 151.
26. N. Moitessier, P. Englebienne, D. Lee, J. Lawandi and C. R. Corbeil, *Br. J. Pharmacol.*, 2008, **153**, S7.
27. C. Bissantz, G. Folkers and D. Rognan, *J. Med. Chem.*, 2000, **43**, 4759.
28. J. C. Cole, C. W. Murray, J. W. Nissink, R. D. Taylor and R. Taylor, *Proteins*, 2005, **60**, 325.
29. S. Cotesta, F. Giordanetto, J. Y. Trosset, P. Crivori, R. T. Kroemer, P. F. Stouten and A. Vulpetti, *Proteins*, 2005, **60**, 629.
30. M. D. Cummings, R. L. DesJarlais, A. C. Gibbs, V. Mohan and E. P. Jaeger, *J. Med. Chem.*, 2005, **48**, 962.
31. D. A. Evans and S. Neidle, *J. Med. Chem.*, 2006, **49**, 4232.
32. A. Evers, G. Hessler, H. Matter and T. Klabunde, *J. Med. Chem.*, 2005, **48**, 5448.
33. R. A. Friesner, J. L. Banks, R. B. Murphy, T. A. Halgren, J. J. Klicic, D. T. Mainz, M. P. Repasky, E. H. Knoll, M. Shelley, J. K. Perry, D. E. Shaw, P. Francis and P. S. Shenkin, *J. Med. Chem.*, 2004, **47**, 1739.
34. T. A. Halgren, R. B. Murphy, R. A. Friesner, H. S. Beard, L. L. Frye, W. T. Pollard and J. L. Banks, *J. Med. Chem.*, 2004, **47**, 1750.
35. E. Kellenberger, J. Rodrigo, P. Muller and D. Rognan, *Proteins*, 2004, **57**, 225.
36. M. Kontoyianni, L. M. McClellan and G. S. Sokol, *J. Med. Chem.*, 2004, **47**, 558.
37. E. Perola, W. P. Walters and P. S. Charifson, *Proteins*, 2004, **56**, 235.
38. G. P. Vigers and J. P. Rizzi, *J. Med. Chem.*, 2004, **47**, 80.
39. J. S. Duca, V. S. Madison and J. H. Voigt, *J. Chem. Inf. Model.*, 2008, **48**, 659.
40. J. A. Feng and G. R. Marshall, *J. Comput. Chem.*, 2010, **31**, 2540.
41. J. H. Voigt, C. Elkin, V. S. Madison and J. S. Duca, *J. Chem. Inf. Model.*, 2008, **48**, 669.
42. N. Metropolis, A. W. Rosenbluth, M. N. Rosenbluth, A. H. Teller and E. Teller, *J. Chem. Phys.*, 1953, **21**, 1087.
43. R. J. Rosenfeld, E. D. Garcin, K. Panda, G. Andersson, A. Aberg, A. V. Wallace, G. M. Morris, A. J. Olson, D. J. Stuehr, J. A. Tainer and E. D. Getzoff, *Biochemistry*, 2002, **41**, 13915.
44. V. Mohan, A. C. Gibbs, M. D. Cummings, E. P. Jaeger and R. L. DesJarlais, *Curr. Pharm. Des.*, 2005, **11**, 323.
45. J. D. Durrant and J. A. McCammon, *J. Chem. Inf. Model.*, 2010, **50**, 1865.
46. J. G. Mandell, V. A. Roberts, M. E. Pique, V. Kotlovyi, J. C. Mitchell, E. Nelson, I. Tsigelny and L. F. Ten Eyck, *Protein Eng.*, 2001, **14**, 105.
47. A. Tovchigrechko and I. A. Vakser, *Nucleic Acids Res.*, 2006, **34**, W310.
48. I. A. Vakser, *Proteins*, 1997, suppl. 1, 226.

49. R. Chen and Z. Weng, *Proteins*, 2002, **47**, 281.
50. F. J. Solis and R. J.-B. Wets, *Math. Oper. Res.*, 1981, **6**, 19.
51. A. R. Conn, N. I. M. Gould and P. L. Toint, *SIAM J. Numer. Anal.*, 1991, **28**, 545.
52. S. Kirkpatrick, C. D. Gelatt, Jr. and M. P. Vecchi, *Science*, 1983, **220**, 671.
53. D. E. Goldberg, *Genetic Algorithms in Search, Optimization and Machine Learning*, Addison-Wesley Longman, Boston, , MA, 1989.
54. J. H. Holland, *Adaptation in Natural and Artificial Systems*, The MIT Press, Cambridge, , MA, 1992.
55. Z. Michalewicz, *Genetic Algorithms + Data Structures = Evolution Programs*, Springer, Berlin, 1996.
56. O. Trott and A. J. Olson, *J. Comput. Chem.*, 2010, **31**, 455.
57. H. M. Chen, B. F. Liu, H. L. Huang, S. F. Hwang and S. Y. Ho, *J. Comput. Chem.*, 2007, **28**, 612.
58. V. Namasivayam and R. Gunther, *Chem. Biol. Drug Des.*, 2007, **70**, 475.
59. O. Korb, T. Stutzle and T. E. Exner, *J. Chem. Inf. Model.*, 2009, **49**, 84.
60. F. S. Kuhl, G. M. Crippen and D. K. Friesen, *J. Comput. Chem.*, 1984, **5**, 24.
61. C. Hetenyi and D. van der Spoel, *Protein Sci.*, 2002, **11**, 1729.
62. C. Hetenyi and D. van der Spoel, *Protein Sci.*, 2011, **20**, 880.
63. C. Hetenyi and D. van der Spoel, *FEBS Lett.*, 2006, **580**, 1447.
64. R. Harris, A. J. Olson and D. S. Goodsell, *Proteins*, 2008, **70**, 1506.
65. M. Vaque, A. Arola, C. Aliagas and G. Pujadas, *Bioinformatics*, 2006, **22**, 1803.
66. G. Jones and P. Willett, *Curr. Opin. Biotechnol.*, 1995, **6**, 652.
67. C. de Graaf, C. Oostenbrink, P. H. Keizers, T. van der Wijst, A. Jongejan and N. P. Vermeulen, *J. Med. Chem.*, 2006, **49**, 2417.
68. D. J. Diller and R. Li, *J. Med. Chem.*, 2003, **46**, 4638.
69. A. Evers and T. Klabunde, *J. Med. Chem.*, 2005, **48**, 1088.
70. B. K. Shoichet, S. L. McGovern, B. Wei and J. J. Irwin, *Curr. Opin. Chem. Biol.*, 2002, **6**, 439.
71. http://www.ccdc.cam.ac.uk/products/csd_system/mogul/ (last accessed 2011).
72. I. J. Bruno, J. C. Cole, M. Kessler, J. Luo, W. D. Motherwell, L. H. Purkis, B. R. Smith, R. Taylor, R. I. Cooper, S. E. Harris and A. G. Orpen, *J. Chem. Inf. Comput. Sci.*, 2004, **44**, 2133.
73. G. J. Kleywegt, M. R. Harris, J. Y. Zou, T. C. Taylor, A. Wahlby and T. A. Jones, *Acta Crystallogr., D: Biol. Crystallogr.*, 2004, **60**, 2240.
74. M. J. Hartshorn, M. L. Verdonk, G. Chessari, S. C. Brewerton, W. T. Mooij, P. N. Mortenson and C. W. Murray, *J. Med. Chem.*, 2007, **50**, 726.
75. http://relibase.ccdc.cam.ac.uk (last accessed 2011).
76. http://www.rcsb.org/pdb (last accessed 2011).
77. http://www.bindingmoad.org (last accessed 2011).

78. R. D. Smith, L. Hu, J. A. Falkner, M. L. Benson, J. P. Nerothin and H. A. Carlson, *J. Mol. Graphics Model.*, 2006, **24**, 414.
79. N. Guex and M. C. Peitsch, *Electrophoresis*, 1997, **18**, 2714.
80. http://www.expasy.org/spdbv (last accessed 2011).
81. I. W. Davis, L. W. Murray, J. S. Richardson and D. C. Richardson, *Nucleic Acids Res.*, 2004, **32**, W615.
82. S. C. Lovell, I. W. Davis, W. B. Arendall III, P. I. W. de Bakker, J. M. Word, M. G. Prisant, J. S. Richardson and D. C. Richardson, *Proteins*, 2003, **50**, 437.
83. P. Labute, *Proteins*, 2009, **75**, 187.
84. S. Forli and A. J. Olson, *J. Med. Chem.*, 2012, **55**, 623.
85. M. L. Raymer, P. C. Sanschagrin, W. F. Punch, S. Venkataraman, E. D. Goodman and L. A. Kuhn, *J. Mol. Biol.*, 1997, **265**, 445.
86. http://www.bch.msu.edu/labs/kuhn/software.html (last accessed 2011).
87. C. J. Cramer and D. G. Truhlar, *J. Comput. Aided Mol. Des.*, 1992, **6**, 629.
88. G. D. Hawkins, C. J. Cramer and D. G. Truhlar, *J. Phys. Chem. B*, 1998, **102**, 3257.
89. B. Q. Wei, W. A. Baase, L. H. Weaver, B. W. Matthews and B. K. Shoichet, *J. Mol. Biol.*, 2002, **322**, 339.
90. C. J. Illingworth, G. M. Morris, K. E. Parkes, C. R. Snell and C. A. Reynolds, *J. Phys. Chem. A*, 2008, **112**, 12157.
91. P. J. Goodford, *J. Med. Chem.*, 1985, **28**, 849.
92. C. A. Sotriffer, H. Gohlke and G. Klebe, *J. Med. Chem.*, 2002, **45**, 1967.
93. S. Radestock, M. Bohm and H. Gohlke, *J. Med. Chem.*, 2005, **48**, 5466.
94. C. W. Murray, C. A. Baxter and A. D. Frenkel, *J. Comput. Aided Mol. Des.*, 1999, **13**, 547.
95. P. Cozzini, G. E. Kellogg, F. Spyrakis, D. J. Abraham, G. Costantino, A. Emerson, F. Fanelli, H. Gohlke, L. A. Kuhn, G. M. Morris, M. Orozco, T. A. Pertinhez, M. Rizzi and C. A. Sotriffer, *J. Med. Chem.*, 2008, **51**, 6237.
96. K. Gunasekaran and R. Nussinov, *J. Mol. Biol.*, 2007, **365**, 257.
97. H. Alonso, A. A. Bliznyuk and J. E. Gready, *Med. Res. Rev.*, 2006, **26**, 531.
98. J. H. Lin, A. L. Perryman, J. R. Schames and J. A. McCammon, *J. Am. Chem. Soc.*, 2002, **124**, 5632.
99. http://ambermd.org/ (last accessed 2002).
100. D. A. Case, T. E. Cheatham III, T. Darden, H. Gohlke, R. Luo, K. M. J. Merz, A. Onufriev, C. Simmerling, B. Wang and R. J. Woods, *J. Comput. Chem.*, 2005, **26**, 1668.
101. J. C. Phillips, R. Braun, W. Wang, J. Gumbart, E. Tajkhorshid, E. Villa, C. Chipot, R. D. Skeel, L. Kale and K. Schulten, *J. Comput. Chem.*, 2005, **26**, 1781.
102. J. R. Schames, R. H. Henchman, J. S. Siegel, C. A. Sotriffer, H. Ni and J. A. McCammon, *J. Med. Chem.*, 2004, **47**, 1879.

103. H. Claussen, C. Buning, M. Rarey and T. Lengauer, *J. Mol. Biol.*, 2001, **308**, 377.

104. M. L. Verdonk, G. Chessari, J. C. Cole, M. J. Hartshorn, C. W. Murray, J. W. Nissink, R. D. Taylor and R. Taylor, *J. Med. Chem.*, 2005, **48**, 6504.

105. M. L. Verdonk, J. C. Cole, M. J. Hartshorn, C. W. Murray and R. D. Taylor, *Proteins*, 2003, **52**, 609.

106. F. Österberg, G. M. Morris, M. F. Sanner, A. J. Olson and D. S. Goodsell, *Proteins*, 2002, **46**, 34.

107. Z. Li and T. Lazaridis, *J. Phys. Chem. B*, 2006, **110**, 1464.

108. J. Luccarelli, J. Michel, J. Tirado-Rives and W. L. Jorgensen, *J. Chem. Theory Comput.*, 2010, **6**, 3850.

109. M. Gastreich, M. Lilienthal, H. Briem and H. Claussen, *J. Comput. Aided Mol. Des.*, 2006, **20**, 717.

110. C. A. Lipinski, F. Lombardo, B. W. Dominy and P. J. Feeney, *Adv. Drug Delivery Rev.*, 2001, **46**, 3.

111. M. M. Hann and T. I. Oprea, *Curr. Opin. Chem. Biol.*, 2004, **8**, 255.

112. J. Gasteiger and M. Marsili, *Tetrahedron Lett.*, 1978, **34**, 3181.

113. M. Rarey, B. Kramer and T. Lengauer, *Proc. Int. Conf. Intell. Syst. Mol. Biol.*, 1995, **3**, 300.

114. S. Forli and M. Botta, *J. Chem. Inf. Model.*, 2007, **47**, 1481.

115. A. Laederach and P. J. Reilly, *Proteins*, 2005, **60**, 591.

116. C. Mulakala, W. Nerinckx and P. J. Reilly, *Carbohydr. Res.*, 2006, **341**, 2233.

117. W. M. Rockey, A. Laederach and P. J. Reilly, *Proteins*, 2000, **40**, 299.

118. J. R. Greenwood, D. Calkins, A. P. Sullivan and J. C. Shelley, *J. Comput. Aided Mol. Des.*, 2007, *24*, 591.

119. M. S. Park, C. Gao and H. A. Stern, *Proteins*, 2011, **79**, 304.

120. J. C. Shelley, A. Cholleti, L. L. Frye, J. R. Greenwood, M. R. Timlin and M. Uchimaya, *J. Comput. Aided Mol. Des.*, 2007, **21**, 681.

121. http://zinc.docking.org/ (last accessed 2011).

122. http://www.molport.com/ (last accessed 2011).

123. http://www.emolecules.com/ (last accessed 2011).

124. D. Weininger, *Daylight Theory Manual*, Daylight Chemical Information Systems, Aliso Viejo, , CA, 2008.

125. J. Gasteiger, C. Rudolph and J. Sadowski, *Tetrahedron Comput. Methodol.*, 1992, **3**, 537.

126. http://www.chemcomp.com/software.htm (last accessed 2011).

127. A. Dalby, J. G. Nourse, W. D. Hounshell, A. K. I. Gushurst, D. L. Grier, B. A. Leland and J. Laufer, *J. Chem. Inf. Comput. Sci.*, 1992, **32**, 244.

128. http://www.molecular-networks.com/online_demos/corina_demo.html. (last accessed 2011).

129. http://www.uku.fi/thassine/projects/ghemical/ (last accessed 2011).

130. http://www.uiowa.edu/ghemical/ghemical.shtml (last accessed 2011).

131. T. Hassinen and M. Peräkylä, *J. Comput. Chem.*, 2001, **22**, 1229.

132. A. W. Schuttelkopf and D. M. van Aalten, *Acta Crystallogr., Sect. D: Biol. Crystallogr.*, 2004, **60**, 1355.

133. D. M. van Aalten, R. Bywater, J. B. Findlay, M. Hendlich, R. W. Hooft and G. Vriend, *J. Comput. Aided Mol. Des.*, 1996, **10**, 255.

134. http://davapc1.bioch.dundee.ac.uk/programs/prodrg/ (last accessed 2011).

135. http://www.eyesopen.com/quacpac (last accessed 2011).

136. C. Bissantz, B. Kuhn and M. Stahl, *J. Med. Chem.*, 2010, **53**, 5061.

137. R. S. Bohacek and C. McMartin, *Nat. Med.*, 1995, **1**, 177.

138. R. S. Bohacek, C. McMartin and W. C. Guida, *Med. Res. Rev.*, 1996, **16**, 3.

139. J. L. Medina-Franco, K. Martínez-Mayorga, M. A. Giulianotti, R. A. Houghten and C. Pinilla, *Curr. Comput. Aided Drug Des.*, 2008, **4**, 322.

140. P. J. Hajduk and J. Greer, *Nat. Rev. Drug Discovery*, 2007, **6**, 211.

141. C. R. Guimaraes and M. Cardozo, *J. Chem. Inf. Model.*, 2008, **48**, 958.

142. M. K. Haider, H. O. Bertrand and R. E. Hubbard, *J. Chem. Inf. Model.*, 2011, **51**, 1092.

143. R. J. Rosenfeld, D. S. Goodsell, R. A. Musah, G. M. Morris, D. B. Goodin and A. J. Olson, *J. Comput. Aided Mol. Des.*, 2003, **17**, 525.

144. A. M. Ruvinsky and A. V. Kozintsev, *J. Comput. Chem.*, 2005, **26**, 1089.

145. A. M. Ruvinsky and A. V. Kozintsev, *Proteins*, 2006, **62**, 202.

146. A. M. Ruvinsky, *J. Comput. Aided Mol. Des.*, 2007, **21**, 361.

147. A. L. Hopkins, C. R. Groom and A. Alex, *Drug Discovery Today*, 2004, **9**, 430.

148. H. Köppen, *Curr. Opin. Drug Discovery Dev.*, 2009, **12**, 397.

149. M. A. Johnson and G. M. Maggiora, *Concepts and Applications of Molecular Similarity*, Wiley, New York, 1990.

150. P. Willett, *Drug Discovery Today*, 2006, **11**, 1046.

151. A. Pozzan, *Curr. Pharm. Des.*, 2006, **12**, 2099.

152. P. C. Hawkins, A. G. Skillman and A. Nicholls, *J. Med. Chem.*, 2007, **50**, 74.

153. B. K. Shoichet, *Nature*, 2004, 432, 862.

154. B. Chen, R. F. Harrison, G. Papadatos, P. Willett, D. J. Wood, X. Q. Lewell, P. Greenidge and N. Stiefl, *J. Comput. Aided Mol. Des.*, 2007, **21**, 53.

155. J. L. Melville, E. K. Burke and J. D. Hirst, *Comb. Chem. High Throughput Screening*, 2009, **12**, 332.

156. S. J. Edgar, J. D. Holliday and P. Willett, *J. Mol. Graphics Model.*, 2000, **18**, 343.

157. N. Triballeau, F. Acher, I. Brabet, J. P. Pin and H. O. Bertrand, *J. Med. Chem.*, 2005, **48**, 2534.

158. J. F. Truchon and C. I. Bayly, *J. Chem. Inf. Model.*, 2007, **47**, 488.

159. R. P. Sheridan, S. B. Singh, E. M. Fluder and S. K. Kearsley, *J. Chem. Inf. Comput. Sci.*, 2001, **41**, 1395.

160. R. P. Sheridan and S. K. Kearsley, *Drug Discovery Today*, 2002, **7**, 903.

161. P. Willett, *QSAR Comb. Sci.*, 2006, **25**, 1143.

162. C. M. R. Ginn, P. Willett and B. Bradshaw, *Perspect. Drug Discovery Des.*, 2000, **20**, 1.
163. M. Whittle, V. J. Gillet, P. Willett and J. Loesel, *J. Chem. Inf. Model.*, 2006, **46**, 2206.
164. M. Whittle, V. J. Gillet, P. Willett and J. Loesel, *J. Chem. Inf. Model.*, 2006, **46**, 2193.
165. S. W. Muchmore, D. A. Debe, J. T. Metz, S. P. Brown, Y. C. Martin and P. J. Hajduk, *J. Chem. Inf. Model.*, 2008, **48**, 941.
166. S. L. Swann, S. P. Brown, S. W. Muchmore, H. Patel, P. Merta, J. Locklear and P. J. Hajduk, *J. Med. Chem.*, 2011, **54**, 1223.
167. J. Hert, P. Willett, D. J. Wilton, P. Acklin, K. Azzaoui, E. Jacoby and A. Schuffenhauer, *J. Med. Chem.*, 2005, **48**, 7049.
168. R. D. Brown and Y. C. Martin, *SAR QSAR Environ. Res.*, 1998, **8**, 23.
169. R. D. Brown and Y. C. Martin, *J. Chem. Inf. Comput. Sci.*, 1996, **36**, 572.
170. R. D. Brown and Y. C. Martin, *J. Chem. Inf. Comput. Sci.*, 1997, **37**, 1.
171. A. C. Good and T. I. Oprea, *J. Comput. Aided Mol. Des.*, 2008, **22**, 169.
172. M. L. Verdonk, V. Berdini, M. J. Hartshorn, W. T. Mooij, C. W. Murray, R. D. Taylor and P. Watson, *J. Chem. Inf. Comput. Sci.*, 2004, **44**, 793.
173. N. Huang, B. K. Shoichet and J. J. Irwin, *J. Med. Chem.*, 2006, **49**, 6789.
174. S. G. Rohrer and K. Baumann, *J. Chem. Inf. Model.*, 2009, **49**, 169.
175. http://www.pharmchem.tu-bs.de/lehre/baumann/MUV.html (last accessed 2011).
176. Y. Wang, E. Bolton, S. Dracheva, K. Karapetyan, B. A. Shoemaker, T. O. Suzek, J. Wang, J. Xiao, J. Zhang and S. H. Bryant, *Nucleic Acids Res.*, 2010, **38**, D255.
177. P. J. Ballester and W. G. Richards, *J. Comput. Chem.*, 2007, **28**, 1711.
178. P. J. Ballester, I. Westwood, N. Laurieri, E. Sim and W. G. Richards, *J. R. Soc., Interface*, 2010, **7**, 335.
179. M. S. Armstrong, G. M. Morris, P. W. Finn, R. Sharma and W. G. Richards, *J. Mol. Graphics Model.*, 2009, **28**, 368.
180. M. S. Armstrong, G. M. Morris, P. W. Finn, R. Sharma, L. Moretti, R. I. Cooper and W. G. Richards, *J. Comput. Aided Mol. Des.*, 2010, **24**, 789.
181. http://www.inhibox.com/dud/ (last accessed 2011).
182. http://dud.docking.org/ (last accessed 2011).

CHAPTER 8

Binding Free Energy Calculation and Scoring in Small-Molecule Docking

CLAUDIO N. CAVASOTTO

School of Biomedical Informatics, University of Texas Health Science Center at Houston, 7000 Fannin Ste. 690, Houston, TX 77030, USA
E-mail: Claudio.N.Cavasotto@uth.tmc.edu

8.1 Introduction

The correct functioning of the biochemical machinery within a cell depends upon the non-covalent molecular association in processes such as enzyme catalysis, molecular transport, and signal transduction. Until recently, high-throughput screening (HTS) has been the dominant technique within pharmaceutical companies to identify and optimize new drug lead compounds.[1] HTS involves combinatorial chemistry and the experimental screening of large chemical libraries against a relevant therapeutic target. It is thus an expensive and time-consuming process, in spite of recent progresses to improve its efficiency.[1]

The use of 3D structures of protein–ligand complexes has been used for many years to guide drug lead optimization, aiming to improve potency or selectivity, thus giving rise to a more rational approach. The natural continuation of this process was the development of *in silico* methods, both as a means to predict protein–ligand interactions and for the computational screen of chemical libraries through protein–ligand docking,[2] a process in

RSC Drug Discovery Series No. 23
Physico-Chemical and Computational Approaches to Drug Discovery
Edited by F. Javier Luque and Xavier Barril

which the ligand is positioned within the binding site and its binding energy estimated. This computational mimic of HTS provided a more rational, more economic, and faster alternative to the traditional HTS. It can be said that currently structure-based drug lead discovery and design is a key first step in the lengthy, expensive, and unpredictable process of developing new drugs.[2–4] The reader is referred to Congreve *et al.*[5] for a comprehensive collection list of marketed drugs developed using structure-based approaches.

The development of *in silico* tools to study protein–ligand interactions and to dock chemical libraries began in the late 1970s[6] and has been in constant progress ever since (see refs. 2, 7, and 8 for a comprehensive discussion on docking methods). The main advantages of docking-based methods compared to ligand-based methods are the structural novelty of the hits discovered (not based on pre-existing, known ligands) and the possibility to model the binding mode of potential ligands within the binding site. Although high-throughput docking (HTD) is plagued with both false positives and false negatives, the low set-up cost, high computational speed, potential structural novelty of discovered ligands, and the possibility to incorporate, at any stage of the process, filters accounting for drug likeness, offset its limitations.

In computational docking studies, three main ideal purposes could be identified: (1) The correct characterization of the binding pose (or the thermodynamic equilibrium ensemble) of a known ligand within the binding site [in principle, this involves the degrees of freedom (DOF) of both the ligand and the receptor]; (2) the accurate calculation of the ligand binding free energy, or the relative free energies of a series of molecules; (3) the prediction of the pose and estimation of the binding free energy of large virtual chemical libraries in a high-throughput fashion (HTD).

The prediction of the dominant binding pose, or the equilibrium ensemble that characterizes the ligand–protein complex, can be determined using only the potential energy surface of the complex, *e.g.* from molecular dynamics (MD) or Monte Carlo (MC) simulations.[9,10] However, in order to use docking in a high-throughput fashion, several approximations should be introduced to reduce the number of DOF to be sampled. Thus, in HTD the receptor is usually considered rigid (or very few protein DOF considered), the solvent is represented in a continuum fashion, and hard DOF (bond lengths and planar angles of the ligand) are frozen. In spite of these approximations, ligand posing through docking has shown a reasonable degree of success,[2,11] and is not the main limitation step in the application of HTD to drug discovery.[2,4] It should be remarked that other factors may also condition docking accuracy, such as the quality of the receptor structure used (see Davis *et al.*[12] for a review about the relevance of crystallographic data in structure-based drug design), the type of receptor used [bound (holo) or unbound],[13] the inclusion of crystallographic water molecules, and the correct assignment of the protonation and tautomerization states of the ligand and binding site residues. These latter factors are not approximations themselves, but reflect uncertainties in the description of the molecular system.

This chapter will open with a review of the theory of binding free energy and its computational calculation. The description of the different types of scoring functions in HTD will follow, giving consideration to comparison studies, assessing their limitations and the facts that hinder their improvement. The post-screening process will be described next, with special emphasis in the use of advanced simulation methods to re-score top-ranking hits. The conclusions and suggested future directions will close the chapter.

8.2 Binding Free Energy

8.2.1 Physico-Chemical Aspects of Protein–Ligand Binding

In the non-covalent association of a ligand and a protein in aqueous solution, the binding free energy is expressed as:

$$G = H - TS \qquad (8.1)$$

where G is the Gibbs free energy, H the enthalpy, S the entropy, and T is the absolute temperature.

The standard binding free energy ($\Delta G^\circ = \Delta H^\circ - T\Delta S^\circ$) is related to the association (K_i) and dissociation constants ($K_d = 1/K_i$) as:

$$G^\circ = -RT \ln K_i = RT \ln K_d \qquad (8.2)$$

where R is the gas constant and the superscript "$^\circ$" reflects the fact that the binding free energy is evaluated at standard conditions.

It is clear from eqn (8.1) that the free energy of binding is a delicate balance between the change in enthalpy and entropy. In a study of more than 100 ligand–protein complexes for which Isothermal Titration Calorimetry (ITC) data for ΔH and ΔS were available, a strong correlation was observed between those two magnitudes (enthalpy–entropy compensation), though, in general, a poor correlation is found between ΔG and ΔH, and ΔG and ΔS, with few exceptions.[14] This highlights the risk of using either enthalpy or entropy as a substitute to estimate ΔG.

The enthalpic component of the binding free energy in eqn (8.1) reflects the change in intra- and intermolecular energy upon binding, involving the ligand and the protein in a fully solvated environment. These account for all type of interactions, such as electrostatic, hydrogen bonds, van der Waals (attractive, dispersive, and repulsive), weakly polar involving π-systems (N–H···π, O–H···π, C–H···π), aromatic–aromatic, cation–π, those involving halogens (C–F···C=O, C–F···H–X with X = O or N, C–X···π with X a halogen), citing those more commonly found in small-molecule–protein complexes. A detailed analysis of these types of interaction will not be presented here, and the reader is referred to reviews in this topic.[15,16] The entropic component is related to the change in the number of accessible states upon binding. At a qualitative level, enthalpy–entropy compensation can be rationalized by the fact that the stronger the ligand–protein interaction (lower ΔH), the fewer the accessible states of the system (lower ΔS), or in other words, bonding opposes motion, and *vice versa*.[17] However, the specific nature of this compensation is system dependent.

Upon binding, ligand solvated polar groups might be buried, totally or partially, within the binding site. If those groups did not have a comparable polar interaction inside the protein, the desolvation energy would strongly oppose binding. On the contrary, the association of hydrophobic partners is usually favored, since it reduces the exposed surface to water, thus decreasing the number of ordered water molecules surrounding those surfaces, and bringing a positive entropic contribution to binding. However, not all processes involving hydrophobic ligands are entropy driven. In the mouse major urinary protein (MUP), it was shown that poor solvation of the binding site leads to an enthalpy-driven process with large ligand desolvation and minor desolvation of the binding site.[18] This effect was also observed in other systems,[19] and its signature seems to be the non-optimal hydration of the binding site in the unbound state.[20] The large decrease in affinity upon a hydrogen substitution of an isopropyl group in a factor Xa inhibitor was due not too much because of the loss of a favorable hydrophobic interaction within the binding site, but especially due to the sub-optimal solvated state of the latter.[21]

The bound conformation of the ligand does not necessarily correspond to the global energy minimum observed in bulk solvent; however, this enthalpic penalty between the bound and unbound state is much lower than what has been usually assumed.[22,23]

Upon binding, there is also a reduction in the number of accessible states, with the corresponding decrease in entropy. Several cases have been reported where improved binding free energies were achieved through lowering the entropic change upon binding by constraining certain degrees of freedom of the ligand.[24] It should be remarked that the loss of ligand conformational entropy, accounting for the change in the number of stable conformers upon binding, does not scale linearly with the number of torsional DOF (as is usually assumed); rather, it has a logarithmic scaling. Among others, calculations on amprenavir[25] and tetrapeptides confirm this fact.[26]

8.2.2 Binding Free Energy Calculation

The accurate *in silico* prediction of binding free energy would be not only extremely valuable in computer-aided drug lead discovery and design,[3] but also to study protein–protein association,[27] and in the understanding of molecular recognition.[28] Although the theoretical foundation of binding free energy is well established,[29,30] its application to biomolecular systems is not straightforward, since it depends on the quality of the energy function, a thorough exploration of the energy landscape, and ultimately is the small difference of large numbers. Methods with which chemical accuracy can be reached in some systems, such as free energy perturbation (FEP)[31] and thermodynamic integration (TI),[32] or those based on the potential of mean force (PMF),[33,34] account for explicit solvent representation and *de facto* include conformational flexibility. These methods can be used to predict the absolute or relative binding affinities of a series of compounds; however, they constitute a rather specialized field of research,[35] and the simulation time hinders their use in a high- or even mid-throughput manner. In end-point methods, such as the Molecular Mechanics–Poisson–Boltzmann Surface Area (MM/PB-SA),[36,37] and the Molecular Mechanics–Generalized Born Surface Area (MM/GB-SA),[38,39] the binding free energy is expressed as the difference between the bound and unbound states of protein and ligand, according to:[26,36,37,39]

$$G^\circ = \langle E_{MM} \rangle + G_{solv} - TS_{bind} \tag{8.3}$$

where $\langle ... \rangle$ is the ensemble average, E_{MM} is the vacuum potential energy, $\Delta G_{solv} = \Delta \langle W \rangle$, where W is the effective solvation energy (accounting for the solvent degrees of freedom), and the third term accounts for the entropy change of the solute.

End-point methods are founded in statistical mechanics[40–44] and also account for conformational flexibility of both ligand and receptor. In spite of the fact that calculation of binding free energies with these approaches showed a mix of successful results[45–48] and failures,[26,34,35,49] and that they still cannot be used in a high-throughput fashion, these methods are appealing for re-scoring purposes, and they will be considered with others in that context in Section 8.4.1.

Needless to say, if the binding free energy could be accurately calculated in a timely manner for large chemical libraries, the HTD problem would be solved. Since this is not yet possible, for docking large chemical libraries in a high-throughput fashion, aiming at the prioritization of molecules for bioevaluation,[3] different types of scoring functions were introduced, which aim to rank the docked molecules in a time-efficient way.

8.3 Scoring Functions in High-Throughput Docking

8.3.1 General Considerations

In HTD, two stages should be distinguished:[2] (1) the docking process itself, or the *in silico* prediction of the pose of a molecule within the binding site; (2) the scoring stage, where the *in silico* generated complexes are assigned a score which attempts to predict the likelihood that a molecule will actually bind to the target.

As has been said above, the prediction of the binding pose can be calculated using the potential energy surface of the ligand-binding complex, usually by searching for the global minimum of the potential energy ("docking energy"), which should correspond to the native pose. To score and rank a set of molecules, however, other contributions should be taken into account according to eqn (8.1), such as the free energy of the unbound molecules, the entropic change, and desolvation effects, all factors which should be taken into account in the "docking score". Thus, the docking energy discriminates among poses of the same ligand, while the docking score discriminates among the best pose (of few poses) per ligand. In many docking programs, however, the docking score serves for both purposes.

Scoring functions should be examined in view of their actual expectations and intrinsic limitations. The main aim of HTD is to populate the hit list with potential binders. It should be highlighted that the main purpose is not to rank potential binders according to binding free energy, nor to extract from the chemical library all possible binders.[4] As far as some new molecules are found, false positives can be tolerated. Moreover, scoring functions have intrinsic limitations that render them unable to provide a universal acceptable correlation with binding free energies (Section 8.3.4).

High-throughput docking and scoring of large virtual chemical libraries should be considered, in fact, as a structural filter imposed on the chemical library to prioritize those molecules which have steric and electrostatic complementarity to the binding site, coupled with an adequate balance of solvation and desolvation effects and a favorable enthalpy–entropy compensation. A post-docking re-scoring stage, with more time-consuming (and more accurate) methods, may follow, in which the relative binding energies of a smaller sub-library could be estimated.

8.3.2 Scoring Functions

As has been said, scoring functions serve the purpose of ranking chemical libraries docked to a receptor with an adequate balance of accuracy and speed. If the scoring and docking energy are different functions, the score is calculated on the best-energy conformation of each ligand (or the few best ones). Scoring functions can be classified as force field based, empirical, and knowledge based. A comprehensive list of scoring functions and their type is shown on Table 8.1. The main characteristics of the three groups of scoring functions are described below, while the reader is referred elsewhere[50,51] for a more detailed analysis and comparison of scoring functions.

8.3.2.1 Force Field-Based Scoring Functions

These functions are based on the non-bonded interaction terms of classical mechanics force fields (FFs), such as AMBER,[92] CHARMM,[93] or OPLS.[94] They usually contain a pairwise van der Waals and electrostatic interaction terms, with additional contributions accounting for hydrogen bonding and, eventually, the internal torsional energy of the ligand. The expression for the basic FF based scoring function is:

Table 8.1 Most common scoring functions.

Scoring function	Type	Ref.
AIScore	Empirical	52
ChemScore	Empirical	53
eHITS SF	Empirical	54
FlexX SF	Empirical	55
Fresno	Empirical	56
GlideScore	Empirical	57
Hammerhead	Empirical	58
HINT	Empirical	59
ICM	Empirical	60
Kroemer's set	Empirical	61
LigScore	Empirical	62
PHOENIX	Empirical	63
PLANTS	Empirical	64
PLP	Empirical	65
Protein Alpha Shape (PAS)	Empirical	66
RankScore	Empirical	67
RF-Score	Empirical	68
SCORE1	Empirical	69
SCORE2	Empirical	70
SCORE 3.0	Empirical	71
ScreenScore	Empirical	72
SFCscore	Empirical	73
SLIDE SCORE	Empirical	74
Surflex-Dock	Empirical	75
X-Score	Empirical	76
AutoDock sf	FF	77
DockScore	FF	78
EADock2	FF	79
GOLDScore	FF	80
QXP	FF	81
ASP	Knowledge based	82
BLEEP	Knowledge based	83
DFIRE	Knowledge based	84
DrugScore[CSD]	Knowledge based	85
DrugScore[PDB]	Knowledge based	86
M-Score	Knowledge based	87
PMF	Knowledge based	88,89
SMoG	Knowledge based	90,91

$$E_{score}^{FF} = \sum_{i \in rec} \sum_{j \in lig} \left[\frac{A_{ij}}{r_{ij}^{n}} + \frac{B_{ij}}{r_{ij}^{m}} + 332.0 \frac{q_i q_j}{\varepsilon(r_{ij}) r_{ij}} \right] \qquad (8.4)$$

where the first two terms represent the van der Waals contribution and the third one the electrostatic contribution, r_{ij} is the distance between atom "i" in the protein and "j" in the ligand, q is the corresponding atomic charge, ε is the internal distance-dependent dielectric constant, and n and m are usually taken as 12 and 6, respectively (for example, in AutoDock[77] scoring function), or 8 and 4 (as in GOLDScore). The factor 332.0 yields kcal mol^{-1} units when q is in atomic units and r in Å.

FF-based scoring functions have been implemented in popular programs such as DOCK,[95] GOLD,[80,96] and AutoDock.[97] The main weakness of these scoring functions is the omission of the entropic term. Thus, larger molecules are likely to get better scores just due to the larger number of interactions.

8.3.2.2 *Empirical Scoring Functions*

Following the seminal work of Böhm in developing the LUDI software,[69] in empirical scoring functions the binding free energy is approximated by a linear combination of terms considered relevant for ligand binding, such as those accounting for hydrogen bonding, ionic, lipophilic, and apolar interactions, metal interactions, and loss of torsional DOF (related to the entropic change). The coefficients of the linear combination are adjusted by multilinear regression using ligand–protein complexes, for which both the structure and the ligand binding energy are available. A general empirical scoring function can be expressed as:

$$
G_{score}^{empirical} = G_0 + G_{HB} \sum_{HB} f(R,\alpha) + G_{ionic} \sum_{ionic} f(R,\alpha)
$$

$$
+ G_{metal} \sum_{metal} f(R,\alpha) + \quad + G_{lipo} \sum_{lipo} f(R) \qquad (8.5)
$$

$$
+ G_{aromatic} \sum_{aromatic} f(R,\alpha) + G_{rot} N_{rot}
$$

where in the hydrogen bonding (HB), ionic, metal, lipophilic (lipo), and aromatic interaction terms, ΔG_x represents the energy of the optimal interaction and f is a penalty function reflecting deviations from the ideal geometry characterized by a distance ΔR and eventually an angle $\Delta\alpha$, ΔG_0 is an independent term, ΔG_{rot} is the energy loss per torsional DOF, and N_{rot} is the ligand rotatable bonds. All ΔG_x are adjusted through linear regression.

The different blends of empirical scoring functions differ in the number and type of terms included in eqn (8.5). One of the most popular scoring functions, ChemScore, is implemented in several docking programs. Other commonly used scoring functions are GlideScore (implemented in Glide[57,98]), FlexX (in FlexX[55]), and PLP (in FRED,[99] DockIt,[100] and LigandFit[101]). The scoring function in ICM docking software[102] has a functional form similar to eqn. (8.5), but using actual FF energy interaction terms from ECEPP/3[103] and MMFF,[104] instead of penalty functions, and where the regression coefficients are fit to maximize the separation between potential binders and non-binders,[60] and not to experimental binding free energies. Since it is not a pure FF function, but has adjustable coefficients, it is classified as empirical.

One limitation of empirical scoring functions is that they rely on currently available structures and binding free energies, which somehow compromises their applicability outside that training set.

8.3.2.3 Knowledge-Based Scoring Functions

In knowledge-based scoring functions, the statistical information about ligand–protein contacts from experimentally solved structures is used to build pairwise atom-type dependent potentials. It is thus implicitly assumed that the more frequent a ligand and a protein atom are found at a given distance range, the more favorable is their interaction. The PMF scoring function[88] is expressed as:

$$E_{\text{score}}^{\text{PMF}} = \sum_{\substack{l \in \text{protein} \\ m \in \text{ligand}}} A_{ij}(r_{lm})$$

$$A_{ij}(r) = -kT \ln \left[f_{\text{vol\_corr}}^{j} g_{ij}(r) \right]$$

(8.6)

where k is Boltzmann constant, T the absolute temperature, i and j are the atom types of atoms l and m, respectively, the sum is performed on those r_{lm} which are within the cutoff distance, g_{ij} is the radial distribution function of a protein atom of type i paired with a ligand atom of type j, and $f_{\text{vol\_corr}}$ is a ligand volume correction factor.

A similar scoring function, DrugScore,[PDB] was developed by Gohlke *et al.*[86] using PDB complexes aiming to rank properly small-molecule docked poses. A newer version, the DrugScore[CSD] scoring function,[85] derived using small-molecule crystal data from the Cambridge Structural Database,[105] gave improved predictions for near-native poses for docked ligands and, interestingly, better binding affinities.

Clearly, the advantage of this type of scoring function is their simplicity, that no fitting to experimental binding free energies is needed, and that desolvation and entropic effects are somehow implicitly accounted for. However, they have been derived from a limited set of static structures and thus some specific interactions are not well parameterized.

8.3.3 Comparison of Scoring Functions

In spite of how valuable for the community are comparative studies on the performance of scoring functions, there are certain considerations which should be taken into account when using these studies.[106] (1) In many cases the program version is not specified, which makes it difficult to track the performance of the same program in different studies. (2) The outcome highly depends on the choice of program settings and the preparation of the receptor structure and molecule libraries (for example, adding hydrogens and missing residues, choosing the correct protonation and tautomerization states), which is linked to the knowledge and familiarity of the scientist with the program and the molecular system under study. (3) It has been shown that neglecting protein flexibility has a higher impact on HTD enrichments than in docking posing.[107,108] (4) In several cases, there is no mention to the average CPU time to dock and score a molecule, critical information to determine the strategy in HTD studies. (5) As noted by Verdonk *et al.*,[109] scoring based on incorrect poses is meaningless, so accuracy at the docking stage is a necessary component (though not sufficient) for a successful HTD. On this issue, a note of concern was raised in the study of Warren *et al.*,[51] where some scoring functions did not require correct poses for an acceptable performance in HTD.

In an interesting study, Moitessier *et al.*[106] analyzed the comparative studies of docking and scoring algorithms published before 2007. Of note among those is the work of Warren *et al.*,[51] who studied 10 docking algorithms and 37 scoring functions on eight unrelated proteins, concluding that no single program performed equally well on all targets. More recently, a study of the scoring functions implemented in DOCK, AMMOS,[110] X-Score, and FRED was performed on five proteins with diverse binding pockets, studying their ability to discriminate active from inactive molecules, using receptor-based focused libraries. The authors showed that the empirical scoring functions used performed better in hydrophobic binding sites, while the FF-based ones were better suited for mixed or polar binding sites.[111] An evaluation of seven docking programs (Surflex,[112] LigandFit, Glide, GOLD, FlexX, eHiTS, and AutoDock) on a set of 1300 protein–ligands complexes from the PDBbind 2007 database[113] showed that the ligand binding conformation can be accurately predicted in most of the cases, while the correlation between docking score and binding affinity was rather weak; in agreement with the study of Warren *et al.*,[51] they also found that the performance of program was system dependent.[114]

The comparative studies mentioned above support the notion that there is no scoring function that consistently outperforms others for all the targets, either at enrichment or at binding free energy prediction. However, in most studies it was found that HTD did enrich the hit list with binders, thus confirming the usefulness of this tool in drug discovery campaigns.

8.3.4 Limitations of Scoring Functions: Can They Be Improved?

Several studies have confirmed the poor correlation between scores and binding free energies.[51,114–116] This is not surprising, considering the number of approximations involved, such as assuming additivity of free energy terms, neglecting protein flexibility, poor handling of entropy, neglecting ligand relaxation,[117] the handling of solvation/desolvation effects,[118] and the use of FF. Although these approximations make scoring functions fast, they add a non-negligible component of noise. Interestingly, although the family of SFCscore functions outperformed most other scoring functions in terms of binding affinity prediction,[73] a clear breakthrough could not be observed. Thus, it seems that within the current representation of the molecular system, and the characterization of the potential energy, scoring functions seem to have reach their intrinsic limit, and achieving an accuracy below 2 pK units seems unreachable.[117] Thus, improvement in scoring should be part of a wider improvement of the whole docking/scoring process.

Free energy is a global property of the system that depends on the volume of the phase space available to the system. Assuming the Hamiltonian of a system to be additive, it can be shown that this implies the enthalpy to be additive, but not the entropy. Thus, free energy cannot be usually expressed as a linear combination of contributions associated with types of interaction (hydrogen bond, van der Waals, *etc.*) or specific molecular groups.[119,120]

The problem of accounting for protein flexibility in HTD has been reviewed by several authors.[11,121,122] Although this is primarily a docking-stage problem, depending on how target flexibility is considered, it may also impact the scoring process.[107] Strategies to overcome this limitation in HTD include the use of soft potentials, averaged grids from different structures, or parallel docking onto a set of conformationally diverse structures, merging the scores/ranks afterwards. In the latter approach, the same scoring function parameterized assuming a single structure per target is usually used. Ideally, it would be desirable to incorporate flexibility on-the-fly by algorithms sampling not only the ligand DOF but also the receptor ones. In this case, a direct FF-based scoring function could be used. However, and due to the computational cost of sampling such a number of DOF, this approach is just feasible as a post-HTD strategy on a reduced molecule library, after a first round of HTD screening (see Section 8.4.1).

The inclusion of water molecules also influences molecule scoring. Usually, water molecules are deleted before HTD, sometimes with the only exception of strongly bound crystallographic waters, to allow the proper docking of molecules that might displace water molecules upon binding. The ideal approach would be to include (or not) water molecules depending on their energy balance, though the complexity of this analysis has been recently highlighted by Michel *et al.*[123] At a practical level, this poses the problem of creating/deleting a molecule at the docking stage, and thus to properly score the presence or absence of water molecules. In GOLD, FITTED,[124] SLIDE,[74,125] and FlexX, water molecules can be switched on and off. In the first three programs there is a penalty function for the displacement of waters (though it should be remarked that there is no single energy value to account for the displacement of a water molecule),[15] while FlexX allows prediction in advance of the favorable locations of water molecules. However, none of these methods has yet been thoroughly validated. In general, empirical and knowledge-based scoring functions should be used under the same conditions with which they were parameterized. For example, if waters were massively deleted during the parameterization process, their effect should be implicitly incorporated in the scoring function, at least partially; thus, adding waters during the docking and scoring may lead somehow to double counting. Recently, Abel *et al.*[21] developed a computational estimator of the contribution of the solvent to the small-molecule binding free energy, evaluating it on a series of congeneric ligands of the factor Xa receptor.

In HTD, ligand entropy associated with rigid DOF (translational and rotational) is usually neglected, though it has been estimated to be in the 3.5–5.0 kcal mol^{-1} range.[126] Vibrational entropy is not considered, and ligand conformational entropy is sometimes accounted for by free torsion counting, as $\Delta S_{conf} = -kTN \ln 3$, where N is the number of free rotatable bonds. However, it was shown that entropy change upon binding does not correlate with the number of free torsions; rather, it is proportional to the logarithm of distinct rotameric states of the molecule. Of course, a more accurate calculation of any of these contributions would make the calculation computationally expensive,[127] and thus their application in HTD would be compromised.

Scoring functions are usually parameterized using databases of ligand–protein complexes for which binding free energies are available, such as PDBbind[113] or BindingDB.[128] It should be ensured that a proper complex preparation regarding the inclusion and orientation of water molecules, assignment of missing hydrogens, and choice of protonation[129,130] and tautomerization[130,131] states is achieved, since these factors may have a dramatic impact in the parameterization of scoring functions. For example, using the PDBbind database, poor results were obtained for all the scoring functions studied when docking to HIV-1 protease[132], while Moitessier *et al.* showed that the omission of a key water molecule and the default automatic assignment of the protonation state to aspartic acids were the reasons for the failure.[106]

Usually, scoring functions are parameterized using experimentally solved structures of proteins complexed with strong binders. Complexes with weak ligands are more rare, since they are difficult to crystallize and are not too appealing for lead optimization purposes. Thus, for most of the time the information of weak ligands and non-binders is not taken into account. In a few approaches, however, both docking ligands and non-binders (decoys) have been considered in the parameterization process, in such a way to maximize the discrimination between them.[60,133] This showed improved results compared to traditional scoring functions trained to reproduce the binding affinity of known ligands.[133] The usefulness of negative data in training scoring functions has been also shown by Pham and Jain.[75] This adds an additional consideration, namely the choice of decoys, which greatly influences the parameterization and evaluation of HTD peformance.[134] Recently, the directory of useful decoys (DUD) was made available to the community.[135] DUD contains ligands and decoys for 40 different targets. For each target, 37 decoy molecules per ligand are selected from the ZINC library[136] to achieve maximal physico-chemical properties similarity with the ligand, and minimal topological similarity with the ligand set.

8.4 Post-Docking Processing in High-Throughput Docking

Given the lack of accuracy of scoring functions in predicting binding affinities, top-scoring compounds are usually subjected to a post-docking processing in an attempt to reduce the number of false positives. False negatives cannot be usually recovered, since they score poorly and are not subject to any further processing.

Top-scoring molecules are usually examined to discard improper binding poses, regardless of their score. For example, it is important to determine if the molecule indeed docked in the binding site, and if its binding pose is meaningful. In particular, they can be examined for key interactions with the receptor known to be conserved throughout known ligand families, such as in protein kinases[137] or GPCRs.[138,139] Clearly, this filtering is plausible of being automated, though visual inspection cannot be substituted. Other filters, such as steric complementarity, can also be applied in an automatic fashion.[140]

Hits can also be clustered according to chemical diversity, its definition depending on the molecular descriptors and metric used. Thus, representative molecules from a wider chemical space (for example, the top-scoring molecule per cluster) can be selected for bio-evaluation.

The assignment to the docked molecules of several scores in a consensus fashion (consensus scoring) has received much attention in recent years.[96,141–143] This emerging technique has been thoroughly review by Feher.[144] Although some early results have been encouraging, a deeper understanding and extensive testing of this technique will help to clarify its potential and limitations. As a necessary condition for a successful performance, each of the individual scoring functions should exhibit high performance and they should all be distinctive.[145] As has been pointed out[8], if correlation exists among scoring functions, consensus scoring may amplify errors rather than cancel them out, thus bringing little improvement to the scoring/ranking process.[146]

Scoring functions are usually trained using a diverse set of ligand–protein complexes. Although, on average, they provide higher enrichments than random selection, this is not valid for all specific targets. Target-specific or tailored scoring functions allow researchers to partially overcome some of the limitations of widely-trained scoring functions.[147–149] The usual approach for tailored scoring functions is to recalibrate existing ones for a given target, to add additional terms, or to complement them with filters.[148] A list of recent developments in this field can be found elsewhere.[148]

Klon *et al.* used 2D fingerprints to train a naïve Bayes classifier based on the scores of top-scoring molecules, and then re-ranking the docked chemical library. This was evaluated on protein tyrosine phosphatase-1B (PTP-1B) and protein kinase B (PKB), showing improved results when compared to plain HTD.[150] A combination with consensus scoring evaluated on the same set of receptors displayed even an enhanced performance.[151] It should be noted that this method was intended to maximize the number of true positives in the top-ranking sub-library, and not to generate more accurate binding poses. Recently, Bourne *et al.* showed that support vector machines, trained by associating sets of individual docking energy terms from the eHiTS[54] program with the available binding affinity of each molecule from high-throughput screening, can enhance the correlation between experimental and calculated binding affinities.[152]

An increasingly used strategy is to re-score top-ranking hits with more computationally expensive scoring methods, aiming in this case for the accurate prediction of the absolute or relative binding free energy.

8.4.1 Rescoring of Top-Ranking Hits Using Advanced Simulation Techniques

8.4.1.1 MM/PB-SA and MM/GB-SA

Studying ligand binding to eight different proteins, Stahl and co-workers showed that the MM/PB-SA approach using only one minimized structure (instead of averaging over the MD trajectories) was enough to discriminate low- from high-affinity binders ($\Delta pIC_{50} \geq 2$–3), but cannot resolve energy differences smaller than those.[153]

A different performance was found by Pearlman studying binding free energy of a congeneric series of 16 ligands to p38 MAP kinase, spanning binding of approximately two orders of magnitude[35]. When compared to TI, OWFEG (one-window free energy grid),[154] ChemScore, PLPScore, and Dock Energy Score, MM/PB-SA performed relatively poorly, remarkably inferior than TI or OWFEG, inferior to Dock Energy Score, and not appreciably better than ChemScore or PLPScore. This might suggest that the performance of MM/PB-SA is system dependent. In line with these results, and using the MM/PB-SA method and the CHARMM FF, a value of -82.8 kcal mol^{-1} was reported for the binding free energy of the Ac-pYEEI peptide to the SH2 domain of human lymphocyte-specific protein tyrosine kinase (LCK), showing a large discrepancy with the experimental value of -9.4 kcal mol^{-1}.[34] The authors also showed that the main source of the discrepancy comes from the enthalpic term.

The binding between avidin and seven biotin analogs was calculated using MM/PB-SA, studying the impact of the FF and the method to generate the complexes.[155] The authors showed that calculations were rather insensitive to the FF of choice, though using different FFs for the geometry generation and the energy calculations was not advisable. Reported mean absolute errors were in the 2.0–4.5 kcal mol^{-1} range, and the standard error in the 1–4 kcal mol^{-1} range, mainly arising from the entropy contribution. In an improved calculation of the entropy evaluated in the binding of seven biotin analogues to avidin, eight (amidinobenzyl)indolecarboxamide inhibitors to factor Xa, and two substrates to cytochrome P450 3A4 and 2C9, the standard error was reduced by a factor of 2–4.[156]

Lyne *et al.* studied the ability of a MM/GB-SA protocol to correctly rank kinase inhibitors after HTD using Glide. This approach performed well in the four cases considered, MAP kinase p38 (p38), Aurora A, cyclin dependent kinase 2 (CDK-2), and c-Jun N-terminal kinase 3 (Jnk-3), always out-performing the HTD scoring function.[157]

Studying ligand specificity on Davies' panel consisting of six protein kinases,[158] p38, cAMP-dependent protein kinase (PKA), glycogen synthase kinase 3β (GSK3β), phosphoinositide-dependent kinase 1 (PDK1), MAP kinase-activated protein kinase 2 (MAPKAP-K2), and LCK, and using MM/PB-SA to estimate binding free energies, Page and Bates obtained equivocal results.[49] Although some correlation with experimental data was obtained, the authors raised questions about the consideration of induced fit and incomplete sampling and, more generally, the validity of FF to study docking and scoring.

Brown and Muchmore decreased by two orders of magnitude the computing time to calculate relative binding free energies with MM/PB-SA using a GB continuum solvent model during MD simulations, an advanced PB solver, and distributed computing. On a set of 18 ligands binding to urokinase, a reasonable correlation with experimental binding free energies was obtained, very similar to the one obtained using the traditional MM/PB-SA method.[159] Recently, this improved methodology was used to estimate binding free energy of 308 small-molecule ligands in complex with the proteins urokinase, PTP-1B, and checkpoint kinase 1 (Chk-1). Statistically significant correlations to experimentally data were reported, with correlation coefficients for the three proteins in the 0.72–0.83 range.[160]

In the solvated interaction energy (SIE) method[161,162] the binding free energy function is expressed as a sum of electrostatic and van der Waals intermolecular interaction terms, with the addition of a solvation energy term calculated using a continuum solvent representation. There is no explicit entropy contribution, but enthalpy–entropy compensation is assumed and represented by an overall scaling factor of ~ 0.1. This method has been parameterized using a training set of 99 different protein complexes and one conformation per complex, showing good agreement with experimental data.

The ability of MM-PBSA rescoring to discriminate between correct and incorrect docking poses was investigated by Thompson *et al.*[163] on a subset of 68 complexes of the CCDC/Astex test set.[164] In this study, calculations were done with the MMFF94s FF,[165,166] the entropy was approximated propor-tional to the number of rotatable bonds, MD simulation was replaced by a simple local energy FF minimization, and the bound conformation of the ligand was used as that corresponding to the ligand in the unbound state. As far as the ability of the rescoring scheme to distinguish between different docked poses of the native conformation of the bound ligand, it was shown that a simple shape-based scoring function was enough. Next, when docking an ensemble of ligand conformations, MM-PBSA did not improve the rescoring over FF rescoring with minimization of the ligand in the field of a rigid protein structure. Performing HTD on the mineralocorticoid receptor

(MR), and using ligands and decoy molecules from the DUD set, MM/PB-SA rescoring showed a marginal improvement for actives in the top 1% of the docked database.

A study of MM/GB-SA rescoring on CDK-2, factor Xa, thrombin, and HIV-RT showed very good agreement with experimental data.[167] MD simulations were substituted by local energy minimization. The authors commented that, while the lack of adequate sampling would not allow the protein to relax and accommodate different scaffolds, this problem should be less important when docking a congeneric series, and that poorly converged simulations add more noise than signal to the calculation. A later contribution showed that replacing the GB-SA protein desolvation by the free energy of displacing water molecules calculated using WaterMap[21,168] provides better results in ranking congeneric series of factor Xa and CDK-2 inhibitors.[169]

Okimoto *et al.* obtained improved enrichment factors by 1.6–4.0 times when rescoring top-ranking compounds with MM/PB-SA on trypsin, HIV protease, acetylcholinesterase (AChE), and CDK-2.[170] However, they noticed that further optimization of the computational protocols is needed to use this methodology in a more high-throughput fashion.

Coveney and co-workers reported MM/PB-SA calculations of lopinavir binding to six HIV-1 protease variants, showing a correlation coefficient of 0.89 and a rmsd of 0.9 kcal mol^{-1}, thus in very good agreement with experimental data.[171]

Some studies have compared the relative performance of MM/PB-SA and MM/GB-SA. It was reported that in calculations with MM/PB-SA and MM/GB-SA on aldose reductase, none of those methods was able to accurately reproduce the absolute binding energies of 28 inhibitors; however, the calculated and experimental values showed good correlation. In terms of estimation of binding free energies, MM/PB-SA performed better than MM/GB-SA.[172] In a recent similar study by the same group, both methods were tested on a series of structurally diverse inhibitors of *Plasmodium falciparum* dihydrofolate reductase (DHFR). Again, calculations with MM/PB-SA and MM/GB-SA showed excellent correlation with experimental values in most of the cases. They also found that this correlation was basically unaffected by using a single minimized complex instead of a structural ensemble from MD simulations, and a continuum solvent model instead of explicit solvent representation, thus saving computing time while keeping the same level of accuracy.[173] A comparison of MM/PB-SA and MM/GB-SA performance on rescoring 45 PTP-1B inhibitors showed good correlation with experimental data, while MM/GB-SA was slightly superior.[174] On the same lines, in a study of 59 ligands and six different proteins, it was shown that the GB performed better than PB in estimating relative binding free energies, while PB performed better to estimate absolute ones. At a methodological level, it was also shown that the calculations were sensitive to the length of the MD simulation and the solute internal dielectric constant.[175] Recently, Linström *et al.* presented a study on several protein–ligand complexes to investigate geometry optimization, pose prediction, and relative binding free energy calculations using the MM/PBSA and MM/GB-SA methods. They found a protocol that in the cases studied performed well both for pose prediction and relative binding free energy prediction.[176]

8.4.1.2   Linear Interaction Energy Calculations

The linear interaction energy (LIE) method[177] uses ensemble averages of the complex and the ligand in isolation in the context of linear response theory. The binding free energy is evaluated as:

$$G_{\text{binding}}^{\text{LIE}} = \alpha \langle U_{l-s}^{\text{vdW}} \rangle + \beta \langle U_{l-s}^{\text{elec}} \rangle + \gamma \qquad (8.7)$$

where $\langle ... \rangle$ represents MC or MD ensemble averages, α, β, and γ are empirical parameters adjusted to reproduce experimental binding free energies, and the Δ represents the difference in the energies in the bound and free state of the ligand.

The γ parameter just depends on the receptor, and thus has no impact when relative binding free energies are sought. Therefore, compared to MM/PB-SA or GB-SA, the LIE requires prior experimental knowledge of ligand binding and fitting parameters.

Huang and Caflisch developed a faster version of the LIE named LIECE, in which the MD sampling is replaced with local energy minimization of the ligand, coupled with a PB description of electrostatics.[178] Jorgensen and co-workers extended the set of energy terms to include surface area contributions and relevant chemical descriptors.[179,180]

Oostenbring *et al.* performed an extensive evaluation of the LIE to predict the binding affinities of docked compounds.[181] Four targets were studied: retinoic acid receptor (RAR) γ, matrix metalloprotease 3, estrogen receptor (ER) α, and dihydrofolate reductase. Both the rms error with experimental binding free energies and the ability to correctly rank molecules were evaluated, comparing the estimated binding free energies with the best out of 10 scoring functions. In all cases, reasonable LIE predictions were obtained in terms of free-energy rms errors, although dihydrofolate reductase proved to be a difficult target both for LIE and the scoring functions. For the other targets, LIE outperformed the scoring functions.

Studying the binding of 22 inhibitors to casein kinase 2 (CK2) with LIE, Bortolato and Moro obtained a correlation of 0.81 and an accuracy of 0.65 kcal mol^{-1} for the training set, while upon validation with 16 external analogs the corresponding values were 0.68 and 0.78 kcal mol^{-1}, respectively.[182]

Åqvist and co-workers compared rescoring with the LIE method to GOLDscore on a set of 43 inhibitors of HIV reverse transcriptase.[183] It was shown that GOLDscore does not reproduce relative binding affinities, while the LIE's standard parameterization reproduced the experimental binding affinities of 39 inhibitors with a correlation coefficient of 0.70 and an average unsigned error (AUE) of 0.8 kcal mol^{-1}.

8.4.1.3 Free Energy Methods

A study on the ranking of congeneric inhibitors of neuraminidase, cyclooxygenase 2, and CDK-2, using Monte Carlo free-energy simulations and a continuum solvent approach, gave very good predictions for the first two receptors, but not for CDK-2. In all cases, the implicit solvent model gave comparable results to simulations with explicit solvent, and were superior to those using empirical scoring functions.[184] Similar results were later obtained from studying the discrimination of 16 ERα ligands from inactive molecules.[185]

The binding of phosphopeptides to the SH2 domain of LCK was investigated by TI. The seven calculated values of relative binding free energy were shown to be in good agreement with experiments.[186] On the same system, in which MM/PB-SA failed by almost an order of magnitude, a PMF approach to calculate the binding free energy of the Ac-pYEEI peptide to the SH2 domain of LCK gave excellent agreement with experimental data.[34]

In the binding energy distribution analysis method (BEDAM) the binding constant is calculated using a weighted integral of the probability distribution of the binding energy obtained in the canonical ensemble, where the solvent is represented in an implicit way using the AGBNP2 model.[187] This method perfectly discriminated binders from decoys in the L99A and L99A/M102Q mutants of T4 lysozyme, showing a reasonable agreement on the prediction of standard binding free energies.[188] Despite the success, the authors acknowledged that further developments are needed to apply this method to more complex targets.

Recently, the relative binding free energy for eight pairs of 3-(amidino-benzyl)-1*H*-indole-2-carboxamide factor Xa inhibitors was calculated using TI. For five pairs the AUE was ~ 0.5 kcal mol^{-1} and for the rest was in the 1.8–4.5 kcal mol^{-1} range, which could be attributed to a different binding mode of the amidinobenzyl group in those molecules.[189] Interestingly, calculations with MM/GB-SA gave slightly better results than those with TI, mainly due to the fact that MM/GB-SA performed better in pairs with a change in net charge. When similar precision was sought, MM/GB-SA exhibited a larger computational overhead.

Metadynamics[190,191] has been also used to study ligand binding in the following complexes: β-trypsin/benzamidine, β-trypsin/chlorobenzamidine, immunoglobulin McPC-603/phosphocholine, and CDK-2/staurosporine. In all cases, this methodology was able to predict correctly the geometry and accurately the free energy of binding.[192] However, there is not yet an exact rule to set up the simulations (number and type of collective variables, size of the hills, total simulation time), although some hints were provided in a recent publication.[193]

8.4.1.4 Quantum Mechanical-Based Rescoring

In 2005, Merz and co-workers showed the feasibility of using a QM-based rescoring method in a set of 165 protein–ligand complexes, both to predict binding affinities and to discriminate binders from non-binders.[194] More recently, using a semi-empirical QM/MM scoring function as implemented in AMBER, and using the DivCon linear-scaling approach,[195] the same group calculated binding free energies on a set of 23 metalloprotein–ligand complexes.[196] The correlation coefficient was 0.64 with a standard deviation of 1.88 kcal mol^{-1} without fitting, which improved to 0.71 and 1.69 kcal mol^{-1}, respectively, when using fitted weighting of the individual scoring terms.

Combining QM/MM with a QM representation of the ligand, and a PB-SA model for solvation, Gräter *et al.* calculated the binding affinities of a set of 47 benzamidine derivatives binding to trypsin. The experimental binding free energies were in the range of -3.9 to -7.6 kcal mol^{-1}. This range was correctly reproduced, with a rms error of 1.2 kcal mol^{-1}. For a separate calculation on ligands of the FK506 protein the rms error was 0.7 kcal mol^{-1}.[197]

Caflisch and co-workers developed a QM version of the LIECE, termed QMLIECE.[198] Both methods were studied on three enzymes, the West Nile virus NS3 serine protease (WNV PR), the aspartic protease of the human immunodeficiency virus (HIV-1 PR), and CDK-2. QMLIECE; all showed an improved performance on 44 peptidic inhibitors of the WNV PR due to the importance of charge polarization in that system. For 24 peptidic inhibitors of HIV-1 PR (20 neutral and four with one formal charge) and for 73 CDK-2 inhibitors (all neutral), both methods performed equally well. The authors concluded the importance of using a QM description of the system when strong charge–charge interactions are present.

Recently, Cavasotto and co-workers introduced a QM-based end-point method for calculating binding free energy in ligand–protein systems termed MM/QM-COSMO (QMSA), which provides an enhanced description of electrostatic interactions while retaining conceptual simplicity and computational efficiency.[26] After extensive conformational sampling through MD simulations, the total energy is re-evaluated using the semi-empirical Hamiltonian PM3[199] within a linear-scaling framework, which allows one to perform consistent QM calculations on the whole system in an affordable time frame. The solvation energy is calculated *via* the COSMO continuum solvent model,[200] for which optimized atomic radii were recently developed.[201] A QM-COSMO energy minimization step precedes the energy re-evaluation of each snapshot. Translational and rotational entropy contributions are calculated through their corresponding configurational integrals,[40,202–204] and the internal (vibrational) entropy is evaluated through normal mode analysis. The binding of a series of tetraphosphopeptides (Ac-pYEEI, Ac-pYAEI, Ac-pYEAI, Ac-pYEEG, Ac-pYEEA) to the SH2 domain of human LCK was studied. End-point methods, MM/PB-SA (using standard[205] and PB-optimized radii[206]), MM/GB-SA, and MM/GB-SA with GB energy minimization prior to rescoring, failed by almost an order of magnitude to accurately reproduce binding free energies (in agreement with an earlier work of Woo and Roux[34]) and to properly rank ligands. Our results confirmed that this large discrepancy arises from the enthalpic contribution. QMSA calculations, however, were in excellent agreement with experimental data (average unsigned error of 0.7 kcal mol^{-1}), with a remarkably good correlation for relative binding free energies. Similar results were obtained studying phosphopeptides binding to the carboxy terminus domain of BRCA1.[207]

The group of Hobza used a fast rescoring of docked complexes based on the semiempirical PM6[208] method, including the DH2 corrections for dispersion energy and hydrogen bonds.[209] The total score consists of the sum of the PM6-DH2 interaction enthalpy, the empirical FF AMBER entropy, and the sum of the deformation and desolvation energies of the ligand, the latter calculated with the SMD[210] method. This methodology builds on the linear-scaling MOZYME[211,212] as implemented in MOPAC[213], and when applied to the HIV-1 protease discriminated well between binders and non-binders, with a reasonable performance in terms of absolute binding free energy prediction. A similar study of 15 CDK-2 ligands[214] showed correlations with experimental binding affinities of 0.87, 0.77, and 0.52, when using only the interaction enthalpy, the interaction enthalpy corrected for ligand desolvation and deformation energies, and also including the entropic term, respectively. Thus, a deterioration is observed upon adding more physical terms to the rescoring function.

8.5   Conclusions

Docking-based screening of chemical libraries in a high-throughput fashion is an established key component in the lengthy process of drug lead discovery. It provides a rational, easy-to-set-up, and economic way to filter large libraries and prioritize a hit-list for bio-evaluation. High-throughput docking is a two-stage process, in which each molecule from the library is posed within the target binding site (docking stage), and that pose (or a few low-energy ones) is assigned a score, which attempts to predict the likelihood of binding to the target (scoring stage). In order to be used in a high-throughput way, docking and scoring hundreds of thousands molecules, scoring functions should trade-off accuracy with speed. Although in most of the numerous studies HTD did enrich the hit list with binders, thus validating the utility of this technique in drug discovery, there is a wide consensus that no scoring function clearly outperforms others for all targets, either at enrichment or at binding affinity calculation. The poor correlation of scoring functions with experimental binding free energies has been shown already by many studies. This fact is quite logical, considering the number of approximations involved, especially regarding the neglect of target flexibility, the poor handling of entropy, and the treatment of solvation/desolvation effects. These approximations are introduced to make scoring functions fast, but they have a non-negligible overhead of noise. It is likely that within the current representation of the biomolecular system, and the characterization of the potential energy, the performance of scoring functions seems to have reach their intrinsic limit, and their improvement should be part of a comprehensive enhancement of the whole docking methodology.

The use of advanced simulation methods at the post-docking stage attempts to correctly predict the binding free energy of the top-ranking library, thus reducing the number of false positives. Since, in principle, these methods are more accurate, their computational overhead is important. However, developments at the algorithmic level, coupled with the wider accessibility to faster computers, open the avenue to routine use of these methods in drug lead discovery and optimization. In this chapter we showed that encouraging results have been obtained using MM/PB-SA, MM/GB-SA, LIE, free energy methods such as FEP and TI, and QM-based approaches, such as the MM/QM-COSMO (QMSA). However, further optimization and validation are needed before they can become reliable computational tools for drug discovery.

References

1. S. S. Phatak, C. C. Stephan and C. N. Cavasotto, *Expert Opin. Drug Discovery*, 2009, **4**, 947.
2. C. N. Cavasotto and A. J. Orry, *Curr. Top. Med. Chem.*, 2007, **7**, 1006.
3. W. L. Jorgensen, *Science*, 2004, **303**, 1813.
4. B. K. Shoichet, *Nature*, 2004, **432**, 862.
5. M. Congreve, C. W. Murray and T. L. Blundell, *Drug Discovery Today*, 2005, **10**, 895.
6. C. R. Beddell, P. J. Goodford, F. E. Norrington, S. Wilkinson and R. Wootton, *Br. J. Pharm.*, 1976, **57**, 201.
7. N. Brooijmans and I. D. Kuntz, *Annu. Rev. Biophys. Biomol. Struct.*, 2003, **32**, 335.
8. D. B. Kitchen, H. Decornez, J. R. Furr and J. Bajorath, *Nat. Rev. Drug Discovery*, 2004, **3**, 935.
9. M. C. Monti, A. Casapullo, C. N. Cavasotto, A. Napolitano and R. Riccio, *ChemBioChem*, 2007, **8**, 1585.
10. M. C. Monti, A. Casapullo, C. N. Cavasotto, A. Tosco, F. Dal Piaz, A. Ziemys, L. Margarucci and R. Riccio, *Chem. Eur. J.*, 2009, **15**, 1155.
11. C. N. Cavasotto and N. Singh, *Curr. Comput. Aided Drug Des.*, 2008, **4**, 221.
12. A. M. Davis, S. J. Teague and G. J. Kleywegt, *Angew. Chem., Int. Ed.*, 2003, **42**, 2718.
13. S. L. McGovern and B. K. Shoichet, *J. Med. Chem.*, 2003, **46**, 2895.
14. C. H. Reynolds and M. K. Holloway, *ACS Med. Chem. Lett.*, 2011, **2**, 433.
15. H. Gohlke and G. Klebe, *Angew. Chem., Int. Ed.*, 2002, **41**, 2644.
16. C. Bissantz, B. Kuhn and M. Stahl, *J. Med. Chem.*, 2010, **53**, 5061.
17. D. H. Williams, E. Stephens, D. P. O'Brien and M. Zhou, *Angew. Chem., Int. Ed.*, 2004, **43**, 6596.
18. N. R. Syme, C. Dennis, S. E. Phillips and S. W. Homans, *ChemBioChem*, 2007, **8**, 1509.

19. B. J. Berne, J. D. Weeks and R. Zhou, *Annu. Rev. Phys. Chem.*, 2009, **60**, 85.

20. S. W. Homans, *Drug Discovery Today*, 2007, **12**, 534.

21. R. Abel, T. Young, R. Farid, B. J. Berne and R. A. Friesner, *J. Am. Chem. Soc.*, 2008, **130**, 2817.

22. K. T. Butler, F. J. Luque and X. Barril, *J. Comput. Chem.*, 2009, **30**, 601.

23. Z. Fu, X. Li and K. M. Merz, Jr., *J. Comput. Chem.*, 2011, **32**, 2587.

24. A. R. Khan, J. C. Parrish, M. E. Fraser, W. W. Smith, P. A. Bartlett and M. N. James, *Biochemistry*, 1998, **37**, 16839.

25. C. E. Chang, W. Chen and M. K. Gilson, *Proc. Natl. Acad. Sci. U. S. A.*, 2007, **104**, 1534.

26. V. M. Anisimov and C. N. Cavasotto, *J. Comput. Chem.*, 2011, **32**, 2254.

27. J. J. Gray, *Curr. Opin. Struct. Biol.*, 2006, **16**, 183.

28. J. A. McCammon, *Curr. Opin. Struct. Biol.*, 1998, **8**, 245.

29. M. K. Gilson, J. A. Given, B. L. Bush and J. A. McCammon, *Biophys. J.*, 1997, **72**, 1047.

30. H. X. Zhou and M. K. Gilson, *Chem. Rev.*, 2009, **109**, 4092.

31. D. L. Beveridge and F. M. DiCapua, *Annu. Rev. Biophys. Biophys. Chem.*, 1989, **18**, 431.

32. T. P. Straatsma and J. A. McCammon, *Methods Enzymol.*, 1991, **202**, 497.

33. S. Izrailev, S. Stepaniants, M. Balsera, Y. Oono and K. Schulten, *Biophys. J.*, 1997, **72**, 1568.

34. H. J. Woo and B. Roux, *Proc. Natl. Acad. Sci. U. S. A.*, 2005, **102**, 6825.

35. D. A. Pearlman, *J. Med. Chem.*, 2005, **48**, 7796.

36. J. Srinivasan, T. E. Cheatham, P. Cieplak, P. A. Kollman and D. A. Case, *J. Am. Chem. Soc.*, 1998, **120**, 9401.

37. Y. N. Vorobjev and J. Hermans, *Biophys. Chem.*, 1999, **78**, 195.

38. M. Feig and C. L. Brooks, III, *Curr. Opin. Struct. Biol.*, 2004, **14**, 217.

39. D. Qiu, P. S. Shenkin, F. P. Hollinger and W. C. Still, *J. Phys. Chem. A*, 1997, **101**, 3005.

40. T. Lazaridis, A. Masunov and F. Gandolfo, *Proteins*, 2002, **47**, 194.

41. J. M. Swanson, R. H. Henchman and J. A. McCammon, *Biophys. J.*, 2004, **86**, 67.

42. M. K. Gilson and H. X. Zhou, *Annu. Rev. Biophys. Biomol. Struct.*, 2007, **36**, 21.

43. H. Luo and K. Sharp, *Proc. Natl. Acad. Sci. U. S. A.*, 2002, **99**, 10399.

44. W. Chen, M. K. Gilson, S. P. Webb and M. J. Potter, *J. Chem. Theory Comput.*, 2010, **6**, 3540.

45. O. Kalid and N. Ben-Tal, *J. Chem. Inf. Model.*, 2009, **49**, 865.

46. B. Kuhn and P. A. Kollman, *J. Med. Chem.*, 2000, **43**, 3786.

47. R. C. Rizzo, S. Toba and I. D. Kuntz, *J. Med. Chem.*, 2004, **47**, 3065.

48. I. Stoica, S. K. Sadiq and P. V. Coveney, *J. Am. Chem. Soc.*, 2008, **130**, 2639.

49. C. S. Page and P. A. Bates, *J. Comput. Chem.*, 2006, **27**, 1990.

50. R. Rajamani and A. C. Good, *Curr. Opin. Drug Discovery Dev.*, 2007, **10**, 308.

51. G. L. Warren, C. W. Andrews, A. M. Capelli, B. Clarke, J. LaLonde, M. H. Lambert, M. Lindvall, N. Nevins, S. F. Semus, S. Senger, G. Tedesco, I. D. Wall, J. M. Woolven, C. E. Peishoff and M. S. Head, *J. Med. Chem.*, 2006, **49**, 5912.

52. S. Raub, A. Steffen, A. Kamper and C. M. Marian, *J. Chem. Inf. Model.*, 2008, **48**, 1492.

53. M. D. Eldridge, C. W. Murray, T. R. Auton, G. V. Paolini and R. P. Mee, *J. Comput. Aided Mol. Des.*, 1997, **11**, 425.

54. Z. Zsoldos, I. Szabo, Z. Szabo and A. P. Johnson, *J. Mol. Struct. (THEOCHEM)*, 2003, **666**, 659.

55. M. Rarey, B. Kramer, T. Lengauer and G. Klebe, *J. Mol. Biol.*, 1996, **261**, 470.

56. D. Rognan, S. L. Lauemoller, A. Holm, S. Buus and V. Tschinke, *J. Med. Chem.*, 1999, **42**, 4650.

57. R. A. Friesner, J. L. Banks, R. B. Murphy, T. A. Halgren, J. J. Klicic, D. T. Mainz, M. P. Repasky, E. H. Knoll, M. Shelley, J. K. Perry, D. E. Shaw, P. Francis and P. S. Shenkin, *J. Med. Chem.*, 2004, **47**, 1739.

58. A. N. Jain, *J. Comput. Aided Mol. Des.*, 1996, **10**, 427.

59. P. Cozzini, M. Fornabaio, A. Marabotti, D. J. Abraham, G. E. Kellogg and A. Mozzarelli, *J. Med. Chem.*, 2002, **45**, 2469.

60. M. Totrov and R. Abagyan, in *RECOMB '99: Proceedings of the Third Annual International Conference on Computational Molecular Biology*, ed. S. Istrail, P. Pevzner and M. Waterman, ACM Press, Lyon, 1999.

61. F. Giordanetto, S. Cotesta, C. Catana, J. Y. Trosset, A. Vulpetti, P. F. Stouten and R. T. Kroemer, *J. Chem. Inf. Comput. Sci.*, 2004, **44**, 882.

62. A. Krammer, P. D. Kirchhoff, X. Jiang, C. M. Venkatachalam and M. Waldman, *J. Mol. Graph. Model.*, 2005, **23**, 395.

63. Y. T. Tang and G. R. Marshall, *J. Chem. Inf. Model.*, 2011, **51**, 214.

64. O. Korb, T. Stutzle and T. E. Exner, *J. Chem. Inf. Model.*, 2009, **49**, 84.

65. D. K. Gehlhaar, G. M. Verkhivker, P. A. Rejto, C. J. Sherman, D. B. Fogel, L. J. Fogel and S. T. Freer, *Chem. Biol.*, 1995, **2**, 317.

66. K. Tondel, E. Anderssen and F. Drablos, *J. Comput. Aided Mol. Des.*, 2006, **20**, 131.

67. N. Moitessier, E. Therrien and S. Hanessian, *J. Med. Chem.*, 2006, **49**, 5885.

68. P. J. Ballester and J. B. Mitchell, *Bioinformatics*, 2010, **26**, 1169.

69. H. J. Böhm, *J. Comput. Aided Mol. Des.*, 1994, **8**, 243.

70. H.-J. Böhm, *J. Comput. Aided Mol. Des.*, 1998, **12**, 309.

71. J. Pei, Q. Wang, Z. Liu, Q. Li, K. Yang and L. Lai, *Proteins*, 2006, **62**, 934.

72. M. Stahl and M. Rarey, *J. Med. Chem.*, 2001, **44**, 1035.

73. C. A. Sotriffer, P. Sanschagrin, H. Matter and G. Klebe, *Proteins*, 2008, **73**, 395.

74. V. Schnecke and L. A. Kuhn, *Perspect. Drug Discovery Des.* 2000, **20**, 171.
75. T. A. Pham and A. N. Jain, *J. Med. Chem.*, 2006, **49**, 5856.
76. R. Wang, L. Lai and S. Wang, *J. Comput. Aided Mol. Des.*, 2002, **16**, 11.
77. G. M. Morris, D. S. Goodsell, R. S. Halliday, R. Huey, W. E. Hart, R. K. Belew and A. J. Olson, *J. Comput. Chem.*, 1998, **19**, 1639.
78. E. C. Meng, B. K. Shoichet and I. D. Kuntz, *J. Comput. Chem.*, 1992, **13**, 505.
79. A. Grosdidier, V. Zoete and O. Michielin, *Proteins*, 2007, **67**, 1010.
80. G. Jones, P. Willett, R. C. Glen, A. R. Leach and R. Taylor, *J. Mol. Biol.*, 1997, **267**, 727.
81. C. McMartin and R. S. Bohacek, *J. Comput. Aided Mol. Des.*, 1997, **11**, 333.
82. W. T. Mooij and M. L. Verdonk, *Proteins*, 2005, **61**, 272.
83. J. B. O. Mitchell, R. A. Laskowski, A. Alex and J. M. Thornton, *J. Comput. Chem.*, 1999, **20**, 1165.
84. C. Zhang, S. Liu, Q. Zhu and Y. Zhou, *J. Med. Chem.*, 2005, **48**, 2325.
85. H. F. Velec, H. Gohlke and G. Klebe, *J. Med. Chem.*, 2005, **48**, 6296.
86. H. Gohlke, M. Hendlich and G. Klebe, *J. Mol. Biol.*, 2000, **295**, 337.
87. C. Y. Yang, R. Wang and S. Wang, *J. Med. Chem.*, 2006, **49**, 5903.
88. I. Muegge and Y. C. Martin, *J. Med. Chem.*, 1999, **42**, 791.
89. I. Muegge, *J. Med. Chem.*, 2006, **49**, 5895.
90. R. S. DeWitte and E. I. Shakhnovich, *J. Am. Chem. Soc.*, 1996, **118**, 11733.
91. A. V. Ishchenko and E. I. Shakhnovich, *J. Med. Chem.*, 2002, **45**, 2770.
92. D. A. Pearlman, D. A. Case, J. W. Caldwell, W. R. Ross, T. E. Cheatham III, S. DeBolt, D. Ferguson, G. Seibel and P. A. Kollman, *Comput. Phys. Commun.*, 1995, **91**, 1.
93. A. D. MacKerell, D. Bashford, M. Bellott, R. L. Dunbrack, J. D. Evanseck, M. J. Field, S. Fischer, J. Gao, H. Guo, S. Ha, D. Joseph-McCarthy, L. Kuchnir, K. Kuczera, F. T. K. Lau, C. Mattos, S. Michnick, T. Ngo, D. T. Nguyen, B. Prodhom, W. E. Reiher, B. Roux, M. Schlenkrich, J. C. Smith, R. Stote, J. Straub, M. Watanabe, J. Wiorkiewicz-Kuczera, D. Yin and M. Karplus, *J. Phys. Chem. B*, 1998, **102**, 3586.
94. W. L. Jorgensen, D. S. Maxwell and J. Tirado-Rives, *J. Am. Chem. Soc.*, 1996, **118**, 11225.
95. C. M. Oshiro, I. D. Kuntz and J. S. Dixon, *J. Comput. Aided Mol. Des.*, 1995, **9**, 113.
96. M. L. Verdonk, J. C. Cole, M. J. Hartshorn, C. W. Murray and R. D. Taylor, *Proteins*, 2003, **52**, 609.
97. D. S. Goodsell, G. M. Morris and A. J. Olson, *J. Mol. Recognit.*, 1996, **9**, 1.
98. T. A. Halgren, R. B. Murphy, R. A. Friesner, H. S. Beard, L. L. Frye, W. T. Pollard and J. L. Banks, *J. Med. Chem.*, 2004, **47**, 1750.
99. M. R. McGann, H. R. Almond, A. Nicholls, J. A. Grant and F. K. Brown, *Biopolymers*, 2003, **68**, 76.

100. *DockIt*, Metaphorics, Aliso Viejo, CA, http://www.metaphorics.com/products/dockit.html.
101. C. M. Venkatachalam, X. Jiang, T. Oldfield and M. Waldman, *J. Mol. Graph. Model.*, 2003, **21**, 289.
102. *ICM*, version 3.6, MolSoft, La Jolla, CA, 2009.
103. G. Nemethy, K. D. Gibson, K. A. Palmer, C. N. Yoon, M. G. Paterlini, A. Zagari, S. Rumsey and H. A. Scheraga, *J. Phys. Chem.*, 1992, **96**, 6472.
104. T. Halgren, *J. Comput. Chem.*, 1995, **17**, 490.
105. F. H. Allen, *Acta Crystallogr., Sect. B: Struct. Sci.*, 2002, **58**, 370.
106. N. Moitessier, P. Englebienne, D. Lee, J. Lawandi and C. R. Corbeil, *Br. J. Pharmacol.*, 2008, (suppl. 1), S7.
107. C. N. Cavasotto and R. A. Abagyan, *J. Mol. Biol.*, 2004, **337**, 209.
108. C. N. Cavasotto, J. A. Kovacs and R. A. Abagyan, *J. Am. Chem. Soc.*, 2005, **127**, 9632.
109. M. L. Verdonk, V. Berdini, M. J. Hartshorn, W. T. Mooij, C. W. Murray, R. D. Taylor and P. Watson, *J. Chem. Inf. Comput. Sci.*, 2004, **44**, 793.
110. T. Pencheva, D. Lagorce, I. Pajeva, B. O. Villoutreix and M. A. Miteva, *BMC Bioinformatics*, 2008, **9**, 438.
111. T. Pencheva, O. S. Soumana, I. Pajeva and M. A. Miteva, *Eur. J. Med. Chem.*, 2010, **45**, 2622.
112. A. N. Jain, *J. Med. Chem.*, 2003, **46**, 499.
113. R. Wang, X. Fang, Y. Lu, C. Y. Yang and S. Wang, *J. Med. Chem.*, 2005, **48**, 4111.
114. D. Plewczynski, M. Lazniewski, R. Augustyniak and K. Ginalski, *J. Comput. Chem.*, 2011, **32**, 742.
115. J. S. Duca, V. S. Madison and J. H. Voigt, *J. Chem. Inf. Model.*, 2008, **48**, 659.
116. M. Sandor, R. Kiss and G. M. Keseru, *J. Chem. Inf. Model.*, 2010, **50**, 1165.
117. J. Tirado-Rives and W. L. Jorgensen, *J. Med. Chem.*, 2006, **49**, 5880.
118. D. L. Mobley and K. A. Dill, *Structure*, 2009, **17**, 489.
119. K. A. Dill, *J. Biol. Chem.*, 1997, **272**, 701.
120. A. E. Mark and W. F. van Gunsteren, *J. Mol. Biol.*, 1994, **240**, 167.
121. F. Spyrakis, A. Bidon-Chanal, X. Barril and F. J. Luque, *Curr. Top. Med. Chem.*, 2011, **11**, 192.
122. R. E. Amaro and W. W. Li, *Curr. Top. Med. Chem.*, 2010, **10**, 3.
123. J. Michel, J. Tirado-Rives and W. L. Jorgensen, *J. Am. Chem. Soc.*, 2009, **131**, 15403.
124. C. R. Corbeil, P. Englebienne and N. Moitessier, *J. Chem. Inf. Model.*, 2007, **47**, 435.
125. M. I. Zavodszky and L. A. Kuhn, *Protein Sci.*, 2005, **14**, 1104.
126. C. W. Murray and M. L. Verdonk, *J. Comput. Aided Mol. Des.*, 2002, **16**, 741.
127. A. M. Ruvinsky, *J. Comput. Aided Mol. Des.*, 2007, **21**, 361.
128. T. Liu, Y. Lin, X. Wen, R. N. Jorissen and M. K. Gilson, *Nucleic Acids Res.*, 2007, **35**, D198.

129. H. Steuber, P. Czodrowski, C. A. Sotriffer and G. Klebe, *J. Mol. Biol.*, 2007, **373**, 1305.

130. T. ten Brink and T. E. Exner, *J. Chem. Inf. Model.*, 2009, **49**, 1535.

131. F. Milletti, L. Storchi, G. Sforna, S. Cross and G. Cruciani, *J. Chem. Inf. Model.*, 2009, **49**, 68.

132. R. Wang, Y. Lu, X. Fang and S. Wang, *J. Chem. Inf. Comput. Sci.*, 2004, **44**, 2114.

133. P. Englebienne and N. Moitessier, *J. Chem. Inf. Model.*, 2009, **49**, 2564.

134. M. L. Verdonk, V. Berdini, M. J. Hartshorn, W. T. Mooij, C. W. Murray, R. D. Taylor and P. Watson, *J. Chem. Inf. Comput. Sci.*, 2004, **44**, 793.

135. N. Huang, B. K. Shoichet and J. J. Irwin, *J. Med. Chem.*, 2006, **49**, 6789.

136. J. J. Irwin and B. K. Shoichet, *J. Chem. Inf. Model.*, 2005, **45**, 177.

137. C. N. Cavasotto, M. A. Ortiz, R. A. Abagyan and F. J. Piedrafita, *Bioorg. Med. Chem. Lett.*, 2006, **16**, 1969.

138. C. N. Cavasotto, A. J. Orry, N. J. Murgolo, M. F. Czarniecki, S. A. Kocsi, B. E. Hawes, K. A. O'Neill, H. Hine, M. S. Burton, J. H. Voigt, R. A. Abagyan, M. L. Bayne and F. J. Monsma, Jr., *J. Med. Chem.*, 2008, **51**, 581.

139. S. S. Phatak, E. A. Gatica and C. N. Cavasotto, *J. Chem. Inf. Model.*, 2010, **50**, 2119.

140. M. Stahl and H. J. Bohm, *J. Mol. Graph. Model.*, 1998, **16**, 121.

141. P. S. Charifson, J. J. Corkery, M. A. Murcko and W. P. Walters, *J. Med. Chem.*, 1999, **42**, 5100.

142. E. M. Krovat and T. Langer, *J. Chem. Inf. Comput. Sci.*, 2004, **44**, 1123.

143. A. Oda, K. Tsuchida, T. Takakura, N. Yamaotsu and S. Hirono, *J. Chem. Inf. Model.*, 2006, **46**, 380.

144. M. Feher, *Drug Discovery Today*, 2006, **11**, 421.

145. J. M. Yang, Y. F. Chen, T. W. Shen, B. S. Kristal and D. F. Hsu, *J. Chem. Inf. Model.*, 2005, **45**, 1134.

146. P. Englebienne and N. Moitessier, *J. Chem. Inf. Model.*, 2009, **49**, 1568.

147. A. J. Orry, R. A. Abagyan and C. N. Cavasotto, *Drug Discovery Today*, 2006, **11**, 261.

148. M. H. Seifert, *Drug Discovery Today*, 2009, **14**, 562.

149. C. N. Cavasotto and S. S. Phatak, *Methods Mol. Biol.*, 2011, **685**, 155.

150. A. E. Klon, M. Glick, M. Thoma, P. Acklin and J. W. Davies, *J. Med. Chem.*, 2004, **47**, 2743.

151. A. E. Klon, M. Glick and J. W. Davies, *J. Med. Chem.*, 2004, **47**, 4356.

152. S. L. Kinnings, N. Liu, P. J. Tonge, R. M. Jackson, L. Xie and P. E. Bourne, *J. Chem. Inf. Model.*, 2011, **51**, 408.

153. B. Kuhn, P. Gerber, T. Schulz-Gasch and M. Stahl, *J. Med. Chem.*, 2005, **48**, 4040.

154. D. A. Pearlman, *J. Med. Chem.*, 1999, **42**, 4313.

155. A. Weis, K. Katebzadeh, P. Soderhjelm, I. Nilsson and U. Ryde, *J. Med. Chem.*, 2006, **49**, 6596.

156. J. Kongsted and U. Ryde, *J. Comput. Aided Mol. Des.*, 2009, **23**, 63.

157. P. D. Lyne, M. L. Lamb and J. C. Saeh, *J. Med. Chem.* 2006, **49**, 4805.

158. S. P. Davies, H. Reddy, M. Caivano and P. Cohen, *Biochem. J.*, 2000, **351**, 95.
159. S. P. Brown and S. W. Muchmore, *J. Chem. Inf. Model.*, 2007, **47**, 1493.
160. S. P. Brown and S. W. Muchmore, *J. Med. Chem.*, 2009, **52**, 3159.
161. M. Naïm, S. Bhat, K. N. Rankin, S. Dennis, S. F. Chowdhury, I. Siddiqi, P. Drabik, T. Sulea, C. I. Bayly, A. Jakalian and E. O. Purisima, *J. Chem. Inf. Model.*, 2007, **47**, 122.
162. Q. Cui, T. Sulea, J. D. Schrag, C. Munger, M. N. Hung, M. Naim, M. Cygler and E. O. Purisima, *J. Mol. Biol.*, 2008, **379**, 787.
163. D. C. Thompson, C. Humblet and D. Joseph-McCarthy, *J. Chem. Inf. Model.*, 2008, **48**, 1081.
164. J. W. M. Nissink, C. Murray, M. Hartshorn, M. L. Verdonk, J. C. Cole and R. Taylor, *Proteins*, 2002, **49**, 457.
165. T. A. Halgren, *J. Comput. Chem.*, 1996, **17**, 616.
166. T. A. Halgren, *J. Comput. Chem.*, 1999, **20**, 720.
167. C. R. Guimaraes and M. Cardozo, *J. Chem. Inf. Model.*, 2008, **48**, 958.
168. T. Beuming, R. Farid and W. Sherman, *Protein Sci.*, 2009, **18**, 1609.
169. C. R. Guimaraes and A. M. Mathiowetz, *J. Chem. Inf. Model.*, 2010, **50**, 547.
170. N. Okimoto, N. Futatsugi, H. Fuji, A. Suenaga, G. Morimoto, R. Yanai, Y. Ohno, T. Narumi and M. Taiji, *PLoS Comput. Biol.*, 2009, **5**, e1000528.
171. S. K. Sadiq, D. W. Wright, O. A. Kenway and P. V. Coveney, *J. Chem. Inf. Model.*, 2010, **50**, 890.
172. A. M. Ferrari, G. Degliesposti, M. Sgobba and G. Rastelli, *Bioorg. Med. Chem.*, 2007, **15**, 7865.
173. G. Rastelli, A. Del Rio, G. Degliesposti and M. Sgobba, *J. Comput. Chem.*, 2010, **31**, 797.
174. X. Zhang, X. Li and R. Wang, *J. Chem. Inf. Model.*, 2009, **49**, 1033.
175. T. Hou, J. Wang, Y. Li and W. Wang, *J. Chem. Inf. Model.*, 2011, **51**, 69.
176. A. Lindström, L. Edvinsson, A. Johansson, C. D. Andersson, I. E. Andersson, F. Raubacher and A. Linusson, *J. Chem. Inf. Model.*, 2011, **51**, 267.
177. J. Aqvist, C. Medina and J. E. Samuelsson, *Protein Eng.*, 1994, **7**, 385.
178. D. Huang and A. Caflisch, *J. Med. Chem.*, 2004, **47**, 5791.
179. R. C. Rizzo, J. Tirado-Rives and W. L. Jorgensen, *J. Med. Chem.*, 2001, **44**, 145.
180. Y. Tominaga and W. L. Jorgensen, *J. Med. Chem.*, 2004, **47**, 2534.
181. E. Stjernschantz, J. Marelius, C. Medina, M. Jacobsson, N. P. Vermeulen and C. Oostenbrink, *J. Chem. Inf. Model.*, 2006, **46**, 1972.
182. A. Bortolato and S. Moro, *J. Chem. Inf. Model.*, 2007, **47**, 572.
183. J. Carlsson, L. Boukharta and J. Åqvist, *J. Med. Chem.*, 2008, **51**, 2648.
184. J. Michel, M. L. Verdonk and J. W. Essex, *J. Med. Chem.*, 2006, **49**, 7427.
185. J. Michel and J. W. Essex, *J. Med. Chem.*, 2008, **51**, 6654.
186. P. W. Fowler, S. Geroult, S. Jha, G. Waksman and P. V. Coveney, *J. Chem. Theory Comput.*, 2007, **3**, 1193.

187. E. Gallicchio and R. M. Levy, *J. Comput. Chem.*, 2004, **25**, 479.
188. E. Gallicchio, M. Lapelosa and R. M. Levy, *J. Chem. Theory Comput.*, 2010, **6**, 2961.
189. S. Genheden, I. Nilsson and U. Ryde, *J. Chem. Inf. Model.*, 2011, **51**, 947.
190. A. Laio and M. Parrinello, *Proc. Natl. Acad. Sci. U. S. A.*, 2002, **99**, 12562.
191. A. Laio, A. Rodriguez-Fortea, F. L. Gervasio, M. Ceccarelli and M. Parrinello, *J. Phys. Chem. B*, 2005, **109**, 6714.
192. F. L. Gervasio, A. Laio and M. Parrinello, *J. Am. Chem. Soc.*, 2005, **127**, 2600.
193. X. Biarnes, S. Bongarzone, A. V. Vargiu, P. Carloni and P. Ruggerone, *J. Comput. Aided Mol. Des.*, 2011, **25**, 395.
194. K. Raha and K. M. Merz, Jr., *J. Med. Chem.*, 2005, **48**, 4558.
195. S. L. Dixon and K. M. Merz, Jr., *J. Chem. Phys.*, 1996, **104**, 6643.
196. S. A. Hayik, R. Dunbrack and K. M. Merz, Jr., *J. Chem. Theory Comput.*, 2010, **6**, 3079.
197. F. Gräter, S. M. Schwarzl, A. Dejaegere, S. Fischer and J. C. Smith, *J. Phys. Chem. B*, 2005, **109**, 10474.
198. T. Zhou, D. Huang and A. Caflisch, *J. Med. Chem.*, 2008, **51**, 4280.
199. J. J. P. Stewart, *J. Comput. Chem.*, 1989, **10**, 209.
200. A. Klamt and G. Schüürmann, *J. Chem. Soc., Perkin Trans. 2*, 1993, 799.
201. V. M. Anisimov and C. N. Cavasotto, *J. Phys. Chem. B*, 2011, **115**, 7896.
202. T. Lazaridis and M. Karplus, *Proteins*, 1999, **35**, 133.
203. M. S. Lee and M. A. Olson, *Biophys. J.*, 2006, **90**, 864.
204. J. M. J. Swanson, R. H. Henchman and J. A. McCammon, *Biophys. J.*, 2004, **86**, 67.
205. A. Bondi, *J. Phys. Chem.*, 1964, **68**, 441.
206. J. M. J. Swanson, S. A. Adcock and J. A. McCammon, *J. Chem. Theory Comput.*, 2005, **1**, 484.
207. V. M. Anisimov, A. Ziemys, S. Kizhake, Z. Yuan, A. Natarajan and C. N. Cavasotto, 2011, submitted.
208. J. J. P. Stewart, *J. Mol. Model.*, 2007, **13**, 1173.
209. J. Rezac, J. Fanfrlik, D. Salahub and P. Hobza, *J. Chem. Theory Comput.*, 2009, **5**, 1749.
210. A. V. Marenich, C. J. Cramer and D. G. Truhlar, *J. Phys. Chem. B*, 2009, **113**, 6378.
211. J. J. P. Stewart, *Int. J. Quantum Chem.*, 1996, **58**, 133.
212. J. J. P. Stewart, *J. Mol. Model.*, 2009, **15**, 765.
213. J. J. P. Stewart, *MOPAC2009*, Stewart Computational Chemistry (www.OpenMOPAC.net), Colorado Springs, CO, 2009.
214. P. Dobes, J. Fanfrlik, J. Rezac, M. Otyepka and P. Hobza, *J. Comput. Aided Mol. Des.*, 2011, **25**, 223.

CHAPTER 9

Accounting for Target Flexibility During Ligand–Receptor Docking

SIMON LEIS AND MARTIN ZACHARIAS*

Technische Universität München, Physik-Department T38, James Franck Str. 1, D-85748 Garching, Germany
*E-mail: martin.zacharias@ph.tum.de

9.1 Introduction

An essential goal of computational drug design is to reduce expenses within the designing process, either by the selection of putative lead structures from databases of drug-like chemical compounds or by *de novo* design of active substances.[1] Target-based drug design uses available three-dimensional (3D) structural information of the receptor to dock compound libraries (target-based virtual screening) or *de novo* designed ligands that specifically bind to the target. In recent years, the number of 3D structures of protein molecules has increased significantly. Additionally, for a growing number of protein target sequences with sequence similarity to a known protein structure (template), comparative modeling methods allow for building fairly accurate model structures (depending on the degree of target–template sequence similarity). The rapid growth of structural knowledge on biomolecular drug target molecules forms the basis for the increasing applicability of structure-based ligand–receptor docking and drug design applications. The ultimate goal of structure-based docking and drug design is the identification of putative ligands, the prediction of the binding geometry, and the prediction of the binding affinity to a given receptor molecule.

RSC Drug Discovery Series No. 23
Physico-Chemical and Computational Approaches to Drug Discovery
Edited by F. Javier Luque and Xavier Barril
© The Royal Society of Chemistry 2012
Published by the Royal Society of Chemistry, www.rsc.org

The affinity and specificity of binding reactions is determined by the structural and physico-chemical properties of the binding partners and the solution environment. The basis for this specificity was first investigated by Emil Fischer in 1894.[2] He addressed the foundations of specific binding by introducing the well-known lock-and-key analogy to describe enzyme–substrate interactions. The basic idea of this concept is that the substrate has to fit specifically like a key into the binding site (lock) of an enzyme. The lock-and-key concept was developed further by Koshland, who proposed that a global conformational change of the enzyme hexokinase is necessary to adapt to its substrate.[3] He developed the idea of "induced fit" recognition, meaning that both partners can structurally differ in their unbound (apo) and bound (holo) conformations. During the association process, the interacting molecules induce conformational changes that are required to achieve high affinity and specificity of binding. It has also been recognized that, in principle, all possible molecular binding processes require a certain degree of conformational adaptation.

Binding reactions can be accompanied by a variety of conformational changes. The magnitude of such changes can range from alterations of side-chain conformations[4] at the binding site to global changes of domain arrangements[5–7] and can even involve refolding of protein segments upon association.[8] Based on ideas from statistical physics, the induced-fit concept has been extended, suggesting a pre-existing ensemble of several interconvertible conformational states that are in equilibrium.[9] These states include conformations close to unbound but also near-bound forms. Binding of a partner molecule to the near-bound form stabilizes this structure and shifts the equilibrium towards the bound form.[10] Computational approaches to realistically model and predict ligand–receptor binding geometries should preferably include such conformational changes during receptor–ligand docking simulations. It is the focus of the present chapter to give an overview on available computational strategies to include receptor conformational changes during docking methods and to discuss their strengths and weaknesses.

While considerable progress has been achieved in modeling the conformational flexibility of ligands during docking over the last decade, inclusion of receptor flexibility is still an unsolved problem, especially in case of significant changes in the protein backbone structure upon binding. In the following, we will present recent efforts and progress on including receptor flexibility during docking, with special emphasis on global and semi-global receptor conformational changes in protein–ligand docking calculations and on aspects of computational efficiency. Various levels of flexibility have been considered in docking approaches. A schematic illustration of the different levels of protein flexibility that can play a role during association is given in Figure 9.1. Since the great majority of drug targets are proteins, the focus will be on the flexibility of proteins. However, most aspects of conformational flexibility are of general importance also for other types of receptor biomolecules.

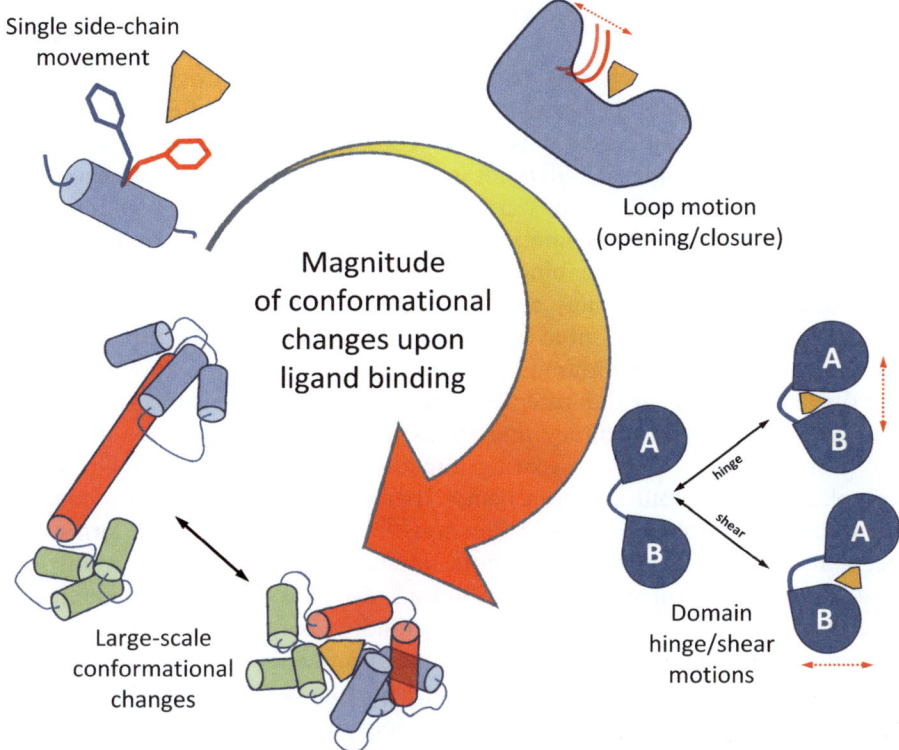

Figure 9.1   Schematic overview of different conformational changes that can occur in a receptor protein upon ligand binding, with increasing magnitude following the large arrow from the upper to the lower left. The place and direction of conformational changes are illustrated in red, potential ligand molecules are shown as orange triangular structures, and protein domains are labeled with the letters A and B.

9.2   Thermodynamic Driving Forces of Binding and the Scoring Problem

Thermodynamics tells us that the driving force for binding of two molecules is given by the free energy change associated with the binding reaction. A free energy change indicates that there are entropic and energetic (enthalpic) components involved in binding reactions.[11–14] On the energetic part, the association process can involve changes in electrostatic interactions, van der Waals interactions and hydrogen bonding. These types of interactions can include contributions within and between the interacting biomolecules, but in addition also between the biomolecules and surrounding solvent molecules (usually water and possibly ions). Besides internal energy or enthalpy, the binding process can also involve changes in the entropy of the biomolecules and the surrounding solvent molecules.[15] For example, binding of biomolecules can result in a change in flexibility of the binding partners affecting the conformational entropy of both molecules. Association can also lead to a change in the ordering of solvent molecules (resulting in changes of enthalpy and entropy) around the binding partners that influences the association process (hydrophobic effect).

There are two important aspects of computational methods to predict receptor–ligand complexes. Firstly, it is desirable to predict an arrangement as close to reality as possible and, secondly, to score it appropriately with respect to alternative complex geometries. Since several enthalpic and entropic contributions influence the binding affinity, the scoring function needs to realistically account for many contributions to binding affinity. At the same time, docking approaches need to be fast enough to allow for rapid docking and evaluation of many thousand putative complexes. Computational speed is a critical issue for ligand–receptor docking, which requires a reasonable compromise between accuracy and speed to calculate a score for a ligand–receptor complex. Current scoring functions to evaluate ligand–receptor complexes range from simple schemes that just account for steric complementarity to complete force field functions.[16–18] These can include terms that account for steric and electrostatic interactions but also account for desolvation and hydrophobic contributions to ligand–receptor interaction.[15] The change in conformational entropy of binding partners also influences the binding affinity. However, so far only a few approaches have tackled the issue of including conformational entropy effects during docking scoring.[19,20]

Instead of scoring functions based on a molecular mechanics force field, it is also possible to design knowledge-based potentials to evaluate complexes that are extracted from known receptor–ligand complexes. The basic idea of such knowledge-based scoring is to relate the observed frequency of atom–atom (or group–group) contacts to the expected contact frequency at receptor–ligand interfaces to extract favorable and unfavorable atom–atom interactions.[21] Here, expected contact frequency means contacts that are obtained if atoms would be distributed randomly at the interface.

The design of an appropriate scoring function for the realistic evaluation of ligand–receptor complexes is still an unresolved issue. The problem of designing a realistic scoring function for computational docking is not the focus of the current chapter. However, its relation to the problem of how to account efficiently for conformational changes is of interest. Frequently, the difficulties of realistic docking (in terms of minimal deviation from an experimental complex structure) and scoring are considered as separate issues. However, since scoring functions are often exclusively parameterized using experimental structures of protein–ligand complexes, it is important to generate docking models already in the sampling phase that come as close as possible to the native complex structure. Appropriate treatment of conformational flexibility during docking is, therefore, tightly connected to the improvement of realistic scoring of docked complexes. The design of more rigorous and more specific scoring functions requires at the same time an improvement of the prediction accuracy of binding modes in terms of deviation from the experimental binding geometry. Highly accurate scoring of a ligand placement requires that the complex geometry has also been predicted with high precision. *Vice versa*, the more errors a scoring function may tolerate, the less specific it becomes. Consequently, there is a direct relation between the robustness or softness of a scoring function and the number of false positives obtained in a virtual screen.

9.3 Molecular Motions of Proteins and Ligands

For a reliable prediction of ligand–receptor binding geometries, it is desirable to include possible conformational changes of both the ligand and the receptor structure. However, depending on the number of added degrees of freedom, the computational demand can increase dramatically during ligand–receptor docking. Hence, it is important to analyze in detail what types of conformational changes can occur in proteins upon ligand binding. In many cases, only minor conformational changes have been found when comparing crystal structures of unbound and bound protein conformations.[22] However, in general, conformational changes can range from local side-chain adjustments to global motions of entire subunits, as indicated in Figure 9.1. Gerstein *et al.* hierarchically classified motions according to their size, into fragment, domain, and subunit motions.[23] The motion of fragments smaller than domains commonly refers to the motion of surface loops, but also the motion of secondary structures. The amplitude of the motion is proportional to its characteristic time-scale; as a consequence, domain motions are still in many cases out of the scope of detailed computer simulation methods such as molecular dynamics simulations.[24]

Domain and fragment motions often involve portions of the protein closing around a binding site. Hence, they make up for specific mechanisms of induced fit. Both hinge and shear motions can contribute to up to 70% of the overall motions in both the fragment and the domain class.[23,25] A simpler way of classification is to discriminate between side-chain and backbone motions. In the light of the aforementioned variety of putative independent small-scale and concerted large-scale motions, one might expect huge conformational changes on different levels at least for those proteins exhibiting large fluctuations in the absence of a binding partner. Luque and Freire found that binding sites can feature regions of high flexibility (often most flexible regions of the entire protein), but also regions of high rigidity.[26] Regarding the atomic displacements involved during conformational changes, however, Najmanovich *et al.* concluded after an extensive database analysis of protein–ligand complexes that in 85% of cases only three or less residues actually change their conformation upon binding.[4] It was also found that certain amino acids exhibit significantly more flexibility than others, *e.g.* Lys is on average more flexible than Phe. Besides, 94% of χ_1 (first side-chain dihedral angle) and 96% of χ_2 dihedral angles remain in their native conformation upon complex formation. Additionally, the authors observed that only in 12% of the complexes does a backbone displacement of more than 2 Å take place upon binding, whereas in 75% of the complexes a backbone motion of less than 1 Å was found. Consequently, the authors concluded that compared to side-chain flexibility, backbone flexibility is of minor importance. However, it has also been observed that receptor backbone conformational changes in the range of 1 Å can significantly affect receptor–ligand interaction.[27] In this regard, one could as well conclude that, in more than 25% of all cases observed, backbone motions might play a decisive role in complex formation. Considering that the set of available experimental structures today is most probably biased towards less flexible proteins (since often proteins with highly flexible parts are more difficult to crystallize), the number of cases where conformational changes of the receptor is of importance might be even higher.

Another important aspect of considering target flexibility during docking is the increased availability of protein model structures generated based on homology to a known structure. Frequently, such homology models are used in ligand–receptor docking studies.[28–30] Depending on the degree of sequence similarity to a known template, modeled structures can contain various types of deviations from the native protein conformation. Hence, an appropriate modeling of receptor flexibility is of particular importance during docking to homology models in order to achieve realistic predictions.

9.4   Accounting for Conformational Flexibility During Docking

9.4.1   Inclusion of Ligand Flexibility

Basically all popular docking approaches include ligand flexibility, which is reasonable, because even small changes in dihedral torsion angles around rotatable bonds can significantly change the shape and electrostatic potential around a ligand molecule. Many methods have been developed and applied, including soft docking,[31,32] ligand conformational ensembles,[33] molecular dynamics,[34] or Monte Carlo simulations,[35] and various smart advanced sampling strategies.[36] Several current efficient docking programs employ "build-up" or incremental construction approaches.[37,38] The incremental construction scheme involves splitting the ligand into fragments that are assumed to be conformationally rigid. Subsequently, a starting fragment (anchor) is docked into the target site and the complete ligand is generated by attaching fragments to a growing ligand in the binding pocket. The connection between fragments allows for conformational adjustment of the ligand to optimally fit into the binding site. The efficient inclusion of ligand flexibility during docking has already been extensively reviewed.[39]

9.4.2   Accounting Implicitly for Receptor Flexibility

As discussed in the previous paragraphs, the analysis of available protein structures indicates in several cases only small differences of receptor structures in a complex with a ligand (bound/holo form) compared to the apo (unbound) form. However, even in such cases, docking of ligands to rigid receptor structures may not be successful due to the sensitivity of steric interactions with respect to even small conformational adjustments in a receptor binding pocket. The application of a soft interaction scoring function represents one possibility to still keep a rigid receptor structure during docking but allowing for some steric overlap between receptor and ligand. Various functional forms of soft scoring functions have been suggested that allow for various degrees of steric overlap between ligand and receptor atoms.[31,32,40,41] Intuitively, such an approach is reasonable since, even in a lock-and-key mechanism of ligand–receptor binding, small atom displacements in the receptor pocket might be possible (*e.g.* due to thermal fluctuations). The thermal fluctuations can effectively reduce steric overlap between ligand and receptor atoms, which can be approximately described by a softening of the steric interactions. A common method to implicitly account for limited conformational changes in the receptor structure is a broadening or shift of the repulsive part of the van der Waals interaction potential (soft-core potential). Thus, slightly increased overlap of atoms upon docking is permitted. One should keep in mind, however, that atomic motions that may lead to a reduction of atomic overlap are usually strongly coupled motions with defined direction, meaning that a motion of one atom to reduce overlap also affects the position and interaction of neighboring atoms. In contrast, simple softening of interactions "reduces" possible overlap independent of neighboring atoms. Nevertheless, soft scoring functions are widely used during docking searches.[42] However, one should keep in mind that uniform softening of steric interactions can greatly reduce the specificity of interactions, resulting in false-positive docking solutions or unrealistic placement of the ligand in the binding pocket.[43]

9.4.3 Ligand–Receptor Docking Using Molecular Dynamics Simulations

In principle, it is possible to perform ligand–receptor docking searches allowing for conformational changes of both the receptor and ligand structure in all Cartesian or bond rotation degrees of freedom. This can be achieved by using energy minimization (EM), molecular dynamics (MD) or Monte Carlo (MC) simulations (or related simulation methods). Such simulations are typically based on a molecular mechanics force field describing all intra- and intermolecular interactions of receptor and ligand molecules.[17,44] In MD simulations, Newton's equations of motion are solved numerically in small time steps (1–2 fs; 1 fs $= 10^{-15}$ s), allowing in principle for full ligand and receptor flexibility.[45] In the case of docking a ligand into a receptor pocket, one typically starts the simulation from various start placements of the ligand near the expected binding site. For computational efficiency, the flexible parts of the protein are frequently restricted to the vicinity of the binding site and the rest of the protein is kept rigid.[46] However, owing to the presence of energy barriers, MD simulations can be trapped in unrealistic docking sub-states and may require long and computationally demanding simulations to reach a realistic complex structure (conformational sampling problem). Di Nola *et al.* partly solved this problem by applying a different coupling scheme for the heat bath. In their method, the center of mass of the ligand moves at a considerably higher temperature than the remaining system.[34] Mangoni *et al.* expanded the approach by allowing for internal motions of the receptor, which was coupled to a low-temperature bath.[47] Alternatively, sampling of ligand placement and receptor conformational states can be improved by simulated annealing methods[48] or scaling/rescaling methods of the ligand–receptor interaction potential during MD simulations.[49,50] Such techniques allow a more rapid convergence to an optimal interface structure with simultaneous adjustment of both side-chain and backbone interface structures. More recently, advanced MD sampling methods have been combined with docking searches.[51,57] These approaches show promising results on test systems but require further testing on larger sets of ligand–receptor pairs to comprehensively prove their applicability for ligand–receptor docking.

The calculation of the ligand binding affinity to a known receptor binding site and the evaluation of ligand modifications on binding affinity is another application area of MD-based approaches. A prominent example is the MM/PBSA (Molecular Mechanics/Poisson–Boltzmann, Surface Area) method or the MM/GBSA method. The latter method employs the Generalized Born approach to calculate electrostatic interactions instead of the more time consuming PB approach.[52] In both cases, an ensemble of conformations of the receptor–ligand complex generated by MD simulations is evaluated, based on a continuum model (Poisson–Boltzmann or Generalized Born model) for the surrounding solvent. The application to the complex as well as to the isolated receptor and ligand molecules results in an estimate of the ligand–receptor interaction. Although computationally much more demanding than using standard docking scoring functions, the approach has been employed also in virtual compound screening efforts.[53] Other even more demanding methods are based on thermodynamic perturbation or thermodynamic integration, where a ligand or parts of a ligand are created or annihilated during MD simulation (reviewed in refs. 54 and 55). Although mostly applied for the evaluation of a selected number of ligands or modifications of ligands,[50] the development of more efficient free energy simulation methods and increasing computer performance may allow for a growing use of such approaches to evaluate ligand–receptor binding affinity.[56]

A major drawback of MD simulations applied to ligand–receptor docking is the large computational demand due to the many flexible degrees of freedom of both solute and solvent molecules. The steady increase of computer power and the implementation of smart sampling methods will undoubtedly broaden the applicability of MD simulation methods for docking searches. However, even in the case of implicit solvent models and a known receptor binding site, the computational effort can be too large to systematically dock multiple ligands. Therefore, MD methods are currently more applied to lead structure optimization or refinement of a small set of pre-selected docking poses. Indeed, multistep docking approaches containing an MD-based refinement step are becoming increasingly popular.[54,57]

9.4.4 Treatment of Side-Chain Flexibility and Local Protein Backbone Changes

In the case of any systematic exploration of ligands and possible binding placements, it is desirable to restrict the flexibility to the fewer degrees of freedom that correspond to the most important variables for the binding adjustment. The variation of bond lengths and angle vibrations makes up only for a comparably small contribution to conformational changes. The motions that contribute most are movements of dihedrals (bond rotation). Often, conformational changes upon ligand–protein association are limited to changes in the side chains that form the binding site. In such cases, the inclusion of side-chain reorientations during docking can result in drastic improvements of the docking performance.[58] It is well known that side chains possess conformational preferences for a discrete number of dihedral states.[59] Dunbrack and Karplus[59] and others[60–65] compiled the most common values for side-chain dihedral angle states from a large database of protein structures. Initially, these libraries were used for assigning side-chain conformations to a given backbone structure in comparative modeling. The first investigation of the applicability of side-chain rotamer states from a library during docking was undertaken by Leach in 1994.[66] There exist backbone dependent and independent libraries, small libraries that cover only the most prominent states, and exhaustive libraries providing residence probabilities for each state.[67] Several available docking programs can include side-chain conformational changes at a proposed ligand binding site, either at the level of searching for optimal discrete side-chain dihedral angle combinations (rotamers) or upon minimization of side-chain dihedrals during docking. The dead-end elimination method and the A* algorithm were tested that allow for the selection of a best combination of possible side-chain rotamers in the spatial vicinity of binding cavities.[68,69] Alternative approaches based on a multi-greedy strategy or a branch-and-cut algorithm have also been used.[70] Several two-stage docking methods allow for a relaxation of side-chain conformations, allowing for continuous dihedral changes either by energy minimization or Monte Carlo at a final stage of docking for a limited set of selected complexes.[71] Wang *et al.* proposed a possible refinement of the rotamer method called Rotamer Trials and Minimization (RTMIN) for protein–protein docking.[72] This method allows for sampling of different rotameric states coupled with a subsequent continuous side-chain adjustment and was applied extensively to protein–ligand docking.[73] Side-chain optimization during docking can also be achieved by energy minimization or employing Monte Carlo methods, allowing continuous adjustment of side-chain dihedral angles and local adjustment of the protein main chain.[74–77] A combination of loop structure prediction and docking was described by Sherman *et al.*, employing several loop copies which were generated in a pre-sampling run.[77] During docking, a particular copy that exhibits the most favorable interactions with the ligand was selected.

9.5 Conformational Ensemble Methods

To account efficiently for larger conformational changes of the main chain and side chains near the binding site, it is also possible to represent the binding site by an ensemble of protein conformations.[78,79] For many proteins of biological or pharmaceutical interest, experimental crystal structures in the apo form, and often also in complex with different ligands, are available. In such cases, the ensemble of target conformations can be formed by the available experimental structures. Alternatively, computational methods such as MD simulations can be used to generate conformational ensembles.[54] Others have generated ensembles using a combination of loop fragments.[80] Distance geometry approaches, as for example implemented in tCONCOORD,[81] can also be used to obtain conformational ensembles. The structures are generated by fulfilling a set of upper and lower inter-atomic distances, where the difference between these upper and lower boundaries depends on the estimated interaction strength and steric hindrance. The resulting structures are usually analyzed by principal component analysis (PCA) to identify a maximally diverse collection of possible conformations. A set of methods, including MD simulations with different solvents and tCONCOORD, has been compared and applied to identify flexible ligand binding pockets on the surface of proteins.[82] The CONCOORD method has also been used to generate conformations in reasonable agreement with bound structures, based on data extracted from unbound structure combined with the radius of gyration of the bound ligand–receptor complex structure.[83] In the CONCOORD/PBSA approach the generated ensembles are successfully used to efficiently sample protein flexibility when predicting free energy and protein stability changes upon protein–protein binding.[84]

Several groups made use of principle components of motion[85] or normal modes to deform a starting receptor structure and thus generate an ensemble. Mustard and Ritchie, for example, employed distance constraint essential dynamics eigenstructures as input for their protein–protein docking program Hex.[86] Cavasotto *et al.* used relevant normal modes to generate an ensemble for the cAMP dependent kinase.[87] These authors also introduced a measure of relevance to select a set of mid-frequency modes to be able to cover localized backbone motions.

Irrespective of the way an ensemble is obtained, the question arises whether the structural snapshots sufficiently cover the protein's conformational space as a whole. It is not clear how many different structures are necessary. Similar to rotamer library and loop copy approaches, there is one nontrivial problem: if there is no "correct" or near correct conformation included in the set of structures, there is no guarantee for improvement over single receptor docking, and the prediction accuracy might even drop below the level of docking using a single rigid receptor.

9.5.1 Employing Structural Ensembles in Docking Calculations

Once an ensemble has been obtained, ligands can for example be docked successively against every receptor structure. For instance, Pang and Kozikowski[88] used 69 snapshots from a short MD simulation in a docking study; Barril and Morley employed a large set of experimentally derived X-ray structures for successive docking;[89] Moreno and Leon proposed to generate a binding site descriptor from a set of protein conformations rather than from a single structure for input in the program DOCK.[90] Similarly, the relaxed complex scheme by McCammon and co-workers utilizes a set of MD snapshots for docking searches in combination with the AutoDock program as docking engine.[91] This concept was extended by Carlson *et al.*, who proposed the "dynamic pharmacophore model",[92,93] where a pharmacophore can be modeled for each structure in the ensemble. In turn, the intersection of all pharmacophore models is called the dynamic pharmacophore, which can then be utilized to identify putative new ligands. The authors tested their approach on HIV integrase and on HIV protease. For multiple protein structures from an MD simulation, it was possible to distinguish true inhibitors from drug-like non-inhibitors.[94] However, docking to each individual structure of an ensemble can become highly computationally demanding and requires evaluation of a large set of alternative structures which may increase the chance of obtaining many false positive docking solutions.

One possibility is to calculate a mean field from all the structures in the entire ensemble. For example, discrete sets of side-chain and protein loop conformations can be efficiently combined in a mean field to self-consistently optimize the ligand–receptor interface structure during the docking calculation (Figure 9.2). In such a mean field approach, each alternative side-chain or main-chain conformation (conformational copy) is given an initial weight based on the internal energy and the interaction with a ligand (usually a Boltzmann weight). During docking, the weights on each copy can change such that in the finally docked structure the optimal side-chain/main-chain structure can be identified as the copy with highest weight.

Many structure-based drug design and docking programs employ pre-calculated grids for representation of the receptor–ligand interactions. The energy function of the receptor atoms in the binding site are first mapped onto a regular grid and interactions with ligand atoms are calculated by interpolation from neighboring grid points. A number of researchers have tried to map not only a single receptor structure onto such grids. Early attempts by Knegtel *et al.* computed composite energy-weighted or geometry-weighted grids, from a set of experimental protein structures.[95] For energy-weighting, the authors computed grids for each structure and subsequently

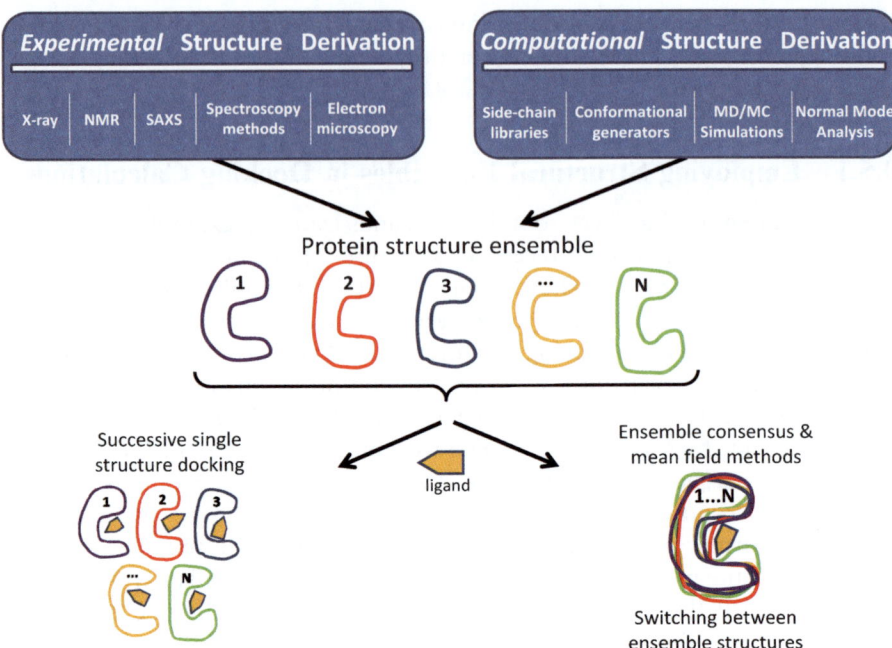

Figure 9.2 Schematic overview of ensemble docking methods and experimental or computational sources for ensemble protein structures. Experimental or computational methods contribute several (ranging from just two to hundreds) of structures in different conformational states. The ensemble is then used for either separate docking to each ensemble structure (*lower left*) or methods that combine or switch between the conformers within the ensemble (*lower right*).

averaged over several grids; thereby, at each grid point, grids with high negative values get high weights, grids with high positive values get small weights. In the case of geometry weighting, first an average structure is calculated, thereby using the mean structure for regions with low flexibility, whereas regions with high flexibility are retained as independent conformational copies. Finally, a grid is calculated for this average structure and used during docking and scoring. Österberg *et al.* reported similar findings: when testing four different ways of combining grids, energy weighting or Boltzmann weighting works well, whereas simple averaging can lead to an increase of incorrect docking solutions.[96] To further improve the efficiency, it is possible to represent part of the structural ensemble by a single rigid structure and to treat only the most flexible part of the binding site by explicit alternative structures.[97] Instead of docking ligands to each conformation in an ensemble, it is possible to use mean field grid representations to reduce the computational expense.[95] Several variants have been presented, mainly differing in the way of averaging the interaction energy contribution.[96,98]

More sophisticated ensemble docking protocols have recently been suggested to avoid the increased computational effort in sequential docking all putative ligands to all target conformers (reviewed by Totrov and Abagyan[99]). Huang and Zou suggested an ensemble docking method that allows simultaneous optimization of placement and receptor conformation out of a set of protein structures and found significant improvement compared to docking to single structures.[100] Besides using different experimental structures of a protein, it is also possible to employ ensembles of homology models.[101]

An interesting new approach for grid-based ensemble docking is the so called 4D-docking method.[102] Here, all structures are sorted by their conformational deviations. During docking, the integer index of the sorted stack is used as a fourth dimension. Hence, the search space can be reduced by consideration of limited conformational deviation, according to the actually considered structure (index neighbors). This allows a faster convergence towards optimal docking placement. Although the use of ensembles of structures can improve docking results, it has also been found that the success depends critically on the choice of the ensemble.[17,103] The underlying assumption of an ensemble docking approach is that conformations sufficiently close to the bound form are included in the ensemble. As indicated in the previous paragraph on generating ensembles, the absence of a receptor structure close to the bound form may deteriorate the docking performance. Recent efforts have therefore also focused on methods to obtain optimal conformational ensembles.[104,105]

9.6 Use of Collective Modes to Describe Global Motions

The application of ensemble approaches, as discussed above, is one promising route to account for the various conformational changes that can occur during ligand binding. However, it has the drawback of a discrete representation of predicted flexibility. In recent years the analysis of protein motions by MD simulations has indicated that most of the conformational fluctuations can be described by a few collective degrees of freedom. These degrees of freedom can be obtained by the PCA of the covariance of atomic fluctuations during MD simulations (also termed essential dynamics analysis).[106] Alternatively, normal mode analysis, that is the analysis of the curvature of the potential energy function around an energy minimum, can also be used to extract collective degrees of large mobility in proteins and other biomolecules. It requires the diagonalization of the second derivative matrix of the energy with respect to the atomic coordinates, yielding harmonic or normal (mass-weighted) modes of the biomolecule. To use the softest modes from a normal mode analysis as variables during flexible docking was suggested by Zacharias and Sklenar and first applied to DNA in complex with a minor groove binding ligand.[107] It allowed for fast minimization of the receptor structure during docking and for an estimation of the receptor deformation energy. The calculation of an energy associated with the deformation is based on the eigenvalue of the corresponding mode. The degree of deformation of the receptor in the soft degrees of freedom can be controlled by a penalty potential to limit deformations. This is only an approximation to the receptor internal energy change; however, it is sufficient for detecting possible ligands and putative binding sites. It avoids the computationally costly explicit calculation of the internal receptor energy at every docking minimization step. Compared to docking methods that employ ensembles of discrete (rigid) receptor structures, the receptor conformation can change continuously during docking (in the pre-calculated soft normal modes) and has therefore an increased capacity for induced fit adaptation.

Normal mode analysis has also been used to study induced fit binding in the case of integrase[108] from human immunodeficiency virus (HIV) and in combination with MD to study ligand binding to HIV protease.[109] However, pre-calculation of harmonic modes of a large receptor molecule requires very extensive energy minimization and is usually performed in the absence of solvent. Energy minimization under these conditions can lead to large deviations from the realistic (experimental) receptor structure and the calculated soft harmonic modes may not correspond to realistic soft degrees of freedom.

The calculation of flexible degrees of freedom of a protein molecule can be performed under more realistic conditions using a PCA of motions obtained during a MD simulation including surrounding waters and ions.[106] Flexible (essential) modes from the MD simulation can then be used as additional variables during docking as described above. This has been explored for the immunosuppressant FK 506 to an "unbound" conformation of a FK506 binding protein, FKBP, using the program PCrelax.[110] Accounting for relaxation in the pre-calculated soft modes of the receptor significantly improved the docking performance compared to docking to a rigid "unbound" FKBP receptor structure at a modest increase in computational demand.[110] The approach can in principle lead to a dramatic reduction of the computational complexity to account for receptor flexibility during docking. This may form the basis for systematic docking, including approximately for global conformational changes in the receptor. The contribution of global motions obtained from MD simulations has also been investigated to support protein–protein docking efforts.[98]

One drawback of the calculation of flexible degrees of freedom of a protein receptor molecule is the high computational demand of calculating either harmonic modes at atomic resolution or principal components of MD trajectories. Also, calculation of principal components of motion from an MD simulation can significantly depend on the simulation length and convergence.[111,112] Elastic network models (ENM) of protein motion[113,114] are based on simplified spring models of proteins and on the assumption that the mobility of a protein region is determined by the local density (or the locally available free space).[94,95] Harmonic mode analysis of this simple energy function allows us to identify possible flexible (soft) collective degrees of freedom (soft modes) of the protein within a few minutes of computer time.[114] Tama and Sanejouand found that such approximate mode calculations resulted in predicted soft modes that showed significant overlap with observed conformational changes in proteins determined experimentally under different conditions (*e.g.* different crystal forms or apo *vs.* bound form of a protein molecule).[115] In some cases, over 50% of the conformational difference between two structures of a protein, determined for example by X-ray analysis of two crystal forms or as ligand-free and bound forms, could be assigned to a single approximate soft mode of the protein.[115] In Figure 9.3, this is demonstrated for the enzyme protein kinase A. In this case, just 10 softest modes obtained from a ENM analysis applied to the apo form of the enzyme are sufficient to approximate the backbone conformational changes observed between apo and an inhibitor bound form to >50%.

It is also possible to use ENM analysis to identify rigid or flexible units in a protein and to define hinge regions.[116] Sandak *et al.* proposed the concept of hinge-domain-bending motion to account for global domain motion during docking with promising results.[117,118] It was shown that, with this approach, relatively large displacements of the receptor backbone structure can be achieved during docking in directions that overlapped significantly with experimentally observed changes.[113–115,119] Inclusion of normal mode minimization during docking was systematically explored by May and Zacharias for protein–protein complexes.[44,120,121] Inclusion of up to five softest modes for protein partners during docking improved the docking results at a very modest increase of computer time by a factor of 2–3 compared to rigid

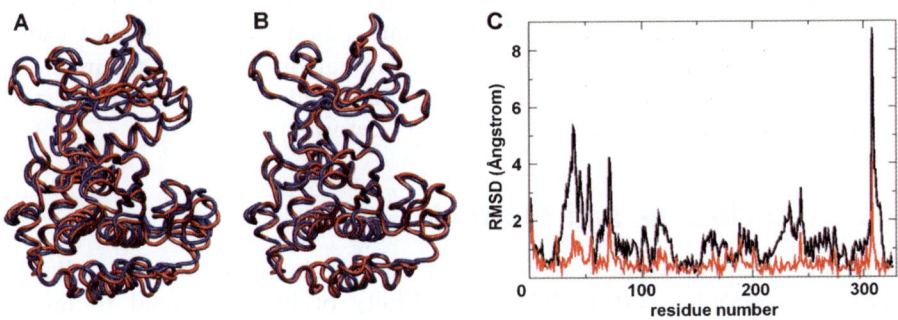

Figure 9.3   (A) Conformational difference between bound (pdb1STC; blue tube representation; complex with stauroporine, ligand not shown) and unbound (pdb 1J3H; red tube) structures of the catalytic domain of protein kinase A (PKA). The backbone rmsd between the two structures is 1.7 Å. (B) Superposition of the bound PKA and the apo structure deformed in the 10 softest normal modes obtained from an elastic network calculation on the apo form to give the smallest possible deviation from the bound form. (C) Backbone rmsd of apo and bound PKA (black line) *vs.* residue number compared to the backbone rmsd for the apo structure optimally deformed in the 10 softest normal modes (red line). The rmsd between the optimally deformed structure and the bound PKA structure is 0.9 Å.

docking. However, the systematic test on several protein–protein complexes also indicated that in order to achieve realistic docking predictions, both side-chain flexibility and global flexibility need to be accounted for simultaneously during docking.[44] One promising route for future developments might be a combined treatment of global flexibility employing soft global degrees of freedom together with several copies of binding site loops for the protein main chain and representation of side-chain flexibility by discrete rotameric states. Such methodology was used to dock several inhibitors to the protein kinase CDK2 (cyclin-dependent kinase 2). Application to different X-ray structures in the apo and various bound forms allowed an evaluation by cross-docking of different inhibitor molecules.[120] Application of the pre-calculated soft modes as flexible variables both with and without inclusion of side-chain flexibility resulted in improved ranking as well as ligand placement during docking. Interestingly, accounting for flexibility in the soft modes alone gave overall similar docking performance as in case of including side-chain flexibility (alone or in combination with normal mode minimization), but at significantly reduced computational costs. Abagyan and co-workers recently applied normal mode deformations to generate conformational variants of receptor structures and used the conformers in a Monte Carlo search during ligand docking.[122] For this application it is necessary to represent each receptor conformation by a potential grid.

Kazemi *et al.* recently presented an interesting variant of the grid representation to account for structural flexibility.[123] The idea of the approach is to deform the geometry of potential grids associated with the receptor structure following possible global conformational changes. The deformed grid can still serve as a basis for calculating interactions with the ligand using interpolation from nearest grid points. In the initial application of the method, however, it was necessary to know the initial and final structures of the receptor in order to determine the necessary deformation.

9.7   Conclusions and Outlook

Ultimately, ligand–receptor docking approaches should be able to reliably predict placement, conformation, and affinity of ligands bound to target receptor molecules. For rigid receptor structures, at least the prediction of a near native binding geometry is often possible. However, many proteins, including some of the most prominent drug targets like protein kinases and HIV protease, undergo significant conformational changes upon complex formation with substrates or inhibitors. Significant progress has been achieved in recent years in particular to better account for changes in side-chain conformation during structure-based drug design and docking. However, efficient and sufficiently accurate inclusion of local but especially more global backbone conformational changes remains a challenge.[17,124] Since one is typically interested in screening large numbers of putative ligands, it is important to find an optimal compromise between required accuracy and feasibility in terms of currently available computational resources. Current methods to tackle this problem range from brute force and time consuming but very detailed MD simulation approaches to conformational ensemble methods and methods that try to restrict the flexible degrees of freedom to a subset of most relevant degrees of freedom. It should be emphasized that even if computational resources are available, the realistic modeling of conformational changes and adaptation is also limited by the accuracy of the underlying force field. Hence, inclusion of too many degrees of freedom may also degrade the performance of docking and scoring approaches. Methods that are based on identifying relevant soft degrees of freedom of a given protein receptor structure are computationally rapid and may allow for approximate inclusion of global flexibility during screening of large databases of putative ligands. Combinations of such approaches with modeling of side-chain flexibility on top of continuous soft mode backbone motion could be promising routes for future developments. Another promising effort is the design of carefully prepared conformational ensembles either based on experimental or modeled structures. Such methods could be valuable for rapid screening of large numbers of putative ligands followed by subsequent refinement steps. As discussed above, there is also a close relation between the accuracy of scoring a ligand–receptor complex and the inclusion of conformational changes during docking. A soft scoring function that tolerates inaccurate placement of a ligand in a binding pocket or allows a large degree of overlap between ligand and receptor atoms can only be of limited specificity. Future efforts to improve scoring of ligand receptor complexes may also include entropic contributions due to changes of receptor and ligand flexibility during the binding process.[18,19]

Acknowledgements

We thank the Deutsche Forschungsgemeinschaft (DFG) for financial support (grant ZA153/5-3).

References

1. G. Schneider and U. Fechner, *Nat. Rev. Drug Discovery*, 2005, **4**, 649.
2. E. Fischer, *Ber. Dt. Chem. Ges.*, 1894, **27**, 2985.
3. D. E. Koshland, Jr., W. J. Ray, Jr. and M. J. Erwin, *Fed. Proc.*, 1958, **17**, 1145.
4. R. Najmanovich, J. Kuttner, V. Sobolev and M. Edelman, *Proteins*, 2000, **39**, 261.
5. M. Gerstein, A. M. Lesk and C. Chothia, *Biochemistry*, 1994, **33**, 6739.
6. A. M. Lesk and C. Chothia, *J. Mol. Biol.*, 1984, **174**, 175.
7. M. Gerstein and C. Chothia, *J. Mol. Biol.*, 1991, **220**, 133.
8. R. B. Best and G. Hummer, *Science*, 2009, **323**, 593.
9. C. J. Tsai, S. Kumar, B. Ma and R. Nussinov, *Protein Sci.*, 1999, **8**, 1181.

10. H. A. Carlson and A. McCammon, *Mol. Pharmacol.*, 2000, **57**, 213.
11. A. V. Finkelstein and J. Janin, *Protein Eng.*, 1989, **3**, 1.
12. G. Klebe and H. J. Bohm, *J. Recept. Signal. Transduct. Res.*, 1997, **17**, 459.
13. W. P. Jencks, *Proc. Natl. Acad. Sci. U. S. A.*, 1981, **78**, 4046.
14. P. R. Andrews, D. J. Craik and J. L. Martin, *J. Med. Chem.*, 1984, **27**, 1648.
15. A. R. Leach, *Molecular Modelling, Principles and Applications*, Pearson, Dorchester, UK, 2001.
16. S. A. Adcock and J. A. McCammon, *Chem. Rev.*, 2006, **106**, 1589.
17. C. B. Rao, J. Subramanian and S. D. Sharma, *Drug Discovery Today*, 2009, **14**, 394.
18. S.-Y. Huang and X. Zou, *Int. J. Mol. Sci.*, 2010, **11**, 3016.
19. N. Trbovic, J.-H. Cho, R. Abel, R. A. Friesner, M. Rance and A. G. Palmer, *J. Am. Chem. Soc.*, 2008, **131**, 615.
20. S. Y. Huang and X. Zou, *J. Chem. Inf. Model.*, 2010, **50**, 262.
21. I. Muegge, *J. Med. Chem.*, 2006, **49**, 5895.
22. K. Gunasekaran and R. Nussinov, *J. Mol. Biol.*, 2007, **365**, 257.
23. M. Gerstein and W. Krebs, *Nucleic Acids Res.*, 1998, **26**, 4280.
24. T. Hansson, C. Oostenbrink and W. van Gunsteren, *Curr. Opin. Struct. Biol.*, 2002, **12**, 190.
25. W. G. Krebs, V. Alexandrov, C. A. Wilson, N. Echols, H. Yu and M. Gerstein, *Proteins*, 2002, **48**, 682.
26. I. Luque and E. Freire, *Proteins*, 2000, suppl. 4, 63.
27. I. Halperin, B. Ma, H. Wolfson and R. Nussinov, *Proteins*, 2002, **47**, 409.
28. A. Schafferhans and G. Klebe, *J. Mol. Biol.*, 2001, **307**, 407.
29. A. Hillisch, L. F. Pineda and R. Hilgenfeld, *Drug Discovery Today*, 2004, **9**, 659.
30. C. N. Cavasotto and S. S. Phatak, *Drug Discovery Today*, 2009, **14**, 676.
31. F. Jiang and S. H. Kim, *J. Mol. Biol.*, 1991, **219**, 79.
32. J. Apostolakis and A. Caflisch, *Comb. Chem. High Throughput Screening*, 1999, **2**, 91.
33. D. M. Lorber and B. K. Shoichet, *Protein Sci.*, 1998, **7**, 938.
34. A. Di Nola, D. Roccatano and H. J. Berendsen, *Proteins*, 1994, **19**, 174.
35. D. S. Goodsell and A. J. Olson, *Proteins*, 1990, **8**, 195.
36. Z. Huang, C. F. Wong and R. A. Wheeler, *Proteins*, 2008, **71**, 440.
37. A. Steffen, A. Kämper and T. Lengauer, *J. Chem. Inf. Model.*, 2006, **46**, 1695.
38. G. Schneider and H.-J. Böhm, *Drug Discovery Today*, 2002, **7**, 64.
39. N. Brooijmans and I. D. Kuntz, *Annu. Rev. Biophys. Biomol. Struct.*, 2003, **32**, 335.
40. R. Abagyan and M. Totrov, *Curr. Opin. Chem. Biol.*, 2001, **5**, 375.
41. C. N. Cavasotto and R. A. Abagyan, *J. Mol. Biol.*, 2004, **337**, 209.
42. I. D. Kuntz, J. M. Blaney, S. J. Oatley, R. Langridge and T. E. Ferrin, *J. Mol. Biol.*, 1982, **161**, 269.

43. A. M. Ferrari, B. Q. Wei, L. Costantino and B. K. Shoichet, *J. Med. Chem.*, 2004, **47**, 5076.
44. A. May and M. Zacharias, *Proteins*, 2008, **70**, 794.
45. M. Karplus and J. A. McCammon, *Nat. Struct. Biol.*, 2002, **9**, 646.
46. Z. R. Wasserman and C. N. Hodge, *Proteins*, 1996, **24**, 227.
47. M. Mangoni, D. Roccatano and A. Di Nola, *Proteins*, 1999, **35**, 153.
48. J. A. Given and M. K. Gilson, *Proteins*, 1998, **33**, 475.
49. N. R. Riemann and M. Zacharias, *Protein Eng.*, 2005, **18**, 465.
50. Y. Pak and S. Wang, *J. Phys. Chem. B*, 2000, **104**, 354.
51. N. Nakajima, J. Higo, A. Kidera and H. Nakamura, *Chem. Phys. Lett.*, 1997, **278**, 297.
52. W. C. Still, A. Tempczyk, R. C. Hawley and T. Hendrickson, *J. Am. Chem. Soc.*, 1990, **112**, 6127.
53. N. Okimoto, N. Futatsugi, H. Fuji, A. Suenaga, G. Morimoto, R. Yanai, Y. Ohno, T. Narumi and M. Taiji, *PLoS Comput. Biol.*, 2009, **5**, e1000528.
54. C. F. Wong and J. A. McCammon, *Annu. Rev. Pharmacol. Toxicol.*, 2003, **43**, 31.
55. Y. Deng and B. Roux, *J. Phys. Chem. B*, 2009, **113**, 2234.
56. J. Michel and J. W. Essex, *J. Comput. Aided Mol. Des.*, 2010, **24**, 639.
57. H. Alonso, A. A. Bliznyuk and J. E. Gready, *Med. Res. Rev.*, 2006, **26**, 531.
58. J. A. Erickson, M. Jalaie, D. H. Robertson, R. A. Lewis and M. Vieth, *J. Med. Chem.*, 2004, **47**, 45.
59. R. L. Dunbrack, Jr. and M. Karplus, *J. Mol. Biol.*, 1993, **230**, 543.
60. J. W. Ponder and F. M. Richards, *J. Mol. Biol.*, 1987, **193**, 775.
61. M. De Maeyer, J. Desmet and I. Lasters, *Fold. Des.*, 1997, **2**, 53.
62. S. C. Lovell, J. M. Word, J. S. Richardson and D. C. Richardson, *Proteins*, 2000, **40**, 389.
63. P. Tuffery, C. Etchebest, S. Hazout and R. Lavery, *J. Biomol. Struct. Dyn.*, 1991, **8**, 1267.
64. Z. Xiang and B. Honig, *J. Mol. Biol.*, 2001, **311**, 421.
65. Z. Xiang, P. J. Steinbach, M. P. Jacobson, R. A. Friesner and B. Honig, *Proteins*, 2007, **66**, 814.
66. A. R. Leach, *J. Mol. Biol.*, 1994, **235**, 345.
67. R. L. Dunbrack, Jr. and F. E. Cohen, *Protein Sci.*, 1997, **6**, 1661.
68. J. Desmet, M. D. Maeyer, B. Hazes and I. Lasters, *Nature*, 1992, **356**, 539.
69. A. R. Leach and A. P. Lemon, *Proteins*, 1998, **33**, 227.
70. E. Althaus, O. Kohlbacher, H. P. Lenhof and P. Müller, *J. Comput. Biol.*, 2002, **9**, 597.
71. L. Schaffer and G. M. Verkhivker, *Proteins*, 1998, **33**, 295.
72. C. Wang, O. Schueler-Furman and D. Baker, *Protein Sci.*, 2005, **14**, 1328.
73. J. Meiler and D. Baker, *Proteins*, 2006, **65**, 538.
74. R. Abagyan, M. Totrov and D. Kuznetsov, *J. Comput. Chem.*, 1994, **15**, 488.

75. M. Totrov and R. Abagyan, *Proteins*, 1997, **1**, 215.
76. J. Fernandez-Recio, M. Totrov and R. Abagyan, *Proteins*, 2003, **52**, 113.
77. W. Sherman, T. Day, M. P. Jacobson, R. A. Friesner and R. Farid, *J. Med. Chem.*, 2006, **49**, 534.
78. K. Bastard, A. Thureau, R. Lavery and C. Prevost, *J. Comput. Chem.*, 2003, **24**, 1910.
79. K. Bastard, C. Prevost and M. Zacharias, *Proteins*, 2006, **62**, 956.
80. A. Shehu, C. Clementi and L. E. Kavraki, *Proteins*, 2006, **65**, 164.
81. D. Seeliger, J. Haas and B. L. de Groot, *Structure*, 2007, **15**, 1482.
82. S. Eyrisch and V. Helms, *J. Comput. Aided Mol. Des.*, 2009, **23**, 73.
83. D. Seeliger and B. L. de Groot, *PLoS Comput. Biol.*, 2010, **6**, e1000634.
84. A. Benedix, C. M. Becker, B. L. de Groot, A. Caflisch and R. A. Böckmann, *Nat. Methods*, 2009, **6**, 3.
85. M. L. Teodoro, G. N. Phillips, Jr. and L. E. Kavraki, *J. Comput. Biol.*, 2003, **10**, 617.
86. D. Mustard and D. W. Ritchie, *Proteins*, 2005, **60**, 269.
87. C. N. Cavasotto, J. A. Kovacs and R. A. Abagyan, *J. Am. Chem. Soc.*, 2005, **127**, 9632.
88. Y. P. Pang and A. P. Kozikowski, *J. Comput. Aided Mol. Des.*, 1994, **8**, 669.
89. X. Barril and S. D. Morley, *J. Med. Chem.*, 2005, **48**, 4432.
90. E. Moreno and K. Leon, *Proteins*, 2002, **47**, 1.
91. J. H. Lin, A. L. Perryman, J. R. Schames and J. A. McCammon, *J. Am. Chem. Soc.*, 2002, **124**, 5632.
92. H. A. Carlson, K. M. Masukawa and J. A. McCammon, *J. Phys. Chem. A*, 1999, **103**, 10213.
93. H. A. Carlson, K. M. Masukawa, K. Rubins, F. D. Bushman, W. L. Jorgensen, R. D. Lins, J. M. Briggs and J. A. McCammon, *J. Med. Chem.*, 2000, **43**, 2100.
94. K. L. Meagher and H. A. Carlson, *J. Am. Chem. Soc.*, 2004, **126**, 13276.
95. R. M. Knegtel, I. D. Kuntz and C. M. Oshiro, *J. Mol. Biol.*, 1997, **266**, 424.
96. F. Österberg, G. M. Morris, M. F. Sanner, A. J. Olson and D. S. Goodsell, *Proteins*, 2002, **46**, 34.
97. H. Claussen, C. Buning, M. Rarey and T. Lengauer, *J. Mol. Biol.*, 2001, **308**, 377.
98. G. R. Smith, M. J. E. Sternberg and P. A. Bates, *J. Mol. Biol.*, 2005, **347**, 1077.
99. M. Totrov and R. Abagyan, *Curr. Opin. Struct. Biol.*, 2008, **18**, 178.
100. S. Y. Huang and X. Zou, *Proteins*, 2007, **66**, 399.
101. E. M. Novoa, L. R. de Pouplana, X. Barril and M. Orozco, *J. Chem. Theory Comput.*, 2010, **6**, 2547.
102. G. Bottegoni, I. Kufareva, M. Totrov and R. Abagyan, *J. Med. Chem.*, 2009, **52**, 397.

103. I. R. Craig, J. W. Essex and K. Spiegel, *J. Chem. Inf. Model.*, 2010, **50**, 511.
104. M. Rueda, G. Bottegoni and R. Abagyan, *J. Chem. Inf. Model.*, 2010, **50**, 186.
105. S.-J. Park, I. Kufareva and R. Abagyan, *J Comput. Aided Mol. Des.*, 2010, **24**, 459.
106. A. Amadei, A. B. Linssen and H. J. Berendsen, *Proteins*, 1993, **17**, 412.
107. M. Zacharias and H. Sklenar, *J. Comput. Chem.*, 1999, **20**, 287.
108. G. M. Keseru and I. Kolossvary, *J. Am. Chem. Soc.*, 2001, **123**, 12708.
109. R. Tatsumi, Y. Fukunishi and H. Nakamura, *J. Comput. Chem.*, 2004, **25**, 1995.
110. M. Zacharias, *Proteins*, 2004, **54**, 759.
111. M. A. Balsera, W. Wriggers, Y. Oono and K. Schulten, *J. Phys. Chem.*, 1996, **100**, 2567.
112. O. F. Lange and H. Grubmüller, *J. Phys. Chem. B*, 2006, **110**, 22842.
113. I. Bahar, A. R. Atilgan and B. Erman, *Fold. Des.*, 1997, **2**, 173.
114. K. Hinsen, *Proteins*, 1998, **33**, 417.
115. F. Tama and Y. H. Sanejouand, *Protein Eng.*, 2001, **14**, 1.
116. M. Shatsky, R. Nussinov and H. J. Wolfson, *J. Comput. Biol.*, 2004, **11**, 83.
117. B. Sandak, R. Nussinov and H. J. Wolfson, *J. Comput. Biol.*, 1998, **5**, 631.
118. B. Sandak, J. Wolfson and R. Nussinov, *Proteins*, 1998, **32**, 159.
119. D. Tobi and I. Bahar, *Proc. Natl. Acad. Sci. U. S. A.*, 2005, **102**, 18908.
120. A. May and M. Zacharias, *J. Med. Chem.*, 2008, **51**, 3499.
121. A. May and M. Zacharias, *Biochim. Biophys. Acta*, 2005, **1754**, 225.
122. M. Rueda, G. Bottegoni and R. Abagyan, *J. Chem. Inf. Model.*, 2009, **49**, 716.
123. S. Kazemi, D. M. Krüger, F. Sirockin and H. Gohlke, *ChemMedChem* 2009, **4**, 1264.
124. X. Barril and X. Fradera, *Expert Opin. Drug Discovery*, 2006, **1**, 335.

CHAPTER 10

COMparative BINding Energy (COMBINE) Analysis as a Structure-Based 3D-QSAR Method†

ANTONIO MORREALE*[a] AND FEDERICO GAGO*[b]

[a] Unidad de Bioinformática, Centro de Biología Molecular Severo Ochoa, CSIC-UAM, c/ Nicolás Cabrera 1, Campus UAM, E-28049 Tres Cantos, Madrid, Spain; [b] Departamento de Farmacología, Universidad de Alcalá, E-28871 Alcalá de Henares, Madrid, Spain
*E-mail: amorreale@cbm.uam.es; federico.gago@uah.es

10.1 Introduction

About 15 years before the seminal "lock and key" concept was introduced by Emil Fischer to visualize enzyme–substrate interactions, and decades before the concept of receptor was formulated by John Newport Langley and Paul Ehrlich, two less well-known Scottish scientists, the organic chemistry Professor Alexander Crum Brown and the physician Thomas Richard Fraser, published their account *"On the connection between chemical constitution and physiological action"*.[1] This groundbreaking report showed that some convulsivant natural alkaloids acquired the muscle relaxant properties characteristic of *d*-tubocurarine upon formation of the corresponding ammonium bases following treatment with "iodide of methyl". In their

† Dedicated to our late colleague and friend Angel R. Ortiz.
RSC Drug Discovery Series No. 23
Physico-Chemical and Computational Approaches to Drug Discovery
Edited by F. Javier Luque and Xavier Barril

own words, "*In fact, a condition exactly the reverse of that produced by strychnia was caused by iodide of methyl-strychnium. In place of violent spasmodic convulsions and muscular rigidity, the appearances were those of paralysis, with a perfectly flaccid condition of all the muscles.... The action of strychnia, brucia, thebaia, codeia, morphia, and nicotia, is evidently greatly diminished in degree and, at the same time, completely changed in character... We may conclude from these facts that when a nitrile base possesses a strychnia-like action, the salts of the corresponding ammonium bases have an action identical with that of curare.*" They concluded their paper with the statement "*This investigation has done little more than merely introduce us into a vast field of inquiry, but it has justified us in expecting that important fruits may be obtained by further and careful cultivation*".

Although flawed because of the oversimplification of relating a specific biological effect (curariform activity) to the presence of a particular chemical functionality (the quaternary ammonium group, later known to be also present in the neurotransmitter acetylcholine), this pioneering work indicated the possibility of establishing general laws between chemical structure and pharmacological activity, and it certainly led to further developments, as foreseen by these authors.

In the early 1900s, Hans Horst Meyer, a German pharmacologist from Marburg, and the English-born Charles Ernest Overton working in Zürich, independently observed a good correlation between the potency of a substance as an anesthetic and its lipid solubility, which was measured as an oil/gas partition coefficient.[2,3] Explanations for this apparently non-structure-specific action focused on unspecific physical effects on the lipid membranes that were intensely studied for decades. However, these long-lasting theories were seriously challenged when it was demonstrated by Nicholas P. Franks and William R. Lieb in London that the activity of soluble firefly luciferase could be inhibited by 50% at concentrations of inhalational agents such as halothane or chloroform, aliphatic and aromatic alcohols, ketones, ethers or alkanes that were normally reached in anesthetized experimental animals.[4] Despite the obvious chemical and structural diversity of these molecules, they all were shown to compete with the luciferin substrate of this enzyme for binding to the active site. This truly original observation opened up the possibility that anesthetics could act in mammalian brains by likewise competing with endogenous ligands for binding to specific receptors on neurons. This prescient thinking was later confirmed by abundant additional evidence and found structural support recently when the binding sites for the general anesthetics propofol and desflurane in a bacterial ion channel homologous to mammalian pentameric ligand-gated ion channels were revealed.[5]

Thus, an eventually erroneous "lipoid theory of narcosis" spurred an interest in lipid solubility for drug action that lingers to this day. In fact, partition coefficients, particularly those calculated for the distribution of a molecule between *n*-octanol and water, have dominated the scene and much of the thinking in medicinal chemistry, and many hydrophobic parameters derived from them (the so-called π constants) were estimated and compiled over the years by Corwin Hansch and co-workers at Pomona College in California.[6] Another set of electronic parameters, σ, calculated from the ionization constants of benzoic acid derivatives and representative of the relative strength of the electron-withdrawing or electron-donating properties of the substituents, was successfully used by Louis Hammett to derive linear free-energy relationships between chemical structure and rates of acid-catalyzed reactions.[7] However, application of the same "extrathermodynamic" approach (no thermodynamic principle states that the relationships should be true) to biological systems was unsuccessful, and this was taken as an indication that other structural descriptors were necessary. When Hansch and others incorporated the π constants representing lipophilicity as additional descriptors, the quality of the correlations with biological activity for congeneric series of compounds improved dramatically[8] and this moment is widely recognized as the birth of quantitative structure–activity relationships (QSAR).[9]

The next major step in QSAR took place when Dick Cramer, at that time working in SmithKline & French in the United States, introduced a technique that entailed the superposition of a set of compounds in the centre of a three-dimensional (3D) cubic lattice and the computation, for each individual molecule, of interaction fields (MIF) at every grid point with a probe resembling an sp^3-hybridized carbon atom endowed with a positive unit charge. This new method was termed CoMFA, an acronym for comparative molecular field analysis.[10] The resulting calculated fields were entered as descriptors in a matrix that also contained a column with measurements of binding affinity and/or experimental activity for all the molecules studied (Figure 10.1).

Partial least squares (PLS) analysis was then used to correlate fields with activity and to detect those regions around the molecules where putative interactions with the receptor would have an important positive or negative impact on activity. The influence of CoMFA in medicinal chemistry has been profound and this is still a method of choice when the 3D structure of the receptor is unavailable. An approach consisting of computing property fields based on the similarity indices of suitably aligned ligands and then subjecting them to PLS analysis was later developed as an alternative by Gerhard Klebe and associates at BASF (Ludwigshafen, Germany) and termed Comparative Molecular Similarity Indices Analysis (CoMSIA).[11] In this case, the resulting CoMSIA maps highlight those regions in the surroundings of the ligands that require a particular physico-chemical property for activity.

10.2 COMparative BINding Energy (COMBINE) Analysis

10.2.1 A Historical Perspective

Since the end of the 1950s, when John Kendrew and associates successfully determined the structure of myoglobin at the Cavendish Laboratory in

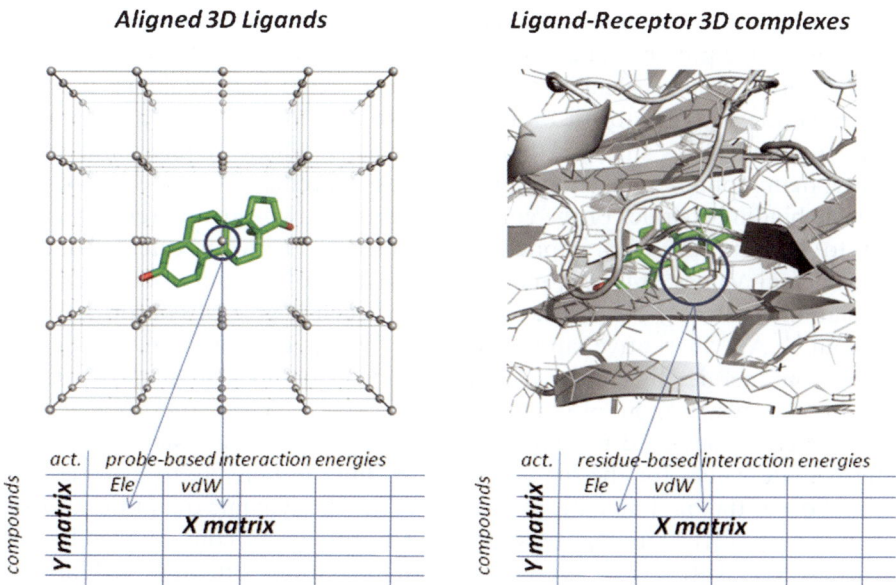

Figure 10.1   Comparison between a ligand-based 3D-QSAR method (CoMFA, *left*) and the COMBINE analysis approach (*right*) that relies on ligand–receptor 3D complexes. In CoMFA, a set of superimposed compounds (only one is shown here for clarity) is placed at the centre of a 3D cubic lattice, and then, for each individual molecule, steric and electrostatic interaction energies with a probe are computed at every grid point. The resulting energy values orderly fill in a data matrix. In COMBINE, the energy values are calculated as pairwise interaction energies between each ligand and every target residue. In both cases, one or more columns contain the relevant experimental information (measures of binding affinity or activity) for all the compounds. Once the matrices have been built, the statistical techniques that will be used are common to both approaches (see Figure 10.2).

Cambridge (UK),[12] the increasing availability of high-resolution crystal structures of biological macromolecules and their complexes with ligands has allowed the visualization of binding sites in proteins with an unprecedented level of detail. As a consequence, researchers started to benefit from the utilization of this 3D structural information not only in qualitative ways[13] but also quantitatively by applying different tools with the long-term aim of being able to predict the binding affinity (or biological activity) of any prospective ligand in advance of synthesis. The range of techniques that can be potentially used is very wide and they vary in complexity and applicability. As a rule of thumb, the more rigorous and accurate, the less suitable for the study of a large series of compounds due to the large computational overload; on the other hand, the simpler the approximation, the more generally applicable and less computationally demanding, although the price to be paid is the lack of precision. Thus, by relating the binding constants of 200 drugs and enzyme inhibitors to the presence of specific functional groups, Peter Andrews and co-workers in Australia calculated some average values that could be used to determine the goodness-of-fit of a drug to its receptor. Charged groups were found to bind more strongly than polar groups, which in turn bind more tightly than nonpolar groups, reportedly as expected. To the sum of the intrinsic binding energies of the component groups, two entropy-related terms had to be subtracted to improve the agreement between calculated and experimental values: 14 kcal mol^{-1} for the loss of overall rotational and translational entropy and 0.7 kcal mol^{-1} for each degree of conformational freedom.[14] An alternative approach is to try to calculate protein–ligand interaction energies by means of computational methods that make use of quantum chemistry and molecular mechanics (MM).[15] The main problem here is that energies obtained from a single configuration of the system may yield approximate values for the binding enthalpy but are unlikely to provide any estimation of the binding entropy, and both thermodynamic quantities are an intrinsic part of true binding free energies. This problem is compounded by the fact that very often enthalpy–entropy compensations are observed in the binding process.[16] To account, at least to some extent, for these entropic contributions, the calculated binding enthalpies are complemented with solvation terms that are typically estimated either from changes in buried nonpolar surface area or from differences in the solvation energies of ligands and binding sites upon complex formation measured using continuum electrostatic methods.[17] Of course, one can also employ molecular dynamics or Monte Carlo simulations that take full advantage of free energy perturbation or thermodynamic integration theory to study the nonphysical stepwise conversions between pairs of rather similar ligands in the free and bound states.[18,19] Less computationally costly, the linear interaction energy method requires simulations only at the endpoints of the mutations and determines the free energy of binding from a linear combination of the differences in the average inhibitor–environment interaction energies in the bound and the unbound states.[20] These high-end methods have a great potential for providing physical insight about the reasons that account for the differences in binding affinities between pairs of very similar ligands, but they are clearly impractical for the comparative study of even small series of receptor–ligand complexes, as customarily produced in most structure-based drug design endeavors.

Straightforward correlations of experimental measures of activity (*e.g.* K_i, IC$_{50}$) or binding affinity (*e.g.* K_d) with plain MM interaction energies, on the other hand, have been found only in a few favorable cases dealing with complexes of a given target with a congeneric series of ligands.[21,22] However, even in these instances, which usually require the manual or automated docking of some or many of the ligands into the binding site of the target protein, the linear regression models obtained for the training set of compounds are seldom accurate enough for predicting the potency of yet unsynthesized compounds that were not included in model derivation. Realizing these limitations, and also aware that enzyme inhibition is often found to be a function of the variance of certain physico-chemical properties at specific sites,[23] Angel R. Ortiz, a PhD student in Alcalá University under the supervision of Federico Gago that was following the PhD programme at the European Molecular Biology Laboratory under the auspices of Rebecca Wade, and in collaboration with Mayte Pisabarro, another PhD student, introduced the idea of analyzing the relationship between biological activities and weighted pairwise interaction energies between each ligand (or parts of it) and selected target residues. The method was termed Comparative Binding Energy (COMBINE) analysis, in clear allusion to CoMFA, because it made use of the same statistical tools but replaced the MIF around the ligands as regressors in the QSAR equation by ligand–receptor interaction energies calculated directly on the modeled 3D complexes (Figure 10.1). It was perceived that the weighting procedure could serve to filter out some of the inaccuracies of potential energy functions and modeling errors in such a way that the analysis could systematically separate true "energetic signals" from "background noise" and therefore derive a model based on a subset of properly weighted interaction energy terms. This novel approach was expected to be more predictive than traditional QSAR analysis because it incorporated more structure-based relevant information about the energetics of the ligand–receptor interaction, even though it was recognized that the selected "effective" energies could themselves act as statistical descriptors of other physically important interactions.[24] One added purported advantage of the new method was that the electrostatic contributions to the desolvation of ligand and receptor could be easily calculated[25] and incorporated in the interaction energy matrix as additional descriptors.

10.2.2 The Statistics Behind COMBINE

The core of COMBINE is a matrix containing structure-related energy descriptors (X variables) obtained from the set of ligand–receptor complexes (see Figure 10.1) that will be used to try and correlate with the binding affinities or the biological activities of the ligands (Y variable). The chemometric method used for this purpose will be briefly outlined.

10.2.2.1 Construction of the X Matrix

The X matrix contains the entire set of descriptor variables that describe the interaction energies between each ligand and every protein residue for all the complexes. Usually these comprise (1) a van der Waals term calculated using a MM force field, and (2) an electrostatic term calculated using point charges for ligand and protein atoms and either Coulomb's law (as implemented in the same force field) or the more elaborate and accurate Generalized Born (GB) or Poisson–Boltzmann (PB) methods.[26] In addition, desolvation energy terms for both receptor and ligand can also be incorporated as additional descriptors ("external variables").

10.2.2.2 Pretreatment of the X Matrix

It is obvious that the interaction energies calculated for protein residues far away from the ligand will tend to be zero and that others will tend to be almost identical for the whole set of ligands. For these reasons, and in order to reduce the total number of variables while keeping all the relevant information within the X matrix, it is customary to eliminate those interaction energy values with zero or negligible values and also those with a standard deviation below a user-defined cut-off because it can be safely assumed that they do not contribute to the overall variance in activity. On the other hand, the observation of positive energy values should make the user aware of possible force field inconsistencies (*e.g.* a certain degree of van der Waals overlap due to very strong electrostatic attraction) or modeling errors that need to be corrected. Once checked, these positive values can be either accepted or truncated to zero. In common with other 3D-QSAR procedures,[10,27] variable scaling can be performed, if desired, by following one of two approaches: (i) standard scaling, where the mean value over the whole set of variables is subtracted from each variable and divided by the standard deviation; or (ii) block scaling, where the mean value of the variables is subtracted from the one being scaled and divided by the standard deviation of these variables.

10.2.2.3 Partial Least Squares (PLS) Regression

This technique, also known as Projection to Latent Structures,[28] combines and generalizes features from Principal Component Analysis (PCA)[29] and Multiple Linear Regression (MLR) in the sense that not only orthogonal Principal Components (PC) or Latent Variables (LV) are extracted, as in PCA, but also a fitting procedure is performed to describe the activities of the compounds, as in MLR. There are two initial matrices in a COMBINE analysis (Figure 10.2): (i) the X matrix (eqn 10.1), which contains the independent variables (interaction energies and, possibly, additional variables such as desolvation energies), and (ii) the Y matrix (column vector) containing the dependent variable (affinities/activities) (eqn 10.2).

$$X = \begin{pmatrix} E_1^1 & E_2^1 & \cdots & E_M^1 & V_1^1 & V_2^1 & \cdots & V_M^1 & A_1^1 & \cdots & A_S^1 \\ E_1^2 & E_2^2 & \cdots & E_M^2 & V_1^2 & V_2^2 & \cdots & V_M^2 & A_1^2 & \cdots & A_S^2 \\ \cdots & \cdots & \cdots & \cdots & \cdots & \cdots & \cdots & \cdots & \cdots & \cdots & \cdots \\ E_1^N & E_2^N & \cdots & E_M^N & V_1^N & V_2^N & \cdots & V_M^N & A_1^N & \cdots & A_S^N \end{pmatrix} \quad (10.1)$$

where E_j^i, V_j^i, and A_j^i are the electrostatic, van der Waals, and additional variables, respectively, N is the number of compounds, M is the number of residues in the protein, and S is the number of additional variables.

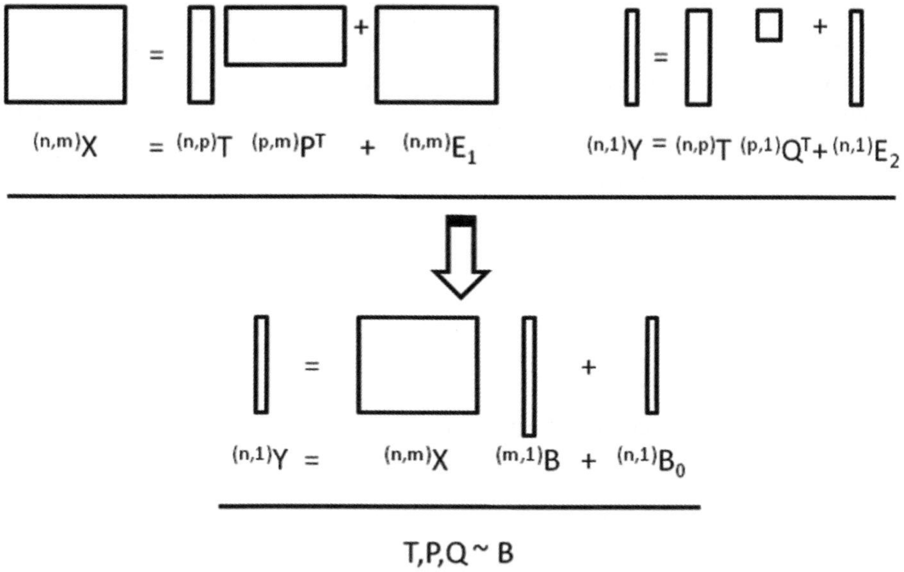

$$^{(n,m)}X \quad = \quad ^{(n,p)}T \quad ^{(p,m)}P^T \quad + \quad ^{(n,m)}E_1 \qquad\qquad ^{(n,1)}Y = {}^{(n,p)}T \; {}^{(p,1)}Q^T + {}^{(n,1)}E_2$$

$$^{(n,1)}Y \quad = \quad ^{(n,m)}X \quad ^{(m,1)}B \quad + \quad ^{(n,1)}B_0$$

$$T, P, Q \sim B$$

Figure 10.2 Graphical representation of the PLS method.[28] n, number of compounds; m, number of independent variables; p, number of principal components; T, score matrix; P, Q, loading matrices; E_1, E_2, B_0, error matrices.

$$Y = \begin{pmatrix} y_1 \\ y_2 \\ \dots \\ y_N \end{pmatrix} \qquad\qquad (10.2)$$

where y_i is the individual activity of compound i. The PLS analysis starts by decomposing the X and Y matrices into one score matrix, T, and two different loading weight matrices, P and Q (eqn 10.3), using the non-linear iterative partial least squares (NIPALS) algorithm:

$$X = TP^T \quad Y = TQ^T \qquad\qquad (10.3)$$

The loading weight matrices P and Q contain information about the variables in the so-called LV or PC space. These are orthogonal vectors obtained as linear combinations of the original variables contained in the X matrix. The coefficients in a given PC provide information on the relative weight of the different terms and can be used to deduce the relevance of each individual ligand–residue interaction for explaining the variance in activity/affinity. On the other hand, the score matrix T contains information about the compounds, which are described in terms of their projections onto the PCs. The PC space is normalized and has a mean of zero, so compounds with high scores should be checked as they could behave as outliers. In addition, clusters of compounds can be detected. Plots of the theoretically calculated activity/ affinity values *versus* the corresponding experimental values and of the evolution of the regression coefficient (r^2, eqn 10.4) as the number of LV increases allow the user to visualize the quality of the fit for the training set compounds, and also for the excluded and/or test compounds.

$$r^2 = \frac{\left[\sum_{i=1}^{N}(y_i-\bar{y})(\hat{y}_i-\langle\hat{y}\rangle)\right]^2}{\sum_{i=1}^{N}(y_i-\bar{y})^2 \sum_{i=1}^{N}(\hat{y}_i-\langle\hat{y}\rangle)^2} \tag{10.4}$$

where $\hat{y}$ is the estimated activity and $\langle\hat{y}\rangle = \dfrac{\sum_{i=1}^{N}\hat{y}_i}{N}$.

10.2.2.4   Cross Validation

This method is used to check that the derived correlation is not spurious and to assess the robustness of the resulting statistical model. It consists of predicting the dependent variable for some complexes that were not included in model derivation. The activities for the excluded compounds will be estimated from the models derived from the remaining complexes. Thus, at the end of the process, a list of predicted activities for all the compounds will be obtained. The performance is then quantified by the q^2 cross-validated correlation coefficient (eqn 10.5):

$$q^2 = 1 - \frac{\sum_{i=1}^{N}(y_i-\hat{y}_i)^2}{\sum_{i=1}^{N}(y_i-\bar{y})^2} \tag{10.5}$$

where $\bar{y}$ is the average value of the activity ($\bar{y} = \dfrac{\sum_{i=1}^{N}y_i}{N}$).

Simply stated, this metric describes the amount of variance in the dataset that is explained by the model. Besides, a standard deviation of error of predictions (*SDEP*, eqn 10.6) and an average absolute error (*AAE*, eqn 10.7) are also calculated.

$$SDEP = \sum_{i=1}^{N}\sqrt{\frac{(\hat{y}_i-y_i)^2}{N}} \tag{10.6}$$

$$AAE = \sum_{i=1}^{N}\frac{|\hat{y}_i-y_i|}{N} \tag{10.7}$$

It can be argued that, despite the cross-validation procedure, the resulting model fits the data just by chance due to the selection of a fortuitous equation out of the large amount of different PLS regression models that can be constructed with the thousands of variables contained in the X matrix. To check against this possibility, the affinities/activities of the compounds can be scrambled or replaced with random numbers to prove the point that in these cases it is not possible to derive a meaningful and statistically robust model.[24]

10.2.2.5 Selection of the Best Model

As successive components are extracted from the X matrix, a check is made to estimate the amount of variance that is recovered. Although there is not a strict rule to select the best model resulting from a PLS analysis, this choice is usually guided by the evolution of both the cross-validated correlation coefficient (q^2, eqn 10.5) and the standard deviation of the errors in prediction (*SDEP*, eqn 10.6). The optimal dimensionality is normally that corresponding to the peak q^2 value beyond which a decrease in this figure of merit is also accompanied by an increase in the number of LV extracted and therefore a less parsimonious model.

10.2.2.6 External Validation

The best way to validate a PLS model is to challenge it *a posteriori* with a set of new complexes that were not used in model derivation by comparing the predictions for this new series of bound ligands with the affinity/activity values obtained from experiment.

10.2.3 gCOMBINE: the Graphical User Interface to COMBINE

To improve on the original COMBINE command-line style implementation and encourage its use by the non-experts, we and our co-workers developed a graphical user interface (GUI) called gCOMBINE.[30] The GUI was written in Java to ensure portability to different operating systems and was released free of charge under a scientific/academic non-profit and non-commercial license agreement. gCOMBINE was designed bearing in mind the need for a simple and user-friendly chemometric tool. Thus, the GUI allows plenty of interactivity, facilitates the handling of data and input/output files, and provides multiple plots representing the energy descriptors entering the analysis, the PLS scores and loading weights, experimental *vs.* predicted regression lines, and the evolution of parameters such as r^2, q^2, and *SDEP* as the number of extracted LV increases (Figure 10.3). Other representative features include the implementation of a sigmoidal dielectric function for electrostatic energy calculations, alternative cross-validation procedures (leave-one-out, leave-N-out, and random groups), drawing of confidence ellipses, and the possibility to carry out several additional tasks such as

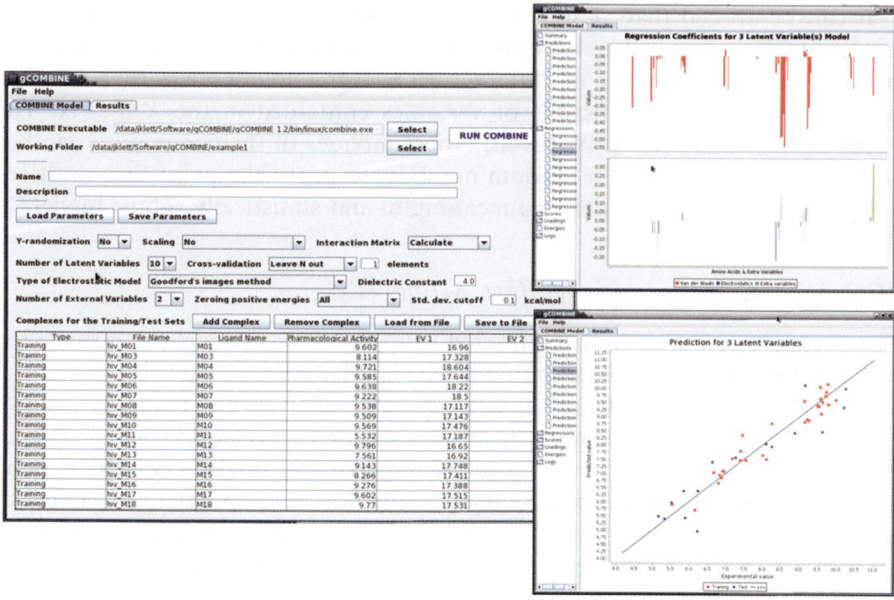

Figure 10.3 Selected screenshots from the gCOMBINE program.[30] *Left*: gCOMBINE main window; *upper right*: PLS weights assigned to the residue-based van der Waals (*red*) and electrostatic (*blue*) interactions energies and also to the desolvation energies (*green*) in a typical COMBINE model;[63] *lower right*: experimental *versus* predicted activity plot.

optional truncation of positive interaction energy values and generation of ready-to-use PDB files containing information related to the importance for activity of individual protein residues. This information can be displayed and color coded using a standard molecular graphics program like PyMOL (http://www.pymol.org/). It is our expectation that, thanks to this GUI, more groups will be encouraged to use COMBINE analysis in their QSAR and drug design research programs, thus expanding its applicability.

10.3 3D-QSAR Models in Virtual Screening

3D-QSAR regression models can be seen as tailor-made scoring functions to be employed in virtual screening (VS) studies for ranking the docked poses of the putative ligands. Despite this potential, only a limited number of such studies were reported by the year 2008, as reviewed by Hillebercht and Klebe.[31] Fortunately, the situation has changed lately, and the number of retrospective (assessment of models) and prospective studies (with experimentally confirmed hits) employing 3D-QSAR regression models as queries to search databases has notably increased. Recent literature in the field (since 2008) can be broadly classified as (a) retrospective studies (no new hits are produced), (b) prospective studies (new hits are found and confirmed), and (c) 3D-QSAR pharmacophores (a 3D-QSAR method is used to select the best pharmaco-phore among multiple possibilities). In the first two classes, 3D-QSAR models are combined with pharmacophore hypotheses either in parallel or sequentially to improve the screening of molecular databases. As for the third class, the field of pharmacophore modeling and applications in drug discovery has been reviewed recently,[32] and the most notable examples are those that employed the HypoGen module of the Catalyst program.

10.3.1 Retrospective Studies

A large proportion of 3D-QSAR work still follows conventional protocols and has model derivation and assessment as its final goal. Most often, these models are combined with a pharmacophore hypothesis to search for putative ligands in molecular databases. Although some hits are always proposed, they are generally not tested experimentally and therefore the model cannot be considered to be fully validated. Some examples are:

(a) GABA$_A$ α_3 receptor modulators.[33] A molecular field analysis (MFA) was performed using 63 known ligands as a training set and 13 as a test set. The derived model accounted for 81% and 83% of the variance in activity, respectively. Then 13 compounds representative of both sets were employed to build a pharmacophore by means of the HipHop program in Catalyst. The Maybridge catalog (60 000 molecules) was searched with the derived pharmacophore and some hits were selected.

(b) New N1-subtype human influenza virus type A (N1hA) neuraminidase inhibitors.[34] A homology model of the receptor was first constructed and then 27 inhibitors were docked with FlexX. The best docking poses were then used in a CoMFA analysis that yielded r^2 and q^2 values of ~ 0.8 and ~ 0.6, respectively.

(c) New phosphoinositide 3-kinase (PI3K) p110a inhibitors.[35] A pharma-cophore was built and used for inhibitor alignment (52 for the training set and 14 for the test set), and the ensuing 3D-QSAR analysis yielded r^2 and q^2 values of 0.95 and 0.88, respectively.

(d) Chemokine receptor type 5 (CCR5) antagonists.[36] Starting from a known X-ray crystal structure, 68 known molecules (60 for the training set and eight for the test set) were docked. The best poses were then subjected to GRid-INdependent descriptors (GRIND)[37]/PLS analysis ($r^2 = 0.9$ and $q^2 = 0.6$). A structure-based pharmacophore was then built and used to screen a database of 3650 compounds, out of which 80% of the actives were retrieved. A subsequent scan of the Maybridge Screening Collection (considered as decoys and seeded with 69 known inhibitors) employing different shape-based VS methods yielded a respectable AUC value of ~ 0.9.

(e) HIV-1 reverse transcriptase (RT) inhibitors.[38] Following a CoMFA-like QSAR analysis using a genetic algorithm on 36 molecules in the training set and five in the test set, a model was produced ($r^2 = 0.87$ and $q^2 = 0.83$) and a pharmacophore was built (24 molecules as the training set and 140 as the test set; $r^2 = 0.86$). A subsequent search within a custom-made database of 10 000 virtual compounds with this pharmacophore and assessment with the 3D-QSAR model yielded six hits.

(f) Blood–brain barrier choline transporter (BBB CHT) inhibitors.[39] A CoMFA analysis was performed on a set of 27 designed, synthesized, and experimentally assayed blockers (very close analogues to other compounds already studied by the same group); r^2 and q^2 values of 0.8 and 0.4 were obtained, respectively.

(g) Enoyl acyl carrier protein (ACP) reductase (ENR) inhibitors. The alignment rule for a combined CoMFA/CoMSIA approach was obtained through the definition of a pharmacophore from a single X-ray structure for 32 molecules in the training set and eight in the test set ($r^2 = 0.98$ and $q^2 = 0.73$).

(h) Binding of aminoglycosides to bacterial 16S ribosomal RNA.[40] A structure-based pharmacophore-driven alignment was performed and used as input for a CoMFA/CoMSIA study (36 and 11 molecules in training and test sets, respectively; $r^2 = 0.95$ and $q^2 = 0.68$). The ZINC and Maybridge fragment databases and the ZINC whole-molecule database were screened using the pharmacophore and 21 molecules were selected on the basis of their predicted activities.

10.3.2 Prospective Studies

The most complete studies are those in which the 3D-QSAR analysis is used as a predictive tool for selecting candidates from VS that are evaluated experimentally. Some examples are:

(a) Nonsteroidal inhibitors of the androgen receptor (AR).[41] Putative ligands from the ASINEX commercial database (http://www.asinex.com/Libraries.html) were docked with GOLD and 10 binding modes per molecule were saved. The poses were then ranked using a CoMSIA model. Six new inhibitors were found with activities in the micromolar range.

(b) Phenylmethylene hydantoins as prostate cancer inhibitors.[42] A total of 636 virtual compounds were predicted on the basis of results from a CoMFA study; 15 were synthesized and five hits were confirmed.

(c) Inhibitors of the ENR FabK from *Streptococcus pneumoniae*.[43] The SPECS database was docked with FlexX and the best scored 5000 molecules were pre-selected. Their activities were predicted according to a CoMFA model derived from 22 molecules in the training set and two in the test set ($r^2 = 0.9$ and $q^2 = 0.5$). Then 87 compounds were selected using as criteria their docking score, predicted pIC_{50} values, binding modes, and chemical diversity as assessed by a clustering analysis. As a result, six molecules were selected for further studies.

(d) *Mycobacterium tuberculosis* ENR inhibitors.[44] Firstly, CoMFA and CoMSIA studies were undertaken for 29 molecules in the training set and eight in the test set using a docking-guided (FlexX) alignment ($r^2 = 0.9$ and $q^2 =$

0.6). Secondly, a pharmacophore was built from an available X-ray structure. The Maybridge database was then filtered using the derived pharmacophore, the selected molecules were docked, and their activities were predicted using the previously generated CoMFA and CoMSIA models. The 20 best ranking molecules were selected for further analysis.

(e) 18 kDa TranSlocator PrOtein (TSPO) inhibitors.[45] In this case a pharmacophore (PHASE) was used as an alignment rule for a GRID[46]/ GOLPE[47]-type of 3D-QSAR analysis (144 molecules, $r^2 = 0.95$ and $q^2 = 0.70$). The Maybridge database was scanned with the pharmacophore and the activities predicted by means of the 3D-QSAR model, which resulted in seven hits with experimental activities in the low nanomolar range.

(f) Epidermal Growth Factor Receptor (EGFR) inhibitors.[48] A docking-based (GOLD) alignment was used as input for a GRID/GOLPE analysis (142 and 64 molecules in training and test sets, respectively; $r^2 = 0.87$, $q^2 = 0.76$). Additionally, a model based on ROCS alignments (http://www.eyesopen.com/ rocs) was developed ($r^2 = 0.87$, $q^2 = 0.76$) and used to search the Maybridge database. The best compounds were docked and their activities were predicted according to the previously derived GRID/GOLPE model. Out of the 10 compounds that were purchased, seven were active in the low micromolar range;

(g) Sex Hormone Binding Globulin (SHBG) ligands.[49] A homology-based model of zebrafish SHBG was built and used for docking in order to align the molecules and perform CoMFA/CoMSIA studies ($r^2_{pred} = 0.6$). The ZINC database was virtually screened and the pK_i for the best scoring compounds was predicted according to the CoMFA and CoMSIA models. Finally, 42 molecules were tested experimentally and six hits were obtained with affinities in the low micromolar range;

(h) Constitutive Androstane Receptor (CAR) agonists.[50] Both homology (receptor view) and pharmacophore (ligand view) models were used to search the LeadQuest database and this resulted in the experimental identification of 11 and 16 compounds, respectively, as true agonists. All the compounds were then docked with GOLD and the best poses selected as input for a GRID/ GOLPE analysis for further validation of the model ($r^2 = 0.95$, $q^2 = 0.73$).

10.3.3 3-D QSAR Pharmacophores

This type of study has been typically carried out using the HypoGen module in the Catalyst program (http://accelrys.com/products/pipeline-pilot/). HypoGen tries to formulate pharmocophoric hypotheses with features found in active molecules but not present in the set of inactives. Then, a 3D-QSAR analysis is conducted for each hypothesis and those yielding the best correlation between the 3D features and the activity/affinity of the ligands are retained for further analysis. Some examples are:

(a) Heat Shock Protein 90 (HSP90) inhibitors.[51] Sixteen molecules were used to build a pharmacophore ($r^2 = 0.91$ and $q^2 = 0.93$) that was used for screening the Maybridge and Scaffold databases. The best 1150 molecules were docked into the target protein and 36 were finally selected to be tested experimentally.

(b) Inhibitor kappa B kinase β (IKKβ) inhibitors.[52] A pharmacophore was built using 23 molecules as the training set out of a total of 159 compounds collected from the literature ($r = 0.93$, $q = 0.77$). The ChemDiv database (~700 000 molecules) was then filtered using this model to select 1806 for docking. From the top-scoring ones, 29 were purchased and experimentally tested and one hit was found in the low micromolar range.

(c) Corticotropin-Releasing Factor 1 (CRF1) antagonists.[53] A pharmacophore was used to screen a virtual library of 15 000 molecules and 60 of them were selected for synthesis and experimental testing.

(d) Human carbonic anhydrase (hCA) isozymes I and II.[31] Hillebrecht and Klebe performed a systematic study on the suitability of CoMFA/CoMSIA to screen large chemical libraries in comparison to 2D fingerprints (MACCS MDL) and 1D properties related to the van der Waals surface area (octanol/water partition coefficient, molar refractivity, and partial charges). A total of 138 sulfonamide-type inhibitors made up the training set and 663 molecules were included in the test set. Affinity predictions were assessed for hCAII, while selectivity studies were conducted with both isoenzymes. Automatic building and alignment of the ligands were performed with the FlexS program.[54] To evaluate the performance of the method they used sensitivity (true positives rate), specificity (true negatives rate), and precision (the hit rate) as the main metrics. The three methods performed equally well ($r^2 \approx 0.9$ and $q^2 \approx 0.8$) in terms of internal predictions, but the predictions for the test set decreased steadily ($r^2 \approx 0.5$). The general trend found was that CoMSIA afforded the best predictive models followed by CoMFA, MACCS, and finally the method based on van der Waals surfaces. On the other hand, the 2D and 1D methods were better at selecting actives and worse at discarding inactives than CoMFA or CoMSIA, which also outperformed 1D and 2D approaches in terms of precision. The low values obtained in the test set predictions were in part ascribed to its composition, as it was based on the experimental data available rather than on chemical diversity.

10.4   3D-QSAR Modeling of Ligand Selectivity

Ideally, if a good regression model is obtained for a series of ligands bound to a common receptor, it should be possible to extend the study to other related receptors provided that the needed affinity/activity data are available for the same set of ligands and the other targets. The goal here is to gain insight into the origins of selectivity, an issue of paramount importance in the field of drug design, and 3D-QSAR computational approaches such as CoMFA and CoMSIA have been employed successfully in this respect.[55] When modeling selectivity, it is customary to analyze it between pairs of targets using as the dependent variable either the difference or the ratio between some experimental measures of biological activity or affinity for a series of ligands. The resulting *selectivity fields* point to ways of increasing the differential binding affinity. Some recent published examples follow:

(a) Weber *et al.* performed CoMFA and CoMSIA analysis to study the selectivity of 86 carbonic anhydrase (CA) inhibitors towards CA I, CA II, and CA IV isoforms.[56] The inhibitors belonged to six different chemical scaffolds, which ensured diversity, and were aligned and energy minimized using three X-ray crystal structures (one for each target) as templates. The resulting ligand coordinates served as input for the calculation of the CoMFA and CoMSIA fields. Selectivity was accounted for by taking the differences in pK_i values of each ligand for a given pair of isoenzymes (CA I–CA II, CAI –CA IV, and CA II–CA IV) as the independent variable. Individual models (CA I, CA II, and CA IV) using the entire set of inhibitors afforded reasonable r^2 (0.8–0.9) and q^2 (0.5) values. The same was obtained when the set of inhibitors was further divided into a training set [60 inhibitors, internal validation ($r^2 \approx 0.6$–0.9, $q^2 \approx 0.3$–0.6)] and a test set [27 inhibitors, external validation ($r^2 \approx 0.6$–0.7)]. In terms of selectivity, only the CA I–CA II pair produced a statistically significant model ($q^2 = 0.56$). The steric (from CoMFA) and the hydrogen bond acceptor (from CoMSIA) fields best accounted for selectivity and some key interacting residues were highlighted. Certain observed activity trends were analyzed and successfully rationalized in terms of these fields, demonstrating the validity of the models obtained.

(b) The selectivity of bisarylmaleimide derivatives as inhibitors of GSK3, CDK2, and CDK4 targets was studied by Dessalew and Bharatam, applying CoMFA.[57] They also employed the differences in activity of the compounds towards pairs of targets (GSK3–CDK2, GSK3–CDK4, and CDK2–CDK4) as the dependent variable. Altogether, 29 and seven compounds were used as training and test sets, respectively. Publicly available X-ray crystal structures were used for inhibitor building and alignment. Individual models performed quite well ($r^2 \approx 0.9$, $q^2 \approx 0.5$–0.7, and $r^2_{pred} \approx 0.7$–0.9) and accounted for the activity tendencies among the inhibitors. On the other hand, when the targets were considered in pairs, reasonably good selectivity models ($r^2 \approx 0.9$, $q^2 \approx 0.8$, and $r^2_{pred} \approx 0.8$–0.9) were obtained and they contained, on average, a lower number of principal components than the former, thus making their interpretation easier. The key selectivity-related residues found were used to guide the design of 11 new inhibitors directed towards GSK3. These compounds were docked in the ATP-binding site of GSK3 using FlexX and scores in agreement with the CoMFA predictions were obtained.

(c) Gilbert *et al.* applied CoMFA and CoMSIA to explore the selectivity of 51 analogs of vanoxerine (a promising candidate for the treatment of cocaine addiction) towards dopamine and serotonin transporters (DAT and SERT, respectively).[58] Owing to the flexibility of the ligands (eight torsional angles), conformational search and hierarchical clustering analysis were used to find the most likely conformations to be used in the alignment prior to CoMFA and CoMSIA field calculations. Six different conformations were employed to align the molecules and these were further divided into a training set (22 analogs) and a test set (six analogs). Selectivity was accounted by using the ratio between the pK_i values of each ligand to both targets, instead of the difference as shown in the previous two examples. The best model (two principal components) afforded r^2 and q^2 values of 0.68 and 0.51, respectively. Key features for DAT/SERT selectivity emerged from the CoMFA contour maps, allowing the authors to propose a new family of compounds with enhanced selectivity.

(d) Inhibitor selectivity towards dipeptidyl peptidases 4, 8, and 9 (DPP4, DPP8, and DPP9, respectively) was examined by Kang *et al.*, applying CoMFA.[59] X-Ray structures of DPP4 co-crystallized with two inhibitors were used to build the ligands. DPP8 and DPP9 receptor structures were obtained from homology modeling. The inhibitors were docked in each binding site and scored. The best pose for each one was selected and used as the conformation for the molecular superposition in three parallel CoMFA calculations after splitting the whole set into training (43) and test (10) subsets. No combination of activities was explored as the dependent variable but good predictive models were obtained for the three independent models [internal test ($r^2 > 0.9$ and $q^2 \approx 0.6$–0.7) and external test ($r^2_{pred} \approx 0.5$–0.6)]. Based on a comparative study of the CoMFA contour maps, two new compounds were designed that were synthesized and tested *in vitro* and displayed enhanced selectivity towards DPP4, in agreement with the theoretical predictions.

(e) Parallel CoMSIA models were employed by Ran *et al.* in their study on the selectivity of 304 selective ATP-competitive inhibitors towards the mammalian target of rapamycin (mTOR) and phosphatidylinositol-3-kinase-alpha (PI3Kα).[60] The mTOR receptor structure was built using homology modeling techniques and the ligands were constructed using as a template the structure of an inhibitor co-crystallized with PI3Kα. The best docking poses for each compound were selected as the conformations to be aligned for CoMSIA. The ligands were divided into training (70 compounds) and test (234) sets, although no combination of activities was used to explore selectivity. Statistically significant 5- and 7-PC models were obtained, respectively, for mTOR [internal ($r^2 = 0.93$ and $q^2 = 0.66$) and external ($r^2_{pred} = 0.84$)] and PI3Kα [internal ($r^2 = 0.97$ and $q^2 = 0.54$) and external ($r^2_{pred} = 0.72$)]. The most influential contributions to selectivity were found to be mainly steric and electrostatic, whereas similar hydrophobic and hydrogen bonding fields were observed.

10.5   Current Trends in COMBINE Analysis as a 3D-QSAR Technique

Since its inception in 1995, when the original approach was developed to account for the differences in activity of a series of human synovial fluid phospholipase A_2 inhibitors developed and tested by scientists at Menarini Pharmaceuticals,[24] the COMBINE method has been applied to the study of other small molecules binding to different protein targets, mostly enzymes, as either substrates (*e.g.* haloalkane dehalogenases DhlA,[61] human cytochrome P450 1A2[62]) or inhibitors (*e.g.* HIV-1 protease,[63] human neutrophil elastase,[64] HIV-1 reverse transcriptase,[65] acetylcholinesterase,[66,67] and LinB[68]) and also of peptide–protein,[69,70] protein–protein,[71] and protein–DNA interactions.[72] The reader is referred to these publications for additional details, as well as to recent reviews by Wade *et al.*,[73,74] Damborsky *et al.*,[68] and Lushington *et al.*[75] In the next subsections some recent enhancements and extensions to this methodology will be briefly discussed, namely: (a) linking to a docking algorithm, as shown by Murcia and Ortiz,[76] when screening virtual libraries, to score and select the most reliable bound conformations of the putative ligands, improve the predictive ability of the regression models, and increase the enrichment factors; (b) incorporation of more than one structure of the same target into the analysis to allow the introduction, at least in part, of target flexibility and/or ligand orientation;[76,77] (c) joint study of affinity and selectivity by use of different protein targets belonging to the same family, as shown by Wang and Wade,[78] Murcia *et al.*,[79] and Henrich *et al.*,[80] which may provide important guidelines for drug design; (d) replacement of the MM descriptors with energy values obtained from a pairwise energy decomposition using a semiempirical quantum mechanical (QM) method,[81] as implemented in the so-called SE-COMBINE (SEmiempirical COMBINE).[82]

10.5.1 COMBINE Analysis as a Scoring Function for Virtual Screening

Results from a COMBINE analysis were coupled to a docking algorithm for the first time by Murcia and Ortiz, with the aim of improving the outcome of a VS campaign using coagulation factor Xa (fXa) as the target.[76] In this study, a two-step fully automatic procedure was designed that linked the results obtained by utilizing a flexible docking algorithm (to generate the 3D ligand–receptor complexes from three different X-ray crystal structures of fXa) with the COMBINE analysis methodology (to rationalize quantitative structure–activity relationships and predict inhibitory activities). The training set consisted of 133 fXa inhibitors that contained a common 3-amidino-1*H*-indole-2-carboxamide scaffold and displayed an activity range of 4 logarithmic units. The ligands were built either from scratch or by using the available X-ray crystal structures as templates. Atom types were assigned according to the AMBER force field (parm99) and semiempirical AM1 ESP charges were located at atomic centres. A conformational analysis was then undertaken to build up a *flexibase* for each ligand. The three receptor structures were prepared with the *protonate* module of AMBER 7.0, assuming standard protonation states for all titratable amino acid side chains. Each inhibitor was docked into each of the three fXa targets using a grid approximation and an exhaustive search algorithm. Finally, all the complexes were subjected to MM energy minimization to allow minor structural rearrangements to take place in the receptor's binding site while keeping the ligand frozen. A first COMBINE analysis was performed using two interaction energy terms, van der Waals (a Lennard-Jones 12-6 potential) and electrostatic (a Coulombic term with a distance-dependent dielectric constant [$4r_{ij}$]), calculated for the 286 protein residues common to all receptor structures, which resulted in 572 X variables. Experimentally measured inhibitory constants (pK_i) were used as the Y variables. To test for the robustness of the models, scrambling tests and cross-validation procedures were carried out. Low q^2 values (≤ 0.3) obtained using leave-one-out for the three sets of 133 complexes were associated with three outliers that displayed automated docking poses inconsistent with the known binding mode. These were identified following an iterative and self-consistent approach that revealed that they had unsigned errors of prediction greater than 1.2 pK_i units. To simulate external validation, 13 complexes were randomly excluded from the training set and a COMBINE model was built for the remaining 120 complexes. This operation was repeated 20 times; in each case, the resulting model was challenged for prediction of the activities of the 13 excluded inhibitors ($q^2 \approx 0.5$ using between 3 and 9 LV) and also used to screen a small in-house virtual library of 99 ligands that contained 86 inactive compounds (ligands extracted from fXa-unrelated complexes in the PDB) plus 13 fXa inhibitors not included in the previous COMBINE study (taken from the literature or other PDB entries). A threshold pK_i value of 5.5 was used to separate active from inactive compounds and the active/inactive list was ranked according to either the *naked* AMBER energy function employed by the docking program or the affinity predicted by the COMBINE analysis. When the screening results were analyzed using Receiver Operating Characteristic (ROC) plots[83] it was shown that the COMBINE analysis models improved the ability to identify structurally related molecules in external sets over the MM force field, although, on the other hand, pK_i predictions for structurally unrelated inhibitors were not significantly better than random. In the best case, a recognition rate of $\sim 80\%$ of known binders was achieved at a $\sim 15\%$ false positives rate. Enrichment factors were also superior for the COMBINE-based method.

10.5.2 COMBINE Analysis for Modeling Target Flexibility and/or Ligand Orientation

The dependence of deriving a successful COMBINE model on the choice of receptor structure employed for ligand docking has been documented above. Incorporation of receptor flexibility into COMBINE beyond ensemble docking, *i.e.* docking every ligand to an ensemble of multiple X-ray crystal structures of the target protein,[84] has not been attempted so far, to the best of our knowledge. On the other hand, in the case of symmetrical binding sites as present, for example, in HIV-1 protease, only one common orientation for each inhibitor is usually considered, but surely an additional set of complexes with the inhibitors docked in a reversed orientation would appear to be needed for a full representation of the binding event. In fact, we could obtain a COMBINE model with good predictive ability ($r^2 = 0.90$, $q^2 = 0.69$) for 48 congeneric inhibitors[63] that displayed their structural variability on the same one-half of the pseudosymmetrical binding cavity (because of the different protonation states of the catalytic aspartates), *i.e.* pockets S1' and S2'. However, this model showed a tendency to underpredict slightly the biological activity of external molecules that displayed structural variation on both sides. By duplicating each row in the matrix of energy descriptors and swapping the duplicated rows between subunits we were able to produce a new model of similar quality ($r^2 = 0.89$, $q^2 = 0.64$) but lacking the tendency to underpredict the activity of the compounds in the external set. By following this procedure, equivalent residues in both subunits were assigned equivalent weights and the resulting model was insensitive to ligand orientation and easier to interpret.[77]

10.5.3 COMBINE Analysis for Modeling Ligand Selectivity

Although these studies can be conducted in parallel (*N* models for *N* targets), or in combination (one model for the entire set of targets), many more examples have been reported for the former than for the latter. Three key studies will be reviewed here: (1) the binding of sialic and benzoic acid analogues to N2 and N9 subtypes of neuraminidase type A (NA),[78] and inhibitor binding to three different serine proteases: (2) either trypsin (Try), thrombin (Thr), and factor Xa (fXa),[79] or (3) Try, Thr, and urokinase-type plasminogen activator (uPA).[80]

(1) The NA study represented the first COMBINE application where more than one receptor structure were entered into the analysis, even though the main goal of the study was not selectivity itself but rather to provide guidelines to modify current inhibitors and optimize their activities. The study was performed on 45 NA–inhibitor complexes belonging to the N2 and N9 subtypes: 24 wild-type N9, 14 wild-type N2, and seven active-site-mutant N9. The vast majority of the complexes were taken directly from the PDB, whereas the rest were built by best-fit superposition of the ligands and/or automated docking with AutoDock. All the complexes were refined by energy minimization using the AMBER force field and the interaction energies between each inhibitor and the protein in the resulting complexes were decomposed into van der Waals and electrostatic terms to build the COMBINE matrix. A PLS analysis was then performed with the GOLPE program[27] using these energy descriptors. The best model ($r^2 = 0.89$ and $q^2 = 0.78$) contained 4 LV and explained most of the variance in affinity by assigning weights to the interaction energies involving 12 key residues and a water molecule. On the basis of this information, a sort of pharmacophore was built that consisted of the interacting features deduced from the COMBINE analysis, covered different parts of the binding site, and provided some hints for the design of new inhibitors.

(2) Murcia *et al.* presented a fully automated computational approach that enabled the simultaneous handling of an arbitrary number of receptors of the

same protein family.[79] The method consists of structurally aligning the multiple targets using MAMMOTH,[85] ligand docking with CGRID/ CDOCK,[76] interaction energy decomposition, and PLS modeling to carry out the COMBINE analyses. The test series were 88 benzamidine derivatives with varying affinities for Thr, Trp, and fXa. Seven different COMBINE models were derived: (a) a three-receptor model (Thr-fXa-Trp), (b) three two-receptor models (Thr-fXa, Thr-Trp, and fXa-Trp), and (c) three single-receptor models (Thr, Trp, and fXa). The electrostatic interactions were described by means of a Coulombic term using either a constant or a distance-dependent dielectric constant. Electrostatic contributions to the desolvation energy of ligands and receptors were calculated by either solving the PB equation[17] or employing a much faster implicit solvation model (ISM).[86] Statistically significant PLS models were obtained for Thr-fXa-Trp ($q^2 = 0.69$, LV = 7), Thr-fXa ($q^2 = 0.74$, LV = 6), and fXa-Trp ($q^2 = 0.62$, LV = 5), and, less satisfactorily, for Thr-Trp ($q^2 = 0.49$, LV = 5), Thr ($q^2 = 0.44$, LV = 7), and Trp ($q^2 = 0.34$, LV = 1). On the other hand, derivation of a predictive model was unsuccessful for fXa inhibitors alone ($q^2 = -0.04$, LV = 1). Thus, use of multiple receptors appears to lead to improved regression models but at the same time an increase in their complexity, as shown by the higher numbers of LV needed to explain the variations in activity/affinity. Overall, these results were consistent with previous computational studies and experimental evidence regarding selectivity for these serine proteases.

(3) Along the same lines, Henrich *et al.* also addressed the selectivity issue by focusing on Thr, Try, and uPA.[80] They used the structures of those complexes in the PDB for which experimentally determined inhibition constants, K_i, for these serine proteases were available as part of the training set and one of the test sets (the so-called "pseudo" test set), or without known 3D structures but with experimental inhibition constants for the other test set (the so-called "real" test set). For each enzyme, a single structure was selected as representative of the target. An X-ray superposition-based alignment rule was employed and all the complexes were minimized within the AMBER force field. Energy partitioning was carried out per receptor residue and these van der Waals and electrostatic interaction energy terms, together with desolvation energies for ligand and receptor, were used to correlate with the affinity using PLS. One COMBINE model for each single enzyme and another model that merged the three enzymes (the so-called three-in-one) were built. In all instances, r^2 and q^2 values above 0.8 and 0.6, respectively, were obtained, although more latent variables were needed to account for the co-variance in the case of the three-in-one model. The essential interaction residues (as highlighted by the PCA analysis) and the interpretation of the PLS weights were found to be in agreement with other theoretical and experimental studies, hence supporting the validity of the COMBINE analysis results. To further explore target selectivity, three new test sets were built: (I) Try *vs.* Thr (44 ligands), (II) Thr *vs.* uPA (150 ligands), and (III) uPA *vs.* Try (40 ligands). Here, ligand alignment relied on docking into the three binding sites, and a

single-target COMBINE model was used to predict the pK_i values. The dependent variable was the ratio between the predicted and experimental K_i values. For test sets I and II, the respective higher selectivities for Try and Thr were well predicted, but no selectivity towards uPA could be discerned in test set III. When a final analysis was performed by plotting predicted *vs.* experimental ratios, r^2 values of 0.66 and 0.18 were found for test sets I and II, respectively, whereas the correlation coefficient for test III was negative. When only those ligands for which the difference between experimental and estimated ΔG values was below 3 kcal mol^{-1} were selected, the selectivity prediction improved, at least for test I ($r^2 = 0.79$). In summary, if good docking poses and accurate affinity/activity data are available, COMBINE models can account for selectivity among targets, provided that these are different enough in their interaction patterns. Results obtained using Protein Interaction Property Similarity Analysis (PIPSA)[87,88] confirmed that the Thr/Try pair is more dissimilar than the Thr/uPA and Try/uPA couples. Furthermore, the electrostatic potential of the Thr active site is anti-correlated to those of Try and uPA. Taken together, these findings suggest that it should be easier to design selective ligands for Thr *vs.* Try and uPA than ligands discriminating between Try and uPA.

10.5.4 SemiEmpirical COMBINE (SE-COMBINE) Analysis

All the methods discussed so far rely entirely on MM force fields and, as a consequence, they suffer from the disadvantages inherent in this methodology. In particular, the reductionist approach of locating fixed partial charges at atomic positions, although convenient, neglects any effects due to charge transfer and polarization. Consequently, replacement of the electrostatic part in MM with a QM description of the interaction under study was expected to solve some of the problems associated with MM calculations. However, the applicability of accurate QM methods to biologically relevant macromolecular systems has been hampered by the long computer times required to perform the calculations, even with semiempirical (SE) methods that provide a reasonable compromise between speed and accuracy. To circumvent this problem, some alternatives have emerged such as the divide-and-conquer (D&C) approach[89,90] and the pairwise energy decomposition (PWD) scheme,[81] among others. These methods make it possible to deal with a set of ligand–receptor complexes and decompose the interaction energies in a manner consistent with the COMBINE analysis method in an affordable time frame. In fact, the PWD scheme has been used in the so-called SE-COMBINE, which was first tested on the same set of 88 trypsin–inhibitor complexes described above.[82] The X matrix (eqn 10.1) was built from the PWD terms that represent ligand–residue electrostatic interaction energies following a previous energy minimization in the AMBER force field. The 3-amidinophenylalanine inhibitors were then divided into three fragments in an attempt to better understand their individual contributions to the binding affinities. SE-COMBINE was able to explain 95% of the variance in the X matrix and yielded r^2 and q^2 values of ~ 0.7 and ~ 0.6, respectively, in both cases. However, when fragments of the inhibitors were used in place of the whole molecules, seven rather than just three LVs were needed. By plotting the exchange repulsion terms between ligands or ligand fragments and protein residues, it was straightforward to identify those residues that contribute the most to the inhibitory potencies observed experimentally. These are the so-called intermolecular interaction maps (IMM) that are especially relevant when the ligands are considered as their constituent fragments. Finally, a comparison of results from this SE-COMBINE analysis and previously reported CoMFA and CoMSIA on the same set of compounds revealed a poorer performance for SE-COMBINE *versus* the other two methods. The likely explanation put forward by the authors was that the semiempirical approach only takes into account electrostatic interactions, and some other terms are needed for a more accurate description of the binding event.

A similar strategy was followed by Zhang *et al.*, who carried out linear scaling, semiempirical D&C single-point calculations on a training set of AMBER-refined 15 inhibitor–protein kinase B (PKB) complexes using the PM3 and AM1 Hamiltonians.[91] Thirty additional ligands with reported inhibitory activities were used as the test set, after exclusion of others containing the sulfonamide group. Atom-to-atom pairwise interaction energies either *in vacuo* or in solvent were then calculated and converted to MM-PWD and QM-PWD terms and, following autoscaling, various combinations of these were evaluated using PLS. A calculated electrostatic energy term representing solvation was disregarded, as it was found to contribute little to the predictive ability of the models. Interaction and SAR maps were constructed to highlight, respectively, only the attractive interactions between residues (*x*-axis) and ligands (*y*-axis) or the most important overall residue–ligand interaction terms derived from the PLS analysis. The SAR map is scaled based on the variance across the series of ligands and tends to highlight protein residues that contribute the most to discriminating between potent and weak inhibitors. It was found that the optimal pure QM and mixed QM/MM models (2–6 LV) characterized the training set better than their purely MM counterpart and also that the AM1 and PM3 Hamiltonians yielded comparable results ($r^2 > 0.9$ and $q^2 \approx 0.8$).

10.6 Future Challenges for COMBINE Analysis

The electrostatic contributions to the desolvation of ligand and receptor are routinely calculated and incorporated as additional descriptors in the interaction energy matrix. The value corresponding to the receptor could be replaced by a collection of individual residue contributions[67] or, as recently shown for a number of serine proteases,[92] by providing estimates of the thermodynamic properties (enthalpy and entropy) of any water molecules solvating the ligand-binding site. This can be particularly important in cases where displacement of water from specific subpockets by the binding of a slightly modified ligand (*e.g.* methylated or halogenated) is accompanied by a substantial increase in potency.

The growing interest in incorporating the much needed protein flexibility into all aspects of molecular recognition and drug discovery[93,94] is manifested in the development and testing of significant methodological advances, such as the relaxed complex scheme,[95,96] which has been advocated for docking and VS purposes but has not yet been applied, to the best of our knowledge, to the field of 3D-QSAR. Having multiple representations of each complex in a COMBINE analysis has not been attempted, but we know that model complexity increases and performance usually decreases as the number of targets grows.

On the other hand, the potential of QM methods to provide a more accurate representation of electronic effects in protein–ligand interactions offers the prospect of capturing crucial charge transfer and polarization events that are outside the current capabilities of most current MM force fields. The fact that methods exist to partition quantum energies into pairwise contributions has made it possible to develop the "second-generation" SE-COMBINE and QM-PWD approaches. However, it still remains to be seen whether the calculated energy values should be combined with other energy terms to compose the global picture of the binding event.

10.7 Conclusions

As more and more 3D structures of ligand–protein complexes are experimentally determined and the quality of homology-built and docking models continues to improve, it seems appropriate to try to move away, whenever possible, from ligand alignment uncertainties and MIF calculations in an empty space, as routinely done in CoMFA and CoMSIA, and focus more on the real protein surroundings, as done in COMBINE analysis. Pairwise decomposition schemes exist for both MM- and QM-derived intermolecular interaction energies and chemometric methods such as PLS are particularly well suited, unlike MLR, to deal with matrices that contain many more descriptors than observations. This means that it should be possible to find the fundamental relations between the calculated energy values and the variance in biological response in the new space provided by the extracted latent variables, as has been successfully demonstrated in a considerable number of cases.

COMBINE analysis stands out as a useful computational tool that can exploit the information contained in a series of 3D ligand–receptor complexes, not necessarily involving just a single target structure but possibly more than one representation of the same target or even multiple targets. The approach that has been reviewed here for the study of selectivity in serine proteases can, in principle, be extended to an arbitrary number of receptors from the same or different protein families.

We anticipate that the success or failure of the COMBINE method will rest to a large extent on the quality of the modeled 3D structures for the whole series of complexes and our ability to compute the interaction energies accurately. The observed dependencies of some of the COMBINE models presented here on protein conformation highlight the importance of incorporating receptor flexibility into the analysis in order to improve the predictions. In favorable cases the resulting model will provide predicted binding affinities or activities for new compounds as well as guidelines for future ligand modification. Nonetheless, it is important to realize that structural modifications on the ligand outside the regions explored by the training set will likely give rise to new interactions that are not properly weighted and therefore have not yet been related to any changes in the actual binding energy. In fact, when COMBINE models have been used as a scoring function in VS, they were better than plain MM results at selecting structurally similar compounds but less proficient at retrieving more dissimilar hits. Likewise, the effect of replacing an amino acid at a given location in the ligand-binding site, *e.g.* through site-directed mutagenesis or by natural selection under drug pressure in the case of antivirals,[65] cannot be anticipated unless the structural variation in the ligands has provided distinct information about that particular region. If that is not the case, alternative strategies have to be employed.[97]

We expect that the current availability of gCOMBINE (https://ub.cbm.uam.es/software/info/gCOMBINE.html) will facilitate a larger dissemination and widespread application of this methodology, both in academic and industrial settings.

Acknowledgements

We are greatly indebted to our former and present collaborators for their enthusiasm, hard work, and many stimulating discussions. Partial support for this research was provided by the Spanish Comisión Interministerial de Ciencia y Tecnología (BIO2008-04384 to A.M. and SAF2006-12713-C02-02 and SAF2009-13914-C02-02 to F.G.) and Comunidad de Madrid (S-BIO/0214/2006 to both authors). A.M. acknowledges the Comunidad Autónoma de Madrid for financial support through the AMAROUTO program to the Fundación Severo Ochoa.

References

1. A. C. Brown and T. R. Fraser, *J. Anat. Physiol.*, 1868, **2**, 224.
2. H. H. Meyer, *Arch. Exp. Pathol. Pharmacol.*, 1899, **42**, 109.
3. C. E. Overton, *Studien uber die Narkose Zugleich ein Beitrag zur Allgemeinen Pharmakologie*, Gustav Fischer, Jena, 1901, p. 195.
4. N. P. Franks and W. R. Lieb, *Nature*, 1984, **310**, 599.
5. H. Nury, C. van Renterghem, Y. Weng, A. Tran, M. Baaden, V. Dufresne, J. P. Changeux, J. M. Sonner, M. Delarue and P. J. Corringer, *Nature*, 2011, **469**, 428.
6. A. Leo, C. Hansch and D. Elkins, *Chem. Rev.*, 1971, **71**, 525.

7. L. P. Hammett, *Trans. Faraday Soc.*, 1938, **34**, 0156.
8. C. Hansch and T. Fujita, *J. Am. Chem. Soc.*, 1964, **86**, 1616.
9. C. Hansch, *J. Comput. Aided Mol. Des.*, 2011, **25**, 495.
10. R. D. Cramer, D. E. Patterson and J. D. Bunce, *J. Am. Chem. Soc.*, 1988, **110**, 5959.
11. G. Klebe, U. Abraham and T. Mietzner, *J. Med. Chem.*, 1994, **37**, 4130.
12. J. Kendrew, G. Bodo, H. M. Dintzis, R. G. Parrish, H. Wyckoff and D. C. Phillips, *Nature*, 1958, **181**, 662.
13. C. Hansch and T. E. Klein, *Methods Enzymol.*, 1991, **202**, 512.
14. P. R. Andrews, D. J. Craik and J. L. Martin, *J. Med. Chem.*, 1984, **27**, 1648.
15. A. Ajay and M. A. Murcko, *J. Med. Chem.*, 1995, **38**, 4953.
16. C. H. Reynolds and M. K. Holloway, *ACS Med. Chem. Lett.*, 2011, **2**, 433.
17. B. Honig and A. Nicholls, *Science*, 1995, **268**, 1144.
18. P. Kollman, *Chem. Rev.*, 1993, **93**, 2395.
19. W. L. Jorgensen, *Acc. Chem. Res.*, 2009, **42**, 724.
20. J. Aqvist and J. Marelius, *Comb. Chem. High Throughput Screening*, 2001, **4**, 613.
21. M. C. Menziani, P. G. de Benedetti, F. Gago and W. G. Richards, *J. Med. Chem.*, 1989, **32**, 951.
22. M. K. Holloway, J. M. Wai, T. A. Halgren, P. M. Fitzgerald, J. P. Vacca, B. D. Dorsey, R. B. Levin, W. J. Thompson, L. J. Chen and S. J. deSolms, *J. Med. Chem.*, 1995, **38**, 305.
23. S. P. Gupta, *Chem. Rev.*, 1987, **87**, 1183.
24. A. R. Ortiz, M. T. Pisabarro, F. Gago and R. C. Wade, *J. Med. Chem.*, 1995, **38**, 2681.
25. A. Checa, A. R. Ortiz, B. de Pascual-Teresa and F. Gago, *J. Med. Chem.*, 1997, **40**, 4136.
26. P. Koehl, *Curr. Opin. Struct. Biol.*, 2006, **16**, 142.
27. M. Baroni, G. Costantino, G. Cruciani, D. Riganelli, R. Valigi and S. Clementi, *Quant. Struct.-Act. Relat.*, 1993, **12**, 9.
28. S. Wold, M. Sjöström and L. Eriksson, *Chemom. Intell. Lab. Syst.*, 2001, **58**, 109.
29. S. Wold, K. Esbensen and P. Geladi, *Chemom. Intell. Lab. Syst.*, 1987, **2**, 37.
30. R. Gil-Redondo, J. Klett, F. Gago and A. Morreale, *Proteins*, 2010, **78**, 162.
31. A. Hillebrecht and G. Klebe, *J. Chem. Inf. Model.*, 2008, **48**, 384.
32. S. Y. Yang, *Drug Discovery Today*, 2010, **15**, 444.
33. R. S. Vijayan and N. Ghoshal, *J. Mol. Graph. Model.*, 2008, **27**, 286.
34. Q. Zhang, J. Yang, K. Liang, L. Feng, S. Li, J. Wan, X. Xu, G. Yang, D. Liu and S. Yang, *J. Chem. Inf. Model.*, 2008, **48**, 1802.
35. Y. Li, Y. Wang and F. Zhang, *J. Mol. Model.*, 2010, **16**, 1449.
36. A. Carrieri, V. I. Perez-Nueno, A. Fano, C. Pistone, D. W. Ritchie and J. Teixido, *ChemMedChem*, 2009, **4**, 1153.

37. M. Pastor, G. Cruciani, I. McLay, S. Pickett and S. Clementi, *J. Med. Chem.*, 2000, **43**, 3233.

38. S. Vadivelan, T. N. Deeksha, S. Arun, P. K. Machiraju, R. Gundla, B. N. Sinha and S. A. Jagarlapudi, *Eur. J. Med. Chem.*, 2011, **46**, 851.

39. W. J. Geldenhuys, V. K. Manda, R. K. Mittapalli, C. J. Van der Schyf, P. A. Crooks, L. P. Dwoskin, D. D. Allen and P. R. Lockman, *Bioorg. Med. Chem. Lett.*, 2010, **20**, 870.

40. P. Setny and J. Trylska, *J. Chem. Inf. Model.*, 2009, **49**, 390.

41. A. A. Soderholm, J. Viiliainen P. T. Lehtovuori, H. Eskelinen, D. Roell, A. Baniahmad and T. H. Nyronen, *J. Chem. Inf. Model.*, 2008, **48**, 1882.

42. M. A. Khanfar and K. A. El Sayed, *Eur. J. Med. Chem.*, 2010, **45**, 5397.

43. Q. Zhang, C. Yu, J. Min, Y. Wang, J. He and Z. Yu, *J. Mol. Model.*, 2011, **17**, 1483.

44. A. Kumar and M. I. Siddiqi, *J. Mol. Model.*, 2010, **16**, 877.

45. T. Tuccinardi, S. Taliani, M. Bellandi, F. da Settimo, E. da Pozzo, C. Martini and A. Martinelli, *ChemMedChem*, 2009, **4**, 1686.

46. P. J. Goodford, *J. Med. Chem.*, 1985, **28**, 849.

47. G. Cruciani and K. A. Watson, *J. Med. Chem.*, 1994, **37**, 2589.

48. C. La Motta, S. Sartini, T. Tuccinardi, E. Nerini, F. da Settimo and A. Martinelli, *J. Med. Chem.*, 2009, **52**, 964.

49. N. Thorsteinson, F. Ban, O. Santos-Filho, S. M. Tabaei, S. Miguel-Queralt, C. Underhill, A. Cherkasov and G. L. Hammond, *Toxicol. Appl. Pharmacol.*, 2009, **234**, 47.

50. J. Jyrkkarinne, B. Windshugel, T. Ronkko, A. J. Tervo, J. Kublbeck, M. Lahtela-Kakkonen, W. Sippl, A. Poso and P. Honkakoski, *J. Med. Chem.*, 2008, **51**, 7181.

51. S. Sakkiah, S. Thangapandian, S. John, Y. J. Kwon and K. W. Lee, *Eur. J. Med. Chem.*, 2010, **45**, 2132.

52. S. Nagarajan, H. Choo, Y. S. Cho, K. J. Shin, K. S. Oh and B. H. Lee, *BMC Bioinf.*, 2010, (suppl. 7), S15.

53. Y. Ye, Q. Liao, J. Wei and Q. Gao, *Neurochem. Int.*, 2010, **56**, 107.

54. C. Lemmen, T. Lengauer and G. Klebe, *J. Med. Chem.*, 1998, **41**, 4502.

55. A. R. Ortiz, P. Gomez-Puertas, A. Leo-Macias, P. Lopez-Romero, E. Lopez-Vinas, A. Morreale, M. Murcia and K. Wang, *Curr. Top. Med. Chem.*, 2006, **6**, 41.

56. A. Weber, M. Bohm, C. T. Supuran, A. Scozzafava, C. A. Sotriffer and G. Klebe, *J. Chem. Inf. Model.*, 2006, **46**, 2737.

57. N. Dessalew and P. V. Bharatam, *Eur. J. Med. Chem.*, 2007, **42**, 1014.

58. K. M. Gilbert, T. L. Boos, C. M. Dersch, E. Greiner, A. E. Jacobson, D. Lewis, D. Matecka, T. E. Prisinzano, Y. Zhang, R. B. Rothman, K. C. Rice and C. A. Venanzi, *Bioorg. Med. Chem.*, 2007, **15**, 1146.

59. N. S. Kang, J. H. Ahn, S. S. Kim, C. H. Chae and S. E. Yoo, *Bioorg. Med. Chem. Lett.*, 2007, **17**, 3716.

60. T. Ran, T. Lu, H. Yuan, H. Liu, J. Wang, W. Zhang, Y. Leng, G. Lin, S. Zhuang and Y. Chen, *J. Mol. Model.*, 2011, in press (DOI: 10.1007/s00894-011-1034-3.

61. J. Kmunicek, S. Luengo, F. Gago, A. R. Ortiz, R. C. Wade and J. Damborsky, *Biochemistry*, 2001, **40**, 8905.

62. J. J. Lozano, M. Pastor, G. Cruciani, K. Gaedt, N. B. Centeno, F. Gago and F. Sanz, *J. Comput. Aided Mol. Des.*, 2000, **14**, 341.

63. C. Perez, M. Pastor, A. R. Ortiz and F. Gago, *J. Med. Chem.*, 1998, **41**, 836.

64. C. Cuevas, M. Pastor, C. Perez and F. Gago, *Comb. Chem. High Throughput Screening*, 2001, **4**, 627.

65. F. Rodríguez-Barrios and F. Gago, *J. Am. Chem. Soc.*, 2004, **126**, 2718.

66. J. X. Guo, J. J. Wu, J. B.Wright and G. H. Lushington, *Chem. Res. Toxicol.*, 2006, **19**, 209.

67. S. Martin-Santamaria, J. Munoz-Muriedas, F. J. Luque and F. Gago, *J. Med. Chem.*, 2004, **47**, 4471.

68. J. Damborsky, J. Kmunicek, T. Jedlicka, S. Luengo, F. Gago, A. R. Ortiz and R. C. Wade, in *Enzyme Functionality: Design, Engineering and Screening*, ed. A. Svendsen, Dekker, New York, 2004, p. 79.

69. K. Schleinkofer, U. Wiedemann, L. Otte, T. Wang, G. Krause, H. Oschkinat and R. C. Wade, *J. Mol. Biol.*, 2004, **344**, 865.

70. T. Wang and R. C. Wade, *J. Med. Chem.*, 2002, **45**, 4828.

71. S. Tomic, B. Bertosa, T. Wang and R. C. Wade, *Proteins*, 2007, **67**, 435.

72. S. Tomic, L. Nilsson and R. C. Wade, *J. Med. Chem.*, 2000, 43, 1780.

73. R. C. Wade, S. Henrich and T. Wang, *Drug Discovery Today: Technol.*, 2004, **1**, 241.

74. R. C. Wade, A. R. Ortiz and F. Gago, in *3D-QSAR in Drug Design*, ed. H. Kubinyi, G. Folkers and Y. Martin, Kluwer-ESCOM, Dordrecht, 1998, vol. 2, p. 19.

75. G. H. Lushington, J. X. Guo and J. L. Wang, *Curr. Med. Chem.*, 2007, **14**, 1863.

76. M. Murcia and A. R. Ortiz, *J. Med. Chem.*, 2004, **47**, 805.

77. M. Pastor, C. Perez and F. Gago, *J. Mol. Graph. Model.*, 1997, **15**, 364389.

78. T. Wang and R. C. Wade, *J. Med. Chem.*, 2001, **44**, 961.

79. M. Murcia, A. Morreale and A. R. Ortiz, *J. Med. Chem.*, 2006, **49**, 6241.

80. S. Henrich, I. Feierberg, T. Wang, N. Blomberg and R. C. Wade, *Proteins*, 2010, **78**, 135.

81. K. Raha, A. J. van der Vaart, K. E. Riley, M. B. Peters, L. M. Westerhoff, H. Kim and K. M. Merz, Jr., *J. Am. Chem. Soc.*, 2005, **127**, 6583.

82. M. B. Peters and K. M. Merz, Jr., *J. Chem. Theory Comput.*, 2006, **2**, 383.

83. P. Baldi, S. Brunak, Y. Chauvin, C. A. Andersen and H. Nielsen, *Bioinformatics*, 2000, **16**, 412.

84. S. Y. Huang and X. Zou, *Proteins*, 2007, **66**, 399.

85. D. Lupyan, A. Leo-Macias and A. R. Ortiz, *Bioinformatics*, 2005, **21**, 3255.

86. A. Morreale, R. Gil-Redondo and A. R. Ortiz, *Proteins*, 2007, **67**, 606.

87. N. Blomberg, R. R. Gabdoulline, M. Nilges and R. C. Wade, *Proteins*, 1999, **37**, 379.
88. R. C. Wade, R. R. Gabdoulline and F. de Rienzo, *Int. J. Quantum Chem.*, 2001, **83**, 122.
89. S. L. Dixon and K. M. Merz, Jr., *J. Chem. Phys.*, 1996, **104**, 6643.
90. T. Akama, M. Kobayashi and H. Nakai, *J. Comput. Chem.*, 2007, **28**, 2003.
91. X. Zhang, A. C. Gibbs, C. H. Reynolds, M. B. Peters and L. M. Westerhoff, *J. Chem. Inf. Model.*, 2010, **50**, 651.
92. R. Abel, N. K. Salam, J. Shelley, R. Farid, R. A. Friesner and W. Sherman, *ChemMedChem*, 2011, **6**, 1049.
93. P. Cozzini, G. E. Kellogg, F. Spyrakis, D. J. Abraham, G. Costantino, A. Emerson, F. Fanelli, H. Gohlke, L. A. Kuhn, G. M. Morris, M. Orozco, T. A. Pertinhez, M. Rizzi and C. A. Sotriffer, *J. Med. Chem.*, 2008, **51**, 6237.
94. M. A. Lill, *Biochemistry*, 2011, **50**, 6157.
95. J. H. Lin, A. L. Perryman, J. R. Schames and J. A. McCammon, *J. Am. Chem. Soc.*, 2002, **124**, 5632.
96. R. E. Amaro, R. Baron and J. A. McCammon, *J. Comput. Aided Mol. Des.*, 2008, **22**, 693.
97. F. Rodriguez-Barrios and F. Gago, *J. Am. Chem. Soc.*, 2004, **126**, 15386.

Enhanced Sampling Methods in Drug Design

WALTER ROCCHIA*[a], MATTEO MASETTI[b] AND
ANDREA CAVALLI*[a,b]

[a] Department of Drug Discovery and Development, Istituto Italiano di
Tecnologia, Via Morego 30, I-16163 Genoa, Italy; [b] Department of
Pharmaceutical Sciences, University of Bologna, Via Belmeloro 6,
I-40126 Bologna, Italy
*E-mail: walter.rocchia@iit.it; andrea.cavalli@iit.it

11.1 Introduction

Today, computational methods have permeated all aspects of drug discovery, and those who are most proficient with such tools can have the advantage in delivering new drugs more quickly at a lower cost than their competitors.[1] In particular, computational methods are extensively used in the hit-to-lead and lead optimization phases of the drug discovery process, and, within such steps, a major role has been played by structure-based drug design, particularly in recent times. Structure-based drug design aims at two major objectives: to accurately estimate the drug–protein binding free energy, and to disclose the structural elements responsible for drug binding to a biological target. Docking algorithms are established computational tools, extensively utilized to identify the binding mode of new drug candidates into a protein. However, they sometimes fail in providing an accurate estimation of the drug–protein binding free energy. To overcome this limitation, several different approaches have been employed over the past 20 years. Among the early strategies, we should mention those methods that calculate the free energy of the bound and unbound states separately, such as

RSC Drug Discovery Series No. 23
Physico-Chemical and Computational Approaches to Drug Discovery
Edited by F. Javier Luque and Xavier Barril
© The Royal Society of Chemistry 2012
Published by the Royal Society of Chemistry, www.rsc.org

the MM/PBSA (Molecular Mechanics/Poisson–Boltzmann Surface Area)[2,3] and LIE (Linear Interaction Energy).[4,5] Although these approaches have been applied quite extensively,[6–9] the accuracy of the free energy estimation has mainly been challenged by the fact that the entropy related to the event is either neglected or calculated inaccurately.

A second family of advanced computational approaches is represented by the so-called enhanced sampling methods, which have very recently played an increasingly relevant role in computational drug design, and hold the promise to overcome some of the major limitations of the methodologies previously applied. Enhanced sampling methods are computational approaches that aim at describing the correct statistics of a thermodynamic event, without the strict requirement of extensively sampling all of the possible accessible states. This allows overcoming one of the major limitations of simulative methods (such as Monte Carlo and molecular dynamics), that is, the need to carry out exceedingly long simulations to properly calculate the thermodynamic quantities associated to a certain (bio)chemical event.

In computational drug design, the most widely used enhanced sampling strategies are based on alchemical transformations,[10] where the free energy is estimated using a thermodynamic cycle.[11,12] The most popular approach, making use of such concepts, is the so-called double decoupling,[13–16] which has provided impressive results in terms of free energy estimation of drug–protein binding, and in some cases the computed thermodynamic quantities have been in remarkable agreement with the experimental values.[17,18] Notwithstanding the notable achievements of alchemical methods in drug design, these approaches do not explicitly account for the dynamical events occurring upon drug binding to a protein. These events can be very important from a drug design standpoint, as they can provide key information for the optimization of a drug candidate in terms of potency and efficacy. Furthermore, dynamical simulations of binding and unbinding can be relevant to estimate kinetic parameters[19] and to evaluate the drug residence time within a target.[20,21] Kinetic parameters are nowadays considered key factors, and understanding how binding kinetics can influence the *in vivo* drug efficacy is one of the central issues of modern drug discovery.

Despite a growing number of different enhanced sampling approaches, such as, for instance, the string method,[22,23] temperature-accelerated molecular dynamics,[24] adaptive biasing force,[25] *etc.*, for the time being, umbrella sampling,[26] steered molecular dynamics,[27] and metadynamics[28] can be considered the methods of choice to dynamically investigate the process of drug binding to a target protein.

In umbrella sampling, the process of drug binding/unbinding can be enhanced by adding an artificial biasing "window potential", which confines the motion of the drug within a small interval around a prescribed value, helping to achieve a more efficient configurational sampling in that specific region. The biasing potential can be a harmonic function, and since the sampling is confined to a small region during a given biased simulation (this is the reason why is called "window potential"), the sampling is restricted only to a small piece of the free energy landscape. This requires that different regions along the binding/

unbinding reaction coordinate are sampled and, ultimately, the collected results along the various windows are cured of the bias and then recombined together to obtain the final estimate of the drug–protein binding free energy.

In steered MD, the drug binding or, more frequently unbinding, can be obtained *via* a tunable restraining potential, which forces the drug to move away from its initial configuration (usually belonging to the bound state) to a new one during an MD run.[27] In particular, in the case of an unbinding process, the target configuration may be defined as one where the drug is embedded into the bulk of the solvent and its interactions with the protein are almost null. Steered MD simulations are out of equilibrium and the reconstruction of the drug–protein binding free energy can be obtained *via* the Jarzynski's equality,[29] stating that it is possible to express the free energy associated with a certain event as a suitable average of the nonreversible work profiles. In this scenario, recently Colizzi *et al.*[30] have also shown that steered MD simulations can discern active from inactive enzyme inhibitors by a simple visual inspection of the force profiles required for pulling ligands out of the protein binding site, and without requiring a precise calculation of the binding free energy. In an evolution of this study, Patel *et al.*[31] have shown that the mechanical work profiles can be analyzed, in a user-independent manner, by means of a multidimensional scaling approach. This allows for automating a quite expensive phase, namely the accurate analysis of multiple work profiles, usually required for accurate steered MD-based thermodynamic estimations.

In metadynamics, the drug binding/unbinding is accelerated by using an external history-dependent potential, which allows the drug to escape from local free energy minima (*e.g.* the bound state) where it can be trapped. Furthermore, this potential prevents the drug from revisiting the already sampled minima. This greatly accelerates the sampling of rather complex free energy landscapes, which are commonly generated during the process of drug binding to a protein. In addition, it leads to an estimate of the free energy related to the event under investigation. Metadynamics has proven to be rather effective in studying drug–protein binding interaction and free energy,[32–34] although this approach can be computationally quite demanding.

Umbrella sampling, steered MD, and metadynamics will represent the methodological focus of this chapter. In particular, we first report some theoretical bases of molecular dynamics and enhanced sampling methods. This section, which is pivotal for the entire chapter, intends to provide the theoretical background necessary to understand and properly apply enhanced sampling to computational drug design. Then, we report and comment on some practical examples of application of steered MD and metadynamics to drug-design-related issues. We close the chapter with some conclusions and perspectives on the use of enhanced sampling methods in the era of GPU-based architectures and microsecond long MD simulations. We deal in the final section with a provocative issue: "are enhanced sampling methods still needed?", and we provide our point of view on the potential role of these approaches in the era of "fully unbiased simulations".

Thereby, we are confident that scientists willing to apply enhanced sampling methods in drug discovery will find in this contribution the theoretical background (along with a few examples) they need to properly apply these innovative approaches in computational medicinal chemistry.

11.2   Basics

We will model our system, *i.e.* the protein, the ligand, and the solvent, as a set of N interacting classical particles in a fixed volume V. The state of the system is completely specified by the knowledge of the set of the coordinates $\{r_i, i = 1, ..., N\}$ and momenta $\{p_i^r, i = 1, ..., N\}$. Together, positions and momenta define a point in the so-called phase space (denoted as $\Gamma = (r, p^r)$) of the system and that we will consider as a classical microstate. Computational techniques such as molecular dynamics (MD) and Monte Carlo (MC) aim at generating a set of points in the phase space having a distribution that is consistent with the statistical thermodynamic description of the system compatible with the fixed control variables. In addition to N and V, the most frequent control variables are the total energy E, the thermodynamic temperature T, and the global pressure P. Three combinations of these variables provide the most common *thermodynamic ensembles*, namely the *microcanonical* (NVE), the *canonical* (NVT), and the *isothermal–isobaric* (NPT).

One of the most relevant results of the statistical mechanics discipline is to have provided a framework where macroscopic observables, such as temperature and pressure, can be calculated as time or ensemble averages, under the ergodic assumption hypothesis,[35] of appropriate microscopic functions defined in the phase space of the system. The correctness of this approach is supported by a well-established phenomenological theory, thermodynamics, which expresses the very same observables as partial derivatives of a suitable state function, *i.e.* the thermodynamic potential, with respect to given control variables. Once the microscopic expression of a thermodynamic potential is stated, a link between the macroscopic and the microscopic description of a phenomenon is created and the results of the computational methods, which are expressed in the microscopic context, can be used to predict results pertaining to the macroscopic world. The importance of knowing the correct statistical distribution in the phase space resides in the fact that it allows us to make correct ensemble averages and to finally achieve accurate predictions.

11.2.1   Fixed Energy Systems (NVE)

Let us assume that our system is isolated so that its energy is constant. The microscopic quantity corresponding to the total energy is the Hamiltonian $H(r, p^r)$. It is assumed that it is given by the sum of coordinate-independent kinetic energy (K) and momentum-independent potential energy (v), so that:

$$H(r, p^r) = K(p^r) + v(r) = \sum_{i=1}^{N} \frac{p_i^r \cdot p_i^r}{2m_i} + v(r) \qquad (11.1)$$

Note that, while the kinetic energy depends on the detailed nature of the system only *via* the atomic masses, the form of the potential energy function, which describes the interactions among the particles, is specific to a given system.

The explicit form of the probability density $\rho(\Gamma)$ associated with this dynamics is obtained by assuming that, at equilibrium, all microscopic states compatible with the fixed energy constraint are equally probable. Thus, the probability to find the system in an infinitesimal volume $d\Gamma$ around the point Γ must be zero if the phase space point is such that $H(\Gamma) \neq E$ and constant otherwise. This requirement is satisfied by choosing:

$$\rho_{NVE}(\Gamma) = \frac{\delta E_0 \delta(H(\Gamma) - E)}{\eta N_\eta! h^{3N} \Omega(N,V,E)} \tag{11.2}$$

where $\delta(\cdot)$ is Dirac's delta,[†] h is Planck's constant, δE_0 is the uncertainty in the measurement of energy, η counts over the different chemical species composing the system so that $\sum_\eta N_\eta = N$,[‡] and:

$$(N,V,E) = \frac{\delta E_0}{\eta N_\eta! h^{3N}} \int\limits_{3N} \int\limits_{D(N,V)} \delta(H(\Gamma) - E) d\Gamma \tag{11.3}$$

$\Omega(N,V,E)$, known as the *microcanonical partition function*, can be interpreted as a counter of the microscopic states of the system compatible with the macroscopic NVE constraints. The condition on E resides in the $\delta(\cdot)$ function, those on N and V in the domain $D(N,V)$, which is the set of all the possible values of $\{r\}$ compatible with the NV constraints and is named configurational space; its measure is V^N.

Thermodynamics tells us that the correct thermodynamic potential in the NVE case is entropy $S(N,V,E)$.[36] The bridge between macroscopic and microscopic description was provided by Boltzmann's relationship among the entropy and the partition function:

$$S(N,V,E) = k_B \ln(N,V,E) \tag{11.4}$$

where k_B is Boltzmann's constant.

[†] For the scope of the present work, the Dirac's delta has to be seen just as an integral kernel such that, for a suitable class of test functions $f(x)$, it happens that $\int f(x)\delta(x - x_0)\, dx = f(x_0)$, provided the domain of integration contains the point x_0. The physical dimension of a Dirac's delta is the reciprocal of that of its argument.

[‡] In this concise description, the number N stands for the composition of the system and can be thought either as a synonym of $\{N_\eta\}$ in the NVE expression or as $\sum_\eta N_\eta$ in expressions such as h^{3N} or $\sum_{i=1}^{N}(\cdot)$. The subscript η spans over all the chemical species comprising the system. In the present formulation, two entities are chemically identical if the Hamiltonian is not affected by their exchange; this should not engender any ambiguity in the chemical context where a whole molecule can be considered as a chemical species.

11.2.2 Fixed Temperature Systems (NVT)

We will now focus on the description of a closed system at fixed temperature. This situation is more interesting from the point of view of the connection with experiments since temperature is easier to control than overall energy. The canonical ensemble is modeled by considering a system σ that interacts with a much larger one, the so-called thermal reservoir $\mathfrak{R}$, which is at fixed temperature. The characteristics of the interaction can be summarized as follows:[36]

- the overall system Σ composed of σ and $\mathfrak{R}$, is isolated;
- the number of degrees of freedom of $\mathfrak{R}$ is much larger than that of σ;
- there is only a weak and nonspecific coupling between σ and $\mathfrak{R}$;
- the volume and the number of particles in each subsystem are kept constant as in the microcanonical ensemble.

Under these assumptions, the equilibrium probability density for the microstate represented by the point $\Gamma = (r,p^r)$ in the phase space of σ is the Boltzmann distribution:

$$\rho_{NVT}(\Gamma) = \frac{\exp[-\beta H(\Gamma)]}{\prod_\eta N_\eta! h^{3N} Q(N,V,T)} \tag{11.5}$$

where $\beta = \dfrac{1}{k_B T}$ and Q is the *canonical partition function*, which, similar to the NVE case, is a weighted counter of all the microstates:

$$Q(N,V,T) = \frac{1}{\prod_\eta N_\eta! h^{3N}} \int_{D(N,V)} dr \int_{\mathfrak{R}^{3N}} \exp(-\beta H(r,p^r)) \, dp^r \tag{11.6}$$

Since kinetic energy depends only on momenta and the potential energy only on coordinates, and using the properties of the exponential function, Q can be expressed as the following product:

$$Q = \frac{\prod_{i=1}^{N}(2\pi k_B T m_i)^{3/2}}{\prod_\eta N_\eta! h^{3N}} Z \tag{11.7}$$

where Z is the so-called configurational integral (eqn 11.8):

$$Z = \int_{D(N,V)} \exp(-\beta v(r)) \, dr \tag{11.8}$$

Owing to the form of Q, it can be seen that most of the quantities of interest depend only on Z and the relevant function for their description is the reduced probability density over the *configurational space*, which is the phase space after suitable average of the momenta:

$$\rho^r_{NVT}(\mathbf{r}) = \int_{\Re^{3N}} \rho_{NVT}(\mathbf{r}, \mathbf{p}^r) d\mathbf{p}^r = \frac{\exp(-\beta\upsilon(\mathbf{r}))}{Z(N,V,T)} \qquad (11.9)$$

In the canonical ensemble, the thermodynamic potential is called Helmholtz's free energy (F) and its microscopic identification is:

$$F(N,V,T) = -k_B T \ln(Q(N,V,T)) \qquad (11.10)$$

The free energy is the key quantity to describe the population of the different states in the canonical ensemble.

11.2.3 Isothermal–Isobaric Systems (NPT)

From the experimental standpoint, pressure is easier to control than volume, so the NPT ensemble is probably the best model for the prediction of experimental quantities. NPT and NVT share the same weak interaction with a thermal reservoir, the only difference being the replacement of the volume constraint with the one on the pressure. The thermodynamic potential for the NPT ensemble is the Gibbs' free energy (G), which can readily be obtained by a change of variable in the Helmholtz's free energy and using well-known thermodynamic relationships:[36]

$$G(N,P,T) = F(N,V(P),T) - V(P)\frac{\partial F}{\partial V} = F + VP \qquad (11.11)$$

It is worth noting that in the case of a typical reaction occurring in water in the NPT ensemble, the following relationship holds:

$$\Delta G = \Delta F + P_{atm}\Delta V \qquad (11.12)$$

Since water can be assumed to be an incompressible liquid with very good approximation at atmospheric pressure, and since both the theoretical formulation and the simulation in the NVT ensemble are somewhat simpler, the Helmholtz's free energy difference is often preferred.

11.2.4 Microscopic Estimators of Thermodynamic Observables

Having established the framework of statistical mechanics and shown its connection to thermodynamics, we are, in principle, in the position to use either a time or an ensemble average to obtain measurable quantities. In practice, however, to do so we must still solve two problems. The first one is the definition of the function of phase space that corresponds to a given observable. This can be done either by microscopic considerations or by means of the thermodynamic relationships involving the thermodynamic potential.

We will show some examples of these definitions, most of which are in the canonical ensemble. The second problem is the fact that to compute the averages we must either be able to solve the evolution equations of the system (Boltzmann's approach, reflected by MD) or to sample configurations according to the ensemble probability density (Gibbs' approach, reflected by Monte Carlo methods). Neither of these tasks can be performed analytically except for systems described by very simple Hamiltonians (free particle, harmonic oscillator, and a few others). Both can however be tackled by numerical methods.

The Hamiltonian is both the microscopic estimator for the internal energy in the NVT ensemble and for the enthalpy in the NPT one. Kinetic energy, which depends only on the momenta, can be used to estimate the thermodynamic temperature:

$$\langle K \rangle_{NVT} = \int\int K(\boldsymbol{p}^r)\rho_{NVT}(\boldsymbol{r},\boldsymbol{p}^r)\mathrm{d}\boldsymbol{r}\,\mathrm{d}\boldsymbol{p}^r = \frac{\frac{3}{2}Nk_{\mathrm{B}}T\left(\sqrt{(\pi k_{\mathrm{B}}T)}^3\right)_1^N Nm_i^{3/2}}{\left(\sqrt{(\pi k_{\mathrm{B}}T)}^3\right)_1^N Nm_i^{3/2}}$$

$$= \frac{3}{2}Nk_{\mathrm{B}}T \tag{11.13}$$

A further microscopic estimator, the one for the pressure, can be obtained with a bit more involved algebra:

$$P = \left\{ \frac{1}{3V}\sum_{i=1}^N \sum_{\alpha=1}^3 \left[\frac{\left(p_{i,\alpha}^r\right)^2}{2m_i} + r_{i,\alpha}F_{i,\alpha} \right] \right\} \tag{11.14}$$

where $F_{i,\alpha}$ is the α-th Cartesian component of the total force acting on the i-th particle.

11.2.5   Numerical Integration of the Equations of Motion in Molecular Dynamics

Here, we will focus on MD, one of the most frequently used techniques used to generate points in the configurational space consistent with the required ensemble statistics. The basic idea behind MD is to study the time evolution of a microscopic system by solving Newton's equations of motion:

$$\boldsymbol{f}_i^r = m_i^r \ddot{\boldsymbol{r}}_i^r = \dot{\boldsymbol{p}}_i^r \tag{11.15}$$

where f_i is the net force acting on the i-th atom having mass m_i and acceleration r_i (the n-th over-dot notation stands for n differentiations over time).

Equation (11.15) represents a set of $3N$ coupled second-order differential equations; once initial conditions are specified, positions and velocities for all the particles may be calculated by integration over time. Isolated systems, *i.e.* those belonging to the NVE ensemble, are conservative and their dynamics is said to be Hamiltonian, ruled by the following evolution equations:

$$\begin{cases} \frac{d\mathbf{r}_i}{dt} = \nabla_{\mathbf{p}_i^r} H(\mathbf{r},\mathbf{p}^r) = \nabla_{\mathbf{p}_i^r} K(\mathbf{p}^r) = \frac{\mathbf{p}_i^r}{m_i}, i = 1,...,N \\ \frac{d\mathbf{p}_i^r}{dt} = -\nabla_{\mathbf{r}_i} H(\mathbf{r},\mathbf{p}^r) = -\nabla_{\mathbf{r}_i} \upsilon(\mathbf{r}) \end{cases} \qquad (11.16)$$

Within the classical approximation, the physics underlying the atomic interactions is described by means of a properly parameterized scalar potential energy function, and the model arising by such a representation is referred to as the *force field*. Since forces depend upon the particle's positions, eqn (11.15) is an example of a many-body problem, which can be analytically solved only for extremely simple systems. In any other case, numerical methods must be employed, unavoidably leading to approximate solutions with respect to the analytical trajectory. By using finite difference methods, the integration of the equations of motion is performed at discrete time intervals δt, called *time steps*, by means of algorithms referred to as *integrators*. The basis for any integrator scheme is the Taylor expansion of the coordinates around the time *t*:

$$r_i(t \pm \Delta t) = r_i(t) \pm \dot{r}_i(t)\Delta t + \frac{1}{2}\ddot{r}_i(t)\Delta t^2 \pm \frac{1}{6}\dddot{r}_i(t)\Delta t^3 + O(\Delta t^4) \qquad (11.17)$$

For simplicity, in eqn (11.17) the expansion is truncated at the fourth order. Similarly, the *order* of an integrator indicates the lower term not included in the expansion, and it represents a measure of the precision of the algorithm (higher order schemes introduce a lower amount of numerical error). Despite the fact that it is unfeasible to follow the "real" trajectory for a many-body system, such instability is not as problematic as it might appear, since the ultimate aim is to calculate statistical properties rather than individual evolutions; as long as the numerical trajectory conserves the energy within a certain threshold, the proper ensemble statistics is preserved.

The Verlet algorithm[37] (or position-Verlet) represents the prototype for all the integrators based on the direct solution of the equations of motion. By summing the expressions $(t + \delta t)$ and $(t - \delta t)$ of the Taylor expansion reported in eqn (11.17), one obtains the Verlet equation for advancing positions:

$$r_i(t + \Delta t) = 2r_i(t) - r_i(t - \Delta t) + \ddot{r}_i(t)\Delta t^2 + O(\Delta t^4) \qquad (11.18)$$

As may be noticed from eqn (11.18), velocities are not explicitly handled in the Verlet algorithm. Still, they are needed to calculate the kinetic energy and other properties depending on them. By *subtracting* the expressions of the Taylor expansions of eqn (11.17), it is possible to derive a useful relation to recover velocities:

$$\dot{r}_i(t) = \frac{r_i(t+\Delta t) - r_i(t-\Delta t)}{2\Delta t} + \mathrm{O}(\Delta t^2) \tag{11.19}$$

However, from eqn (11.19) it is clear that velocities at time t are available only once the positions have reached the next time step ($t + \delta t$). Despite its simplicity, the Verlet algorithm has many appealing features: it is time reversible, stable, and moderately robust. The main drawbacks of the Verlet algorithm are both the relatively poor precision and the definitely uncomfortable handling of velocities.

An improvement over the position-Verlet algorithm is the velocity-Verlet algorithm.[38] It is an integrator able to store positions, velocities, and accelerations at the same time t. Actually, velocities are evaluated at each half time step, but since they are evaluated twice per iteration, the synchronization with positions is ensured. The advancement in position directly follows the Taylor expansion of eqn (11.17) truncated at the 3rd order:

$$r_i(t+\Delta t) = r_i(t) + \dot{r}_i(t)\Delta t + \frac{1}{2}\ddot{r}_i(t)\Delta t^2 \tag{11.20}$$

Then, the velocities are propagated according to:

$$\dot{r}_i(t+\Delta t/2) = \dot{r}_i(t) + \frac{1}{2}\ddot{r}_i(t)\Delta t \tag{11.21}$$

and

$$\dot{r}_i(t+\Delta t) = \dot{r}_i(t+\Delta t/2) + \frac{1}{2}\ddot{r}_i(t+\Delta t)\Delta t \tag{11.22}$$

In between the velocities' half steps, the evaluation of the forces at time ($t + \delta t$) is performed. Then, the net velocity update is obtained:

$$\dot{r}_i(t+\Delta t) = \dot{r}_i(t) + \frac{1}{2}[\ddot{r}_i(t) + \ddot{r}_i(t+\Delta t)]\Delta t \tag{11.23}$$

The velocity-Verlet integrator is stable and well behaved. In addition, high-order variants have been developed in order to increase the precision. These Verlet-like integrators are efficient algorithms, involving just one force evaluation per iteration.

The dynamical system considered so far corresponds to the evolution in the microcanonical ensemble, where energy is constant during the time evolution of the thermodynamic system. To simulate other ensembles, *ad hoc* algorithms are used to mimic the effect of the interaction with the reservoir, namely *thermostats* for the NVT and *thermostats* and *barostats* for the NPT. Their description is, however, out of the scope of this contribution.

11.3 Description of the Binding Process

11.3.1 Identification of Sub-States in a Given Thermodynamic Ensemble

Considering a protein–ligand binding process occurring in the NVT or, equivalently, in the NPT ensemble, we can identify two states, namely the *unbound* one and the *bound* one, this latter corresponding to the formed complex. Both states are compatible with the fixed set of control variables and each of them corresponds to a different region of the phase space. A thorough characterization of bound and unbound states is far from trivial (see ref. 13), especially if the protein and the ligand form a loosely bound complex or if there are accessory interaction sites. Basically, the region of the configurational space that must be counted in the bound (B) state should include all the configurations where the protein binding site and the ligand make a strong interaction, while all non-interacting microstates form the unbound state (A). It is important to note, however, that the definition of A and B might depend on many factors, such as the experimental protocols one aims at describing. If, for instance, one considers the signal provided by a FRET[39] experiment, one obtains basically a two-state characterization of the process, where all the microstates involving some ligand–protein interaction occurring out of the binding site are to be assigned to the unbound state A. This latter is a different definition from that given previously and may well provide numerical discrepancies.

From the energetic point of view, we should imagine the unbound state as a very large free energy basin, substantially flat, whereas the bound state is more limited and likely presents a rougher landscape with several different local minima. There are basically no rules concerning the landscape of the remaining part of the configurational space, although a rough profile with plenty of peaks and a few valleys is to be expected.

11.3.2 The Role of Free Energy and the Concept of Reaction Coordinate

If we were able to limit the accessibility of our system to a given region of the phase space, for instance that corresponding to state A or B, and to consider the free energy of the resulting constrained system in each of these states, we would then observe that the probabilities of finding the unconstrained system at equilibrium in the regions pertaining to A or B are closely related to the free energy difference of the corresponding states:

$$\Delta F_{AB} = F(N,V,T,state''B'') - F(N,V,T,state''A'') = -k_B T \ln\left[\frac{\wp(B)}{\wp(A)}\right] \quad (11.24)$$

The described operation of constraining (that is, the imposition of such a sophisticated constraint) can be very difficult to accomplish in an experimental context, but it can be much easier to perform on the computational side, where the individual degrees of freedom are accessible.

Thermodynamic chemistry relates this free energy difference to the association constant K_{ass}, defined as the ratio between the kinetic constants k_{on} and k_{off} of the following bimolecular reaction where protein (P) and ligand (L) produce the complex C:

$$P + L \underset{k_{off}}{\overset{k_{on}}{\rightleftharpoons}} C \tag{11.25}$$

$$\Delta F_{AB} = -k_B T \ln K_{ass} = -k_B T \ln \frac{k_{on}}{k_{off}} \tag{11.26}$$

These relationships tell us that free energy difference, for the calculation of which only a thorough definition of the initial and final states is needed, is sufficient to describe the net association/dissociation rates, but it does not get to the details of the binding and unbinding mechanisms, which are described separately by the k_{on} and k_{off} constants. Interestingly, it seems that the efficacy of a compound as a drug is related to its residence time in the target binding site.[21] In turn, the residence time is directly connected to the k_{off} constant, which accounts specifically for the unbinding process. It descends from the need to predict this specific aspect that in addition to the initial and final states also the progress of the reaction has to be properly described. One way to achieve this goal in the statistical mechanical framework is based on the concept of *reaction coordinate*.

The reaction coordinate, henceforth termed ξ, is a function of the degrees of freedom of the system capable to mark the progress of the reaction from the state of the non-interacting reagents to that of the product, in our case the complex. If we imagine to know the exact form of ξ and to be able to constrain the system so that only the microstates compatible with a given value, or set of values, $\hat{\xi}$ of ξ are accessible, then we can consider the free energy $F(N, V, T, \hat{\xi})$ of that constrained system. In this perspective, it can be shown that a key quantity to obtain free energy differences and profiles is the probability density $\rho^r_{NVT\xi}$ of ξ:

$$\Delta F_{AB} = -k_B T \ln\left[\frac{\wp(B)}{\wp(A)}\right] = -k_B T \ln\left[\frac{\int_{I_B} \rho^r_{NVT\xi}(\hat{\xi}) d\hat{\xi}}{\int_{I_A} \rho^r_{NVT\xi}(\hat{\xi}) d\hat{\xi}}\right] \tag{11.27}$$

where

$$\rho^r_{NVT\xi}(\hat{\xi}) = \int_{D(N,V)} \delta\left(\xi(r) - \hat{\xi}\right) \rho^r_{NVT}(r) dr \tag{11.28}$$

and I_A and I_B are the sets of the $\hat{\xi}$ values that correspond to states A and B, respectively.

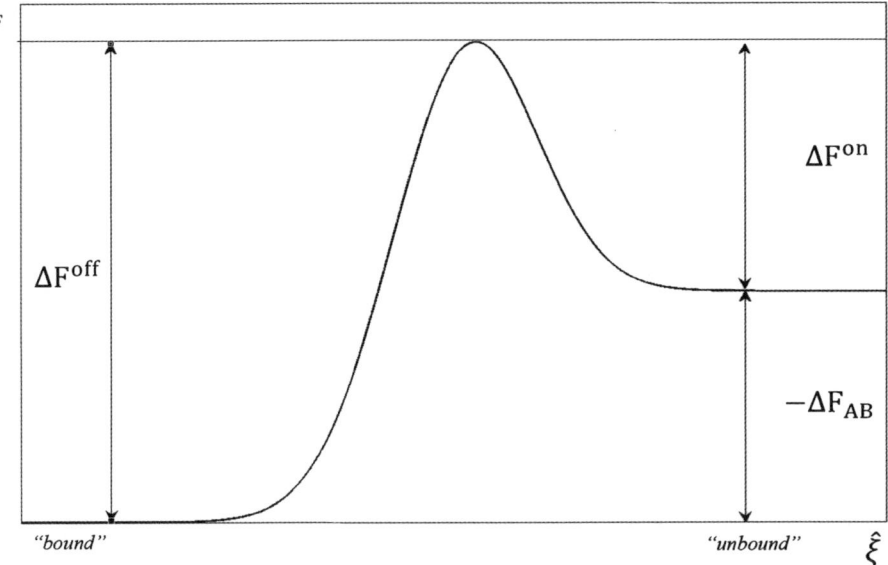

Figure 11.1 Schematic representation of a free energy profile with respect to the reaction coordinate for an activated process such as protein–ligand binding.

The profile of F as a function of $\hat{\xi}$ will be similar to what is shown in Figure 11.1, where the two states are separated by an energetic barrier that affects the kinetics of the process.

$$\Delta F_{AB} = \Delta F^{on} - \Delta F^{off} \tag{11.29}$$

ΔF^{on} is often called "activation energy" and is not only a function of the specific interaction between P and L, but also of their concentration in the solution. In contrast, ΔF^{off} depends only on the forces between the protein and the ligand while they are in the bound state. Moreover, it is responsible for the residence time since it accounts for the mechanisms that characterize the exit of the ligand from the site, and it can also be more feasibly tuned by computational means.

11.3.3 Volumetric Effect on the Unbound State

The profile of the free energy in the previous plot presents a plateau corresponding to the unbound state. If one simulated a ligand unbinding process, and approximated the reaction coordinates with the distance between interaction site and ligand, one would then observe a different, decreasing, asymptotic behavior as the distance increases. The reason for this is that in a "universe" containing only one molecule of ligand and one protein, the space available to them increases as they get farther away and the same applies to the translational entropy of the system. More exactly, one can show that the free energy of the unbound state contains the term $-k_{\mathrm{B}}T\ln\left[V_{\mathrm{avail}}\left(\hat{\xi}\right)\right]$, where $V_{\mathrm{avail}}\left(\hat{\xi}\right)$ is the available volume for a given $\hat{\xi}$. The horizontal asymptotic behavior results from the compensation to the translational entropy given by the presence of a given concentration of ligands that compete for a given target.

11.4 Basics of Enhanced Sampling

The main goal of enhanced sampling techniques is to force the correct sampling of the event of interest, in our case protein–ligand interaction and binding. Unfortunately, a good *a priori* knowledge of the specific phenomenon, ideally of the reaction coordinate, is often needed in order for these techniques to be successful. In fact, albeit there are many different approaches, the vast majority of them relies on the identification of one or more parameters that try to describe either the interaction or the progress of the binding process.

11.4.1 The Importance of Sampling and the "Importance Sampling"

So far, we have pointed out two major difficulties for an efficient description of the binding/unbinding processes, namely the characterization of bound and unbound states and the identification of the reaction coordinate. From the computational standpoint, there is another important challenge, which is evidently related to the previous ones. The tools used to estimate the binding process are based on physical grounds through the "force fields",[40] which are parametric descriptions of intra- and intermolecular classical forces. They lead to an extremely rugged potential energy landscape, which is a function of the different microscopic configurations of the system. Owing to the contact with the thermal reservoir, the system spontaneously explores the whole configurational space with a frequency which is proportional, in the long term, to the Boltzmann probability. The potential that rules this exploration is the free energy, whose complete knowledge is particularly difficult to achieve. Easier to estimate are free energy differences, which are however important from the practical point of view and for which the two already mentioned computational tools (MD and MC) can in principle be used to accomplish this task. Both techniques experience serious challenges when the free energy landscape is characterized by high barriers between minima. In this case, both MD and MC may fail in sampling correctly all the regions that contribute to the desired observable. That is the reason why the binding process is listed among the so-called *rare events*, which are hard to sample correctly. In this scenario, the role and the compelling need of a set of '*importance sampling*' techniques able to drive the exploration of the configurational space to the regions that cannot be neglected emerges.

11.4.2 Collective Variables

The identification of the reaction coordinate in a real case scenario is practically impossible and only approximations can be found; usually these approximations consist of sets of scalar quantities of clear physical meaning, such as the distance between the geometric centers of the ligand and the binding site, which are gathered in a vector and called *collective variables* (CVs). Different choices lead to different mappings from A to B. Any mapping leads to a well-defined $F(N,V,T,\hat{\xi})$, and, upon suitable integration, provides the correct ΔF_{AB}. However, not every mapping is equally capable of describing how the interconversion between A and B occurs, and to provide the correct kinetic parameters; the discipline that studies in detail these aspects is the Transition Path Theory,[41] whose enunciation is beyond the scope of this work. Good agreement with experimental data is the best, if not the only, way to ascertain that a good set of collective variables was chosen.

If the collective variables have the form of a linear combination of the Cartesian coordinates of the system, the free energy of the constrained system turns out to coincide with the so-called Potential of Mean Force (PMF) and to fulfill the following relation:

$$-\nabla_{\xi} F(N,V,T,\hat{\xi}) = \langle \vec{F}_{\xi} \rangle_{NVT\hat{\xi}} \qquad (11.30)$$

where $\vec{F}_{\xi}$ is the force exerted by the constraint $\xi(r) = \hat{\xi}$ over the system.[36]

From the practical viewpoint, once the computational practitioner has identified a promising set of collective variables, she/he needs to use a method to estimate how the free energy changes from A to B by varying the constraint $\hat{\xi}$ in a simulation. At fixed $\hat{\xi}$, it must be ensured that all the non-constrained degrees of freedom, sometimes called orthogonal variables, sample all the compatible microstates according to the Boltzmann distribution. There are a number of methods aimed at this goal, which go under the name of enhanced sampling techniques and some of them will be described in the following section of this work.

11.4.3 Enhanced Sampling Techniques

One can broadly distinguish three different families of enhanced sampling techniques:

- Methods that sample one or several systems in equilibrium (*e.g.* thermodynamic integration, free energy perturbation, umbrella sampling).
- Nonequilibrium sampling techniques (*e.g.* steered MD).[27]
- Methods that introduce additional degrees of freedom, along which the free energy is calculated (*e.g.* λ-dynamics[42,43] and metadynamics[28]).

An exhaustive description of all the methods belonging to these families is beyond the scope of this work. Here, we would like to describe just a few among them, which are particularly meaningful either from the foundational point of view or for their importance in applications.

11.4.3.1 Thermodynamic Integration

Let us imagine having a Hamiltonian where the interaction between P and L can be parametrized by a suitable scalar λ; then, the free energy difference can be expressed as follows:

$$\Delta F_{AB} = \int_{\lambda_A}^{\lambda_B} \frac{\mathrm{d}F(\lambda)}{\mathrm{d}\lambda}\mathrm{d}\lambda = \int_{\lambda_A}^{\lambda_B} \frac{1}{Q}\frac{\mathrm{d}}{\mathrm{d}\lambda}\left(\int_{D(N,V)} \int_{\Re^{3N}} e^{-\beta H(\mathbf{r}, \mathbf{p}^r, \lambda)}\mathrm{d}\mathbf{p}^r \mathrm{d}\mathbf{r}\right)\mathrm{d}\lambda =$$

$$\int_{\lambda_A}^{\lambda_B} \langle \frac{\partial H(\mathbf{r}, \mathbf{p}^r, \lambda)}{\partial \lambda} \rangle_{NVT\lambda}\mathrm{d}\lambda \qquad (11.31)$$

Usually, the lambda parameter only involves the potential energy part, leading to the following simplified expression involving only the potential energy:

$$\Delta F_{AB} = \int_{\lambda_A}^{\lambda_B} \langle \frac{\partial \upsilon(\mathbf{r}, \lambda)}{\partial \lambda} \rangle_{NVT\lambda}\mathrm{d}\lambda \qquad (11.32)$$

This integral can be estimated by means of several simulations, each at fixed λ, where the average energy derivative is calculated. The limitation of this approach consists in the fact that a significant number of simulations are needed to obtain a good approximation of the integral and that each simulation must be long enough to provide an accurate average.

11.4.3.2 Free Energy Perturbation

Similar to thermodynamic integration, also free energy perturbation considers parameterized Hamiltonians:

$$\Delta F_{AB} = -k_B T \ln \frac{Q(N,V,T,\lambda_B)}{Q(N,V,T,\lambda_A)} =$$

$$-k_B T \ln \left(\int_{D(N,V)} \int_{\Re^{3N}} e^{\beta H(\mathbf{r}, \mathbf{p}^r, \lambda_A) - \beta H(\mathbf{r}, \mathbf{p}^r, \lambda_B)} \frac{\exp(-\beta H(\mathbf{r}, \mathbf{p}^r, \lambda_A))}{Q(N,V,T,\lambda_A)}\mathrm{d}\mathbf{p}^r \mathrm{d}\mathbf{r}\right) = \qquad (11.33)$$

$$-k_B T \ln \langle \exp(\beta H(\mathbf{r}, \mathbf{p}^r, \lambda_A) - \beta H(\mathbf{r}, \mathbf{p}^r, \lambda_B)) \rangle_{NVT\lambda_A}$$

Even if it seems that only one simulation is needed in this case, namely that where $\lambda = \lambda_A$, this is not entirely true: if the two Hamiltonians are very different, the sampling of the $H(\mathbf{r}, \mathbf{p}^r, \lambda_B)$ contribution would be poor and lead to inaccurate results. In order to solve for this issue, one needs to divide the $[\lambda_A; \lambda_B]$ interval in several sub-intervals, so as to allow a good overlap of the distributions. This results in a computational cost similar to that required by thermodynamic integration.

11.4.3.3 Umbrella Sampling

One of the most widely used computational techniques used to accelerate the sampling along one, or more, collective variables, is *umbrella sampling*. Given a suitable generator of configurations, for instance MD or MC, it practically biases the generation by means of an additive *biasing potential*, $w\left(\xi(\mathbf{r}),\hat{\xi}\right)$, *i.e.* the umbrella, so that the corresponding collective variable values stay close to a given target. The most common biasing potential is a harmonic restraint, $w\left(\xi(\mathbf{r}),\hat{\xi}\right) = \dfrac{K}{2}\left(\xi(\mathbf{r})-\hat{\xi}\right)^{2}$, corresponding to a parabolic bias. This procedure alters the probability density of the generated configurations as follows:

$$\rho_i^{bias}\left(\hat{\tilde{\xi}}\right) = \int_{D(N,V)} \delta\left(\xi(\mathbf{r})-\hat{\tilde{\xi}}\right) \frac{\exp\left\{-\beta\left[\upsilon(\mathbf{r})+w\left(\xi(\mathbf{r}),\hat{\xi}_i\right)\right]\right\}}{\int_{D(N,V)} \exp\left\{-\beta\left[\upsilon(\mathbf{r})+w\left(\xi(\mathbf{r}),\hat{\xi}_i\right)\right]\right\}d\mathbf{r}}\,d\mathbf{r} =$$

$$(11.34)$$

$$\frac{\exp\left\{-\beta w\left(\hat{\tilde{\xi}},\hat{\xi}_i\right)\right\}}{\left\langle \exp\left\{-\beta w\left(\xi(\mathbf{r}),\hat{\xi}_i\right)\right\}\right\rangle_{NVT}}\,\rho^{unbias}\left(\hat{\tilde{\xi}}\right) = \exp\left\{-\beta w\left(\hat{\tilde{\xi}},\hat{\xi}_i\right)+\beta F_i^{bias}\right\}\rho^{unbias}\left(\hat{\tilde{\xi}}\right)$$

In simpler words, the statistics obtained applying the *i*-th umbrella is the original one windowed around $\hat{\xi}_i$ and multiplied by a factor, $exp\{\beta F_i^{bias}\}$, which is related to the free energy of the umbrella itself. It is evident that this relation is not invertible and that many different umbrellas have to be applied to cover the desired $\hat{\xi}$ interval. It is therefore assumed that the desired distribution can be obtained by a suitable combination of N_w simulated biased distributions:

$$\rho^{unbias}\left(\hat{\tilde{\xi}}\right) = \sum_{i=1}^{N_w} \alpha_i\left(\hat{\tilde{\xi}}\right)\rho_i^{bias}\left(\hat{\tilde{\xi}}\right) \qquad (11.35)$$

Equation (11.34) provides a constraint over the coefficients, namely that:

$$\sum_{i=1}^{N_w} \alpha_i\left(\hat{\tilde{\xi}}\right)\exp\left\{-\beta w\left(\hat{\tilde{\xi}},\hat{\xi}_i\right)+\beta F_i^{bias}\right\} = 1 \qquad (11.36)$$

Further conditions are imposed by the minimization of the statistical error made in each windowed sampling.[44] This is equivalent to the minimization of the variance along different sampling processes and leads to the following expression for the reconstructed unbiased distribution:

$$\rho^{unbias}\left(\hat{\tilde{\xi}}\right) = \frac{\displaystyle\sum_{i=1}^{N_w} n_i\rho_i^{bias}\left(\hat{\tilde{\xi}}\right)}{\displaystyle\sum_{i=1}^{N_w} n_i\exp\left\{-\beta w\left(\hat{\tilde{\xi}},\hat{\xi}_i\right)+\beta F_i^{bias}\right\}} \qquad (11.37)$$

where n_i is the number of samples collected during the *i*-th simulation.

The $\{F_i^{\text{bias}}\}$ constants can be derived following the WHAM approach[44] by suitably recasting their definition:

$$\exp\{-\beta F_i^{\text{bias}}\} = \langle \exp\{-\beta w(\xi(r),\hat{\xi}_i)\}\rangle_{NVT}$$

$$= \int d\hat{\xi} \int_{D(N,V)} \delta(\xi(r)-\hat{\xi}) \frac{\exp\{-\beta[v(r)+w(\xi(r),\hat{\xi}_i)]\}}{Z_{NVT}} dr \quad (11.38)$$

$$= \int \exp\{-\beta w(\hat{\xi},\hat{\xi}_i)\}\rho^{\text{unbias}}(\hat{\xi})d\hat{\xi}$$

The rationale for avoiding the original average in the NVT ensemble resides in the fact that in realistic cases a robust generator of points in the configurational space is not available, which is the main reason to perform umbrella sampling and similar methods. On the other hand, the final expression depends recursively on the set of $\{F_i^{\text{bias}}\}$ via the unbiased distribution; this issue is faced by means of a self-consistent iterative approach.

An alternative method to WHAM is umbrella integration, whose details are well described,[45] while a totally different approach is based on the integration of the mean force acting on the restraint and proves to be particularly simple when the restraint is a linear combination of the Cartesian coordinates of the system.

11.4.3.4   Steered Molecular Dynamics

Steered molecular dynamics is a technique where an external force is applied to the system so as to force the variation of the collective variable $\hat{\xi}$ in a relatively fast manner. The resulting statistics is not necessarily Boltzmann-like since it is collected from out of equilibrium trajectories; however, Jarzynski[29] showed the following relationship between the free energy difference and a suitable average of the nonreversible work exerted during the simulations:

$$e^{-\beta\Delta F_{AB}} = \langle e^{-\beta W_{AB}}\rangle \quad (11.39)$$

Consistently with what it is known from thermodynamics, this relationship indicates that the lower work profiles are more similar to the actual free energy surface. Often, the external force is modeled as a spring connected to a dummy atom, which drives the system to the desired state at constant velocity.

11.4.3.5   Metadynamics

In metadynamics, an external history-dependent bias potential, which is a function of $\hat{\xi}$, is added to the Hamiltonian of the system. This potential is expanded as a sum of Gaussians that are deposited while the system wanders in the phase space. The aim of these Gaussians depending on $\hat{\xi}$ is to refrain the system from revisiting configurations that have already been sampled. The bias is applied continuously during the MD simulation, either through an extended Lagrangian formalism[46] or by acting directly on the microscopic coordinates of the system.[47] At time t, the metadynamics bias potential can be expressed as follows:

$$w(\xi,t) = \int_0^t \frac{\omega}{\tau_G} \exp\left\{ -\frac{[\xi(r(t)) - \xi(r(\tau))]^2}{2\sigma^2} \right\} d\tau \qquad (11.40)$$

where ω is the height of the Gaussian and τ_G is the deposition stride.

The basic effect of this accumulation of Gaussians is to accelerate the exit from any basin where the system could become trapped and to explore realistic reactive pathways between basins since the point of exit is expected to be around the lowest free-energy saddle point. Moreover, after a transient, the bias potential $w(\xi,t)$ provides an unbiased estimate of the underlying free energy:

$$w(\xi,t) \xrightarrow[t \to \infty]{} -F(\xi) + C \qquad (11.41)$$

where C is a constant that cancels out in any free energy difference expression.

Metadynamics is not completely free of many problems that characterize other enhanced sampling techniques; in particular, in a single run the additional bias tends to oscillate around the actual free energy plus a constant, leading to an overfilling of the underlying free energy surface and making critical the decision of when to stop a simulation. However, in this respect, one can say that, as a general rule, if metadynamics is used to find the closest saddle point, it should be stopped as soon as the system exits from the minimum. Otherwise, if one is interested in reconstructing a free energy surface (FES), it should be stopped when the motion of the CVs becomes diffusive in the region of interest.[48] Two interesting improvements over the original method consist in the linear scaling "multiple walkers" method[49] and in well-tempered metadynamics.[50] In the first, multiple replicas of the system are introduced, each with an associated "walker". All the replicas depose Gaussians, simultaneously contributing to the same history-dependent bias potential. As in ordinary metadynamics, the sum of the Gaussians laid by all the walkers provides an unbiased estimate of the free energy and the resulting statistics are combined by means of a weighted histogram analysis. In the well-tempered variant of metadynamics, the biasing potential is made adaptive so to provide control over convergence and error; moreover, this approach offers the possibility of controlling the regions of FES that are physically meaningful to explore. These improvements as well as others are well described in the work of Leoni *et al.*[51] and references therein.

11.5 Enhanced Sampling Applications to Protein–Ligand Binding

In the following, we will briefly describe some applications and challenges pertaining to two among the most widely used enhanced sampling techniques to the problem of protein–ligand binding, namely steered molecular dynamics (SMD) and metadynamics. This is not intended to be an exhaustive review, but aims at providing some insights into this relatively new field.

11.5.1   Steered Molecular Dynamics

SMD has been extensively used in biomolecular simulations, accompanying conventional MD simulations, and in particular in simulating small molecule undocking from a target binding site.

In this context, several studies have been performed in the Jiang's group in the Shanghai Institute of Materia Medica. We will start by mentioning the nanosecond-long simulation of binding and unbinding of a-APA, a nonnucleoside reverse trascriptase inhibitor, from HIV-1 reverse transcriptase.[52] Then, two works focused on the binding to acetylcholinesterase (AChE); the first[53] unveiled some mechanistic details of entrance and exit of huperzine A from the AChE gorge, while Niu *et al.*[54] studied the unbinding mechanism of E2020 from AChE with simulations on the order of nanoseconds. They tried different velocities of the dummy atom. The simulation results based on the SMD using a pulling velocity of 0.005 Å ps^{-1} have revealed many dynamic features of the inhibitor unbinding process and provided insights into the inhibition mechanism of E2020. One year later the same group published a study[55] concerning the blocking of the ion channel pore of the nicotinic acetylcholine receptor by the neuroleptic chlorpromazine. They were able to estimate the free energy of binding and to identify the interaction site. Moreover, their simulations led to the conclusion that chlorpromazine may not only block ion transport, but also restrict the conformational transition of the receptor channel, which is also a requisite for this latter to exert its physiological functions.

Li *et al.*[56] probed the possible testosterone exit channels in cytochrome P450 2B1. They built a homology model, docked the substrate, and pulled this out through three putative channels using SMD. The results were able to rule out one of the three possibilities due to the large rupture force and backbone motion. More recently, the Jiang group also published a study on the dynamical mechanism of transport of long-chain fatty acids across FadL, a fatty acid transport protein across membranes, where distinct preferred pathways for different fatty acids have been hypothesized. Mechanistic details concerning ligand dissociation from thyroid hormone receptor are presented in the work of Martinez *et al.*,[57] where the flexible region more prone to allow ligand dissociation was identified.

SMD was used in 2007[58] to analyze the exit pathways of cytochrome c2 from the reaction center of *Rhodobacter sphaeroides*. However, the MM/PBSA[59] approach was preferred in order to achieve a quantitative estimation of the binding free energy. SMD was also utilized in an actual docking process together with three other docking tools by Fairchild *et al.*[60] to see how the VX nerve agent binds human serum paroxonase 1. This latter was modeled by homology and then several different mutual positions were prepared as starting points for the SMD procedure. The binding was induced by steering the distance of three atoms of VX from the active site. The resulting bound pose turned out to be quite similar to the best predicted by the DOCK program. The authors concluded that, unlike methods that consider the binding site as a rigid structure, which need to start from open conformations to achieve good results, SMD binding simulations can be useful for obtaining binding predictions even when starting with a protein structure in a closed conformation.

More complex collective variables were adopted by Isberg *et al.*,[61] who simulated the activation of a 5-HT$_{2A}$ model by an agonist ligand. Collective variables were based on knowledge concerning hydrogen bonds of interhelical, ligand, and G protein interactions gathered from mutagenesis and X-ray crystallographic data. The simulation included a DPPC lipid membrane and TIP4P explicit water solvent. Unbiased MD assessed the stability of the achieved conformations.

Lipids have also been studied by SMD, especially cholesterol; for instance, Murcia *et al.*[62] undocked cholesterol from the steroidogenic acute regulatory protein-related lipid transfer domains of two homolog proteins. Their results show that cholesterol can adopt a similar conformation in the binding cavity in both cases and that the main contribution to the protein–ligand interaction energy derives from hydrophobic contacts. However, hydrogen-bonding and water-mediated interactions appear to be important in the fine tuning of the binding affinity and the position of the ligand.

Similar conclusions were attained by Canagarajah *et al.*,[63] who studied the cholesterol exchange mechanism in the one oxysterol-binding protein-related protein (ORP) family member, yeast Osh4. By means of SMD, they observed an opening mechanism of a lid covering the cholesterol-binding tunnel. Relevant microscopic factors for the dissociation are water-mediated hydrogen bonds and van der Waals contacts within the binding pocket, opening of a lid covering the binding pocket, and the breakage of transient cholesterol contacts with the rim of the pocket and hydrophobic residues on the interior face of the lid.

Interestingly, SMD was also used in the CAPRI protein–protein docking challenge. Heifetz *et al.*[64] developed a fully solvated SMD protocol and used it to identify near-native complexes, and to refine the complexes obtained with other larger scale methods allowing for backbone flexibility. They applied this approach combined with other techniques in CAPRI rounds 6–12 and were able to provide acceptable or medium decoys in three out of four rounds, showing better results in cases lacking a large conformational rearrangement.

Finally, we would like to mention an interesting new protocol for ranking within series of congeneric small molecules. Colizzi *et al.*[30] combined docking, cluster analysis, and SMD simulations with the ligand-active site distance as steered variable; by means of SMD, they obtained approximate potential of mean force profiles and exploited the differences corresponding to the rupture of the anchoring interactions during the unbinding events to discriminate active from inactive ligands. In order to make this comparison, a high accuracy as the one required by the Jarzynski equality it is not needed, making the approach computationally affordable. With their protocol, the authors were able to rationalize the structure–activity relationship of a flavonoid series, and to design and predict the biological activity of a novel inhibitor of the *Plasmodium falciparum* FabZ, a dehydratase playing a vital role in the life cycle of the parasite.

In summary, it can be observed that SMD is rarely used to actually calculate the free energy profiles according to the original Jarzynski formula, probably due to the computationally prohibitive number of the required trajectories, but rather to discover mechanistic details of the binding process that can be instrumental for the improvement of hit compounds.

11.5.2 Metadynamics

The recognition of tetramethylammonium by the peripheral anionic site and the subsequent penetration into the human AChE were studied by Branduardi and co-workers.[32] In that work, one of the biggest challenges of enhanced sampling techniques is evident, namely the effect of an incomplete choice for the collective variable. The authors show that accelerating the sampling of the distance between ligand and site is not sufficient to guarantee convergence of the metadynamics run and to achieve a reliable description of the process. In that case, the definition of a new collective variable describing the cation–π interaction, and accelerating also the opening of a gate made of aromatic residues, led to results in good agreement with experiments.

Four flexible docking simulations, namely β-trypsin/benzamidine, β-trypsin/chlorobenzamidine, immunoglobulin McPC-603/phosphocholine, and cyclin-dependent kinase 2/staurosporine were performed by Gervasio *et al.*[65] using as CVs the angle between the major inertia axis of each ligand and the line connecting the ligand to the centroid of the receptor, and the distance of the ligand from an atom in the binding site or the distance of the ligand from the centroid of the receptor. In these cases, the method was able to predict the docked geometry of the complexes even without assuming their previous knowledge.

Another example of application to docking is provided by Masetti *et al.*,[33] where the following protein–ligand complexes were studied: glycogen synthase kinase-3β in complex with a selective ureidic inhibitor, estrogen receptor-r complexed with 4-hydroxy tamoxifen, and *Escherichia coli* enoyl-ACP reductase complexed with triclosan. First, a set of 2D collective variables, *i.e.* the usual distance and orientation angle, were chosen and then compared with the inclusion of a third variable, namely a dihedral angle between two reference points on the protein, the ligand center, and the major inertia axis of the rigid moiety of the ligand. The addition of a third variable slowed down the computation and provided a significant improvement only in the case of triclosan, supporting the idea that the "right" CVs vary from case to case. Therefore, it is difficult to define "universal" collective variables for docking and undocking simulations via metadynamics.

The availability of high-resolution X-ray crystal structures of non-rhodopsin GPCRs for diffusible hormones and neurotransmitters was exploited by Provasi *et al.*[66] to apply microsecond-scale metadynamics to probe the interaction between homology models of opioid receptors, and a nonselective antagonist, naxolone. Interestingly, a DPPC/cholesterol lipid bilayer was included in the simulation because of its modulatory role in the function and structure of a number of GPCRs. The results of these simulations, corrected to improve ligand sampling in the bulk region, provided equilibrium constant values for the final bound state of naxolone very close to experimental results.

In the work of Limongelli *et al.*,[34] the full unbinding pathway of the highly selective inhibitor SC-558 of cyclooxygenase-2 (COX-2) was studied by means of all-atom MD simulations in explicit solvent and metadynamics calculations, using a distance and a torsional angle as collective variables. These unbinding simulations were coupled to explicit simulations of large scale motions of two α-helices that allowed greatly reducing hysteresis. Using this approach, the authors found a putative alternative binding mode to that experimentally known, and they conjectured that the ability of SC-558 to bind COX-2 in two different ways could explain the increased residence time in this enzyme isoform, likely at the basis of SC-558 toxicity. Consistently, the same procedure applied to the other isoform of the cyclooxygenase, namely COX-1, showed only one binding pose, which agrees remarkably well with the X-ray structure of COX-1 complexed with celecoxib, an analogue of SC-558, which was very recently reported.[67]

Metadynamics was also used to study the translocation of two antibiotics, namely moxifloxacin and ampicillin, through the OmpF protein channel.[68] In this study, in addition to the most obvious collective variable, *i.e.* the progress along the direction of the channel, more informative reaction coordinates were included, accounting explicitly for antibiotic flexibility and interactions with the channel. This approach showed two different molecular paths for the two antibiotics and outlined the relevance of the interaction made at the constriction region.

Metadynamics has also been used in studying the folding process. For instance, Bonomi *et al.*[69] analyzed p-S8, a small peptide belonging to the HIV-1 protease, which is thought to have a prominent role in the folding of the protein. The authors combined metadynamics and parallel tempering, and used a measure of the water present at the interface and the formation of hydrogen bonds across the molecules as collective variables. They simulated the interaction of p-S8 and another, less structured, segment of the protein under the hypothesis that their docking is responsible of the creation of the folding nucleus of the protease. As a result, they found three hydrogen bonds that stabilize the interaction, giving a clue for the design of anti-AIDS drugs. This approach was further extended in 2010,[70] where the whole folding process was considered. The authors found that in this case the formation of a hydrophobic core is a milestone event, while the rest of the protein can reach its native state following different pathways and order of assembling.

Overall, metadynamics and its more recent improvements provide a number of powerful tools to describe complex biomolecular phenomena.

11.6 Conclusions and Perspectives

The ultimate goal of molecular simulations is to complement experimental observations, and possibly to predict properties that are hardly accessible to measurements. Clearly, among these properties, free energy differences have always played a role of major interest. However, since the pioneering computational studies, efforts addressed in this direction have been extensively frustrated primarily due to the limitations in computational power compared to the complexity entailed by the problems such as protein–ligand binding. To overcome these difficulties, two different strategies have been adopted. The most obvious solution is the "technological" route: as the computational power increases, longer time scales become accessible to simulations, with the consequence of gaining better chances to achieve the convergence for a given sampling. Notably, since the early MC and MD studies, the computational power has increased by more than six orders of magnitude: from 1000 floating point operations per second (Flops) of the late 1950s to the hundreds of TFlops achieved in recent years. In this context, active research has primarily dealt with the development of specialized and highly performing hardware.[71] A different approach is the "methodological" route, which is focused on developing methods aimed at escaping from a complete Boltzmann sampling while retaining a correct Boltzmann statistics of the visited states (enhanced sampling methods). Nowadays, enhanced sampling methods have become of age, and they are widely used in several fields of computational research, ranging from materials science to medicinal chemistry. In particular, a number of these methods have been either developed or massively employed, especially during the last decade, when the commercial cost of parallel computing architectures became progressively accessible to a larger extent of the scientific community. This increase in average available computational power corresponded to a parallel increased demand for enhancing sampling, as the nature of the problems investigated became more and more ambitious. From a pharmaceutical standpoint, the estimation of the protein–ligand binding free energy, which in the past has often been termed as the "holy grail" in computational drug design, became finally accessible at an acceptable level of description (all atoms, explicit water molecules, and long-range electrostatic forces accounted for). Notwithstanding the difficulties often encountered leading to a lack of generalization (*i.e.* no enhanced sampling method can be blindly applied to every kind of problem with an absolute chance of success), and in some cases the involvedness of the theoretical framework they rely upon (most of the enhanced sampling methods require a high level of expertise to be handled), an encouraging agreement with experimental data has been in general achieved. In spite of these undoubting successes, and in light of the amazing performances recently shown by some specialized architectures as well as GPU-based computational clusters,[19] it is worth wondering whether there is still room for enhancing MC or MD sampling. Judging from the recent trends emerging from the literature, it might seem that the decline of the enhanced sampling era has finally come. Nowadays, simulations lasting hundreds of microseconds have become reality, and although these timescales are not affordable yet for the vast majority of the scientific community, it is reasonable to imagine that they will be in the near future. Despite these tremendous technological advances, in our opinion, enhanced sampling methods are not to be completely abandoned yet. Indeed, there is some matter of concern, often overlooked, that is worth addressing:

1. First of all, it must be considered that the protein–ligand binding event takes place in a time scale ranging from microseconds up to milliseconds (and possibly longer). The legitimate excitement caused by the possibility to extend MD simulations to the unprecedented duration of hundreds of microseconds should not distract attention from the fact that achieving a binding event in an unbiased manner does not directly imply the possibility to calculate the free energy or the kinetics of binding.[72,73] Rather, an ensemble of many unbiased binding events must be collected to achieve a proper statistical description of the binding process. This subject has not been explicitly addressed in the literature yet, although it will be probably covered in the future to strengthen the technique or to assist further advances.

2. Closely related with the previous point is the possibility to take advantage of the aforementioned exceptional computational power to drive enhanced sampling simulations towards an effective convergence. Indeed, even though this issue is hidden or simply not addressed in many applications, the use of these methods has always suffered from the limitations due to an incomplete convergence of sampling. From a practical perspective, this results in poor precision in free energy estimation and/or hysteresis in free energy profiles. Several diagnostic techniques and *ad hoc* procedures have been developed to monitor and limit this problem. However, it is clear that all the enhanced sampling methods would strongly benefit from a massive increment in computational power.

3. The more you get, the more you want. The computational demand depends not only on the timescales one wishes to simulate, but it is also a function of the system size and of the level of description employed to treat the problem. Researchers have just started to investigate very big systems, as the need to understand biological processes such as protein–protein interactions, protein–nucleic acid interactions, allostery, *etc.*, intrinsically requires larger size scales compared to those commonly used to study protein–ligand binding. In order to let these simulations affordable, the increase in computational cost is usually paid by a decrease in the level of description. That is, instead of using all atom force fields, simpler functional forms have been devised to treat the

potential energy terms (coarse-grained force fields). On the one side, the combined use of a huge computational power together with the help of enhanced sampling methods would finally open the possibility to study larger systems and larger motions at a fully atomistic level. On the other side, limitations in the current widespread additive force fields for condensed-phase simulations have long been recognized. From this standpoint, given a moderate system size, the opportunity offered by a combined use of enhanced sampling methods and the recently available computational power would allow a routinely use of demanding polarizable force fields, as their superior accuracy has been proven to be substantial in case of binding free energy estimates.[74]

In conclusion, the importance and the usefulness of enhanced sampling methods are largely demonstrated by their widespread use in the community. The growth in computational power does not preclude *per se* the use of these methods, as we have shown that the "technological" and "methodological" routes are not mutually exclusive. In contrast, a smart combination of the two would increase both the precision and the accuracy of molecular simulations and, ultimately, would strengthen their predictive power.

References

1. W. L. Jorgensen, *Science*, 2004, **303**, 1813.
2. J. M. J. Swanson, R. H. Henchman and J. A. McCammon, *Biophys. J.*, 2004, **86**, 67.
3. M. S. Lee and M. A. Olson, *Biophys. J.*, 2006, **90**, 864.
4. B. O. Brandsdal, F. Österberg, M. Almlöf, I. Feierberg, V. B. Luzhkov, J. Åqvist and D. Valerie, *Adv. Protein Chem.*, 2003, **66**, 123.
5. J. Åqvist, C. Medina and J.-E. Samuelsson, *Protein Eng.*, 1994, **7**, 385.
6. J. Marelius, M. Graffner-Nordberg, T. Hansson, A. Hallberg and J. Åqvist, *J. Comput. Aided Mol. Des.*, 1998, **12**, 119.
7. J. Åqvist and T. Hansson, *J. Phys. Chem.*, 1996, **100**, 9512.
8. G. Grazioso, A. Cavalli, M. De Amici, M. Recanatini and C. De Micheli, *J. Comput. Chem.*, 2008, **29**, 2593.
9. G. Rastelli, G. Degliesposti, A. Del Rio and M. Sgobba, *Chem. Biol. Drug Des.*, 2009, **73**, 283.
10. L. J. William, J. K. Buckner, B. Stephane and T. R. Julian, *J. Chem. Phys.*, 1988, **89**, 3742.
11. Y. Deng and B. Roux, *J. Chem. Theory Comput.*, 2006, **2**, 1255.
12. J. Wang, Y. Deng and B. Roux, *Biophys. J.*, 2006, **91**, 2798.
13. M. K. Gilson, J. A. Given, B. L. Bush and J. A. McCammon, *Biophys. J.*, 1997, **72**, 1047.
14. Y. Deng and B. Roux, *J. Chem. Theory Comput.*, 2006, **2**, 1255.
15. D. L. Mobley, J. D. Chodera and K. A. Dill, *J. Chem. Phys.*, 2006, **125**, 084902.

16. D. L. Mobley, J. D. Chodera and K. A. Dill, *J. Chem. Theory Comput.*, 2007, **3**, 1231.
17. D. L. Mobley, A. P. Graves, J. D. Chodera, A. C. McReynolds, B. K. Shoichet and K. A. Dill, *J. Mol. Biol.*, 2007, **371**, 1118.
18. S. E. Boyce, D. L. Mobley, G. J. Rocklin, A. P. Graves, K. A. Dill and B. K. Shoichet, *J. Mol. Biol.*, 2009, **394**, 747.
19. I. Buch, T. Giorgino and G. De Fabritiis, *Proc. Natl. Acad. Sci. U. S. A.*, 2011, **108**, 10184.
20. R. A. Copeland, D. L. Pompliano and T. D. Meek, *Nat. Rev. Drug Discovery*, 2006, **5**, 730.
21. D. C. Swinney, *Nat. Rev. Drug Discovery*, 2004, **3**, 801.
22. W. E. Ren, W. Ren and E. Vanden-Eijnden, *J. Phys. Chem. B*, 2005, **109**, 6688.
23. L. Maragliano, A. Fischer, E. Vanden-Eijnden and G. Ciccotti, *J. Chem. Phys.*, 2006, **125**, 024106.
24. L. Maragliano and E. Vanden-Eijnden, *Chem. Phys. Lett.*, 2006, **426**, 168.
25. J. Hénin and C. Chipot, *J. Chem. Phys.*, 2004, **121**, 2904.
26. B. Roux, *Comput. Phys. Comm.*, 1995, **91**, 275.
27. S. Park and K. Schulten, *J. Chem. Phys.*, 2004, **120**, 5946.
28. A. Laio and M. Parrinello, *Proc. Natl. Acad. Sci. U. S. A.*, 2002, **99**, 12562.
29. C. Jarzynski, *Phys. Rev. Lett.*, 1997, **78**, 2690.
30. F. Colizzi, R. Perozzo, L. Scapozza, M. Recanatini and A. Cavalli, *J. Am. Chem. Soc.*, 2010, **132**, 7361.
31. J. S. Patel, D. Branduardi, M. Masetti, W. Rocchia and A. Cavalli, *J. Chem. Theory Comput.*, in press.
32. D. Branduardi, F. L. Gervasio, A. Cavalli, M. Recanatini and M. Parrinello, *J. Am. Chem. Soc.*, 2005, **127**, 9147.
33. M. Masetti, A. Cavalli, M. Recanatini and F. L. Gervasio, *J. Phys. Chem. B*, 2009, **113**, 4807.
34. V. Limongelli, M. Bonomi, L. Marinelli, F. L. Gervasio, A. Cavalli, E. Novellino and M. Parrinello, *Proc. Natl. Acad. Sci. U. S. A.*, 2010, **107**, 5411.
35. J. R. Waldram, *The Theory of Thermodynamics*, Cambridge University Press, New York, 1985.
36. D. A. McQuarrie, *Statistical Mechanics*, Harper & Row, New York, 1973.
37. L. Verlet, *Phys. Rev.*, 1967, **159**, 98.
38. C. S. William, C. A. Hans, H. B. Peter and R. W. Kent, *J. Chem. Phys.*, 1982, **76**, 637.
39. T. Förster, *Ann. Phys.*, 1948, **437**, 55.
40. D. Frenkel and B. Smit, *Understanding Molecular Simulation*, Academic Press, San Diego, , CA, 2002.
41. E. Vanden-Eijnden, in *Computer Simulations in Condensed Matter Systems: From Materials to Chemical Biology* (*Lectures Notes in Physics*, vol. 703), ed. M. Ferrario, G. Ciccotti and K. Binder, Springer, Berlin, 2006, p. 453.
42. J. L. Knight and C. L. Brooks, III, *J. Comput. Chem.*, 2009, **30**, 1692.

43. B. Tidor, *J. Phys. Chem.*, 1993, **97**, 1069.
44. M. Souaille and B. Roux, *Comput. Phys. Comm.*, 2001, **135**, 40.
45. J. Kästner, *WIREs Comput. Mol. Sci.*, 2011, in press.
46. M. Iannuzzi, A. Laio and M. Parrinello, *Phys. Rev. Lett.*, 2003, **90**, 238302.
47. A. Laio, A. Rodriguez-Fortea, F. L. Gervasio, M. Ceccarelli and M. Parrinello, *J. Phys. Chem. B*, 2005, **109**, 6714.
48. A. Barducci, M. Bonomi and M. Parrinello, *WIREs Comput. Mol. Sci.*, 2011, **1**, 826.
49. P. Raiteri, A. Laio, F. L. Gervasio, C. Micheletti and M. Parrinello, *J. Phys. Chem. B*, 2005, **110**, 3533.
50. A. Barducci, G. Bussi and M. Parrinello, *Phys. Rev. Lett.*, 2008, **100**, 020603.
51. V. Leone, F. Marinelli, P. Carloni and M. Parrinello, *Curr. Opin. Struct. Biol.*, 2010, **20**, 148.
52. L. Shen, J. Shen, X. Luo, F. Cheng, Y. Xu, K. Chen, E. Arnold, J. Ding and H. Jiang, *Biophys. J.*, 2003, **84**, 3547.
53. Y. Xu, J. Shen, X. Luo, I. Silman, J. L. Sussman, K. Chen and H. Jiang, *J. Am. Chem. Soc.*, 2003, **125**, 11340.
54. C. Niu, Y. Xu, Y. Xu, X. Luo, W. Duan, I. Silman, J. L. Sussman, W. Zhu, K. Chen, J. Shen and H. Jiang, *J. Phys. Chem. B*, 2005, **109**, 23730.
55. Y. Xu, F. J. Barrantes, J. Shen, X. Luo, W. Zhu, K. Chen and H. Jiang, *J. Phys. Chem. B*, 2006, **110**, 20640.
56. W. Li, H. Liu, E. E. Scott, F. Grater, J. R. Halpert, X. Luo, J. Shen and H. Jiang, *Drug. Metab. Dispos.*, 2005, **33**, 910.
57. L. Martinez, P. Webb, I. Polikarpov and M. S. Skaf, *J. Med. Chem.*, 2005, **49**, 23.
58. T. V. Pogorelov, F. Autenrieth, E. Roberts and Z. A. Luthey-Schulten, *J. Phys. Chem. B*, 2006, **111**, 618.
59. P. A. Kollman, I. Massova, C. Reyes, B. Kuhn, S. Huo, L. Chong, M. Lee, T. Lee, Y. Duan, W. Wang, O. Donini, P. Cieplak, J. Srinivasan, D. A. Case and T. E. Cheatham, *Acc. Chem. Res.*, 2000, **33**, 889.
60. S. Fairchild, M. Peterson, A. Hamza, C.-G. Zhan, D. Cerasoli and W. Chang, *J. Mol. Model.*, 2010, **17**, 97.
61. V. Ìsberg, T. Balle, T. Sander, F. S. Jørgensen and D. E. Gloriam, *J. Chem. Inf. Model.*, 2011, **51**, 315.
62. M. Murcia, J. D. Faráldo-Gómez, F. R. Maxfield and B. Roux, *J. Lipid Res.*, 2006, **47**, 2614.
63. B. J. Canagarajah, G. Hummer, W. A. Prinz and J. H. Hurley, *J. Mol. Biol.*, 2008, **378**, 737.
64. A. Heifetz, S. Pal and G. R. Smith, *Proteins: Struct., Funct., Bioinf.*, 2007, **69**, 816.
65. F. L. Gervasio, A. Laio and M. Parrinello, *J. Am. Chem. Soc.*, 2005, **127**, 2600.
66. D. Provasi, A. Bortolato and M. Filizola, *Biochemistry*, 2009, **48**, 10020.

67. G. Rimon, R. S. Sidhu, D. A. Lauver, J. Y. Lee, N. P. Sharma, C. Yuan, R. A. Frieler, R. C. Trievel, B. R. Lucchesi and W. L. Smith, *Proc. Natl. Acad. Sci. U. S. A.*, 2009, **107**, 28.
68. E. Hajjar, A. Kumar, P. Ruggerone and M. Ceccarelli, *J. Mol. Model.*, 2009, **16**, 1701.
69. M. Bonomi, F. L. Gervasio, G. Tiana, D. Provasi, R. A. Broglia and M. Parrinello, *Biophys. J.*, 2007, **93**, 2813.
70. M. Bonomi, A. Barducci, F. L. Gervasio and M. Parrinello, *PLoS ONE*, 2010, **5**, e13208.
71. J. L. Klepeis, K. Lindorff-Larsen, R. O. Dror and D. E. Shaw, *Curr. Opin. Struct. Biol.*, 2009, **19**, 120.
72. R. O. Dror, A. C. Pan, D. H. Arlow, D. W. Borhani, P. Maragakis, Y. Shan, H. Xu and D. E. Shaw, *Proc. Natl. Acad. Sci. U. S. A.*, 2011, **108**, 13118.
73. Y. Shan, E. T. Kim, M. P. Eastwood, R. O. Dror, M. A. Seeliger and D. E. Shaw, *J. Am. Chem. Soc.*, 2011, **133**, 9181.
74. D. Jiao, P. A. Golubkov, T. A. Darden and P. Ren, *Proc. Natl. Acad. Sci. U. S. A.*, 2008, **105**, 6290.

CHAPTER 12

Expanding the Target Space: Druggability Assessments

PETER SCHMIDTKE, DANIEL ALVAREZ-GARCIA, JESUS SECO AND XAVIER BARRIL*

ICREA & Departament de Fisicoquímica, Facultat de Farmàcia, Universitat de Barcelona, Av. Joan XXIII s/n, E-08028 Barcelona, Spain
*E-mail: xbarril@ub.edu

12.1 Introduction

In recent decades, drug discovery has mainly followed a target-based approach, where relationships are sought between a pathological situation and the biological activity of individual macromolecules. Once such a link is established (and confirmed at the target validation stage)[1] the goal of drug discovery is to identify organic molecules capable of modulating the activity of the target. As expressed by Hopkins and Groom, a potential target is only druggable (*i.e.* pharmacological tractable) if it can *"bind a small molecule with the appropriate chemical properties at the required binding affinity".*[2] However, either due to an overly optimistic confidence in the ability to discover active compounds or to the inexistence of methods to reliably predict druggability, this property is often presumed or crudely extrapolated from the functional classification of the target. Considering that 50% of drug discovery programs fail to produce viable leads even for a privileged target class such as enzymes,[3] it is evident that druggability prediction methods can make a major economical impact, directing the use of resources towards the most promising targets.

RSC Drug Discovery Series No. 23
Physico-Chemical and Computational Approaches to Drug Discovery
Edited by F. Javier Luque and Xavier Barril
© The Royal Society of Chemistry 2012
Published by the Royal Society of Chemistry, www.rsc.org

Perhaps more importantly, druggability assessment methods will become fundamental in guiding drug discovery away from the "comfort zone" provided by privileged target families such as kinases, nuclear receptors or G-Protein Coupled Receptors (GPCRs).[4] Indeed, the whole target-based approach is nowadays under intense scrutiny because its dominance coincides with a productivity crisis in the pharmaceutical industry,[5,6] with high attrition rates in clinical trials, mostly due to lack of efficacy.[7,8] Furthermore, even during that period, the most novel drugs were often discovered by means of phenotypic screenings, which were used far less frequently.[6] The limited success of target-based approaches may partly be due to its reductionist approach, but they also offer both practical and conceptual advantages and, even when the initial hits are discovered by phenotypic screening, identification of the molecular target becomes desirable.[9] This suggests that the way in which targets have been traditionally selected may bear more responsibility on the ultimate low productivity than the approach itself. For instance, in an overwhelming majority of cases, target-based approaches aim at displacing the endogenous ligand of a protein, while natural compounds or those identified by phenotypic screening exploit a more diverse array of targets and molecular mechanisms of action (MMoA).[6] In this context, druggability prediction methods have the potential to extend the target space beyond the current limits, while still increasing the probability of success.

12.2 The Target Space

In order to be useful as a pharmacological target, a biological component must fulfil the double condition of being disease modifying and amenable to modulation by a therapeutic strategy. This includes a potentially large proportion of the molecular machinery of human cells and pathogenic species. Owing to the diversity of roles played by proteins, they are by far the most common targets.[4,10] Accordingly, this review is exclusively focused on proteins, but it should be noted that many drugs, particularly in oncology and antimicrobial classes, act on DNA or RNA.[10] Our choice is also supported by initial studies on the druggability of the ribosome, which suggest that the structural determinants may be quite distinct from proteins.[11] A precise definition of the target space is precluded by many factors, including the fact that many drugs act on multiple targets and the mechanism of action of a small but significant proportion is still unknown.[10] Furthermore, it should also be noted that a good target is not only druggable, but should also satisfy many other conditions.[12] However, an analysis of current drug targets affords some clues to understand the importance of druggability and offers an insight into the future evolution of the target space.

12.2.1 Privileged Target Classes

An inspection at the distribution of drug targets by protein family (Figure 12.1)[4] suggests that enzymes, receptors and transporters are druggable

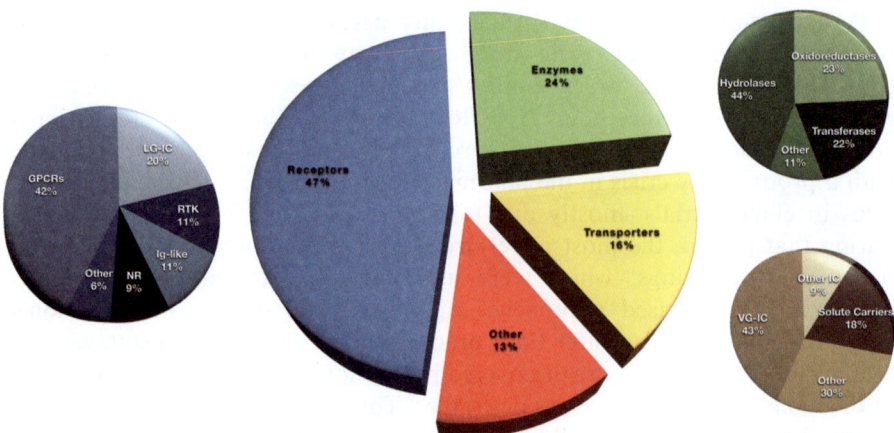

Figure 12.1 Protein targets of small-molecule drugs, classified by protein type. Receptors, enzymes and transporters are sub-divided into protein families. Data obtained from ref. 4. Abbreviations: GPCRs, G-protein coupled receptors; LG-IC, ligand-gated ion channels; RTK, receptor tyrosine kinases; Ig-like, immunoglobulin-like receptors; NR, nuclear receptors; VG-IC, voltage-gated ion channels.

targets, while other protein classes are difficult or undruggable. This is generally a good assumption because these protein classes have evolved to interact with small organic molecules (substrates, hormones, neurotransmitters, *etc.*), which means that drugs can compete in equal terms and achieve high affinity for the binding site. Depending on the type of natural substrate, their difficulty as targets will also vary: proteins binding *bona fide* small molecules (*e.g.* class A GPCRs, kinases) are more druggable than those binding non-drug-like ligands, such as peptides (*e.g.* classes B and C in GPCRs, proteases). The influential paper by Hopkins and Groom and other works estimating the size of the so-called "druggable genome"[2,13] relied on protein domain annotations (*e.g.* from PFAM[14]) to predict the number of proteins containing domains experimentally known to be targeted by small-molecule drugs. A support vector machine method has also been developed to predict protein druggability based on an amino acid sequence independent of sequence similarity.[15] However, these methods rely on the assumption that similar proteins display similar druggability, which is likely to be true at a statistical level, but not necessarily for a particular target of interest. Some large and eminently druggable protein families, such as nuclear receptors, ion channels, GPCRs, serine proteases or kinases, have attracted particular attention from the pharmaceutical research industry, as they usually bind molecules in the same chemical space, and the design of focused libraries facilitates the identification of active compounds.[16] Selectivity across family members is the main limitation of this approach and, paradoxically, the development of allosteric modulators that target alternative binding sites is a growing trend.[17]

This demonstrates the interest in exploiting mechanisms of action that are not linked to a conserved sequence or domain topology, both within and outside the main target classes.

12.2.2 Under-Exploited Molecular Mechanism of Action

Although most drugs acting on privileged target families compete with other small organic molecules (substrates, ligands), other molecular mechanisms of action are also very common. Interestingly, these mechanisms are exploited much more frequently by first-in-class drugs discovered through phenotypic screening,[6] and some of them are also valid for protein classes that do not normally bind small molecules. Although the presence of druggable cavities allowing the exploitation of alternative molecular mechanism of action (*e.g.* allosterism, conformational trapping) may be probabilistically low for each individual structure, the ubiquity of the phenomenon and the abundance of diverse biological components (for instance, current estimates for the human interactome are 130 000 protein–protein pairs[18]) warrants many new therapeutic opportunities.

12.3 Properties of Drug Binding Sites

Most biomacromolecules have a tendency to interact with other organic molecules (proteins, nucleic acids, metabolites, *etc.*), but this property is unevenly distributed over the protein surface. Surface patches with a larger interaction potential are known as "hot spots", a concept originating from alanine scanning experiments that probed the interface between protein–protein complexes. These studies revealed that most of the binding free energy is contributed by a few residues only.[19,20] The O-ring theory, initially introduced by Bogan and Thorn, suggests that hot-spot residues are easily desolvated because the local environment induces solvent exclusion.[21–23] Although structurally different, ligand binding sites display the same behaviour: some regions in the binding area interact very favourably with particular functional groups, while the rest may provide the right shape and solvent exclusion capacity.[24] The presence of hot spots is, therefore, necessary for binding to occur, but what are the distinct properties of druggable binding sites? This has been investigated in parallel with the development of druggability prediction methods.

As expected from the properties of drug-like ligands, closed and lipophilic binding sites are more likely to be druggable. This is supported by the usefulness of related parameters to obtain predictive models following inductive[25,26] or deductive[27] reasoning. Binding site curvature relates to the necessity to maximize the contact surface area between the ligand and the protein, while the positive correlation of apolar surface area with druggability would presumably suggest that binding potency is entirely due to hydrophobic interactions. This was justified on the basis that electrostatic interaction and desolvation energies act in opposition, resulting in an overall negligible contribution.[25,27] However, this contradicts the empirical observation that polar interactions often constitute anchoring points, featuring predominantly in pharmacophoric models of binding sites. In fact, the contribution of polar interactions is context dependent, and a single hydrogen bond can contribute up to 1.8 kcal mol^{-1},[28] nearly twice as much as a hydrophobic gain provided by a methyl group.[29] The increased proportion of apolar surface area in druggable binding sites compared to non-druggable ones (70% *vs.* 50%) can, actually, be reconciled with the importance of polar interactions: polar atoms in druggable binding sites are less solvent exposed and have a predominantly hydrophobic environment, resulting in a lower dielectric environment that strengthens electrostatic interactions. This effect has been quantified in proteins, demonstrating that hydrogen bonds can be up to 1.2 kcal mol^{-1} stronger in hydrophobic environments.[30] This clearly indicates that, beyond the obvious gain in hydrophobic potential, a decrease in the polar surface ratio can have the paradoxical effect of increasing the hydrogen bonding potential of the binding site.

An intriguing property of druggable binding sites is that, in spite of being mostly buried, polar atoms protrude more from the cavity surface than apolar atoms.[26] In such disposition they are readily available to interact with incoming ligands, providing anchoring or selectivity points. It has been suggested that this type of environment can also transform polar atoms into kinetic traps. The molecular mechanism consists in a simple decoupling of the association and dissociation processes, which cannot happen in a concerted-like manner due to the steric impediments imposed by the local environment. In such circumstances, a transition state emerges and, in consequence, diffusion is slowed down.[31] This could stabilize transient encounter complexes, facilitating their mutual recognition[32] and, once formed, lock the binding mode of the interacting molecules.

12.4   Druggability: Term Definition and Controversy

"Druggability" is a neologism that appeared first in abstracts of scientific papers in 1999.[33] However, the "Druggable Genome" paper by Hopkins and Groom[2] made the first comprehensive review on chemical tractability in 2002. Since that review, the usage of the term became widespread among different areas in drug discovery and can designate different concepts related to the suitability of a target for drug discovery. In fact, despite the widespread usage of this neologism in today's literature, the exact scope of the term remains fuzzy and varies substantially from paper to paper.

Through this chapter, druggability is associated with a binding site, taking the definition that a pocket is druggable if it can bind a molecule similar to orally available marketed drugs. The definition of druggability is, therefore, necessarily linked to the notion of drug likeness, which is not unequivocally defined. From a purely structural point of view, the usage of this term is too vague because the structural features responsible for the interaction between a drug and the protein only constitute a part of all features required for a protein to be druggable. Structure-based techniques do not assess the disease relation of proteins and whether the binding of a small molecule has an impact on the function of the protein. As stated by Robert Sheridan and co-workers at Merck: "The term druggable implies too much. Getting to an actual drug involves many hurdles that are almost impossible to predict in advance and involve properties of small molecules as well as the target".[34] Alternative terms such as "chemical tractability", "bindability" or "ligandability" have also been proposed and may be more accurate; however, "druggability" is more concise and appealing. More importantly, the fuzziness of the term reflects the datasets on which druggability prediction methods have been trained and the use of the term is not problematic as long as the underlying assumptions are understood.

12.5 Experimental Determination of Druggability

12.5.1 High-Throughput Screening

In theory, the most rigorous form of assessing druggability is to test the ability of drug-like compounds to modify the biological activity of the target of interest. In this regard, examining the rate of success of high-throughput screening (HTS) results is very informative. In 2006, Macarron published a retrospective analysis of HTS campaigns at GlaxoSmithKline, grouping the results by target families. The analysis is particularly interesting because success was defined as the ability to produce a confirmed hit (*i.e.* activity in a biologically relevant assay with a tractable chemical structure and an initial indication of SAR such that a chemical optimization effort can begin). As expected, some target families offer very good results (*e.g.* a lead could be identified from HTS in $> 70\%$ of nuclear hormone receptors and ion channels), whereas the success rate was only 33% for targets not belonging to the main classes.[3] Gupta *et al.*, from AstraZeneca, also published a retrospective analysis of HTS on 22 enzymes (identified by function), but in this case the hit rate (*i.e.* percentage of compounds with read-outs about a certain threshold) was reported as a measure of success.[35] The values reported range from 0.06% to 3.85% for a common collection of 37 275 compounds. As pointed out by Macarron, one of the limitations of these retrospective analyses is that it is not possible to know if failure happens because the target is undruggable or because the collection of compounds tested does not cover the adequate chemical space: 30% of targets that failed when tested on a subset of the historical collections turned out to be tractable when tested against the unified GSK collection.[3] Considering the immensity of the drug-like chemical space (estimated at 10^{20}–10^{24} synthetically accessible compounds),[36] this is an important issue and suggests that success with novel target types is partly limited by the composition of current historical collections, which have focused on specific families of receptors, enzymes or transporters. It also raises questions about the usefulness of hit rates as druggability predictions. One should also be aware of the limitations of the specific assay; for instance, a binding assay may not be the most suitable to identify allosteric modulators. Although useful as a retrospective analysis, the costs of HTS make it absolutely impractical as a druggability assessment method.

12.5.2 Fragment Screening

Fragment screening was initially described in 1996,[37] adopting new detection methods and becoming extremely popular as a hit identification strategy in the past decade.[38–41] Its main advantage is the superior ability to detect binders because it explores much simpler compounds than HTS.[42] Considering that the number of possible chemical compounds grows as a quadratic function with the number of atoms,[43] even if the number of compounds tested is usually three orders of magnitude smaller than HTS it can in fact explore a much larger proportion of the corresponding chemical space. Accordingly, the fragment screening hit rates may be more informative about the druggability of a given protein than those coming from HTS. Abbot, Vernalis and more recently AstraZeneca have published data for 23, 12 and 36 targets, respectively,[25,44,45] and in all cases there is a good correlation between poor hit rates and the difficulty in obtaining high affinity ligands. This demonstrates the suitability of fragment screening as a method to detect good binding sites for small molecules. Once the necessary infrastructure and know-how is in place, the cost of fragment screening and the time needed to set up the experiment is much lower than the corresponding HTS assay, so carrying a fragment screening experiment before launching a full drug discovery project may be a wise and feasible approach for both small and large pharmaceutical companies. One potential limitation of this approach is that it is difficult to predict the drug-likeness of future ligands based on the chemical structure of the fragment hits. In other words, the fragment screening hit rate predicts "ligandability" better than druggability, and a good hit rate does not warrant that advanced leads can be developed. In a few cases, lead compounds were developed at AstraZeneca for targets with low fragment hit rates, which leads them to suggest that poor "ligandability" carries a substantial risk increase, but should not preclude the pursuit of the target using alternative approaches if the potential reward is sufficiently high.[45]

12.5.3 Multiple Solvent Crystal Structures

Well before the fragment screening era, it was observed that organic solvents have a large propensity to interact with binding sites of proteins both in solution and in crystals.[46,47] This raised the possibility of using "solvent mapping" to detect and characterize binding sites, something that has been achieved for a few systems.[46,48,49] Retrospectively, this can be seen as an extreme form of fragment screening: as the ligands tested are smaller, they are more likely to bind and fewer compounds need to be tested, but more sensitive methods are needed to detect binding.[42] The detection method is precisely the limitation of this approach: few proteins form crystals sufficiently stable to withstand the high concentrations of organic molecules necessary to carry out multiple solvent crystallographic screening. It is, however, conceivable that small libraries of "sub-fragments" (*i.e.* compounds with properties well below the "rule of 3")[50] could be specifically developed and tested with current fragment screening methods for the purpose of predicting druggability.

12.6 Computational Assessment of Druggability

Sequence similarity principles can be applied to infer druggability from closely related proteins. However, a more quantitative prediction requires structural information. In addition, most methods rely on a training set and on pre-existing methods to define the binding site. Accordingly, these two aspects will be reviewed before delving into druggability prediction methods.

12.6.1 Data Sets

The first druggability prediction method was trained to reproduce NMR fragment screening hit rates. The dataset consisted of 28 binding sites on 23 different proteins, for which 10 000 compounds were tested. Using hetero-nuclear NMR, perturbations anywhere on the protein can be detected and ligands with K_d values as high as 5 mM can be identified.[37] The physicochemical properties of the screening library conform to the definition of fragments (average molecular weight of 220 and an average $c \log P$ of 1.5) and, being tested at high concentrations (0.5–1.0 mM), they are highly soluble. The hit rates, ranging from 0.01% to 0.94%, were used as a measure of druggability.[25] As demonstrated in the paper, high correlation is observed between the experimental NMR hit rate and the ability to identify high-affinity ($K_d < 300$ nM) ligands. In line with this approach, researchers at AstraZeneca have used the HTS hit-rates as a measure of druggability. Using a set of 22 undisclosed targets, they obtained predictive models.[35] However, this definition of druggability presents two main limitations. Firstly, ligand drug-likeness implicitly derives from the composition of the screening library, but its physicochemical properties can be very different from typical drugs, particularly in the case of fragments. The second problem is of a practical nature, because screening data is proprietary, expensive to obtain and rarely made publicly available. Additionally, extension of published datasets would require using the same screening library and methodology, limiting its transferability across organizations.

In 2007, Cheng and co-workers presented an alternative view of druggability, defined as the maximal affinity that a drug-like ligand (ideally an orally bioavailable compound) can achieve for a binding pocket.[27] This definition also presents some limitations, such as the fact that the drug-likeness of a compound is sometimes difficult to assess or that the classification of a target may change over time. Obtaining good quality data can also be difficult, particularly when it comes to undruggable binding sites, because they can only be classified as such after substantial research efforts have been invested and negative data are often not published. However, the definition is useful in decision making, because it can distinguish between targets that are likely to have a successful outcome (*i.e.* deliver an orally bioavailable lead) and those that are more likely to prove very challenging and may require other approaches (*e.g.* a pro-drug strategy). Subsequent druggability prediction methods have mostly adhered to this definition.

In order to facilitate further developments and to establish a benchmark that could be used in prediction performance, the initial set of 27 targets presented by Cheng *et al.* was extended by Schmidtke and Barril[26] with 1070 structures representing 70 different targets. The set was obtained by crossing a list of oral drugs with information from the PDB[51] and the DrugBank,[52] followed by visual inspection. The unified catalogue is publicly available as the Druggable Cavity Directory (http://fpocket.sourceforge.net/dcd), a resource that can also be used to extend the dataset or to reassess target classification in a collaborative manner.

The most extensive dataset to date has been published by Brenk and co-workers.[53] Known as the non-redundant set of druggable and less druggable binding sites (NRDLD), the set contains 115 structures, 71 of which are considered druggable and the remaining 44 are "less druggable". The set of druggable cavities was obtained from bibliographic searches, DrugBank,[52] the Cheng *et al.* dataset[27] and selected structures from the Astex diverse set (a collection of protein–ligand complexes used to assess docking performance).[54] The set of "less druggable" structures, on the other hand, consist of proteins for which none of the known ligands fulfils the rule of five, has a $c \log P \geq 2$ and a ligand efficiency ≥ 0.3 kcal mol^{-1} per heavy atom. This set of criteria is designed to ensure that proteins in this category are unable to bind orally bioavailable compounds with high efficiency.

Although larger datasets are always desirable, it would seem that these pioneering works have provided a sufficiently large body of data and, once unified and supplemented with conformational diversity, they will facilitate the development of subsequent druggability predictions methods.

12.6.2   Cavity Detection Algorithms

Owing to the shape complementarity requisite, the binding sites of ligands correspond to protein surfaces with inward curvature. Deep pockets are generally assumed to play a functional role and, in consequence, cavity detection algorithms have long been used to predict ligand binding sites. A large range of computer programs has been developed to identify pockets and to predict their likelihood to act as ligand binding sites (reviewed in ref. 55). The algorithms can roughly be classified into geometric or energetic approaches. In the first class, which is the most common, the protein shape is directly probed to detect void spaces surrounded by protein atoms. In the second approach the interaction energy of chemical probes (ranging from a simple sphere with van der Waals parameters to a diverse set of chemical fragments with van der Waals, hydrogen bonding and electrostatic potentials) is mapped on the three-dimensional space of the protein and ligand binding sites are identified on the basis of interaction energy profiles.

The main objective of those programs is to distinguish the true ligand binding site from the rest of cavities in a protein structure. As ligand binding sites often coincide with the largest protein pocket,[24,56] size alone is a good predictor, but most methods use a combination of parameters to rank the pockets, which in some cases also include information on residue conservation. Success rates for the most recently published methods are close to 70% for the highest ranked pocket and 90% when the top three pockets are considered.[57–60]

Achieving a representation of pockets that matches the space occupied by the ligands in an automated manner is far from trivial, because the ligand binding site is usually part of a larger network of pockets on the protein surface. However, cavity detection algorithms have a long history and have reached a fair level of maturity. At the same time, these programs are evolving to incorporate new functionalities that can be extremely useful in drug design.[55] These include consideration of pocket flexibility, pocket comparison algorithms and pocket druggability, which is discussed in the next section.

12.6.3 Methods Based on Cavity Descriptors

Most druggability prediction methods build on cavity detection algorithms to extract pocket surface descriptors and make their predictions, but they present two main differences with their parent methods: (1) pockets are compared not only within, but also across, protein structures; and (2) instead of distinguishing binding sites from non-binding sites, their goal is to predict the likelihood that the pocket displays high affinity for drug-like ligands. Naturally, they require a completely new parameterization, and the datasets described in Section 12.6.2 were developed with the specific purpose of being used as training sets. In this regard, the work of Sheridan *et al.* deserves a special mention, because the method is not calibrated against targets of known druggability. Instead, they analyze hundreds of thousands of pockets in the PDB and relate cavity descriptors to the probability of binding a drug-like ligand.[34]

The first druggability prediction method was published in 2005 by Hajduk *et al.*, from Abbott Laboratories,[25] the same group that pioneered fragment screening.[37] It is not surprising then that their definition of druggability was based on fragment screening hit rates. Different cavity parameters were compared between the positive and negative sets, but only the apolar surface area was different at a statistically significant level ($\sim 35\%$ lower for non-druggable pockets). Trends in other parameters were also consistent with the view of druggable pockets as larger, more hydrophobic, more complex and more compact than the non-druggable ones. Regression analysis allowed them to obtain a quantitative correlation that could be used to discriminate between druggable and non-druggable binding sites of proteins not used in the training set.[25]

The second druggability method, and possibly the most cited one, was published in 2007 by Cheng *et al.*, at Pfizer.[27] In this case, a model was developed to estimate the maximum affinity that an ideal drug-like molecule could have for a given binding site. Remarkably, correct predictions were obtained on the assumption that the binding affinity of drug-like molecules derives exclusively from the hydrophobic effect. The only pocket descriptors necessary are the radius of curvature and the hydrophobic surface area. A lower radius means that the pocket is more closed and, as expected, correlates with better druggability. The method assumes a linear relationship between hydrophobic surface area and maximal binding affinity, consistent with the view that druggable binding sites are more hydrophobic.

Unfortunately, both Hajduk's and Cheng's methods used a combination of algorithms that included commercial and proprietary software that is not publicly available, neither obvious to reimplement.[25,27] Additionally, Cheng's method relied on the presence of a ligand to define the binding site, precluding its applicability in an unsupervised manner to other cavities. Therefore, the introduction of a druggability score (Dscore) in the SiteMap cavity detection program from Schrödinger was a welcome addition.[61] The scoring function was trained on Cheng's data, but resulted in a different model using three parameters: number of site points (related to cavity size) and the degree of enclosure make a positive contribution towards the score, while pocket hydrophilicity makes a negative contribution.

In 2010, the open source program fpocket[57] was also endowed with a druggability score, in this case trained on an extended dataset.[26] Given that the main goal was to correctly classify pockets into druggable or undruggable categories, the score is based on a two-step logistic function (sigmoid). As the score is based on descriptors extracted from automatically detected cavities and each cavity was represented by multiple crystal structures, the formula also reflects the need to provide robust predictions in spite of the variability introduced by the pocket detection algorithm. The same applies to the specific choice of descriptors, which include the mean local hydrophobic density (a measure of size and spatial distribution of hydrophobic subpockets) and information about the amino acid composition in terms of hydrophobicity and polarity. Although not selected by the learning and validation procedure, this work also showed a different behaviour of polar atoms: in druggable cavities they protrude more from the surface than apolar atoms.[26] This work also was the first to investigate the effect of protein flexibility (through the use of multiple crystallographic structures) on the druggability score. Both the mean values and the values of the top scoring cavities were clearly different between druggable and non-druggable cavities, suggesting that confidence in druggability predictions may increase when multiple structures are considered.[26] Application of the druggability score to rescue druggable cavities from a pool of druggable and undruggable ones demonstrated a similar performance of SiteMap and fpocket. Both programs can be used in an unsupervised manner, but, if applied to a large collection of structures, it should be noted that the Voronoi tessellation shape-based algorithm in fpocket is computationally much more efficient than the grid-based interaction energy algorithm in SiteMap (2–4 seconds compared to several minutes).

Also in 2010, Sheridan *et al.*, at Merck, developed the drug-like density (DLID) metric.[34] As noted above, the training method was substantially different from previous contributions. However, they found a very effective linear equation with three variables (volume, buriedness and hydrophobicity), therefore conceptually resembling the one implemented in SiteMap.[61]

The last publicly available method that we are aware of has been published in 2011 by Brenk and co-workers, who also introduced a larger dataset.[53] In their approach, definition of the cavity volume is based on docking of a large set of diverse molecules. However the performance of this approach is not compared with more common and computationally less expensive pocket detection algorithms. A set of 16 descriptors capturing the polarity, size and compactness of the cavity, plus six descriptors related to the amino acid composition (all of which follow a Gaussian distribution), were subjected to principal component analysis, which could effectively separate pockets according to their degree of druggability. In agreement with previous studies, druggable sites showed a tendency to be larger and less polar. A model with five descriptors, called DrugPred, was derived. In line with previous studies, the selected descriptors relate to the size and the hydrophobicity of the pocket (positive correlation) or to the polarity of the pocket (negative correlation). Comparison with previous methods shows that DrugPred performs better in terms of accuracy, precision and recall despite using similar descriptors. This seems to suggest that the larger training provides a significant advantage.[53]

12.6.4 Methods Based on Interaction Potentials

Computational methods based on the principles of physical chemistry can be used to predict the interaction free energy between a ligand and a protein binding site. As this property is intimately linked to the druggability concept, molecular simulations offer an alternative to empirical approaches. The main difficulty in predicting binding free energies is that they are the end result of multiple terms of large and opposing magnitude. Consequently, accurate predictions are computationally very demanding and extremely hard to achieve.[62] The concept has nevertheless been used successfully in energy-based binding site detection methods, which rely on extremely crude but very fast approximations.[55] With increasingly rigorous approaches, it is theoretically possible to carry out the *in silico* equivalent of experimental druggability prediction methods. For instance, Huang and Jacobson have demonstrated that hit rates in docking-based virtual screening experiments correlate with the experimental hit rates obtained by NMR.[63] Other methods that also combine exhaustive sampling of the ligand–receptor configurational space with severe approximations on the interaction energy predictions have proven useful to identify and characterize the most druggable binding sites of a target protein with a reasonable computational cost.[64,65] Obtaining quantitative predictions, however, requires more rigorous approaches that take into account often neglected terms such as solvation or entropy. This is achieved by the druggability index developed by Seco *et al.*, which predicts the maximal binding affinity that a drug-like compound could achieve for a binding site from molecular simulations based on first principles.[66] Initially, the method reproduces a solvent-mapping experiment, in which the protein is exposed to a certain concentration of an organic solvent. Both NMR and crystallographic experiments have demonstrated that organic solvents tend to localize on binding sites,[46,48,49] which is a natural consequence of the tendency of binding hot-spots to become desolvated (see above). Molecular dynamics simulations using 20% isopropyl alcohol (IPA) as solvent reproduce this behaviour, correctly identifying the experimentally determined IPA binding sites. Knowing that the method provides a correct sampling of the protein–ligand space, the collection of configurations generated by molecular dynamics can be subjected to a statistical treatment, leading to binding site identification and druggability predictions. The process involves the following steps:

(i) A grid encompassing the whole of the simulation box is generated and the number of times that a solvent atom type (IPA-OH, IPA-CH$_3$, Water-O) falls within each grid element is counted. Then, comparing the observed population (N_i) with the expected value (N_o), the associated free energy can be obtained using eqn (12.1):

$$G_i = -k_B T \ln\left(\frac{N_i}{N_o}\right) \qquad (12.1)$$

where k_B is the Boltzmann constant and T the temperature at which the simulation was run.

(ii) The points with the best interaction free energies are identified, taking care that all points are separated by, at least, the distance of a covalent bond.

(iii) Finally, points corresponding to IPA atom types (OH and CH$_3$) are considered transferable to aliphatic and polar neutral features of drug-like compounds, respectively. They are clustered together to form binding sites of maximal binding efficiencies.

Given that state-of-the-art molecular dynamics simulations can be used to correctly determine both the affinity and binding kinetics of organic solvents,[67] small ligands[68] and even drug-like ligands,[69,70] it is reasonable to expect that in the coming years it will be practical to assess druggability of a target based on computational experiments analogous to multiple solvent crystallographic screening or even fragment screening.

12.7 Summary and Outlook

Formal investigation of the causes of druggability has only started in recent years. Sitting at the interface of pharmacokinetics, molecular recognition and biomolecular structure, this incipient knowledge area builds on previous methods and understanding about drug-likeness, binding site identification and structure-based drug design, amongst others. Driven by a real necessity from the pharmaceutical industry, significant progress has been achieved. Of particular note is the existence of a small but diverse set of druggability prediction methods and the creation of a catalogue of systems with various degrees of druggability against which new methods can be trained and tested. Future challenges include explicit consideration of protein flexibility and achieving more quantitative predictions.

Druggability prediction methods are expected to have two seemingly opposed consequences: on the one hand, they will help concentrate on those targets offering better prospects; on the other hand, they will raise awareness about less obvious binding sites that may be used to exert a biological effect through non-standard mechanisms such as protein–protein inhibition,[71] protein–protein stabilization,[72] target chaperoning,[73] conformational trapping[74] and allosteric sites in general.[75] The exploitation of more diverse molecular mechanisms of action is one of the unresolved limitations of target-based drug discovery,[6] and druggability prediction methods have the potential to give access to new resources that will alleviate the productivity crisis of pharmaceutical research.

Acknowledgements

This work was financed by the Ministerio de Educación y Ciencia (Grant SAF2009-08811).

References

1. C. Smith, *Nature*, 2003, **422**, 341, 343, 345, passim.
2. A. L. Hopkins and C. R. Groom, *Nat. Rev. Drug Discovery*, 2002, **1**, 727.
3. R. Macarron, *Drug Discovery Today*, 2006, **11**, 277.
4. M. Rask-Andersen, M. S. Almen and H. B. Schioth, *Nat. Rev. Drug Discovery*, 2011, **10**, 579.
5. F. Sams-Dodd, *Drug Discovery Today*, 2005, **10**, 139.
6. D. C. Swinney and J. Anthony, *Nat. Rev. Drug Discovery*, 2011, **10**, 507.
7. H. Kubinyi, *Nat. Rev. Drug Discovery*, 2003, **2**, 665.
8. I. Kola and J. Landis, *Nat. Rev. Drug Discovery*, 2004, **3**, 711.
9. C. P. Hart, *Drug Discovery Today*, 2005, **10**, 513.
10. P. Imming, C. Sinning and A. Meyer, *Nat. Rev. Drug Discovery*, 2006, **5**, 821.

11. H. David-Eden, A. S. Mankin and Y. Mandel-Gutfreund, *Nucleic Acids Res.*, 2010, **38**, 5982.
12. I. Gashaw, P. Ellinghaus, A. Sommer and K. Asadullah, *Drug Discovery Today*, 2011, in press.
13. A. P. Russ and S. Lampel, *Drug Discovery Today*, 2005, **10**, 1607.
14. R. D. Finn, J. Mistry, J. Tate, P. Coggill, A. Heger, J. E. Pollington, O. L. Gavin, P. Gunasekaran, G. Ceric, K. Forslund, L. Holm, E. L. Sonnhammer, S. R. Eddy and A. Bateman, *Nucleic Acids Res.*, 2010, **38**, D211.
15. L. Y. Han, C. J. Zheng, B. Xie, J. Jia, X. H. Ma, F. Zhu, H. H. Lin, X. Chen and Y. Z. Chen, *Drug Discovery Today*, 2007, **12**, 304.
16. C. J. Harris, R. D. Hill, D. W. Sheppard, M. J. Slater and P. F. Stouten, *Comb. Chem. High Throughput Screening*, 2011, **14**, 521.
17. R. Eglen and T. Reisine, *Pharmacol. Ther.*, 2011, **130**, 144.
18. K. Venkatesan, J. F. Rual, A. Vazquez, U. Stelzl, I. Lemmens, T. Hirozane-Kishikawa, T. Hao, M. Zenkner, X. Xin, K. I. Goh, M. A. Yildirim, N. Simonis, K. Heinzmann, F. Gebreab, J. M. Sahalie, S. Cevik, C. Simon, A. S. de Smet, E. Dann, A. Smolyar, A. Vinayagam, H. Yu, D. Szeto, H. Borick, A. Dricot, N. Klitgord, R. R. Murray, C. Lin, M. Lalowski, J. Timm, K. Rau, C. Boone, P. Braun, M. E. Cusick, F. P. Roth, D. E. Hill, J. Tavernier, E. E. Wanker, A. L. Barabasi and M. Vidal, *Nat. Methods*, 2009, **6**, 83.
19. T. Clackson and J. A. Wells, *Science*, 1995, **267**, 383.
20. W. L. DeLano, *Curr. Opin. Struct. Biol.*, 2002, **12**, 14.
21. A. A. Bogan and K. S. Thorn, *J. Mol. Biol.*, 1998, **280**, 1.
22. F. Rodier, R. P. Bahadur, P. Chakrabarti and J. Janin, *Proteins*, 2005, **60**, 36.
23. I. Halperin, H. Wolfson and R. Nussinov, *Structure*, 2004, **12**, 1027.
24. C. Sotriffer and G. Klebe, *Farmaco*, 2002, **57**, 243.
25. P. J. Hajduk, J. R. Huth and S. W. Fesik, *J. Med. Chem.*, 2005, **48**, 2518.
26. P. Schmidtke and X. Barril, *J. Med. Chem.*, 2010, **53**, 5858.
27. A. C. Cheng, R. G. Coleman, K. T. Smyth, Q. Cao, P. Soulard, D. R. Caffrey, A. C. Salzberg and E. S. Huang, *Nat. Biotechnol.*, 2007, **25**, 71.
28. A. R. Fersht, *Trends Biochem. Sci.*, 1987, **12**, 301.
29. P. A. Karplus, *Protein Sci.*, 1997, **6**, 1302.
30. J. Gao, D. A. Bosco, E. T. Powers and J. W. Kelly, *Nat. Struct. Mol. Biol.*, 2009, **16**, 684.
31. P. Schmidtke, F. J. Luque, J. B. Murray and X. Barril, *J. Am. Chem. Soc.*, 2011, **133**, 18903.
32. C. Tang, J. Iwahara and G. M. Clore, *Nature*, 2006, **444**, 383.
33. H. Labischinski and L. Johannsen, *Drug Resist. Updates*, 1999, **2**, 319.
34. R. P. Sheridan, V. N. Maiorov, M. K. Holloway, W. D. Cornell and Y. D. Gao, *J. Chem. Inf. Model.*, 2010, **50**, 2029.
35. A. Gupta, A. K. Gupta and K. Seshadri, *J. Comput. Aided Mol. Des.*, 2009, **23**, 583.

36. P. Ertl, *J. Chem. Inf. Comput. Sci.*, 2003, **43**, 374.
37. S. B. Shuker, P. J. Hajduk, R. P. Meadows and S. W. Fesik, *Science*, 1996, **274**, 1531.
38. M. L. Verdonk, V. Berdini, M. J. Hartshorn, W. T. Mooij, C. W. Murray, R. D. Taylor and P. Watson, *J. Chem. Inf. Comput. Sci.*, 2004, **44**, 793.
39. D. A. Erlanson, J. A. Wells and A. C. Braisted, *Annu. Rev. Biophys. Biomol. Struct.*, 2004, **33**, 199.
40. C. Dalvit, *Drug Discovery Today*, 2009, **14**, 1051.
41. S. Perspicace, D. Banner, J. Benz, F. Muller, D. Schlatter and W. Huber, *J. Biomol. Screening*, 2009, **14**, 337.
42. M. M. Hann, A. R. Leach and G. Harper, *J. Chem. Inf. Comput. Sci.*, 2001, **41**, 856.
43. T. Fink, H. Bruggesser and J. L. Reymond, *Angew. Chem. Int. Ed.*, 2005, **44**, 1504.
44. I. J. Chen and R. E. Hubbard, *J. Comput. Aided Mol. Des.*, 2009, **23**, 603.
45. F. N. Edfeldt, R. H. Folmer and A. L. Breeze, *Drug Discovery Today*, 2011, **16**, 284.
46. E. Liepinsh and G. Otting, *Nat. Biotechnol.*, 1997, **15**, 264.
47. C. Mattos and D. Ringe, *Nat. Biotechnol.*, 1996, **14**, 595.
48. A. C. English, C. R. Groom and R. E. Hubbard, *Protein Eng.*, 2001, **14**, 47.
49. C. Mattos, C. R. Bellamacina, E. Peisach, A. Pereira, D. Vitkup, G. A. Petsko and D. Ringe, *J. Mol. Biol.*, 2006, **357**, 1471.
50. M. Congreve, R. Carr, C. Murray and H. Jhoti, *Drug Discovery Today*, 2003, **8**, 876.
51. H. M. Berman, J. Westbrook, Z. Feng, G. Gilliland, T. N. Bhat, H. Weissig, I. N. Shindyalov and P. E. Bourne, *Nucleic Acids Res.*, 2000, **28**, 235.
52. D. S. Wishart, C. Knox, A. C. Guo, D. Cheng, S. Shrivastava, D. Tzur, B. Gautam and M. Hassanali, *Nucleic Acids Res.*, 2008, **36**, D901.
53. A. Krasowski, D. Muthas, A. Sarkar, S. Schmitt and R. Brenk, *J. Chem. Inf. Model.*, 2011, **51**, 2829.
54. M. J. Hartshorn, M. L. Verdonk, G. Chessari, S. C. Brewerton, W. T. Mooij, P. N. Mortenson and C. W. Murray, *J. Med. Chem.*, 2007, **50**, 726.
55. S. Henrich, O. M. Salo-Ahen, B. Huang, F. F. Rippmann, G. Cruciani and R. C. Wade, *J. Mol. Recognit.*, 2010, **23**, 209.
56. S. J. Campbell, N. D. Gold, R. M. Jackson and D. R. Westhead, *Curr. Opin. Struct. Biol.*, 2003, **13**, 389.
57. V. Le Guilloux, P. Schmidtke and P. Tuffery, *BMC Bioinf.*, 2009, **10**, 168.
58. A. Tripathi and G. E. Kellogg, *Proteins*, 2010, **78**, 825.
59. A. Volkamer, A. Griewel, T. Grombacher and M. Rarey, *J. Chem. Inf. Model.*, 2010, **50**, 2041.
60. P. Schmidtke, C. Souaille, F. Estienne, N. Baurin and R. T. Kroemer, *J. Chem. Inf. Model.*, 2010, **50**, 2191.
61. T. A. Halgren, *J. Chem. Inf. Model.*, 2009, **49**, 377.

62. C. Chipot and A. Pohorille, *Free Energy Calculations: Theory and Applications in Chemistry and Biology*, Springer, Berlin, 2007.
63. N. Huang and M. P. Jacobson, *PLoS One*, 2010, **5**, e10109.
64. R. Brenke, D. Kozakov, G. Y. Chuang, D. Beglov, D. Hall, M. R. Landon, C. Mattos and S. Vajda, *Bioinformatics*, 2009, **25**, 621.
65. M. Clark, F. Guarnieri, I. Shkurko and J. Wiseman, *J. Chem. Inf. Model.*, 2006, **46**, 231.
66. J. Seco, F. J. Luque and X. Barril, *J. Med. Chem.*, 2009, **52**, 2363.
67. D. Huang and A. Caflisch, *ChemMedChem*, 2011, **6**, 1578.
68. I. Buch, T. Giorgino and G. De Fabritiis, *Proc. Natl. Acad. Sci. U. S. A.*, 2011, **108**, 10184.
69. Y. Shan, E. T. Kim, M. P. Eastwood, R. O. Dror, M. A. Seeliger and D. E. Shaw, *J. Am. Chem. Soc.*, 2011, **133**, 9181.
70. R. O. Dror, A. C. Pan, D. H. Arlow, D. W. Borhani, P. Maragakis, Y. Shan, H. Xu and D. E. Shaw, *Proc. Natl. Acad. Sci. U. S. A.*, 2011, **108**, 13118.
71. M. R. Arkin and J. A. Wells, *Nat. Rev. Drug Discovery*, 2004, **3**, 301.
72. Y. Pommier and J. Cherfils, *Trends Pharmacol. Sci.*, 2005, **26**, 138.
73. P. Leandro and C. M. Gomes, *Mini Rev. Med. Chem.*, 2008, **8**, 901.
74. G. M. Lee and C. S. Craik, *Science*, 2009, **324**, 213.
75. J. E. Lindsley and J. Rutter, *Proc. Natl. Acad. Sci. U. S. A.*, 2006, **103**, 10533.

CHAPTER 13

Computational Strategies and Challenges for Targeting Protein–Protein Interactions with Small Molecules

DANIELA GRIMME, DOMINGO GONZÁLEZ-RUIZ AND
HOLGER GOHLKE*

Department of Mathematics and Natural Sciences, Heinrich-Heine-University,
Universitätsstr. 1, D-40225 Düsseldorf, Germany
*E-mail: gohlke@uni-duesseldorf.de

13.1 Introduction

In biological systems, function and malfunction ultimately originates from
interacting rather than isolated molecules. As such, most cellular processes are
carried out by multiprotein complexes,[1] and the organization of the ensemble
of expressed proteins into functional units results in complex protein
interaction networks.[2] Thus, protein–protein interactions (PPI) are a large
and important class of potential therapeutic targets.[3,4]

Currently, three classes of protein–protein interaction modulators (PPIM)
are known: (i) therapeutic antibodies, which are highly target-specific and
stable in human serum.[5] As a downside, antibodies are not cell-permeable and
show a lack of oral bioavailability; (ii) peptides derived from protein–protein
interfaces (PPIF), which are applied as dimerization and interaction
inhibitors.[6–11] However, poor metabolic stability and low bioavailability[9]

RSC Drug Discovery Series No. 23
Physico-Chemical and Computational Approaches to Drug Discovery
Edited by F. Javier Luque and Xavier Barril

limit their perspective for drug development; (iii) small-molecule PPIM,[12–15] which are less likely to suffer from the above limitations. Although considered almost impossible only a few years ago,[16] a number of recent studies have successfully demonstrated the application of (drug-like) PPI antagonists that bind *directly* to a PPIF.[3,12,17–19] In addition, inhibitors that influence PPI by binding to *allosteric* sites have emerged as promising alternatives.[20,21]

Approaches towards the identification of small-molecule PPIM can be classified into three general categories.[17] In principle, these approaches resemble those applied to find small-molecule inhibitors of "classical" (*e.g.*, enzyme) targets. First, interface-derived peptides may aid as lead structures in guiding the development of peptidomimetics.[8,9] Second, screening of (combinatorial) chemical libraries has proven successful in identifying small-molecule PPIM in a number of cases,[22–29] taking advantage of novel techniques such as cell-based translocation assays,[30] tethering,[31,32] or fragment-discovery approaches.[33–35] Notably, although structural information about the PPI to target was available in some of the cases, most of these screens were initially performed on non-targeted libraries.[17] Third, virtual screening of databases is a viable approach for finding small-molecule PPIM. However, this has been applied successfully only in a few cases to date (Table 13.1).

Recent advances in the understanding of the energetics and dynamics of PPI and methodological developments in the field of structure-based drug design methods may open new avenues to apply virtual screening and rational design approaches for finding small-molecule PPIM. These developments include (i) computational approaches to dissect binding interfaces in terms of energetic contributions of single residues (to identify "hot spot" residues), (ii) prediction of potential binding sites from unbound protein structures, (iii) recognition of allosteric binding sites as alternatives to directly targeting interfaces, (iv) docking approaches that consider protein flexibility and improved descriptions of the solvent influence on electrostatic interactions, and (v) data-driven docking approaches. In this updated and extended review,[36] we will describe and summarize these recent developments with a particular emphasis on their applicability to screen for or design small-molecule PPIM.

13.2   What Makes Protein–Protein Interfaces Difficult to Target?

The challenge of identifying small-molecule PPIM in general and through structure-based drug design methods owes much to the overall characteristics of PPIF. Compared to active sites of enzymes, PPIF are typically flat and devoid of deep binding sites for small molecules.[37,38] In addition, the majority of protein–protein complex interfaces are approximately 1600 ± 400 Å^2 in size,[39–42] with an average of 22 buried amino acid residues per binding partner.[39] This amount of buried surface area upon protein–protein complex formation greatly exceeds the potential binding area of small molecules.[17]

Table 13.1 Small-molecule PPIM identified in prospective studies by molecular modeling and virtual screening.

Biological system	Methods / results	References
Bcl-x$_L$/2 and Bak-BH3 domains	1.- Target modeled by homology; Virtual Screeing (VS) of MDL/ACD 3D database (193.833 compounds (cmpds.)); DOCK 3.5; 1000 visually examined; 28 selected; found 1 with activity.	288
	2.- Target modeled by homology; refinement by MD; VS of NCI 3D database (206.876 cmpds.); DOCK 4.01; 500 analyzed; 80 selected; 35 tested; 7 inhibit the interaction.	289
	3.- 93 compounds selected from commercial databases on the basis of their similarity to BH3Is ($> 80\%$) and solubility (logP > 6.0). Ligand placed according to CSP. Refinement with TreeDock.	290
	4.- innovative VS strategy with various filters, 45 compounds for biological evaluation, one compound showed inhibitory activity of IC$_{50}$ 2.48 µM (K$_{i\ (calc)}$ 0.38 µM).	291
	5.- in silico screening of compounds from the free database ZINC^{286}, identified 17 sulfonamide derivatives as potential new inhibitors for biological evaluation.	292
CD4/MHC class II	DOCK3.5; ACD screened (~ 150.000 cmpds). 1000 best shape complementarity + 1000 best scored. 3 visual inspections; 41 compounds selected for cell-adhesion experiment; 8 significantly inhibit activity.	293
T cell response – Myelin basic protein	1.- NCI-3D DB screened (~ 150.000 cmpds.); DOCK 4.0.1. 1800 selected. Screening of these with MCDOCK. 150 selected; 106 tested for toxicity; 39 screened in IL-2 functional secretion assay; 7 "leads" identified.	294
XIAP – caspase-9info	Traditional chinese medicinal herbs-3D DB (8221 cmpds). DOCK. Top 1000 re-scored with X-score (consensus scoring). 200 selected; 36 tested; 5 inhibit the interaction. Embelin is the most potent one.	295
Rac GTPase – Guanine nucleotide exchange factors	UNITY search for compounds from NCI DB that fit in the binding site. FlexX to dock and rank the selected ones. 100 inspected; 15 further investigated; 1 inhibits the interaction.	296
PDZ protein interaction domain (PTEN and MAGI3)	From a known protein-peptide complex, scaffolds are derived and a (targeted) virtual library generated. DOCK used to guide the synthesis.	297
PDZ domain of Dishevelled (Dvl)	Structure-based ligand screening (from the NCI) and NMR spectroscopy for final validation of potential inhibitors; found several inhibitors, the best has similar binding affinity like native binding partner, Fz, in vivo validation for blocking Wnt signaling.	298
Gp41 of HIV	ComGenex database (20 000 cmpds.) VS with DOCK 3.5. 200 selected for visual inspection; 20 selected for biological assays.	299
MDM2 - p53	Pharmacophore model was developed on the crystal structure and MD simulations; 50.000 synthetic compounds; evaluation in a flourecence-polarization MDM2 binding assay; 5 small-molecule MDM2 inhibitors; most potent inhibitor K$_i$ of 110 nM.	300

Many PPIF also consist of non-contiguous regions in the primary protein sequence, which makes it difficult to rationally design binding site mimetics.[17] Furthermore, although smaller interfaces (< 2000 Å^2) usually form a single epitope on the surface of each component protein, larger interfaces are generally composed of multiple epitopes.[43] In the latter case, many contacts distributed over a large surface may be required to yield a potent PPI inhibitor, as indicated by a synthetic mimic of a shallow, bowl-shaped protein surface designed by Hamilton and co-workers.[44]

With respect to the amino acid *composition* of PPIF of proteinase–inhibitor or antibody–antigen complexes, no difference was found in terms of polar, nonpolar, and charged proportions compared to the composition of solvent-accessible surface regions in general.[39,40,45] In the case of large interfaces (> 5000 Å^2), hydrophobic residues were more abundant, while polar residues were more abundant in small interfaces (< 1000 Å^2).[46] More specifically, aromatic amino acids (in particular, tyrosine) were found most frequently among the nonpolar residues,[47] whereas arginine was more abundant than aspartate, glutamate, and lysine in the case of charged amino acids.[39] Thus, given the general lack of differentiation in the amino acid *composition* of PPIF compared to other protein surface regions, finding small-molecule PPIM that preferentially bind to the former does not seem feasible at first glance.

The fact that PPIF were found to be mutually complementary with respect to their electrostatic potentials led to the proposal that long-range electrostatic interactions considerably influence binding to these epitopes.[48] This has been confirmed recently by identifying mutants of the β-lactamase inhibitor protein that lie > 7 Å apart from the binding partner, yet show a contribution to binding by improving the overall electrostatic complementarity of the binding partners.[49,50] Likewise, libraries of closely related catalytic antibodies generated by phage display revealed that "second sphere" residues (*i.e.*, outside the active site) contributed favorably to both the binding of a transition state analog and catalytic efficiency.[51] Taking into account these long-range interactions may therefore provide a general means to aid in the design and tuning of binding interactions.

Several recent studies also point to the important role of water-mediated interactions in PPIF.[39,52,53] As such, it was shown that the close-packing found for atoms buried in PPIF can be extended to the majority of all interface atoms if solvent positions are taken into account.[39] Water molecules thus contribute to the close packing of atoms, which insures complementarity between the binding partners. Similarly, by comparison of knowledge-based direct and water-mediated contact potentials, partial solvation was found to be important in stabilizing charged groups in PPIF.[52] In fact, water-mediated polar interactions are as abundant at interfaces as are direct protein–protein hydrogen bonds, although the pattern of hydration varies between interfaces.[53] On the one hand, water molecules that form a ring around "dry" interfaces have been observed; on the other hand, "wet" interfaces may be permeated by water molecules. Overall, these findings indicate that proper accounting for effects of (de-)solvation and specific water-mediated interactions is necessary for computational approaches towards the design and screening of small-molecule PPIM to be successful.

Another important contribution to the observed complementarity in PPIF arises from the inherent flexibility and plasticity of those regions.[54–58] Conformational variability thereby leads to an ensemble of substates around the native protein structure. The *distribution* of populations of these conformational substates depends on binding or other physical influences.[59,60] In view of small molecule-binding to PPIF, this leads to two implications. First, small-molecule PPIM can "induce" striking conformational changes, leading to grooves or pockets in PPIF that were not apparent in the unbound structure. Structural evidence for such a binding site plasticity has been found for Ro264550 binding to interleukin 2 (IL-2)[3] (Figure 13.1). Given that structural and thermodynamic studies further demonstrated that this portion of IL-2 is inherently flexible, one can expect that ligand binding *captured* a low-energy protein conformation instead of *inducing* a high-energy one.[14,61] In principle, it should be possible to detect these potential binding sites in the unbound state by computational means, which may give a hint as to the druggability of these interfaces. Above all, appropriately taking into account

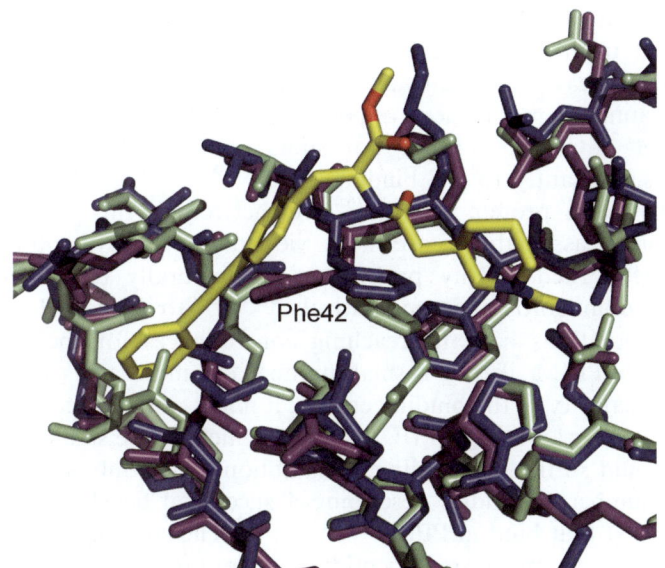

Figure 13.1 Pronounced plasticity observed in the PPI of interleukin 2 (IL2) upon binding of the small-molecule PPIM Ro264550 (*yellow*) (*magenta*: unbound IL2, PDB code 1m47; *gray*: bound IL2, PDB code 1m48). In addition, a conformation extracted from an ensemble of *unbound* IL2 structures is shown (*blue*) that most closely resembles the bound protein structure. The ensemble of unbound IL2 structures was generated by constrained geometric simulation using an enhanced version of FRODA.[161,301]

protein flexibility and plasticity in the process of small-molecule PPIM discovery is mandatory. Second, a prevailing distribution of different receptor conformations is also related to the functional adaptability of PPI partners. Studies have shown that residues participating in interactions with multiple binding partners often show a higher degree of flexibility and plasticity.[54,62,63] This implies that also different PPIM should be able to bind to such regions, leading to the notion that diverse molecule libraries need to be screened for potential binders.[64]

Although some of the challenges imposed by the structural and dynamical properties of PPIF also exist in the case of "classical" targets, the above described characteristics led to the view that the development of small molecules PPIM is difficult.[12] However, a more optimistic viewpoint is provided by energetic data on the stability of macromolecular complexes. Analyses of protein–protein complexes *versus* protein–peptide complexes show very similar thermodynamic behavior, despite considerable differences in the interface sizes.[65] Similarly, a "non-linear free energy" relationship with respect to atomic properties has been found recently in an investigation of the stability of macromolecular complexes, resulting in roughly constant binding free energies of the tightest complexes, independent of the interface size.[66] Both observations support an early view[67] that the actual interfaces that contribute to binding ("functional epitopes") have a similar size; these epitopes need to be distinguished from the "structural epitopes" that are given by the overall buried surface areas, which can vary considerably. Along these lines, extensive "alanine scanning" mutagenesis experiments in PPIF revealed that only a small subset of amino acids ("hot spots") within the overall interface contribute significantly to the binding affinity.[38,50,65,68–75] Recent structural analyses of protein–protein surfaces[76,77] further refined this picture such that protein–protein associations are now viewed as locally optimized, with clustered, networked, highly packed, and structurally conserved residues contributing dominantly and cooperatively[50,70,78,79] to the complex stability.[80]

These observations have far-reaching consequences for the discovery of small-molecule PPIM: if only a small number of amino acids within a PPIF provide the majority of the binding energy, it might not be necessary for small molecules to cover an entire PPIF. Instead, mimicking the smaller "functional epitope" should suffice for binding. This notion is not only supported by an increasing number of studies describing a successful development of small-molecule PPIM that bind in PPIF.[13,14,17,18,29,81] The striking observation that phage-display selections of small peptides[82–85] not targeted for protein–protein inhibition nevertheless bind at the protein hot spot indicates that these interface regions are particularly prone to binding, even for small molecules.

13.3 Computational Hot Spot Detection

Given that convergent binding apparently correlates with hot spots, these regions of PPIF should be considered primary targets for virtual screening or target-oriented combinatorial chemistry.[86] Indeed, small-molecule mimics of hot spots have been found to inhibit PPI, albeit with generally weak affinities.[44,87–89] Although a plethora of experimental alanine scanning data exists for a wide range of PPI,[90,91] it is advisable not to overinterpret the data in terms of specific interactions between residues.[86] This is based on the fact that alanine mutations are perturbations to the free energy surfaces of the unbound state of the protein, the bound state, or both. If it can be assured that the mutation only influences the conformational ensemble of the complex (*i.e.*, it has no effect on the unbound state), measured binding free energy differences between mutant and wild-type protein can be related to specific contact differences. It is for this reason that computational approaches for detecting hot spots become increasingly important.[92] Furthermore, these approaches can be applied to predict hot spots also for (modeled or structurally known) protein–protein complexes for which no experimental mutagenesis data is available. A database, HotSprint, that collects computationally predicted hot spots has been introduced recently.[93]

As for techniques based on first principles, computational alanine scanning has been performed using the MM-PBSA approach[94] to estimate the individual contribution of each residue to the binding.[95] Here, explicit molecular mechanical energies are combined with continuum model-based solvation free energies (see below) and estimates of vibrational entropy changes to probe PPI. These terms are averaged over configurations of the molecular system obtained from high-quality molecular dynamics (MD) simulations. Applied to the "classical" example of experimental alanine scanning, the human growth hormone–receptor complex, the average unsigned error of calculated binding free energy differences obtained by this approach was ~ 1 kcal mol^{-1}.[96] If a proper thermodynamic cycle is employed, a full description of the structural and energetic consequences of a mutation upon the unbound and bound state is obtained.[96] However, this requires repeating the MD sampling of the unbound mutant protein and the mutated complex for each amino acid of interest, which is computationally expensive. Hence, approximations to generate the mutant ensembles have been introduced that only require the simulation of the wild-type proteins, which are then post-processed to introduce either mutations to alanine[95] or larger amino acids.[97] These methods provide a computationally inexpensive way of screening a large variety of possible modifications on either side of the interface. An approach along these lines is presented by Moreira *et al.*[98] To identify hotspot residues in PPIF, the molecular mechanics parmm94 force field and a continuum solvation approach with different internal dielectric constant values, depending on the type of amino acid, is used. After a MD simulation using a modified Generalized Born (GB) solvation model, the post-processing of the complexes, which follows a single-trajectory protocol, permits us to calculate effective energies of the complex and the interacting monomers. Overall, a success rate of 80% in predicting hot spots (binding free energy difference > 4.0 kcal mol^{-1}), warm spots (binding free energy difference between 4.0 and 2.0 kcal mol^{-1}), and null spots (binding free energy difference < 2.0 kcal mol^{-1}) was reported for three protein–protein complexes with 46 alanine mutations.[98,99] Recently, the CONCOORD/Poisson–Boltzmann surface area (CC/PBSA) approach[100] has been introduced, which mimics the MM-PBSA approach in general. However, in contrast to MM-PBSA, a conformational ensemble of protein–protein complex structures is generated using the CONCOORD program[101] rather than MD simulations. As a main advantage, protein flexibility is considered at a much lower computational cost. When tested on alanine mutants for the interface residues of the protein–protein complex Ras–RalGDS, however, a lower predictive power of CC/PBSA compared to MM/PBSA was found (see also below).[102]

As with experimental alanine scanning, the computational mimic inevitably leads to perturbations of the system under consideration.[86,103] In contrast, non-perturbing alternatives to determine the contribution of each residue to the binding free energy are provided by means of component analysis.[104–107] Here, contributions of molecular mechanical energies and solvation free energies are assigned to those atoms that participate in the respective interaction. Summing over atoms of a residue then yields the contribution to the binding free energy. Most importantly, these values are obtained without the need to make structural modifications in the binding partners. It is noted, however, that while the total binding free energy is a state function, free energy components, in general, are not, and are sensitive to the decomposition scheme chosen.[108–110] We pursued a free energy decomposition for the Ras–Raf and Ras–RalGDS protein–protein complexes recently.[104,111] For the first time, decomposition of the solvation free energy contribution was obtained by applying a GB model (see below). Compared to an analogous decomposition based on Poisson continuum electrostatics,[105,107] the GB approach allowed us to "screen" all residues of the binding partners at once, drastically lowering computational demand. Convincingly, squared correlation coefficients of 0.55 and 0.46 are found for both systems when comparing the calculated contributions to the binding free energy to experimentally determined binding free energy differences for alanine mutants in the PPIF. Thus, the applied decomposition scheme provides a means by which hot spots in PPIF can be determined rapidly and reliably. In addition, by extending the analysis to all residues of the binding partners, significant contributions to the binding free energy can be identified for single residues as far apart as 25 Å from the interface. This clearly indicates the presence of "actions-at-a-distance" in these systems.

Computationally cheaper alternatives to the first principle-based methods have been reported in terms of regression-based scoring functions that allow us to predict binding energy hot spots in PPIF. Here, contributions due to van der Waals and electrostatic energies, hydrogen bonds, water bridges, solvation free energies, and variations of the protein flexibility are combined linearly. The respective weighting factors of the energy terms are parameterized using a dataset of stability changes measured for single mutations in different proteins. In that respect, these functions resemble regression-based scoring functions for protein–ligand interactions, first reported 17 years ago.[112] Widely used approaches in that respect are FoldX[113,114] and Robetta.[115,116] Encouragingly, although parameterized on alanine scanning data of monomeric proteins only, these functions also perform well if applied to predict alanine scanning results on protein interfaces, resulting in average unsigned errors of 0.9[113] and 1.1[116] kcal mol^{-1} between observed and calculated changes in binding energy. Hence, although not explicitly parameterized on protein interfaces, these functions seem to be general enough to also explain hot spot phenomena. These functions have been used to computationally redesign PPI specificity[117] and to validate homology modeled complexes of Ras and effector proteins.[118] The contributions of the single energy terms may be analyzed in more detail to better understand the thermodynamic characteristics of protein–protein recognition. Two results stand out. First, taking into account the fine details of the structure is crucial. In particular, explicitly modeling hydrogen bond strengths in an environment-dependent fashion considerably enhances the accuracy of hot spot predictions over the sole use of Coulomb electrostatics with a distance-dependent dielectric[116,119] (although it is noted that a more sophisticated treatment of electrostatics[120] may change this view). Second, accounting for (changes of) protein flexibility improved predictions for some complexes,[113,116] although the restricted accounts of flexibility clearly show limitations in more dramatic examples of interface plasticity such as the human growth hormone-receptor interface.[116]

As a knowledge-based approach, DrugScore^{PPI} is a fast and accurate method[102] for calculating relative binding free energies of Ala mutants in PPIF with respect to the wild-type complexes. For DrugScore^{PPI}, statistical pair-potentials have been derived from 851 complex structures and have been adapted against 309 experimental alanine scanning results. Available as a user-friendly webservice, DrugScore^{PPI} offers a fast and accurate prediction of hotspot residues in PPIF. When applied to an external test set of 22 alanine mutations in the interface of Ras–RalGDS, DrugScore^{PPI} significantly outperforms the CC/PBSA, FoldX, and Robetta methods with respect to predictive power and performs as good as the MM/GBSA method, which had been applied to a subset of 16 mutations.[102] Similarly, Tuncbag et al.[121] presented an empirical method that determines hot spot residues based on residue conservation, solvent accessibility, and statistical pairwise potentials for interface residues. Adjusted on 150 experimentally determined residues [58 (92) (non-)hot spot residues] and tested on an independent test set of 112 experimentally determined residues [54 (58) (non-)hot spot residues], they observed an accuracy of 70% to match with the experimental hot spot residues. The approach is available as a webserver, HotPoint.[122]

Finally, the machine-learning approach KFC for predicting hot spot residues in protein interfaces has been presented by Darnell et al.[123,124] It uses a combination of two physics-based and knowledge-based models characterizing shape specificity features (atomic density, residue size) and biochemical contacts (atomic contacts, hydrogen bonds, salt bridges). KFC already shows a better predictive power than Robetta, as demonstrated for training and test datasets of 16 protein complexes. Still, the combination of KFC and Robetta's alanine scanning, termed KFCA, results in a statistically significant improvement in the accuracy of hot spot prediction with respect to KFC.

13.4 Predicting Potential Binding Sites in Protein–Protein Interfaces from Unbound Protein States

Although the discovery of interfacial hot spots led to the expectation that small-molecule complements of these regions could attain sufficient binding affinity, in many cases only micromolar inhibitors could be developed.[89] In turn, much more effective small-molecule PPIM have been found to bind to well-defined clefts or grooves in the interface.[3,38,42,89] Methods available for cleft detection are, among others, POCKET,[125] LIGSITE,[126] LIGSITE[CSC],[127] SURFNET,[128] CAST,[129] PASS,[130] PocketPicker,[131] Fpocket,[132] and PocketAnalyzer (C. Pfleger, T. Jimenez Vaquero, H. Gohlke, unpublished results). Unfortunately, these clefts rarely occur in PPIF. However, in some cases, small molecules that bind to clefts not observed in the unbound protein could be identified experimentally.[3,61,133] This clearly demonstrates that inherent flexibility and plasticity is a hallmark of PPIF. Accordingly, detecting clefts in unbound protein interfaces by computational means will provide valuable starting points for the further rational design of small-molecule PPIM.[134]

As recent studies suggest, biomolecular recognition processes and flexibility (or changes of flexibility) of the binding partners are more fundamentally interrelated than acknowledged by the classical models[135] of "lock and key"[136] or "induced fit".[137] As such, the "conformational selection" model[60,138] proposes that proper conformations are "picked" by the binding from the ensembles of rapidly interconverting conformational species of the unbound molecules. This is supported by experimental evidence for the presence of conformational variability of binding partners prior to their association,[139] and yields an explanation as to why a single protein can bind multiple unrelated ligands at the same site.[64] This model also provides the foundation for computational investigations of conformational fluctuations of the unbound protein state, which may reveal conformational states adopted by the bound proteins (see below).

MD simulations offer the most direct computational approach to address the extent that conformational fluctuations of unbound proteins reflect the conformational changes upon association. While initially only rotamers of key side chains were investigated,[140] a recent analysis of 11 proteins by at least 4 ns long simulations has shown that a few key residues in protein interfaces frequently sample their bound state and may be critical in the early recognition of association.[141] In a more extensive study on 41 proteins for which the three-dimensional structures of bound and unbound state are known,[142] about half of the short interface segments of unbound proteins were found to sample the bound state during a 5 ns simulation. These findings are striking because even in the absence of the binding partner, certain conformations of *substructural* parts resemble already known bound states. However, in no case in the latter study[142] do the proteins *as a whole* fluctuate closer to the bound state than the unbound state. This points to a limited sampling of adequate conformations due to insufficient simulation times as a primary reason, and possibly to inaccuracies of the underlying energetic descriptions of the systems. Encouragingly, however, for the "classical" target aldose reductase, complexed conformations of the protein could be identified from MD trajectories of the unbound protein state,[143] so predicting conformations of PPIF competent for binding of small-molecule PPIM might be equally feasible.

As a viable alternative to MD simulations, normal mode analysis (NMA)[144–146] and coarse-grained alternatives[147–149] have re-emerged as powerful methods for analyzing the dynamics of biomolecules from a structural perspective.[150] Here, an analytical solution to the equations of motions yields collective variables (normal modes) that describe the dynamics of the system. It is particularly interesting in view of predicting bound protein states from unbound ones in that usually a small subset of low-frequency normal modes (in many cases, even a single mode is sufficient) reliably describes the observed conformational changes.[151] One interpretation is that protein structures have evolved such that biologically relevant motions near the folded state predominantly occur along the directions of lowest-energy modes. Phrased differently, upon going from an unbound to a bound state, proteins most readily explore directions linked to a smooth energy ascent.[149] This provides an explanation as to why bound conformations may already exist in the ensemble of unbound proteins, as proposed by the "conformational selection" model. From a practical point of view, NMA is accordingly applied to identify potential conformational changes of proteins upon binding.[148,152,153] Macromolecular conformations generated through NMA may then be used in docking algorithms that account for protein flexibility.[154,155] We note, however, that due to the harmonic approximation inherent to NMA, transitions from one local minimum to another one are neglected by the method. This becomes particularly important for more localized motions that have been observed in the case of PPIF plasticity.[3,55,156,157] For this case, hybrid MD/NMA techniques[158] may be valuable, combining low-frequency modes from NMA, which describe the collective motions of the protein, with MD, which in turn accounts for more localized motions. That way, large-scale conformational changes of the system are amplified by the modes, while the MD contribution allows to escape the local minimum near the starting structure.

Constrained geometric simulation (CGS) is a computationally very efficient approach to model large-scale conformational changes in proteins. This approach considers all atoms of the macromolecule and is not restricted to exploring only the minimum near the starting structure.[159–161] CGS is based on flexibility concepts[162,163] that allow for the efficient and accurate location of rigid and flexible regions within a macromolecule from a single, static structure. Here, 3D molecule-like bond networks (where bonds originate from covalent as well as non-covalent interactions in the protein) are analyzed with respect to the bond-rotational degrees of freedom.[164] This concept has been successfully applied to identifying collectively and independently moving regions in a series of proteins[164] or determining the change in protein flexibility upon protein–protein complex formation.[165] Coupled networks of covalent and non-covalent bonds within the protein are then used as input to two computational methods that explore the internal mobility of proteins. Thereby, the coupled bond network in the protein is preserved, and van der Waals overlaps are avoided. In the first method (ROCK),[159,160] correlated motions in flexible protein regions are explored by random-walk sampling of rotatable bonds, thereby leaving rigid regions undisturbed. In the second method (FRODA),[161] rigid protein regions are replaced by so-called ghost templates, which are then used to guide the movements of protein atoms. In both cases, generated protein conformations compare favorably with conformational ensembles determined by NMR.

As an application of CGS to identifying potential binding sites in PPIF, an enhanced version of FRODA was used to generate a conformational ensemble of the *unbound* state of IL-2 within a few hours on a single processor.[301] IL-2 shows a pronounced interface plasticity upon binding of Ro264550.[3] As depicted in Figure 13.1, a conformation very similar to the one found in the *bound* IL-2 can be identified from these simulations. Unbound IL-2 is thus able to sample bound states even in the absence of the ligand. As this example indicates, these types of simulations may provide an efficient starting point for investigating the conformational variability of PPIF. For a successful prospective prediction of druggable clefts, the simulations need to be combined with screening for energetically accessible protein conformations and a geometrical detection of indentations in the interface region.

CONCOORD[101] is another method to predict protein flexibility based on geometrical considerations. Again, covalent and non-covalent interactions within the structure are translated into a set of geometrical constraints, which provides the starting point for the generation of an ensemble of new conformations. Using principal component analysis, the "essential" degrees of freedom of the structure can then be extracted from the ensemble. tCONCOORD[166] is a reimplementation of the CONCOORD method. Recently, Eyrisch *et al.* presented a protocol for identifying transient pockets in PPIF of BCL-X_L, IL-2, and MDM2 by applying the PASS algorithm[130] to MD snapshots.[133] In a second study, Eyrisch *et al.* showed that backbone movements *and* side-chains dynamics are important for transient pocket formation in protein interfaces.[167] When comparing transient pocket detection based on CONCOORD-, tCONCOORD-, and NMA-generated ensembles, only tCONCOORD was able to generate as many and as large pockets as obtained by MD simulations.[167]

13.5 Allosteric Binding Sites as Alternative Targets for Modulating Protein–Protein Interactions

Peptidic and synthetic mimics of protein surfaces and small-molecule PPIM targeting PPIF provide the most direct approach to disrupt PPI. In a more indirect way, allosteric mechanisms have been exploited as promising targets for modulating PPI, especially for cell-surface receptors.[20,38,168] Here, allosteric drugs modulate receptor activity through conformational changes in the receptor protein that are transmitted from the allosteric site to the effector coupling site. More specifically, allosteric modulators enrich certain subsets of conformations available to the protein in the global conformational ensemble[60,138] that differ in their biological binding/signaling properties.[20,169] This reinforces the role of protein dynamics as an entropic carrier of free energy of allostery.[139,170–172]

At least three advantages of allosteric PPIM over "direct" ones have been pointed out:[20] (i) the effect of allosteric modulators is saturable and less prone to overdosing; (ii) allosteric modulators can selectively tune responses only in tissues in which the endogenous agonist exerts its physiological effects; (iii) allosteric modulators have the potential for greater receptor subtype selectivity, based either on a mechanism related to the location of the allosteric sites or different degrees of cooperativity exerted by a modulator at each subtype. In the case of PPI, an additional advantage is that allosteric modulators need not bind at difficult-to-target PPIF regions but can address more pronounced binding sites either between protein subunits or in the interior of a protein.

The discovery of new allosteric sites is a challenging prerequisite to the rational development of allosteric PPIM. Several new allosteric sites have been identified by experimental means,[173] *e.g.* in glycogen phosphorylase,[174] protein tyrosine phosphatase 1B,[175] and HIV-1 reverse transcriptase.[176] The techniques applied include traditional high-throughput screening followed by X-ray crystallography, phage display with crystallography, and tethering.[177]

As an alternative to experiment-led discovery, several computational methods have demonstrated their capability to predict allosteric sites. In evolutionary trace analysis (ETA),[178,179] sequence and structural data are combined to infer the location of functional sites in proteins. Here, members of a protein family are first divided into functional classes based on their sequence identity tree. Then residues that are invariant within each class but vary among them are identified and mapped onto a representative structure. A cluster of these class-specific residues on the protein structure implies an evolutionarily privileged site that is responsible for the functional specificity of the individual family members. The method was applied to the family of regulator of G protein signaling (RGS) proteins,[180] which interact with G protein α subunit (G_α) proteins. A novel functional surface located next to but distinct from the interface between RGS and G_α was identified. Subsequent mutagenesis experiments[181] and crystal structure data[182] confirmed this surface to be the interaction site for binding of the G protein effector subunit PDEγ. Interestingly, since some of the surface residues had profound effects on the regulation of G_α activity by PDEγ but did not directly interact with G_α, a form of allosteric communication among these residues was inferred.

The ETA exploits evolutionary information about *individual* residues to identify functional/allosteric protein sites. However, a hallmark of allosteric interaction is that it occurs between topographically distinct binding sites; this requires a reliable propagation of signals originating at the allosteric site to the functional one. Hence, if long-range "through-space" interactions can be neglected, the signal flow must proceed *via* a physically connected network of residues that link both sites. The statistical coupling analysis (SCA)[79] detects such coupling between two sites in a protein by analyzing the *co-evolution* of these positions in large and diverse multiple sequence alignments of a protein family. Applied to G-protein coupled receptors, the chymotrypsin class of serine proteases, hemoglobin, guanine-nucleotide-binding proteins, and RXR nuclear receptors, evolutionary conserved sparse networks of amino acid interactions could indeed be identified as structural motifs for allosteric communication in proteins retro- and prospectively.[183–185]

For a successful application of SCA, sub-alignments of members of a protein family of sufficient size and diversity are required so that coupling observed between sites reflects evolutionary constraints during evolution and not just historical relationships.[185] In those cases where the sequence information is insufficient for SCA but structural information about the protein is available, cooperative networks of residues within proteins can be predicted by the COREX approach.[186–188] Here, a large number of different conformational states of a protein are generated through the combinatorial unfolding of a set of predefined folding units. The correlation between the folding states of two residues then indicates the mutual susceptibility of each residue to perturbations at every other site, which in turn reveals the energetic coupling between those residues. Taken one step further, correlations between binding sites as a whole and the rest of the protein can be identified in addition to the pairwise residue correlations. Notably, the analysis of energetic couplings in dihydrofolate reductase revealed that perturbations at one site do not necessarily propagate through structure to the other site *via* a series of conformational distortions. Instead, perturbations exert an influence by affecting the distribution of folded and unfolded states in the ensemble, reinforcing the influence of dynamics on allosteric modulation.[189]

Although the described methods are exciting means to rationalize intramolecular communication, only a few allosteric sites predicted *de novo* have been reported so far.[173] Even if structural protein information is available, locating such sites is hampered by the fact that in many cases some degree of conformational adaptation is required for binding the allosteric modulator. Hence, potential allosteric binding sites may not be readily detectable in the unbound receptor from geometric considerations alone. However, additional structural features might provide a hint to the possibility of opening up a binding site. This is demonstrated by the rather extreme example of inhibitor binding to the highly packed core region between helices 11 and 12 of β-lactamase (Figure 13.2).[61] Here, unfavorable φ/ψ angles of Leu220 located at the N-terminal end of helix 11 suggest conformational strain in the unbound structure, which may be relieved upon complex formation and then counterbalances the energetic cost of core disruption. In addition, a COREX analysis predicts the revealed binding site region to be relatively unstable.[56] Together, these observations suggest that helices 11 and 12, although well-packed, are more prone to an induced-fit adaptation than other core regions. Finally, the most compelling evidence for the existence of the cryptic site comes from the binding of a crystallization agent in the same site of a *homologous* β-lactamase structure.[190] Taken together, perhaps the most promising way to predict new allosteric sites is by, first, obtaining suggestions for potential sites from "crystallization artifacts" or already known ligands binding to related protein structures and, subsequently, confirming these potential allosteric sites computationally.

Finally, "interfacial inhibition" through uncompetitive inhibition has been elucidated recently[191–193] as another natural paradigm for interfering with

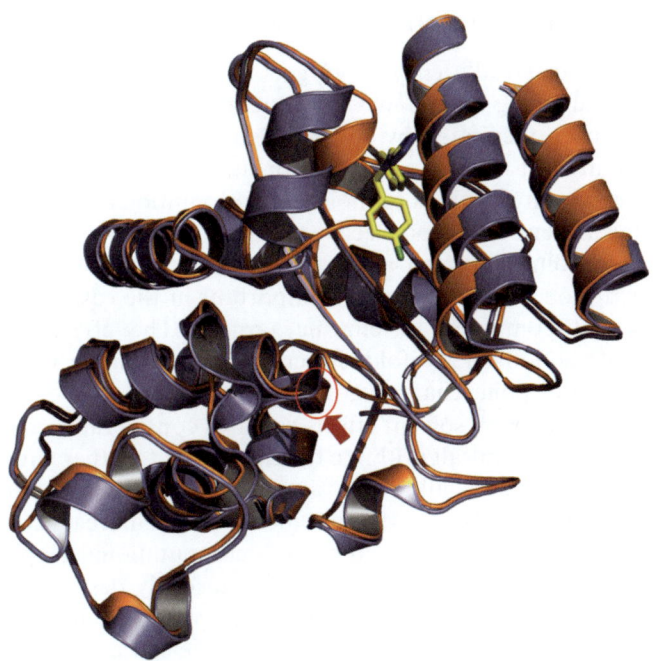

Figure 13.2   Conformational rearrangement of the backbone structure of β-lactamase (*orange*: unbound structure, PDB code 1zg4; *magenta*: bound structure, PDB code 1pzo) upon binding of an allosteric inhibitor (*yellow*). The location of the catalytically active serine is marked by a *red circle*.

macromolecular interactions. Here, targets are captured in dead-end complexes that are unable to complete their biological function. These intermediates display deeply curved surfaces with unbalanced energetic characteristics that are targeted by the inhibitor. However, such conditions are less likely to occur in the unbound proteins or completely bound complexes, and predictions of intermediate complex states as a prerequisite for identifying such sites are generally beyond current computational capabilities.

13.6 Docking for Targeting Protein–Protein Interfaces

Once potential binding sites have been identified, performing docking experiments or virtual screening (VS) is the next step in computer-aided drug development. In particular, this holds for PPI as a new class of targets because no large collections of known ligands may be available to successfully apply ligand-based approaches. Methodologically, current docking approaches face two main difficulties: dealing with solvent effects and protein flexibility. The impact of this situation on targeting PPI along with recent progress is discussed below.

13.6.1 Improved Descriptions of Solvent Effects

In the case of "classical" enzyme targets, a large number of successful applications of docking in VS has been reported.[194] By and large, these successes have been facilitated by steric constraints imposed by well-defined deep cavities that exist in these targets.[195] In such cases, the description of the complex energetic contributions to molecular recognition can be simplified. In fact, neglecting solvent effects on electrostatics did not have a significant effect on the success of some computer-aided drug-design programs.[194,196] However, a proper description of electrostatics is important in the case of PPIF, which are typically flat compared to enzyme targets. The effect of water on electrostatic interactions is twofold: it screens direct charge–charge interactions, and it solvates polar/charged groups.

Dealing thoroughly with solvent effects requires considerable computational resources. However, recent algorithmic developments together with increasing computational power now allow for a much more rigorous treatment of electrostatics. In particular, continuum models, which treat the solvent as being structureless, are now widely applied in computational biophysics.[197] In this approximation, Poisson's equation (PE) rigorously describes the electrostatics of a system consisting of a solute modeled as a distribution of charges in a low dielectric medium immersed in a high dielectric medium (typically water). In general, the PE can only be solved numerically for arbitrarily shaped molecules using, for example, finite difference techniques.[198] In order not to re-solve the PE for every newly generated conformation in docking, Arora and Bashford[199] have introduced the Solvation Energy Density Occlusion (SEDO) approach. Here, the system is represented in terms of a solvation energy density that is pre-computed for receptor and ligands prior to starting the docking simulation. Upon binding, the interacting region of both counterparts changes from a high dielectric medium to a low dielectric one. By neglecting a charge density rearrangement in the remaining high-dielectric region, one then only has to subtract the contribution arising from the newly occluded areas in the complex, which pays off a great gain in efficiency. The methodology was tested on two different data sets: a series of MHC class I protein–peptide complexes, and a congeneric series of HIV-1 protease–ligand complexes. The complexes with the small ligands of the HIV-1 protease yielded slightly better results than the peptides with the MHC class I protein, but all of them were in very good agreement with the results obtained when a non-modified PE approach was followed.

When a docking simulation runs, every new ligand conformation has to be evaluated. Practically, the scoring function complexity and implementation must be efficient. In this regard, the SEED approach by Majeux *et al.*[200] introduces an appealing treatment of the electrostatic contribution to the total binding free energy. Two important approximations are made. First, a simple distance-dependent dielectric model for the screened ligand–receptor interaction is used. Second, for both receptor and ligand, the main contribution to desolvation is considered to come from the removal of the first shell of water.[201] Totrov has estimated that the first solvent layer contributes 66% to the total desolvation energy.[202] Considering this, the receptor and ligand molecules are independently mapped onto a grid, and the corresponding desolvation is pre-computed at the centers of low dielectric probe-spheres rolled over the solvent accessible surface. This computation is done according to the Coulomb approximation of the electric displacement. During docking, only occluded areas in both counterparts have to be detected and summed to assess the total contribution. The approach was validated against solutions of the PE, showing very good correlations for every single contribution and the total electrostatic energy. The oncoprotein MDM2 was targeted to investigate the virtues of this approach. MDM2 binds to the p53 tumor suppressor, keeping it inactive.[203] 1,4-Benzodiazepines and dibenzocyclohexane were found computationally to yield the best binding energies. However, to our knowledge, these findings have not been validated experimentally.

The GB model is an approximation to the PE approach.[204] Here, space is also divided in regions of high (solvent) and low (solute) dielectric, but the reaction field energy is approximated by a pairwise sum over interacting charges. From the original Born theory, the electrostatic contribution to the free energy of solvation of a point charge q located in the center of a spherical cavity of radius R (Born radius) is given by eqn (13.1):

$$\Delta G_{\mathrm{pol}} = -\frac{1}{2}\left(1-\frac{1}{\varepsilon}\right)\frac{q^2}{R} \tag{13.1}$$

where ε is the dielectric permittivity of the medium.

The term "generalized" comes into play when considering more than two point charges and arbitrarily shaped cavities instead of spherical ones. Then, a smooth function that considers charge–charge interactions according to their location in the solute is required. To date, the formulation proposed by Still *et al.*[204] is the most widely used one to estimate solvation energies in this situation. An important parameter in the formulation is the effective Born radius, which measures the burial of an atom in the low-dielectric medium and depends on the atom's intrinsic Born radius and the arrangement of the rest of the atoms in the system. The way of estimating R has implications not only in terms of accuracy, but also in terms of efficiency. The different flavors of GB currently in use arise mainly from the way of defining and computing this parameter. The accuracy and efficiency of a variety of GB models for computing electrostatic solvation energies in comparison with the more rigorous PE model has been assessed by Feig *et al.*[205] The latest GB models yield results comparable to PE, although efficiency is still a concern for those models. A more in-depth review on GB models is available.[206]

Zou *et al.*[207] incorporated a GB model into DOCK.[208] In a further development of the ideas in this work, Liu *et al.*[209] have recently incorporated the more efficient, yet arguably less accurate, pairwise approximation proposed by Hawkins *et al.*[210] that takes into account atomic overlaps to compute Born radii. Despite the limited data set of only three systems used for the evaluation, acceptable results are obtained for the Born radii, considering the gain in efficiency ($\sim 8\%$ of error for the receptor and 4% for the ligand). More interestingly, the GB enhanced scoring scheme performs better than the standard "force field" scoring function in DOCK in a VS experiment. This has, in fact, an important implication as these kinds of computational screenings are now commonplace in the early stages of current drug-design programs and suffer from high rates of false positives. Additionally, this re-implementation of the GB model into DOCK introduces a correction to properly treat possible void formations between the protein and "misaligned" ligands. This is one of the most problematic aspects of GB models: the boundary definition by means of spheres between solvent and solute may result in solvent-inaccessible, yet high-dielectric, voids in the interior of large biomolecules.[206] Interestingly, successful VS studies implementing these improvements in DOCK have been reported for glyceraldehyde-3-phosphate dehydrogenase,[211] lysosomal cysteine proteases,[212] and a PDZ protein interaction domain[213] as targets.

Aside from GB, other implicit solvation models have been adapted in some popular docking programs as a compromise between the required accuracy and the affordable computational effort.[214] In that respect, a particularly efficient implicit solvation model for computing the electrostatic part of the binding free energy in protein–ligand docking has been introduced recently.[215] With a similar performance in accuracy as GB, the mean pose calculation time by this model amounts to about 40 ms. On the other hand, when facing rather large VS experiments, it is now commonplace to follow a hierarchical approach[216] where candidate compounds are pre-screened and selected with simpler scoring functions. This reduced database is subsequently re-ranked with more accurate approaches such as MM-PBSA.[217] Very encouragingly, by employing a modified protocol which includes using a GB model during the MD simulations, a very efficient PE solver, and a computational design based on a distributed-computing paradigm, a high-throughput variant of MM-PBSA has been introduced.[218] With this variant, more than 300 compounds were evaluated against three different targets overnight, using approximately 400 desktop computers.[219] As for the accuracy, statistically significant correlations to experimental data were obtained, with correlation coefficients > 0.72.

13.6.2 Protein Flexibility in Protein–Ligand Docking

Proteins are inherently flexible, which provides the origin for their plasticity and enables them to conformationally adapt to a binding partner. However, current docking-based drug-design approaches generally treat the target protein as a rigid unit. Following this approximation, better results in VS experiments have been obtained when target structures extracted from complexes were used compared to "apo" structures.[220] On the other hand, "holo" structures appear to introduce a bias in the experiment which, in many cases, precludes finding chemically novel ligands.[221] If the adaptability of binding sites in enzymes is a real concern, taking into account protein flexibility appears even more critical in the case of PPIF, as they have been proven to be highly adaptive,[14] as discussed above.

From a practical point of view, challenges of incorporating protein flexibility into docking are twofold: first, one needs to detect what is flexible,[222] and second, this knowledge needs to be transformed into the docking algorithm. With respect to the former, options range from using experimental information to predicting the most relevant movements through computational methods such as MD, NMA, or a graph-theoretical approach.[164] The latter issue is in general far more open to creative approaches but must be guided by the unavoidable concern about efficiency.

13.6.2.1 Determining What is Moving and How

Protein flexibility comprises a range of possible movements, from single side-chains to drastic structural rearrangements as seen in calmodulin.[223] Depending on where the binding site is located, one or more of these movement types will be relevant for docking. Interestingly, a recent study by Zavodszky and Kuhn[224] assessed, for a large set of typical enzymatic targets, to what extent side-chain rotations of amino acids contribute to the flexibility within the binding site. Ligands from 63 different complexes comprising a total of 20 different enzymes were re-docked into the corresponding *apo* structure using their docking program SLIDE[225] that allows for protein side-chain rotations. These side-chain rotations were proven to be necessary to correctly dock 54% of the ligands, but encouragingly, 95% of the rotations were smaller than 45°. The plausibility of every adaptation proposed by SLIDE was then evaluated by comparing the free with the resulting bound structures and by analyzing the geometry with PROCHECK.[226] Only 7% of the conformations were evaluated as unfavorable. Previously, Najmanovich *et al.*[227] had shown that there is no correlation between backbone and side-chain flexibility. They reported as well that rotations in side chains of up to three residues account for ~85% of all the cases where there is a conformational change upon ligand binding. For the case of PPI, it has been shown that the conformational change can be even more important for those residues involved in the interface.[228] Having the whole dynamic picture of a protein in hand would be the optimal situation to deal with the protein's flexibility. Although this is not the case for most of the interesting targets, there is hope that one can predict stable conformers even from the unbound structure due to the fact that the bound conformation is likely to be a pre-existing one in the free state[229] (see Section 13.4).

In summary, it is encouraging to see that with little effort much can be gained: many changes will be related only to side chains. Even more, our understanding of protein dynamics has enabled unbound structures as useful starting points for flexible docking. However, the challenge remains in considering every movement that occurs in the binding, independent of its range. In what follows, three approaches to predict flexible regions in proteins are described: automated conformation exploration restrained by experimental knowledge, derived from MD simulations, and through NMA.

Cavasotto and Abagyan[155] incorporated a receptor-ensemble docking approach in the frame of the IFREDA (ICM-flexible receptor docking algorithm) method for VS. The first step of this involves the *de novo* generation of alternative receptor conformations. Four protein kinases subfamilies were investigated. Forty random configurations of a ligand were generated in the binding site. The ligand, side chains in the binding site, and pre-selected loops known to undergo rearrangements are considered fully flexible. An *in vacuo* minimization is performed followed by a stochastic energy minimization and a final full minimization of the top ranked conformations. In this way, the area considered as flexible is enlarged, which yields plausible receptor conformations that are relevant for the binding site. Yet, the requirement for an exhaustive sampling is avoided. However, a drawback of the method is that flexible protein regions must be known in advance.

Zacharias[230] has proposed to use MD to study the movements of the protein before docking. From the MD one can calculate soft flexible degrees of freedom *via* principal component analysis; these soft modes can be afterwards incorporated into a flexible docking algorithm as additional variables to guide the movement of the protein. As only C_α atoms are considered, side-chain movements are not represented by the precalculated modes. However, the method has still proven useful in VS for pre-filtering databases because it is not very computationally demanding. The author was also able to achieve successful docking results starting from an unbound protein structure where the ligand did not fit sterically. The method has the advantage that the explored protein conformations are fairly realistic, including solvent and counterion effects. In an attempt to also consider side-chain flexibility, Tatsumi combined MD with harmonic dynamics.[231] Collective movements are incorporated into the motion of C_α atoms by means of harmonic modes, whereas the motions of all other atoms are simulated by unbiased MD. This method is theoretically appealing, although inefficient, as one single docking takes 40 days of CPU time (using up to 16 CPUs in parallel).

NMA is a convenient and widespread method to study the dynamics of macromolecules in cases where only one structure is available (see above). Information about low-frequency normal modes are increasingly incorporated into docking procedures.[155,230] The aforementioned IFREDA, which needs information about side chains and backbone movements, was extended in that respect.[155] From NMA, "relevant modes" are selected, and protein structures are subsequently modified following them. The "relevant modes" are a means of focusing the general analysis to the binding site region, such that those modes that influence binding site atoms are selected. Although these are low-frequency modes, they do not correspond exactly with the lowest ones, so that intermediate-scale loop motions that might occur around the binding site are also captured. The methodology was validated by docking ligands into *apo* structures, where rigid-receptor docking had failed. Also, a small-scale VS showed larger enrichment factors than when performed with the rigid receptors.

Recently, we have developed a multi-scale modeling approach that combines concepts from rigidity theory, elastic network theory, and constrained geometric simulations. The approach is able to accurately predict protein conformational changes in three steps.[152,232] In the first step, the molecule is decomposed into rigid clusters using the FIRST approach.[164] Importantly, the composition of these clusters is not limited to residues adjacent in sequence or secondary structure elements. Instead, residues that are distant in primary sequence but close in the 3D structure may also be comprised in one cluster. In the second step, clusters are treated as rigid bodies, and the motions of these clusters are predicted by RCNMA (Rigid Cluster Normal Mode Analysis) using an elastic network representation of the coarse-grained protein. When applied to 10 proteins that show conformational changes upon complex formation, directions and magnitudes of the motions predicted by RCNMA agree well with experimentally determined ones, particularly if the movement is dominated by loop or fragment motions. In the last step, the NMSim module generates new conformers of the macromolecule using low-energy normal mode directions, predicted by RCNMA, and random direction components. The generated conformers are then iteratively corrected regarding steric clashes and constraint violations in order to generate stereochemically allowed conformations that lie preferentially in the subspace spanned by low-frequency normal modes. The NMsim approach was validated on hen egg white lysozyme (HEWL). For this, experimentally determined structures and conformations from state-of-the-art MD simulations[233] were compared to conformations determined by FRODA,[161] CONCOORD,[234] and NMSim. Regarding residue fluctuations, NMSim results show a good agreement with those from MD simulations and experimental structures. With respect to the stereochemical quality, NMSim-generated structures have backbone torsion angle characteristics that are in remarkable agreement with the characteristics of 100 high-resolution experimental structures.

13.6.2.2 *Incorporating Flexibility to Docking Algorithms*

Once the dynamic properties of the protein are known, incorporating this information is the next challenge in algorithmic development. Comprehensive reviews on methods to deal with flexibility in docking have been published.[155,222,235,236] The main concern in this step is computational efficiency. If a number of conformations is known either from experiment or calculation, a trivial solution is to run a parallel docking against every structure.[237] The main potential drawback is that the dynamic picture might not be complete, and some relevant conformations might not be present.

Running parallel dockings with all available structures is generally affordable when the number of structures to consider is small. An interesting alternative to incorporate flexibility in a mean-field sense was accomplished within Autodock[238,239] by Österberg *et al.*[240] Using a grid-based approach for evaluating the interaction energy between ligand and receptor, they followed different strategies to combine representations of multiple target structures by merging their individual interaction energy grids. Taking either minimum values or potential averaged grid values did not perform well. In turn, weighting different grids according to a Boltzmann distribution assumption yielded the best results. The approach was evaluated on 21 complexes of peptidomimetic inhibitors with human immunodeficiency virus type 1 (HIV-1) protease, and they were able to correctly dock 20 of them using the weighted maps. As a drawback of the method, merging grid representations of different receptor conformations into one may lead to the situation where mutually exclusive combinations of receptor conformations are present. It can also be expected that the method reaches its limits in cases of larger protein mobility. The latest AutoDock version, AutoDock4,[241] also allows us to consider the flexibility of receptor side-chains explicitly. When tested in a redocking experiment with 188 diverse protein–ligand complexes, successful dockings were reported for complexes with 10 or less torsional degrees of freedom. In a cross-docking experiment with 87 HIV protease complexes, adding side-chain flexibility overall leads to more successful docking results but also raises problems like increased computational costs and an increased potential for false positives because of the larger search space. Another docking algorithm, FlipDock,[242] performs automated docking of flexible ligands into flexible receptors, using the AutoDock force field. To represent ligand and receptor flexibility, a special data structure, the Flexibility Tree (FT),[243] is used. FlipDock was tested on 400 cross-dockings and showed a docking success rate of 93.5%, which compares favorably to the rate by AutoDock3.0 of 72%.

An alternative to the strategies of AutoDock to cope with protein mobility is to work explicitly with known (or predicted) conformations of the protein, but in a way that is more computationally efficient than trivially running several dockings in parallel. FlexE[244] incorporates a united protein description that handles similarities and differences in the ensemble of conformations. Structures are initially superimposed by backbone atoms. Then, united structures are created by combining the alternative side-chain conformations and backbone parts. A united structure is composed of instances, that is, conformationally different substructures. These structures are then used in an incremental docking approach. The ligand is placed fragment by fragment into the active site and possible interactions between the ligand and all instances are determined. In the final step, contributions from all instances are summed, whereby mutually incompatible instances are discarded in order to retain realistic protein structures. As there are several possible combinations of independent instances at each construction step, finding high-scoring independent sets of instances is the most time consuming step of the whole procedure. The authors report an improvement compared to running parallel dockings and merging the results (67% of the best-ranked solutions below 2.0 Å *vs.* 63%) with a considerable reduction of running time. Probably the most severe limitation of this approach is that it does not take into account changes in internal energy of the different protein conformations. This clearly favors open binding pockets that can accommodate large ligands, which form many favorable intermolecular interactions.

Along these lines, Wei *et al.* have pointed out that the largest impact on the improvement of their VS results was obtained by precisely including this contribution.[221] They have used an in-house modified version of DOCK[245] and an ensemble of experimental structures of the receptor as templates to represent conformational changes. The receptor is decomposed into rigid and flexible regions. For each of the flexible regions, there are several conformational possibilities according to the experimental data. In this sense, the receptor is multicomponent: a rigid component is combined with flexible ones. A depth-first search algorithm is used to scan through all possible conformations for the possibility to dock the ligand without steric clashes. If found, the score is computed by summing the contributions from every component. The best-fit conformation of each flexible receptor region is used to assemble the receptor conformation. They applied the method to identify known ligands of a hydrophobic cavity mutant of T4 lysozyme and the folate-binding pocket of thymidylate synthase from the Available Chemicals Directory. The inclusion of a weighted receptor conformational energy in the scoring function led to an improvement in the enrichments, particularly for lysozyme.

This example clearly shows that incorporating flexibility into a docking algorithm does not only involve sampling protein conformations, but also evaluating them[221,246] with respect to this energy. In this regard, it is worth emphasizing the importance of considering the free energy of the receptor's conformation and the stabilizing influence that a bound ligand can provide.[57]

13.6.3 Data-Driven Docking Approaches

Computational docking approaches fail due to two main reasons: insufficient sampling (which includes considering receptors as rigid) and simplifications in the scoring functions. Although rigorous descriptions of intermolecular interactions are available and could be readily incorporated, this comes together with an increase of computational demands. In fact, for most of the cases, the loss in efficiency does not pay off. To overcome this dilemma, another strategy is possible: supplement the scoring functions with experimental information. This is now a current trend in the field of protein–protein docking, as can be seen from the assessments of the last CAPRI round.[247] From there, a relevant conclusion is that "using prior knowledge of the protein regions that are likely to interact remains an important ingredient for achieving successful docking".

To supplement "pure" docking, there are two possibilities: first, to use the knowledge derived directly from the binding partners and, second, to incorporate experimental information about a particular complex. The first alternative, which has been reviewed by Fradera and Mestres,[248] includes the field of the so-called "tailored scoring functions".[249,250] There, the goal is to adapt general scoring functions to the particular target of interest by including information from known interactions of the protein with similar ligands. In general, one could say that the better the studied system is known, the better results can be expected, if the information is properly incorporated. Because PPI as targets do not enjoy such a wealth of information as is available for typical enzyme targets, the strategy cannot be fully exploited. Incorporating directly measured information from experiments is the alternative.

Methods that use experimental information to drive the docking of biomolecular complexes (typically protein–protein or nucleic acid–protein) have been reviewed by van Dijk *et al.*[251] The information that is currently being used in this field has two main sources: mutagenesis studies and different NMR properties (*i.e.*, chemical shifts perturbation, H/D exchange, residual dipolar couplings, diffusion anisotropy). Data derived from biochemical or biophysical experiments can aid docking on two different levels: first, by identifying binding sites and, second, by restricting the conformational search space, concentrating the sampling around native-like poses. The second objective is more challenging and requires data that incorporate very specific structural information describing the orientation of the interacting counterparts. The use of mutagenesis data, for example, is restricted to the first purpose, as the collected examples show.[251]

Since the very beginning, biomolecular NMR data have been connected to computational processing for structure elucidation. Structures are calculated by means of minimizing a hybrid energy function (eqn 13.2) that incorporates NMR measures (E_{data}, with a weighting factor w_{data}; *e.g.*, inter-proton distances derived from peak intensities arising from NOEs, torsion angles and hydrogen bond restraints from scalar couplings, bond orientations from residual dipolar couplings) and a force-field-based term (E_{ff}):[252]

$$E_{hybrid} = E_{ff} + w_{data}E_{data} \qquad (13.2)$$

The accuracy of the obtained structures using this approach is highly dependent on the amount and the quality of data available, as has been acknowledged by Chen *et al.*[253] Since complete collection and assignments of structural restraints is a daunting task (and in some cases impossible), a look from the "other side of the coin" might prove useful: when experimental data are incomplete or inaccurate, directly deriving biomolecular (complex) structure from it may not be viable. The data may still serve, however, to guide a computational structure prediction tool or distinguish solutions generated by it.

With respect to the whole process, the experimental data can be employed before, at the same time, or after the computational sampling step. When used as a pre-filter, approximate poses are manually generated that are then computationally optimized.[254] However, if the data are incorporated at the search stage, on the one hand, the tedious and non-exhaustive manual generation can be avoided and, on the other hand, native-like configurations are likely to be searched. This last feature is also an advantage over methods that use the experimental data only for post-filtering of proposed solutions.[255]

NMR as a tool for investigating the conformation of bound ligands has been recently reviewed.[256] There, the quantitative analysis of transferred-NOEs and cross-correlated relaxation data are examined from the experimentalist's perspective. From the computational viewpoint, the far more interesting question is which easily obtainable NMR measurements can be incorporated to guide a docking algorithm. In this regard, chemical shift perturbations (CSP) and saturation transfer difference (STD) have already proven promising and will be discussed below. We will not consider here the case of complex structure elucidation using intermolecular NOEs that is applicable to tightly bound ligands, which is the standard protein NMR methodology.[257]

13.6.3.1 Guided Docking Using Chemical Shift Perturbations (CSP)

^1H–^{15}N heteronuclear single quantum correlation (HSQC) NMR is nowadays a well-established experiment. The most well-known application in the field of drug design is the so-called "SAR (structure–activity relationship) by NMR"[33] approach. Upon ligand binding, the chemical shifts (CS) of the interacting partners are affected due to a change in the environment. Provided that no large conformational changes occur during this process, the largest perturbations that can be observed, *e.g.* on the protein side, are due to the binding of the ligand. In the SAR by NMR approach, different low-affinity fragments are used to "explore" the binding site of a ^{15}N-labeled protein by monitoring CSP and mapping them onto the protein surface. Combining the information obtained from fragments that bind at different regions of the binding pocket, these fragments can be chemically linked to yield new molecules with increased binding affinity. ^{1}H–^{15}N HSQC spectra have also been used to determine that sulindac-derived inhibitors of the Ras–Raf interaction[258] bind directly to Ras. This is a relevant result not only with respect to the particular studied system, but it also shows the technique to be suitable for facilitating the task of finding binding sites at the target (see Section 13.4).

The HADDOCK approach[259,260] uses CSP information upon complexation for structure prediction of macromolecular complexes in a paradigmatic way. The underlying idea is that the size of the configurational/conformational search space can be significantly reduced once the residues involved in the intermolecular interactions are known. Information of this kind can be obtained from the analysis of CSPs. CSPs, however, do not reveal which residues interact with each other. At this point, docking comes into play. Here, the experimental information is introduced by means of ambiguous interaction restraints (AIRs), originally proposed by Nilges.[261] An AIR (eqn 13.3) is defined as an upper-bounded intermolecular distance that must be fulfilled upon complex formation. However, it does not require a particular residue pair to fulfill it, but a subset of pre-selected possible pairs.

$$d_{i(A),B}^{\mathrm{eff}} = \left(\sum_{m=1}^{N_{\mathrm{atoms},i(A)}} \sum_{k=1}^{N_{\mathrm{res},B}} \sum_{n=1}^{N_{\mathrm{atoms},k(B)}} \frac{1}{d_{m,n}^{6}} \right)^{-1/6} \tag{13.3}$$

Residues defined as "involved in the interaction" are taken as pairs (i, k), one from each counterpart A and B, respectively. The distance is computed for every atom m in residue i from the first protein to every atom n of residue k in the second protein. In this way, as soon as two atoms are in contact the restraint is satisfied. Schieborr *et al.*[262] have applied HADDOCK to the problem of protein–ligand docking. Owing to the large size difference between ligand and receptor, the authors modified the protocol, such that only strong effects of the ligand on the protein were considered. As a result, the impact of incorporating the AIR in this case is a restriction of the conformational search to the binding site area. However, no information about the mutual orientation of both binding partners is exploited. In fact, the authors acknowledge that the binding site of the protein could have been located also by mapping the strong CSP onto the protein structure. Thus, with this approach, the success in determining the native structure of the complex still depends on the force-field component of the scoring function.

It is stimulating that CSPs have been used successfully in structure refinement.[263–265] This opens a new perspective for the docking field.

Refinement against CSP differs from structure calculations with distance restraints. In the case of CSP refinement, there are no pairwise distance restraints to fulfill, but a scoring function that minimizes differences between observed and calculated CS is required. This implies that an efficient method to compute theoretical CS is available.[266] Contributions to proton CS can be decomposed into local (diamagnetic and paramagnetic) and non-local contributions. In the latter case, effects from nearby aromatic rings (δ_{rc}) and other sources of magnetic anisotropy (δ_{mag}) as well as electrostatic and solvent effects (δ_{el}) are included (eqn 13.4). The local effects are approximated by the observed shifts in short peptides with corresponding secondary structure.

$$\delta_{total} = \delta_{local} + \delta_{rc} + \delta_{mag} + \delta_{el} \tag{13.4}$$

If there is no significant conformational change of the protein upon complexation, the observed CS differences between the free and complex spectra of the protein are due to the ligand. Compared to the SAR by NMR approach that exploits such information only qualitatively, one can now quantify these differences, such that one can deduce additional information about the orientation of different groups of the ligand. Aromatic rings, when present, constitute the main source of the contribution to the total CS difference. Accordingly, the orientation of the ring with respect to the protein can be determined because of the anisotropic nature of this effect. McCoy and Wyss[267] have explored the usefulness of these "ring current effects" within the frame of the program SHIFTS[268] as a post-filtering tool for elucidating the structure of protein–ligand complexes. Although native-like and close-to-native-like conformations were best ranked, the possibility of using such a method for guiding docking remains unclear. There are two concerns that would hamper the approach: first, in many cases errors in the prediction are in the same range as the observed changes upon complex formation and, second, CSP may also arise from conformational changes of the protein upon complex formation. González-Ruiz et al. recently presented a new approach that steers protein–ligand docking with quantitative NMR chemical shift perturbations.[269] This method is based on a hybrid scoring scheme that combines a weighted sum of DrugScore,[270] describing protein–ligand interactions, and Kendall's rank correlation coefficient,[271] which scores the agreement between experimentally determined and computed CSP for a given ligand pose. An efficient empirical model considering only ring-current effects is used for back-calculating CSP for a ligand pose. The hybrid scoring scheme was tested on 70 protein–ligand complexes with computed CSP reference data. Without CSP information, a docking success rate of 71% was achieved. This increased to 99% if CSP information was included. The approach should be helpful for protein–ligand complexes that are computationally difficult to predict, e.g. if a ligand binds to a flat binding site.

Instead of monitoring CSP from the protein side, Wang and Merz used a semiempirical method at the NMDO level developed to predict CS of ligand protons.[272] As an application they ranked manually generated poses of a FKBP12 inhibitor, finding a very good correlation between CSP and structural RMSDs. Despite the higher level of theory applied than in the SHIFTS approach, there are still concerns due to issues of protein flexibility. In addition, although sufficiently fast for a single calculation, a scoring function in a docking algorithm has to be evaluated millions of times, which makes this method unappealing for VS applications.

13.6.3.2   Impact of Data-driven Docking on Computational Targeting of PPI

Most of the current advances in designing small molecules to target PPI come, as expected, from experimental and not theoretical approaches.[273] However, as the docking field develops, more challenging examples can be targeted, which also includes PPI. Mixing experimental information concerning a particular system with current docking algorithms has proven not only feasible but also as the source for large improvements. This holds especially true in the particularly demanding field of protein–protein docking.

At the moment, quantitative information that can be derived from experiment (such as in the cases of CS and STD described) is used in combination with computational approaches to confirm and validate binding modes as a post-processing tool. Still, docking methods that rely on the comparison between computer-predicted and experimentally observed properties remain inefficient, which prevents them from being useful in VS experiments. Nevertheless, further developments of approaches that use experimental data at search time should be pursued. If used in a pre-processing step, using experimental information can help supplement computational approaches by restricting the search space to those regions that involve key interacting residues.

13.7   Summary and Outlook

Recent advances in computational approaches to detect PPI and identify small-molecule PPIM have been reviewed. The number of successful examples of computer-aided identification of agents that modulate PPI is still limited. However, there is significant progress in understanding and modeling molecular recognition properties of PPI, and we expect that in the next few years the influence of computational means on targeting PPI will increase considerably. This is reflected in the hit rates of around 15–20% obtained for four of the virtual screening experiments listed in Table 1,[289,293-295] which are much higher than if non-targeted libraries are screened.[17]

We have primarily focused on structure-based approaches that require knowledge of at least one of the interacting components of the protein–protein complex. Although detailed structural information about the PPIF may not be available in all cases, and further challenges arise from the inherent plasticity of

PPIF, this way seems to be more promising to pursue in our opinion than ligand-based approaches. On the one hand, for the latter methods to be successful, structure–activity relationships or pharmacophore models derived from lead structures obtained by experimental high-throughput screening need to be established. However, a sufficient amount of good-quality data that is usually required for training may not be available in many cases of novel protein–protein targets. On the other hand, structural knowledge and insights into the forces that act at a PPIF are not only critical to the prospective discovery or design of small-molecule PPIM. It will also allow us to understand the reasons for affinity and selectivity towards a given target retrospectively,[274] which will aid in guiding future efforts to improve both properties.

The approaches presented here form two key levels of what may become an integrated approach for finding small-molecule PPIM by computational means (Figure 13.3). Given the structure of the target, the first level comprises the prediction of potential binding sites, which includes hot spot analysis, cleft detection, and allosteric site detection. Subsequently, flexible, data-driven ligand docking with improved scoring will be applied in the frame of VS. Equally important, but not covered in this review, are aspects of target identification, target druggability, and ligand pre-selection. Even in the field of "classical" targets the first two topics have only been touched recently,[275,276] and computational approaches are emerging.[202,277–279] However, they may become even more important in the case of protein–protein targets due to the large variability of PPIF. As such, the interfaces of Bcl-2/Bak and p53/Mdm2 are highly hydrophobic whereas the Ras/Raf-1 kinase interface is largely polar.[274] Experiences gained from one PPIF may thus not be transferable to another case and identifying "dead-end targets" early will save time and expenses.

With respect to ligand pre-selection, computational approaches to filter for drug-like properties of small molecules are already widely used.[280,281] Here, the main focus is on the estimation of ADMET (absorption, distribution, metabolism, excretion, toxicity) properties. Considering, however, that the lack of well defined clefts or grooves in PPIF often results in PPIM with upper-limit potencies in the micromolar range,[89] a potential high degree of promiscuity of these PPIM on other targets is the consequence. Such non-specific, "promiscuous" compounds have been identified among screening hits,[282] lead compounds,[283] and known drugs.[284] A single, aggregation-based mechanism of action has been proposed to explain the observed effects.[285] Accordingly, the panel of approaches to estimate the pharmacological profile of a potential PPIM should be extended to methods[284,286] that allow us to identify and eliminate potentially promiscuous compounds at early stages. Finally, of heightened interest for the case of protein–protein targets would be schemes that pre-select compounds to be used in the virtual screening that bind to specific protein domains.[17] For this, however, a systematic identification of such entities by experiment is required first.

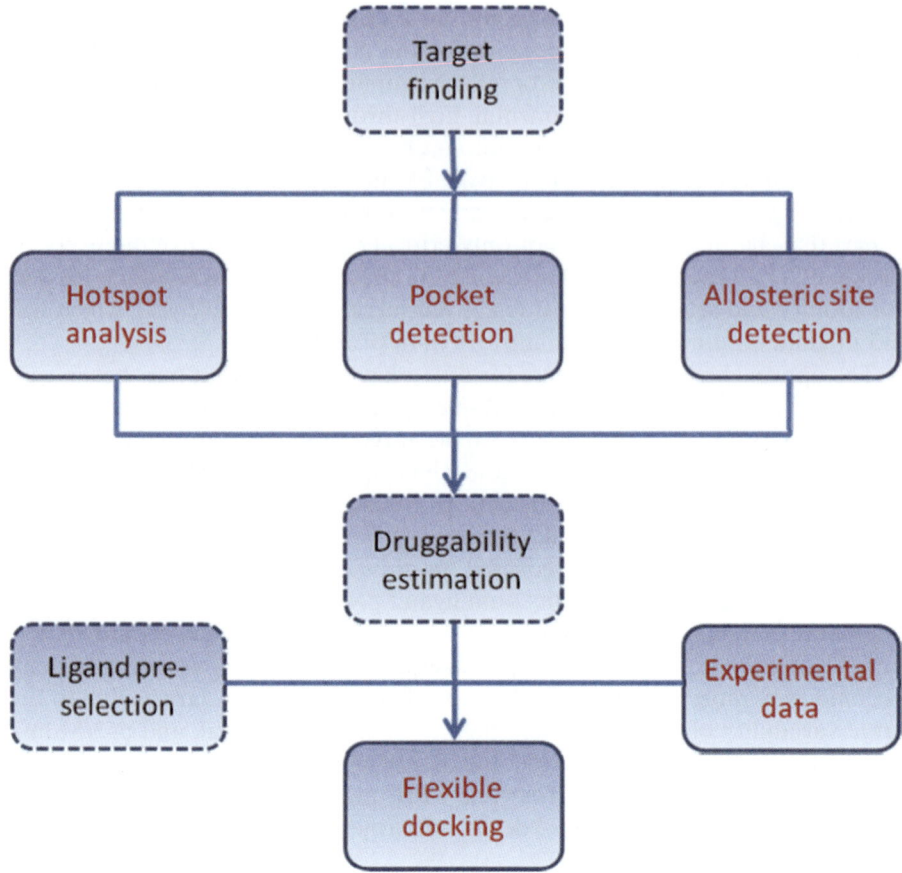

Figure 13.3 Flowchart describing an integrated approach to the computer-aided identification of small-molecule PPIM. Topics depicted in boxes with straight lines have been covered in this review.

Overall, being able to establish such a process will be a major breakthrough in the field, and in combination with progress in various experimental areas, we are awaiting exciting times for modulating PPI.

Acknowledgements

This study was supported by funds from Heinrich-Heine-University.

References

1. A. C. Gavin, M. Bosche, R. Krause, P. Grandi, M. Marzioch, A. Bauer, J. Schultz, J. M. Rick, A. M. Michon, C. M. Cruciat, M. Remor, C. Hofert, M. Schelder, M. Brajenovic, H. Ruffner, A. Merino, K. Klein, M. Hudak,

D. Dickson, T. Rudi, V. Gnau, A. Bauch, S. Bastuck, B. Huhse, C. Leutwein, M. A. Heurtier, R. R. Copley, A. Edelmann, E. Querfurth, V. Rybin, G. Drewes, M. Raida, T. Bouwmeester, P. Bork, B. Seraphin, B. Kuster, G. Neubauer and G. Superti-Furga, *Nature*, 2002, **415**, 141.

2. P. Bork, L. J. Jensen, C. von Mering, A. K. Ramani, I. Lee and E. M. Marcotte, *Curr. Opin. Struct. Biol.*, 2004, **14**, 292.

3. M. R. Arkin, M. Randal, W. L. DeLano, J. Hyde, T. N. Luong, J. D. Oslob, D. R. Raphael, L. Taylor, J. Wang, R. S. McDowell, J. A. Wells and A. C. Braisted, *Proc. Natl. Acad. Sci. U. S. A.*, 2003, **100**, 1603.

4. P. M. Fischer, *Drug Design Rev.*, 2005, 179.

5. L. Stockwin and S. Holmes, *Expert Opin. Biol. Ther.*, 2003, **3**, 1133.

6. R. Zutshi, M. Brickner and J. Chmielewski, *Curr. Opin. Chem. Biol.*, 1998, **2**, 62.

7. Z.-Y. Zhang, R. A. Poorman, L. L. Maggiora, R. L. Heinrikson and F. J. Kezdy, *J. Biol. Chem.*, 1991, **266**, 15591.

8. M. Liuzzi, R. Deziel, N. Moss, P. Beaulieu, A. M. Bonneau, C. Bousquet, J. G. Chafouleas, M. Garneau, J. Jaramillo and R. L. Krogsrud, *Nature*, 1994, **372**, 695.

9. V. Bottger, A. Bottger, S. F. Howard, S. M. Picksley, P. Chene, C. Garcia-Echeverria, H. K. Hochkeppel and D. P. Lane, *Oncogene*, 1996, **13**, 2141.

10. M. Rubinstein and M. Y. Niv, *Biopolymers*, 2009, **91**, 505.

11. J. Eichler, *Curr. Opin. Chem. Biol.*, 2008, **12**, 707.

12. L. Pagliaro, J. Felding, K. Audouze, S. J. Nielsen, R. B. Terry, C. Krog-Jensen and S. Butcher, *Curr. Opin. Struct. Biol.*, 2004, **8**, 442.

13. J. A. Wells and C. L. McClendon, *Nature*, 2007, **450**, 1001.

14. M. R. Arkin and J. A. Wells, *Nat. Rev. Drug Discovery*, 2004, **3**, 301.

15. T. Berg, *Curr. Opin. Drug Discovery Dev.*, 2008, **11**, 666.

16. R. L. Juliano, A. Astriab-Fisher and D. Falke, *Mol. Interv.*, 2001, **1**, 40.

17. T. Berg, *Angew. Chem., Int. Ed.*, 2003, **42**, 2462.

18. A. G. Cochran, *Curr. Opin. Chem. Biol.*, 2001, **5**, 654.

19. H. Yin and A. D. Hamilton, *Angew. Chem., Int. Ed.*, 2005, **44**, 4130.

20. A. Christopoulos, *Nat. Rev. Drug Discovery*, 2002, **1**, 198.

21. G. Fuentes, J. Oyarzabal and A. M. Rojas, *Curr. Opin. Drug Discovery Dev.*, 2009, **12**, 358.

22. A. Degterev, A. Lugovskoy, M. Cardone, B. Mulley, G. Wagner, T. Mitchison and J. Yuan, *Nat. Cell Biol.*, 2001, **3**, 173.

23. D. L. Boger, J. K. Lee, J. Goldberg and Q. Jin, *J. Org. Chem.*, 2000, **65**, 1467.

24. S. Silletti, T. Kessler, J. Goldberg, D. L. Boger and D. A. Cheresh, *Proc. Natl. Acad. Sci. U. S. A.*, 2001, **98**, 119.

25. D. L. Boger, J. Goldberg, W. Jiang, W. Chai, P. Ducray, J. K. Lee, R. S. Ozer and C. M. Andersson, *Bioorg. Med. Chem.*, 1998, **6**, 1347.

26. J. Goldberg, Q. Jin, Y. Ambroise, S. Satoh, J. Desharnais, K. Capps and D. L. Boger, *J. Am. Chem. Soc.*, 2002, **124**, 544.

27. D. V. Erbe, S. Wang, Y. Xing and J. F. Tobin, *J. Biol. Chem.*, 2002, **277**, 7363.
28. W. Jahnke, A. Florsheimer, M. J. Blommers, C. G. Paris, J. Heim, C. M. Nalin and L. B. Perez, *Curr. Top. Med. Chem.*, 2003, **3**, 69.
29. S. Fletcher and A. D. Hamilton, *Curr. Top. Med. Chem.*, 2007, **7**, 922.
30. K. Almholdt, P. O. Arkhammar, O. Thastrup and S. Tullin, *Biochem. J.*, 1999, **137**, 211.
31. D. A. Erlanson, R. S. McDowell, M. M. He, M. Randal, R. L. Simmons, J. Kung, A. Waight and S. K. Hansen, *J. Am. Chem. Soc.*, 2003, **125**, 5602.
32. D. A. Erlanson, J. W. Lam, C. Wiesmann, T. N. Luong, R. L. Simmons, W. L. DeLano, I. C. Choong, M. T. Burdett, W. M. Flanagan, D. Lee, E. M. Gordon and T. O'Brien, *Nat. Biotechnol.*, 2003, **21**, 308.
33. S. B. Shuker, P. J. Hajduk, R. P. Meadows and S. W. Fesik, *Science*, 1996, **274**, 153.
34. A. C. Braisted, J. D. Oslob, W. L. DeLano, J. Hyde, R. S. McDowell, N. Waal, C. Yu, M. R. Arkin and B. C. Raimundo, *J. Am. Chem. Soc.*, 2003, **125**, 3714.
35. H. J. Boehm, M. Boehringer, D. Bur, H. Gmuender, W. Huber, W. Klaus, D. Kostrewa, H. Kuehne, T. Luebbers, N. Meunier-Keller and F. Mueller, *J. Med. Chem.*, 2000, **43**, 2664.
36. D. Gonzalez Ruiz and H. Gohlke, *Curr. Med. Chem.*, 2006, **13**, 2607.
37. F. K. Pettit and J. U. Bowie, *J. Mol. Biol.*, 1999, **285**, 1377.
38. O. Keskin, A. Gursoy, B. Ma and R. Nussinov, *Chem. Rev.*, 2008, **108**, 1225.
39. L. L. Conte, C. Chothia and J. Janin, *J. Mol. Biol.*, 1999, **285**, 2177.
40. J. Janin and C. Chothia, *J. Biol. Chem.*, 1990, **265**, 16027.
41. D. R. Davies, E. A. Padlan and S. Sheriff, *Annu. Rev. Biochem.*, 1990, **59**, 439.
42. P. Chene, *ChemMedChem*, 2006, **1**, 400.
43. P. Chakrabarti and J. Janin, *Proteins*, 2002, **47**, 334.
44. M. W. Peczuh and A. D. Hamilton, *Chem. Rev.*, 2000, **100**, 2479.
45. S. Jones and J. M. Thornton, *Proc. Natl. Acad. Sci. U. S. A.*, 1996, **93**, 13.
46. F. Glaser, D. M. Steinberg, I. A. Vakser and N. Ben-Tal, *Proteins*, 2001, **43**, 89.
47. E. A. Padlan, *Proteins*, 1990, **7**, 112.
48. A. J. McCoy, V. Chandana Epa and P. M. Colman, *J. Mol. Biol.*, 1997, **268**, 570.
49. B. A. Joughin, D. F. Green and B. Tidor, *Protein Sci.*, 2005, **14**, 1363.
50. G. Schreiber and A. R. Fersht, *J. Mol. Biol.*, 1995, **248**, 478.
51. M. R. Arkin and J. A. Wells, *J. Mol. Biol.*, 1998, **284**, 1083.
52. G. A. Papoian, J. Ulander and P. G. Wolynes, *J. Am. Chem. Soc.*, 2003, **125**, 9170.
53. F. Rodier, R. P. Bahadur, P. Chakrabarti and J. Janin, *Proteins*, 2005, **60**, 36.
54. E. J. Sundberg and R. A. Mariuzza, *Structure*, 2000, **8**, R137.

55. W. L. DeLano, M. H. Ultsch, A. M. de Vos and J. A. Wells, *Science*, 2000, **287**, 1279.
56. I. Luque and E. Freire, *Proteins*, 2000, **4**, 63.
57. S. J. Teague, *Nat. Rev. Drug Discovery*, 2003, **2**, 527.
58. C.-S. Goh, D. Milburn and M. Gerstein, *Curr. Opin. Struct. Biol.*, 2004, **14**, 104.
59. B. Ma, H. J. Wolfson and R. Nussinov, *Curr. Opin. Struct. Biol.*, 2001, **11**, 364.
60. C.-J. Tsai, B. Ma and R. Nussinov, *Proc. Natl. Acad. Sci. U. S. A.*, 1999, **96**, 9970.
61. J. R. Horn and B. K. Shoichet, *J. Mol. Biol.*, 2004, **336**, 1283.
62. V. A. Feher and J. Cavanagh, *Nature*, 1999, **400**, 289.
63. L. E. Kay, D. R. Muhandiram, N. A. Farrow, Y. Aubin and J. D. Forman-Kay, *Biochemistry*, 1996, **35**, 361.
64. B. Ma, M. Shatsky, H. J. Wolfson and R. Nussinov, *Protein Sci.*, 2002, **11**, 184.
65. W. E. Stites, *Chem. Rev.*, 1997, **97**, 1233.
66. N. Brooijmans, K. A. Sharp and I. D. Kuntz, *Proteins*, 2002, **48**, 645.
67. B. C. Cunningham and J. A. Wells, *J. Mol. Biol.*, 1993, **234**, 554.
68. L. Jin and J. A. Wells, *Protein Sci.*, 1994, **3**, 2351.
69. T. Clackson and J. A. Wells, *Science*, 1995, **267**, 383.
70. T. Clackson, M. H. Ultsch, J. A. Wells and A. M. de Vos, *J. Mol. Biol.*, 1998, **277**, 1111.
71. A. A. Bogan and K. S. Thorn, *J. Mol. Biol.*, 1998, **280**, 1.
72. Z. Hu, B. Ma, H. Wolfson and R. Nussinov, *Proteins*, 2000, **39**, 331.
73. J. Novotny, R. E. Bruccoleri and F. A. Saul, *Biochemistry*, 1989, **28**, 4735.
74. J. L. Kouadio, J. R. Horn, G. Pal and A. A. Kossiakoff, *J. Biol. Chem.*, 2005, **280**, 25524.
75. I. S. Moreira, P. A. Fernandes and M. J. Ramos, *Proteins*, 2007, **68**, 803.
76. B. Ma, T. Elkayam, H. Wolfson and R. Nussinov, *Proc. Natl. Acad. Sci. U. S. A.*, 2003, **100**, 5772.
77. X. Li, O. Keskin, B. Ma, R. Nussinov and J. Liang, *J. Mol. Biol.*, 2004, **344**, 781.
78. D. E. Otzen and A. R. Fersht, *Protein Eng.*, 1999, **12**, 41.
79. S. W. Lockless and R. Ranganathan, *Science*, 1999, **286**, 295.
80. O. Keskin, B. Ma and R. Nussinov, *J. Mol. Biol.*, 2005, **345**, 1281.
81. A. W. White, A. D. Westwell and G. Brahemi, *Expert Rev. Mol. Med.*, 2008, **10**, e8.
82. S. S. Sidhu, W. J. Fairbrother and K. Deshayes, *ChemBioChem*, 2003, **4**, 14.
83. R. C. Pillutla, K. C. Hsiao, J. R. Beasley, J. Brandt, S. Ostergaard, P. H. Hansen, J. C. Spetzler, G. M. Danielsen, A. S. Andersen, R. E. Brissette, M. Lennick, P. W. Fletcher, A. J. Blume, L. Schaffer and N. I. Goldstein, *J. Biol. Chem.*, 2002, **277**, 22590.

84. N. C. Wrighton, F. X. Farrell, R. Chang, A. K. Kashyap, F. P. Barbone, L. S. Mulcahy, D. L. Johnson, R. W. Barrett, L. K. Jolliffe and W. J. Dower, *Science*, 1996, **273**, 458.

85. S. E. Cwirla, P. Balasubramanian, D. J. Duffin, C. R. Wagstrom, C. M. Gates, S. C. Singer, A. M. Davis, R. L. Tansik, L. C. Mattheakis, C. M. Boytos, P. J. Schatz, D. P. Baccanari, N. C. Wrighton, R. W. Barrett and W. J. Dower, *Science*, 1997, **276**, 1696.

86. W. L. DeLano, *Curr. Opin. Struct. Biol.*, 2002, **12**, 14.

87. P. L. Toogood, *J. Med. Chem.*, 2002, **45**, 1543.

88. D. A. Erlanson, R. S. McDowell and T. O'Brien, *J. Med. Chem.*, 2004, **47**, 3463.

89. Y. Li, Y. Huang, C. P. Swaminathan, S. J. Smith-Gill and R. A. Mariuzza, *Structure*, 2005, **13**, 297.

90. K. S. Thorn and A. A. Bogan, *Bioinformatics*, 2001, **17**, 284.

91. T. B. Fischer, K. V. Arunachalam, D. Bailey, V. Mangual, S. Bakhru, R. Russo, D. Huang, M. Paczkowski, V. Lalchandani, C. Ramachandra, B. Ellison, S. Galer, J. Shapley, E. Fuentes and J. Tsai, *Bioinformatics*, 2003, **19**, 1453.

92. N. Tuncbag, G. Kar, O. Keskin, A. Gursoy and R. Nussinov, *Brief Bioinform.*, 2009, **10**, 217.

93. E. Guney, N. Tuncbag, O. Keskin and A. Gursoy, *Nucleic Acids Res.*, 2008, **36**, D662.

94. J. Srinivasan, T. E. Cheatham, III, P. Cieplak, P. A. Kollman and D. A. Case, *J. Am. Chem. Soc.*, 1998, **120**, 9401.

95. I. Massova and P. A. Kollman, *J. Am. Chem. Soc.*, 1999, **121**, 8133.

96. S. Huo, I. Massova and P. A. Kollman, *J. Comput. Chem.*, 2002, **23**, 15.

97. W. Wang and P. A. Kollman, *J. Mol. Biol.*, 2000, **303**, 567.

98. I. S. Moreira, P. A. Fernandes and M. J. Ramos, *J. Comput. Chem.*, 2007, **28**, 644.

99. I. S. Moreira, P. A. Fernandes and M. J. Ramos, *J. Phys. Chem. B*, 2006, **110**, 10962.

100. A. Benedix, C. M. Becker, B. L. de Groot, A. Caflisch and R. A. Bockmann, *Nat. Methods*, 2009, **6**, 3.

101. B. L. de Groot, D. M. van Aalten, R. M. Scheek, A. Amadei, G. Vriend and H. J. Berendsen, *Proteins*, 1997, **29**, 240.

102. D. M. Kruger and H. Gohlke, *Nucleic Acids Res.*, 2010, **38**, W480.

103. C. K. Vaughan, A. M. Buckle and A. R. Fersht, *J. Mol. Biol.*, 1999, **266**, 1487.

104. H. Gohlke, C. Kiel and D. A. Case, *J. Mol. Biol.*, 2003, **330**, 891.

105. G. Archontis, T. Simonson and M. Karplus, *J. Mol. Biol.*, 2001, **306**, 307.

106. J. Gao, K. Kuczera, B. Tidor and M. Karplus, *Science*, 1989, **244**, 1069.

107. Z. S. Hendsch and B. Tidor, *Protein Sci.*, 1999, **8**, 1381.

108. A. E. Mark and W. F. van Gunsteren, *J. Mol. Biol.*, 1994, **240**, 167.

109. G. P. Brady and K. A. Sharp, *J. Mol. Biol.*, 1995, **254**, 77.

110. S. Boresch and M. Karplus, *J. Mol. Biol.*, 1995, **254**, 801.

111. H. Gohlke and D. A. Case, *J. Comput. Chem.*, 2004, **25**, 238.

112. H. J. Böhm, *J. Comput. Aided Mol. Des.*, 1994, **8**, 243.
113. R. Guerois, J. E. Nielsen and L. Serrano, *J. Mol. Biol.*, 2002, **320**, 369.
114. J. Schymkowitz, J. Borg, F. Stricher, R. Nys, F. Rousseau and L. Serrano, *Nucleic Acids Res.*, 2005, **33**, W382.
115. T. Kortemme, D. E. Kim and D. Baker, *Sci. STKE*, 2004, 2004, pl2.
116. T. Kortemme and D. Baker, *Proc. Natl. Acad. Sci. U. S. A.*, 2002, **99**, 14116.
117. T. Kortemme, L. A. Joachimiak, A. N. Bullock, A. D. Schuler, B. L. Stoddard and D. Baker, *Nat. Struct. Biol.*, 2004, **11**, 371.
118. C. Kiel, S. Wohlgemuth, F. Rousseau, J. Schymkowitz, J. Ferkinghoff-Borg, F. Wittinghofer and L. Serrano, *J. Mol. Biol.*, 2005, **348**, 759.
119. T. Kortemme, A. V. Morozov and D. Baker, *J. Mol. Biol.*, 2003, **326**, 1239.
120. D. Bashford and D. A. Case, *Annu. Rev. Phys. Chem.*, 2000, **51**, 129.
121. N. Tuncbag, A. Gursoy and O. Keskin, *Bioinformatics*, 2009, **25**, 1513.
122. N. Tuncbag, O. Keskin and A. Gursoy, *Nucleic Acids Res.*, 2010, **38**, W402.
123. S. J. Darnell, L. LeGault and J. C. Mitchell, *Nucleic Acids Res.*, 2008, **36**, W265.
124. S. J. Darnell, D. Page and J. C. Mitchell, *Proteins*, 2007, **68**, 813.
125. D. G. Levitt and L. J. Banaszak, *J. Mol. Graphics*, 1992, **10**, 229.
126. M. Hendlich, F. Rippmann and G. Barnickel, *J. Mol. Graphics Model.*, 1997, **15**, 359.
127. B. Huang and M. Schroeder, *BMC Struct. Biol.*, 2006, **6**, 19.
128. R. A. Laskowski, *J. Mol. Graphics*, 1995, **13**, 323.
129. J. Liang, H. Edelsbrunner and C. Woodward, *Protein Sci.*, 1998, **7**, 1884.
130. G. P. Brady, Jr. and P. F. Stouten, *J. Comput. Aided Mol. Des.*, 2000, **14**, 383.
131. M. Weisel, E. Proschak and G. Schneider, *Chem. Cent. J.*, 2007, **1**, 7.
132. V. Le Guilloux, P. Schmidtke and P. Tuffery, *BMC Bioinf.*, 2009, **10**, 168.
133. S. Eyrisch and V. Helms, *J. Med. Chem.*, 2007, **50**, 3457.
134. S. Leis, S. Schneider and M. Zacharias, *Curr. Med. Chem.*, 2010, **17**, 1550.
135. Y. Levy, S. S. Cho, J. N. Onuchic and P. G. Wolynes, *J. Mol. Biol.*, 2005, **346**, 1121.
136. E. Fischer, *Ber. Dtsch. Chem. Ges.*, 1894, **27**, 2985.
137. D. E. Koshland, *Science*, 1967, **156**, 540.
138. C.-J. Tsai, S. Kumar, B. Ma and R. Nussinov, *Protein Sci.*, 1999, **8**, 1181.
139. B. F. Volkman, D. Lipson, D. E. Wemmer and D. Kern, *Science*, 2001, **291**, 2429.
140. S. R. Kimura, R. C. Brower, S. Vajda and C. J. Camacho, *Biophys. J.*, 2001, **80**, 635.
141. D. Rajamani, S. Thiel, S. Vajda and C. J. Camacho, *Proc. Natl. Acad. Sci. U. S. A.*, 2004, **101**, 11287.
142. G. R. Smith, M. J. E. Sternberg and P. A. Bates, *J. Mol. Biol.*, 2005, **347**, 1077.
143. C. A. Sotriffer, O. Krämer and G. Klebe, *Proteins*, 2004, **56**, 52.

144. D. A. Case, *Curr. Opin. Struct. Biol.*, 1994, **4**, 285.

145. S. Hayward and B. L. de Groot, *Methods Mol. Biol.*, 2008, **443**, 89.

146. L. Skjaerven, S. M. Hollup and N. Reuter, *J. Mol. Struct.: THEOCHEM*, 2009, **898**, 42.

147. I. Bahar, A. R. Atilgan and B. Erman, *Fold. Des.*, 1997, **2**, 173.

148. F. Tama, F. X. Gadea, O. Marques and Y.-H. Sanejouand, *Proteins*, 2000, **41**, 1.

149. I. Bahar and A. J. Rader, *Curr. Opin. Struct. Biol.*, 2005, **15**, 586.

150. J. Ma, *Structure*, 2005, **13**, 373.

151. W. Zheng and S. Doniach, *Proc. Natl. Acad. Sci. U. S. A.*, 2003, **100**, 13253.

152. A. Ahmed and H. Gohlke, *Proteins*, 2006, **63**, 1038.

153. F. Tama and Y. H. Sanejouand, *Protein Eng.*, 2001, **14**, 1.

154. M. Zacharias and H. Sklenar, *J. Comput. Chem.*, 1999, **20**, 287.

155. C. N. Cavasotto and R. A. Abagyan, *J. Mol. Biol.*, 2004, **337**, 209.

156. O. Livnah, E. A. Stura, S. A. Middleton, D. L. Johnson, L. K. Jolliffe and I. A. Wilson, *Science*, 1999, **283**, 987.

157. S. Atwell, M. Ultsch, A. M. De Vos and J. A. Wells, *Science*, 1997, **278**, 1125.

158. Z. Zhang, Y. Shi and H. Liu, *Biophys. J.*, 2003, **84**, 3583.

159. M. Lei, M. I. Zavodszky, L. A. Kuhn and M. F. Thorpe, *J. Comput. Chem.*, 2004, **25**, 1133.

160. M. I. Zavodszky, M. Lei, M. F. Thorpe, A. R. Day and L. A. Kuhn, *Proteins*, 2004, **57**, 243.

161. S. Wells, S. Menor, B. M. Hespenheide and M. F. Thorpe, *Phys. Biol.*, 2005, **2**, 1.

162. D. J. Jacobs and M. F. Thorpe, *Phys. Rev. Lett.*, 1995, **75**, 4051.

163. *Rigidity Theory and Applications*, ed. M. F. Thorpe and P. M. Duxbury, Kluwer/Plenum, New York, 1999.

164. D. J. Jacobs, A. J. Rader, L. A. Kuhn and M. F. Thorpe, *Proteins*, 2001, **44**, 150.

165. H. Gohlke, L. A. Kuhn and D. A. Case, *Proteins*, 2004, **56**, 322.

166. D. Seeliger, J. Haas and B. L. de Groot, *Structure*, 2007, **15**, 1482.

167. S. Eyrisch and V. Helms, *J. Comput. Aided Mol. Des.*, 2009, **23**, 73.

168. R. A. Laskowski, F. Gerick and J. M. Thornton, *FEBS Lett.*, 2009, **583**, 1692.

169. J.-P. Changeux and S. J. Edelstein, *Science*, 2005, **308**, 1424.

170. D. Kern and E. R. Zuiderweg, *Curr. Opin. Struct. Biol.*, 2003, **13**, 748.

171. F. Rousseau and J. Schymkowitz, *Curr. Opin. Struct. Biol.*, 2005, **15**, 23.

172. K. Gunasekaran, B. Ma and R. Nussinov, *Proteins*, 2004, **57**, 433.

173. J. A. Hardy and J. A. Wells, *Curr. Opin. Struct. Biol.*, 2004, **14**, 706.

174. N. G. Oikonomakos, V. T. Skamnaki, K. E. Tsitsanou, N. G. Gavalas and L. N. Johnson, *Structure*, 2000, **8**, 575.

175. C. Wiesmann, K. J. Barr, J. Kung, J. Zhu, D. A. Erlanson, W. Shen, B. J. Fahr, M. Zhong, L. Taylor, M. Randal, R. S. McDowell and S. K. Hansen, *Nat. Struct. Mol. Biol.*, 2004, **11**, 730.

176. L. A. Kohlstaedt, J. Wang, J. M. Friedman, P. A. Rice and T. A. Steitz, *Science*, 1992, **256**, 1783.
177. J. A. Hardy, J. Lam, J. T. Nguyen, T. O'Brien and J. A. Wells, *Proc. Natl. Acad. Sci. U. S. A.*, 2004, **101**, 12461.
178. O. Lichtarge, H. R. Bourne and F. E. Cohen, *J. Mol. Biol.*, 1996, **257**, 342.
179. O. Lichtarge and M. E. Sowa, *Curr. Opin. Struct. Biol.*, 2002, **12**, 21.
180. M. E. Sowa, W. He, T. G. Wensel and O. Lichtarge, *Proc. Natl. Acad. Sci. U. S. A.*, 2000, **97**, 1483.
181. M. E. Sowa, W. He, K. C. Slep, M. A. Kercher, O. Lichtarge and T. G. Wensel, *Nat. Struct. Biol.*, 2001, **8**, 234.
182. K. C. Slep, M. A. Kercher, W. He, C. W. Cowan, T. G. Wensel and P. B. Sigler, *Nature*, 2001, **409**, 1071.
183. G. M. Suel, S. W. Lockless, M. A. Wall and R. Ranganathan, *Nat. Struct. Biol.*, 2003, **10**, 59.
184. M. E. Hatley, S. W. Lockless, S. K. Gibson, A. G. Gilman and R. Ranganathan, *Proc. Natl. Acad. Sci. U. S. A.*, 2003, **100**, 14445.
185. A. I. Shulman, C. Larson, D. J. Mangelsdorf and R. Ranganathan, *Cell*, 2004, **116**, 417.
186. V. J. Hilser and E. Freire, *J. Mol. Biol.*, 1996, **262**, 756.
187. E. Freire, *Proc. Natl. Acad. Sci. U. S. A.*, 1999, **96**, 10118.
188. E. Freire, *Proc. Natl. Acad. Sci. U. S. A.*, 2000, **97**, 11680.
189. H. Pan, J. C. Lee and V. J. Hilser, *Proc. Natl. Acad. Sci. U. S. A.*, 2000, **97**, 12020.
190. A. P. Kuzin, M. Nukaga, Y. Nukaga, A. M. Hujer, R. A. Bonomo and J. R. Knox, *Biochemistry*, 1999, **38**, 5720.
191. S. Robineau, M. Chabre and B. Antonny, *Proc. Natl. Acad. Sci. U. S. A.*, 2000, **97**, 9913.
192. L. Renault, B. Guibert and J. Cherfils, *Nature*, 2003, **426**, 525.
193. Y. Pommier and J. Cherfils, *Trends Pharmacol. Sci.*, 2005, **26**, 138.
194. B. K. Shoichet, *Nature*, 2004, **432**, 862.
195. P. Ferrara, H. Gohlke, D. J. Price, G. Klebe and C. L. Brooks, *J. Med. Chem.*, 2004, **47**, 3032.
196. W. L. Jorgensen, *Science*, 2004, **303**, 1813.
197. B. Honig and A. Nicholls, *Science*, 1995, **268**, 1144.
198. M. K. Gilson, K. Sharp, B. Honig, R. Fine and R. Hagstrom, *Biophys. J.*, 1987, **51**, A234.
199. N. Arora and D. Bashford, *Proteins*, 2001, **43**, 12.
200. N. Budin, N. Majeux and A. Caflish, *Biol. Chem.*, 2001, **382**, 1365.
201. M. Schaefer and M. Karplus, *J. Phys. Chem.*, 1996, **100**, 1578.
202. J. An, M. Totrov and R. Abagyan, *Genome Inform.*, 2004, **15**, 31.
203. J. Momand, G. P. Zambetti, D. C. Olson, D. George and A. J. Levine, *Cell*, 1992, **69**, 1237.
204. W. C. Still, A. Tempczyk, R. C. Hawley and T. Hendrickson, *J. Am. Chem. Soc.*, 1990, **112**, 6127.

205. M. Feig, A. Onufriev, M. S. Lee, W. Im, D. A. Case and C. L. Brooks, III, *J. Comput. Chem.*, 2004, **25**, 265.
206. D. Bashford and D. A. Case, *Annu. Rev. Phys. Chem.*, 2000, **51**, 129.
207. X. Zou, Y. Sun and I. D. Kuntz, *J. Am. Chem. Soc.*, 1999, **121**, 8033.
208. J. Krumrine, F. Raubacher, N. Brooijmans and I. D. Kuntz, *Methods Biochem. Anal.*, 2003, **44**, 443.
209. H. Y. Liu, I. D. Kuntz and X. Q. Zou, *J. Phys. Chem. B*, 2004, **108**, 5453.
210. G. D. Hawkins, C. J. Cramer and D. G. Truhlar, *Chem. Phys. Lett.*, 1995, **246**, 122.
211. J. C. Bressi, C. L. Verlinde, A. M. Aronov, M. L. Shaw, S. S. Shin, L. N. Nguyen, S. Suresh, F. S. Buckner, W. C. Van Voorhis, I. D. Kuntz, W. G. Hol and M. H. Gelb, *J. Med. Chem.*, 2001, **44**, 2080.
212. M. Bogyo, S. Verhelst, V. Bellingard-Dubouchaud, S. Toba and D. Greenbaum, *Chem. Biol.*, 2000, **7**, 27.
213. N. Fujii, J. J. Haresco, K. A. P. Novak, D. Stokoe, I. D. Kuntz and R. K. Guy, *J. Am. Chem. Soc.*, 2003, **125**, 12074.
214. I. W. Davis and D. Baker, *J. Mol. Biol.*, 2009, **385**, 381.
215. A. Morreale, R. Gil-Redondo and A. R. Ortiz, *Proteins*, 2007, **67**, 606.
216. D. A. Case, T. E. Cheatham, III, T. Darden, H. Gohlke, R. Luo, K. M. Merz, Jr., A. Onufriev, C. Simmerling, B. Wang and R. J. Woods, *J. Comput. Chem.*, 2005, **26**, 1668.
217. P. A. Kollman, I. Massova, C. Reyes, B. Kuhn, S. H. Huo, L. Chong, M. Lee, T. Lee, Y. Duan, W. Wang, O. Donini, P. Cieplak, J. Srinivasan, D. A. Case and T. E. Cheatham, *Acc. Chem. Res.*, 2000, **33**, 889.
218. S. P. Brown and S. W. Muchmore, *J. Chem. Inf. Model.*, 2007, **47**, 1493.
219. S. P. Brown and S. W. Muchmore, *J. Med. Chem.*, 2009, **52**, 3159.
220. S. McGovern and B. Shoichet, *J. Med. Chem.*, 2003, **46**, 2895.
221. B. Q. Wei, L. H. Weaver, A. M. Ferrari, B. W. Matthews and B. K. Shoichet, *J. Mol. Biol.*, 2004, **337**, 1161.
222. C. B. Rao, J. Subramanian and S. D. Sharma, *Drug Discovery Today*, 2009, **14**, 394.
223. D. Barnard, B. Diaz, L. Hettich, E. Chuang, X. F. Zhang, J. Avruch and M. Marshall, *Oncogene*, 1995, **10**, 1283.
224. M. I. Zavodszky and L. A. Kuhn, *Protein Sci.*, 2005, **14**, 1104.
225. V. Schnecke, C. Swanson, E. Getzoff, J. Tainer and L. A. Kuhn, *Proteins*, 1998, **33**, 74.
226. R. A. Laskowski, M. W. MacArthur, D. S. Moss and J. M. Thornton, *J. Appl. Crystallogr.*, 1993, **26**, 283.
227. R. Najmanovich, J. Kuttner, V. Sobolev and M. Edelman, *Proteins*, 2000, **39**, 261.
228. M. J. Betts and M. J. E. Sternberg, *Protein Eng.*, 1999, **12**, 271.
229. C. S. Goh, D. Milburn and M. Gerstein, *Curr. Opin. Struct. Biol.*, 2004, **14**, 104.
230. M. Zacharias, *Proteins*, 2004, **54**, 759.
231. R. Tatsumi, Y. Fukunishi and H. Nakamura, *J. Comput. Chem.*, 2004, **25**, 1995.

232. A. Ahmed and H. Gohlke, presented at the 1st International Conference on Computational & Mathematical Biomedical Engineering (CMBE09), Swansea, 2009.
233. A. N. Koller, H. Schwalbe and H. Gohlke, *Biophys. J.*, 2008, **95**, L04.
234. B. L. de Groot, G. Vriend and H. J. Berendsen, *J. Mol. Biol.*, 1999, **286**, 1241.
235. S. F. Sousa, P. A. Fernandes and M. J. Ramos, *Proteins*, 2006, **65**, 15.
236. M. Totrov and R. Abagyan, *Curr. Opin. Struct. Biol.*, 2008, **18**, 178.
237. H. A. Carlson, *Curr. Opin. Chem. Biol.*, 2002, **6**, 447.
238. T. Y. Lee, V. D. Le, D. Y. Lim, Y. C. Lin, G. M. Morris, A. L. Wong, A. J. Olson, J. H. Elder and C. H. Wong, *J. Am. Chem. Soc.*, 1999, **121**, 1145.
239. G. M. Morris, R. Huey, W. Lindstrom, M. F. Sanner, R. K. Belew, D. S. Goodsell and A. J. Olson, *J. Comput. Chem.*, 2009, **30**, 2785.
240. F. Österberg, G. M. Morris, M. F. Sanner, A. J. Olson and D. S. Goodsell, *Proteins*, 2002, **46**, 34.
241. G. M. Morris, R. Huey, W. Lindstrom, M. F. Sanner, R. K. Belew, D. S. Goodsell and A. J. Olson, *J. Comput. Chem.*, 2009, **30**, 2785.
242. Y. Zhao and M. F. Sanner, *Proteins*, 2007, **68**, 726.
243. Y. Zhao, D. Stoffler and M. Sanner, *Bioinformatics*, 2006, **22**, 2768.
244. H. Claussen, C. Buning, M. Rarey and T. Lengauer, *J. Mol. Biol.*, 2001, **308**, 377.
245. D. M. Lorber and B. K. Shoichet, *Protein Sci.*, 1998, **7**, 938.
246. X. Barril, C. Aleman, M. Orozco and F. J. Luque, *Proteins*, 1998, **32**, 67.
247. R. Mendez, R. Leplae, M. F. Lensink and S. J. Wodak, *Proteins*, 2005, **60**, 150.
248. X. Fradera and J. Mestres, *Curr. Top. Med. Chem.*, 2004, **4**, 687.
249. J. M. Jansen and E. J. Martin, *Curr. Opin. Chem. Biol.*, 2004, **8**, 359.
250. S. Radestock, M. Bohm and H. Gohlke, *J. Med. Chem.*, 2005, **48**, 5466.
251. K. Brejc, W. J. van Dijk, R. V. Klaasen, M. Schuurmans, J. van der Oost, A. B. Smit and T. K. Sixma, *Nature*, 2001, **411**, 269.
252. R. Abseher and M. Nilges, *J. Mol. Biol.*, 1998, **279**, 911.
253. J. H. Chen, W. Im and C. L. Brooks, *J. Am. Chem. Soc.*, 2004, **126**, 16038.
254. A. P. R. Zabell and C. B. Post, *Proteins*, 2002, **46**, 295.
255. A. Dobrodumov and A. M. Gronenborn, *Proteins*, 2003, **53**, 18.
256. T. Carlomagno, *Annu. Rev. Biophys. Biomol.*, 2005, **34**, 245.
257. G. Martorell, M. J. Gradwell, B. Birdsall, C. J. Bauer, T. A. Frenkiel, H. T. Cheung, V. I. Polshakov, L. Kuyper and J. Feeney, *Biochemistry*, 1994, **33**, 12416.
258. H. Waldmann, I. M. Karaguni, M. Carpintero, E. Gourzoulidou, C. Herrmann, C. Brockmann, H. Oschkinat and O. Muller, *Angew. Chem., Int. Ed.*, 2004, **43**, 454.
259. C. Dominguez, R. Boelens and A. M. J. J. Bonvin, *J. Am. Chem. Soc.*, 2003, **125**, 1731.
260. S. J. de Vries, A. D. van Dijk, M. Krzeminski, M. van Dijk, A. Thureau, V. Hsu, T. Wassenaar and A. M. Bonvin, *Proteins*, 2007, **69**, 726.
261. M. Nilges and S. I. O'Donoghue, *Prog. Nucl. Magn. Reson. Spectrosc.*, 1998, **32**, 107.

262. U. Schieborr, M. Vogtherr, B. Elshorst, M. Betz, S. Grimme, B. Pescatore, T. Langer, K. Saxena and H. Schwalbe, *ChemBioChem*, 2005, **6**, 1891.

263. G. M. Clore and A. M. Gronenborn, *Proc. Natl. Acad. Sci. U. S. A.*, 1998, **95**, 5891.

264. J. Kuszewski, A. M. Gronenborn and G. M. Clore, *J. Magn. Reson., Ser. B*, 1995, **107**, 293.

265. K. Osapay, Y. Theriault, P. E. Wright and D. A. Case, *J. Mol. Biol.*, 1994, **244**, 183.

266. D. S. Wishart and D. A. Case, *Methods Enzymol.*, 2001, **338**, 3.

267. M. A. McCoy and D. F. Wyss, *J. Biomol. NMR*, 2000, **18**, 189.

268. K. Osapay and D. A. Case, *J. Am. Chem. Soc.*, 1991, **113**, 9436.

269. D. Gonzalez-Ruiz and H. Gohlke, *J. Chem. Inf. Model.*, 2009, **49**, 2260.

270. H. Gohlke, M. Hendlich and G. Klebe, *J. Mol. Biol.*, 2000, **295**, 337.

271. M. G. Kendall, *Biometrika*, 1938, **30**, 81.

272. B. Wang, K. Raha and K. M. Merz, Jr., *J. Am. Chem. Soc.*, 2004, **126**, 11430.

273. T. Berg, *Angew. Chem., Int. Ed.*, 2003, **42**, 2462.

274. S. K. Sharma, T. M. Ramsey and K. W. Bair, *Curr. Med. Chem.: Anti-Cancer Agents*, 2002, **2**, 311.

275. A. L. Hopkins and C. R. Groom, *Nat. Rev. Drug Discovery*, 2002, **1**, 727.

276. N. C. Meisner, M. Hintersteiner, V. Uhl, T. Weidemann, M. Schmied, H. Gstach and M. Auer, *Curr. Opin. Chem. Biol.*, 2004, **8**, 424.

277. P. J. Hajduk, J. R. Huth and S. W. Fesik, *J. Med. Chem.*, 2005, **48**, 2518.

278. J. Seco, F. J. Luque and X. Barril, *J. Med. Chem.*, 2009, **52**, 2363.

279. R. P. Sheridan, V. N. Maiorov, M. K. Holloway, W. D. Cornell and Y. D. Gao, *J. Chem. Inf. Model.*, 2010, **50**, 2029.

280. C. A. Lipinski, F. Lombardo, B. W. Dominy and P. J. Feeney, *Adv. Drug Delivery Rev.*, 1997, **23**, 3.

281. D. E. Clark and S. D. Pickett, *Drug Discovery Today*, 2000, **5**, 49.

282. S. L. McGovern, E. Caselli, N. Grigorieff and B. K. Shoichet, *J. Med. Chem.*, 2002, **45**, 1712.

283. S. L. McGovern and B. K. Shoichet, *J. Med. Chem.*, 2003, **46**, 1478.

284. J. Seidler, S. L. McGovern, T. N. Doman and B. K. Shoichet, *J. Med. Chem.*, 2003, **46**, 4477.

285. S. L. McGovern, B. T. Helfand, B. Feng and B. K. Shoichet, *J. Med. Chem.*, 2003, **46**, 4265.

286. B. Y. Feng, A. Shelat, T. N. Doman, R. K. Guy and B. K. Shoichet, *Nat. Chem. Biol.*, 2005, **1**, 146.

287. J. J. Irwin and B. K. Shoichet, *J. Chem. Inf. Model.*, 2005, **45**, 177.

288. J. L. Wang, D. Liu, Z. J. Zhang, S. Shan, X. Han, S. M. Srinivasula, C. M. Croce, E. S. Alnemri and Z. Huang, *Proc. Natl. Acad. Sci. U. S. A.*, 2000, **97**, 7124.

289. I. J. Enyedy, Y. Ling, K. Nacro, Y. Tomita, X. Wu, Y. Cao, R. Guo, B. Li, X. Zhu, Y. Huang, Y. Q. Long, P. P. Roller, D. Yang and S. Wang, *J. Med. Chem.*, 2001, **44**, 4313.

290. A. A. Lugovskoy, A. I. Degterev, A. F. Fahmy, P. Zhou, J. D. Gross, J. Yuan and G. Wagner, *J. Am. Chem. Soc.*, 2002, **124**, 1234.

291. P. Mukherjee, P. Desai, Y. D. Zhou and M. Avery, *J. Chem. Inf. Model.*, 2010, **50**, 906.

292. A. M. Almerico, M. Tutone and A. Lauria, *J. Mol. Model.*, 2009, **15**, 349.

293. S. Li, J. Gao, T. Satoh, T. M. Friedman, A. E. Edling, U. Koch, S. Choksi, X. Han, R. Korngold and Z. Huang, *Proc. Natl. Acad. Sci. U. S. A.*, 1997, **94**, 73.

294. N. K. Koehler, C. Y. Yang, J. Varady, Y. Lu, X. W. Wu, M. Liu, D. Yin, M. Bartels, B. Y. Xu, P. P. Roller, Y. Q. Long, P. Li, M. Kattah, M. L. Cohn, K. Moran, E. Tilley, J. R. Richert and S. Wang, *J. Med. Chem.*, 2004, **47**, 4989.

295. Z. Nikolovska-Coleska, L. Xu, Z. Hu, Y. Tomita, P. Li, P. P. Roller, R. Wang, X. Fang, R. Guo, M. Zhang, M. E. Lippman, D. Yang and S. Wang, *J. Med. Chem.*, 2004, **47**, 2430.

296. Y. Gao, J. B. Dickerson, F. Guo, J. Zheng and Y. Zheng, *Proc. Natl. Acad. Sci. U. S. A.*, 2004, **101**, 7618.

297. N. Fujii, J. J. Haresco, K. A. Novak, D. Stokoe, I. D. Kuntz and R. K. Guy, *J. Am. Chem. Soc.*, 2003, **125**, 12074.

298. D. Grandy, J. Shan, X. Zhang, S. Rao, S. Akunuru, H. Li, Y. Zhang, I. Alpatov, X. A. Zhang, R. A. Lang, D. L. Shi and J. J. Zheng, *J. Biol. Chem.*, 2009, **284**, 16256.

299. A. K. Debnath, L. Radigan and S. Jiang, *J. Med. Chem.*, 1999, **42**, 3203.

300. A. L. Bowman, Z. Nikolovska-Coleska, H. Zhong, S. Wang and H. A. Carlson, *J. Am. Chem. Soc.*, 2007, **129**, 12809.

301. A. Metz, C. Pfleger, H. Kopitz, S. Pfeiffer-Marek, K.-H. Baringhaus and H. Gohlke, *J. Chem. Inf. Model.*, 2012, **52**, 120.

CHAPTER 14

Using Molecular Simulations and Metadynamics to Predict Binding Free Energies and Kinetics: the Case of Cox and CDK2

GIORGIO SALADINO AND FRANCESCO L. GERVASIO*

Computational Biophysics Group, Spanish National Cancer Research Centre (CNIO), calle Melchor Fernadéz Almagro 3, E-28029, Madrid, Spain
*E-mail: flgervasio@cnio.es

14.1 Introduction

Obtaining reliable estimates of the free energy profiles and kinetics constants associated with ligand binding is an ambitious goal of modern computational drug design, with significant potential benefits in lead discovery and lead optimization phases. While several structure-based methods have been developed for virtual screening purposes, they usually focus on pose prediction and quick screening purposes, often failing to predict the correct binding energy order of similar ligands if large conformational changes or buried water molecules are involved.[1,2] In these cases, computationally intensive methods, based on fully atomistic or implicit solvent Monte Carlo and molecular dynamics (MD) simulations, have been shown to be more useful. A typical example is that of the rational design of HIV integrase inhibitors,[3] where a secondary binding site, invisible in the X-ray structure, was found by MD simulations, suggesting a novel approach to design stronger inhibitors. By

RSC Drug Discovery Series No. 23
Physico-Chemical and Computational Approaches to Drug Discovery
Edited by F. Javier Luque and Xavier Barril
© The Royal Society of Chemistry 2012
Published by the Royal Society of Chemistry, www.rsc.org

modeling the full flexibility of the target and of the ligand along with a proper consideration of the solvent effect, these methods overcome most of the concerns of fast-docking methods. Unfortunately, they require much longer computations and often their predictive power is impaired by the impossibility to perform an exhaustive sampling of ligand binding and conformational free energy landscape. This central problem of molecular simulations has been recently systematically addressed by different approaches, ranging from algorithmic advances in MD codes,[4,5] the development of advanced sampling algorithms,[6–8] as well as the use of specialized hardware[9] and distributed computing platforms.[10] Among the advanced sampling techniques, meta-dynamics and its derivatives as parallel tempering metadynamics (PTmetaD),[11] bias-exchange metadynamics,[12] and path collective variables[13] (PCVs) have been shown to be efficient in studying both molecular recognition and large conformational changes.[14–18]

Metadynamics-based methods are able to reconstruct the free energy landscape much faster than non-accelerated MD simulations,[19] but still require very significant computer resources compared to standard docking algorithms. Their advantage is that they provide not only a good estimate of the binding free energy, but also quantitative information on the metastable minima and transition states, which can be used in the lead optimization phase to optimize the underlying ligand binding kinetics to achieve therapeutically safe and differentiated responses.[20]

Recently we have shown how the use of PCVs together with metadynamics allows the calculation of the free energy profile along the ligand binding coordinate very efficiently and can be used to estimate the kinetics.[12,21] PCV-enhanced metadynamics was able to correctly calculate the relative (and absolute) binding free energies ($\Delta\Delta G_{bind}$) and to reconstruct the full docking free energy profile, including the transition states and metastable minima for different systems.

In the following we will review the application of PCV-enhanced metadynamics to two pharmaceutically interesting cases: cyclin-dependent kinase 2 (CDK2) and COX-1 and COX-2 isoforms. After a brief description of the computational techniques employed, we will show how these techniques permitted the prediction of the binding free energy profile and kinetics and to explain the different residence time of a ligand that binds to both COX isoforms.[21,22]

14.2 Methods

14.2.1 Metadynamics

The metadynamics algorithm is based on biasing the normal evolution of an MD simulation by a history-dependent potential. The latter is constructed by adding every few thousand MD time-steps a repulsive Gaussian centered at the current position projected on a set of collective variables (CVs) that approximate the reaction coordinate.

Using a metaphor first introduced by Laio and Gervasio in,[8] metadynamics is able to escape local minima in the free-energy surface by filling it like a person being able to climb out of a deep pool by filling it with sand. The advantage with respect to similar methods[23] is that the time-dependent potential defined by the sum of the Gaussians deposited up to time t provides an unbiased estimate of the free energy in the region explored during the dynamics.[24]

As the sum of the added Gaussians iteratively compensates the underlying free energy, an MD biased with metadynamics tends to escape from free energy basins *via* the lowest saddle point, a property that turns out to be useful in undocking simulations (see below). The computational efficiency, the ease of coding and the recent availability of an open-source plug-in (PLUMED),[25] that works with most of the available MD codes, makes metadynamics a very flexible and widely used tool.[14,15,26–37]

However, being based on CVs, metadynamics requires the preliminary identification of a set of variables that describe the process of interest, something far from trivial in many cases. Typically, to study the free energy associated with the binding of a ligand, after setting up a normal MD run of the target and ligand system with any of several MD codes (Amber, Gromacs, NAMD, LAMMPS, ACEMD, *etc.*), one could take the distance between the center of mass of the ligand and the center of mass of the protein as the CVs.[14] However, as we shall see in the next section, such a simple choice of the CV can adversely affect the efficiency.

14.2.2   The Choice of the CVs

Similarly to other free energy methods, the efficiency of metadynamics is influenced by the choice of CVs. If the CVs are chosen sensibly, the system will quickly find its way over the lowest free energy saddle point and evolve over the next minimum. The simplest types of CV are geometry related, such as distances, angles and dihedrals formed by atoms or group of atoms. Still, choosing the right set of CVs can be difficult. Sometimes it is necessary to proceed by trial and error, attempting several runs with different combinations of variables and checking *a posteriori* if the free energy surfaces converge. Fortunately, the use of special CVs as the vectors of a principal component analysis of an MD trajectory,[38] or the use of PCVs or the combination of metadynamics with parallel tempering,[11] provide a systematic way to avoid trial-and-error attempts with simple combinations of geometric-based CVs.

14.2.3   Path Collective Variables

Often one has the knowledge of the states of interest of the chosen biological system (*e.g.* the ligand-bound and unbound form of a target). In these cases, it is possible to define a guess path in a configurational space from the initial state to the final one. With this path (even if it is a very rough approximation to the real reaction coordinate), two CVs can be defined that are able to describe the position of a point in configurational space relative to the pre-assigned path:[13]

$$s(x) = \lim_{\lambda \to \infty} \frac{\int_0^1 t e^{-\lambda \|\|S(x) - S(t)\|\|^2} dt}{\int_0^1 e^{-\lambda \|\|S(x) - S(t)\|\|^2} dt} \qquad (14.1)$$

$$z(x) = -\frac{1}{\lambda} \lim_{\lambda \to \infty} \int_0^1 e^{-\|S(x) - S(t)\|\|^2} dt \qquad (14.2)$$

where t parameterizes a path $S(t)$ in a high-dimensional CV space and indicates the distance in this space; for any microscopic configuration x, $s(x)$ and $z(x)$ measure the progression along the "guess" path and the distance from it.

In practical applications, the path is discretized with a discrete number of frames $S(l)$, with l varying from 1 to P, so that $S(1) = S_A$ and $S(P) = S_B$, and eqns (14.1) and (14.2) are approximated by finite sums over l. The distance $\|...\|$ in eqns (14.1) and (14.2) can be defined in different ways. It can be the rmsd between the two structures after they are optimally aligned[39] or the difference between the current contact map matrix $S_C(R)$ and the pre-defined contact maps along the path, where $S_C(R)$ is defined as:

$$Sc(R) = \sum_{i,j} \frac{1 - \left(\frac{r_{ij}}{r_0}\right)^n}{1 - \left(\frac{r_{ij}}{r_0}\right)^m} \qquad (14.3)$$

where the sums on i and j run on two sets of atoms, r_{ij} is the distance between the i-th and j-th C_α atoms of the protein backbone, n and m are set to 6 and 10, respectively, and the cutoff distance r_0 is taken to be $r_0 = 8.5$ Å.

The squared distance $\|...\|^2$ between a generic state R and a point $S_C(l)$ along the path is measured in this case as:

$$\|Sc(R) - Sc(l)\|^2 = \sum_{j > i} \left(C_{ij}(R) - C_{ij}(l)\right)^2 \qquad (14.4)$$

where nearest neighbors are excluded from the sum.

While the initial guess on the path can be refined at will,[15] eventually finding a rigorous parameterization of the path, this is usually unnecessary. The best strategy in the case of ligand binding is to pull the ligand out of the target and build a first path, taking equally spaced frames along the pulling trajectory, and then perform subsequent pulls on the path and take the path requiring the least work as the definition of the PCV. Moreover, using the distance from the path $z(R)$ as a biased variable in metadynamics allows us to explore reaction pathways that are far from the initial guess, eventually finding reaction pathways that are significantly different[13] from the guess path. A variant of this approach has been used in the case of CDK2 to obtain an optimal binding reaction coordinate and the free energy profile along it. Its use minimizes human intervention on the choice of CV and drastically decreases the computational resources needed to calculate the binding–unbinding free energy (see Section 14.3.1).

14.2.4 Optimal Collective Variables for Ligand Binding

The need to fully explore the free energy basin of the ligand in the bulk solution makes the use of geometry-based CVs from computationally demanding to outright unfeasible, depending on which approach is used to calculate the free energy corresponding to the "external area". The different possible approaches are presented in Figure 14.1.

In Figure 14.1, the free energy volume that must be filled with bias by a metadynamics simulation using r and θ as CVs is shown in diagram 1. After fully exploring the internal cavity A and filling it with bias, metadynamics must completely fill the "outside" space B, which in this case is delimited by restraining the ligand within a conical area, and return back to the bound state. This approach is very inefficient due to the long time necessary to fill B and to the multiple recrossing of the narrow gate G needed to reach convergence. Moreover, the free energy difference between A and B depends on the volume accessible to the drug in the area B.[40,41] To obtain the absolute binding free energy it must be reweighted according to the standard volume. In a recent paper describing the use of bias-exchange metadynamics, the authors were able to reconstruct the free energy surface associated with docking and undocking of a small peptide to the HIV-protease.[12] Using this approach, notwithstanding the speed-up due to the bias-exchange procedure, the computational resources that were needed to complete the task are out of the reach of a typical pharmaceutical work-flow, limiting its applicability to selected cases.

A different approach was proposed by Masetti *et al.*[16] (Figure 14.1, diagram 2), where only the internal area of the cavity is explored by metadynamics. This approach, called "coarse metadynamics", is not rigorous because it neglects part of the difference in free energy of the unbound state of different ligands. Still, the relative binding free energy of similar inhibitors can be calculated by making the assumption that once the drugs reach the "outside area" their free energy is approximately the same.

A rigorous and efficient approach was introduced by Fidelak *et al.*[21] It is based on PCVs and it should be regarded as the method of choice in most common cases. It provides the free energy profile along the binding reaction coordinate with explicit solvent while drastically decreasing the computational resources needed to calculate the binding–unbinding free energy. The protocol

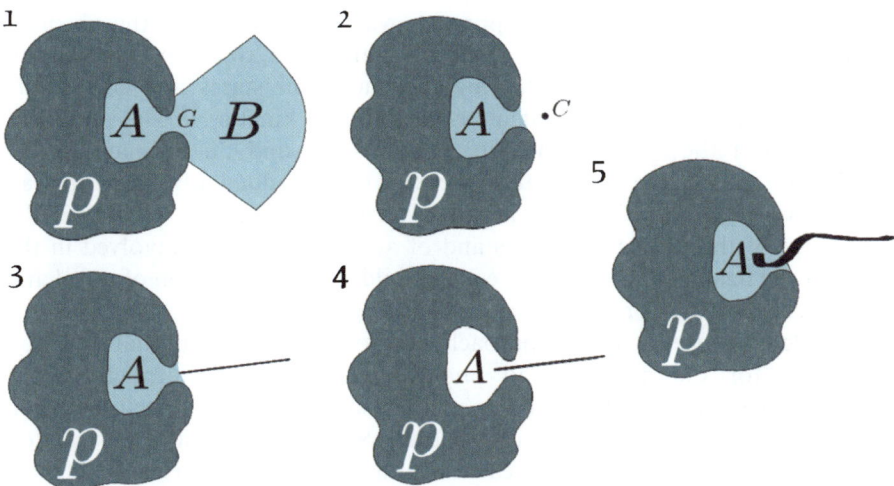

Figure 14.1   Different approaches to calculate binding free energies with metadynamics or umbrella sampling on a path. In the scheme, P is the protein, A is the binding cavity, G the narrow gate, B is the external area, and O is a point just outside the internal cavity. The schemes in the figure correspond to (1) running metadynamics with r and an angle θ as CVs; (2) the approach of Masetti *et al.*;[16] (3) the approach introduced by Fidelak *et al.*;[21] (4) metadynamics or umbrella sampling on a straight line; (5) metadynamics using the optimal path as CV, respectively.

uses metadynamics in the space of the distance (r) and a dihedral angle (ω) to quickly undock the ligand and build a guess path from A to B to be used with PCVs (Figure 14.1, diagram 5). Here state A is the relaxed crystallographic pose of the ligands, while state B is an unbound state several angstroms away from the surface of the target. Clearly this choice must be corrected by taking into account the standard volume of a free ligand in solution if the absolute $\Delta G_{binding}$ is needed.[40,41] This approach requires a much lower computational cost when compared to a fully converged metadynamics with simple geometry-based CVs.

14.3   Results and Discussion

14.3.1   Binding Profiles for a Series of CDK2 Inhibitors

Cyclin-dependent kinase 2 (CDK2) is a member of the CDK family that performs numerous functions in the regulation of the cell cycle.[42] It is a complex target for computational studies as it is known to be very flexible. Crystal structures revealed the existence of an active conformation and at least one inactive conformation. In a recent paper we have studied the binding free energy profile and the mode of inhibition of five different 2-anilino-4-[(hetero)aryl]pyrimidine derivatives of pharmacological relevance, for which experimental data are available.[43]

Metadynamics with the PCV approach was used to predict the absolute binding free energies ($\Delta\Delta G_{binding}$) of all the ligands, and to reconstruct the full docking free energy profile including the transition states and metastable minima that can be used to calculate the binding kinetics.[20] The approach provided both the structure–activity relationships and an estimate of the kinetics at a reasonable computational cost, with the added advantage that an *a priori* choice of the CV is not required. In the specific case of CDK2, the knowledge of the metastable states and of all of the residues involved in the access of the ligands to the binding cavity could lead to the design of novel and more selective drugs for this important oncological target.

In order to use a PCV approach, we obtained an optimal reaction coordinate for the docking and undocking of each inhibitor from the metadynamics trajectories, and we used it as a CV for a well-tempered metadynamics as a function of path variables S and Z. The full reconstruction of the free energy surface for each ligand was reached after 40–200 ns (corresponding to 2–3 days of computations on an HPC or on a GPU machine with a GPU-enable MD code), thanks to the optimized nature of the path used as the CV. For comparison,[44] umbrella sampling calculations performed on a straight path results in the necessity of using an extremely long sampling time (20.5 μs) to converge the free energy profile.

The calculated $\Delta G_{binding}$ are in excellent agreement with experimental data, leading to a correct ranking of the inhibitors (see Table 14.1).

As we obtained the reaction coordinates of binding, we have a good insight of the common and peculiar features of the docking mechanism of the various inhibitors. The protocol requires reduced human intervention with respect to metadynamics and converges the free energy profile with reasonable computational resources, highlighting its relevance for structure-based lead optimization purposes. The proposed approach, as for other pathway methods, by predicting metastable and transition states provides a biological insight that is not available either with docking or with end-point methods. Indeed, optimizing the transition state and the underlying ligand binding

Table 14.1 Comparison between experimental and calculated relative and absolute $\Delta G_{binding}$.[a]

inhibitor	inh1	inh2	inh3	inh4	inh5
$\Delta\Delta G^{exp}$	0.0	-2.0	-2.0	-4.0	-1.5
$\Delta\Delta G^{coarse}$	0.0	-1.5	-2.7	-2.2	-1.8
$\Delta\Delta G^{pathCV}$	0.0	-2.0	-1.4	-4.7	-1.4

[a]The naming convention for the inhibitors and the experimental values for the $\Delta\Delta G_{binding}$ have been obtained from Kontopidis *et al.*[43] Coarse and pathCV are used to refer to the results obtained with "coarse metadynamics"[16] and pathCV metadynamics,[21] respectively.

kinetics could potentially provide an additional mechanism to achieve therapeutically safe and differentiated responses.[20] Finally, the use of metadynamics to explore the binding cavity allows the discovery of additional relevant binding modes, if they are present, something that is not possible even with most of the other pathway methods.

14.3.2 Rationalizing the Different Residence Times of a Ligand in COX Isoforms

The importance of ligand target *residence time* to fine tune *in vivo* efficacy and toxicology is being increasingly recognized.[19,45] The residence time (τ), defined as the period for which a receptor is occupied by a ligand, is simply related to the dissociation rate constant k_{off} ($\tau = 1/k_{off}$). The longer the duration of binding to the intended target(s), the greater is the efficacy, provided that the associated toxicity is minimal. Moreover, the shorter the time the ligand binds to unintended and colateral targets, the safer the effect.

Metadynamics with PCVs has been successfully used to understand the different residence time of a non-steroidal anti-inflammatory ligand binding to COX-1 and COX-2 isoforms.[22] Non-steroidal anti-inflammatory drugs (NSAIDs) are used for the treatment of pain and inflammation. Their mechanism of action is based on blockage of the cyclooxygenase enzymes (COXs).[46] The classical NSAIDs such as aspirin or ibuprofen are nonselective *versus* the various COXs isoforms. However, the inhibition of COX-1 may lead to dangerous side effects, *e.g.* ulcers. As a consequence, a selective inhibition of COX isoforms has been actively sought, leading to a new generation of widely used COX-2 selective drugs such as celecoxib and nimesulide. The kinetics of selective and nonselective inhibitors also seem to be different in different inhibitors, and in recent years a significant effort has been made to understand the kinetics of binding of inhibitors to the COX isoforms.[47–49] In a recent paper it has been shown that a COX inhibitor binding to both COX-1 and COX-2 isoforms has a very different residence time in the two: hours in COX-2 but 30 s in COX-1.[48] This important finding opens the avenue to the design of "selective" inhibitors based on the different residence time (and binding kinetics) in different isoforms instead of the classically exploited difference in potency (equilibrium affinity). However, to fully exploit this possibility, the mechanism by which the ligand attains a different residence time had to be explained.

In this particular case the PCV has been used not to describe the exit path of the ligand, but a conformational change of three α-helices that form a narrow gate through which the ligand exits from the cavity.[50] In order to sample this important slow motion, we used a PCV[21] in the contact map[51] space of the residues that determine the gate flexibility. The docking and undocking of the ligand was described by two additional CVs: its distance from the cavity and a dihedral angle identifying its orientation relative to the enzyme.

Surprisingly, two separate and almost equally favorable minima have been found in COX-2, while only one deep minimum was found in COX-1. The

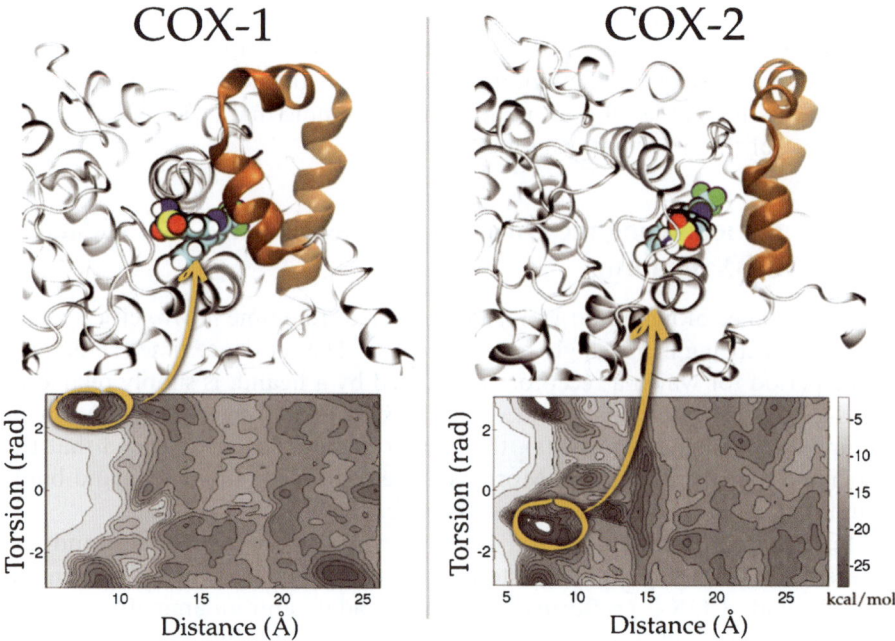

Figure 14.2   Free energy surfaces of SC558 binding to COX-1 and COX-2. The docking and undocking of the ligand was described by the distance from the cavity and a dihedral angle identifying its orientation relative to the enzyme.

existence of two binding modes is compatible with the existence of at least an additional binding mode for many diaryl heterocycles. This fact has been suspected to be the main cause of the peculiar kinetics exhibited by this class of inhibitor.[47–49] With metadynamics it was not only possible to confirm the existence of two different binding modes in COX-2, but also to observe the direct interconversion from one mode to the other.

Comparing the free energy surfaces of SC558 binding to COX-1 and COX-2 (Figure 14.2), one sees that the alternative binding site is only present in COX-2. This explains the COX-2 selectivity of this compound and the higher residence time in COX-2 found for similar compounds.[48]

These results show how metadynamics and PCV-based approaches can be used to understand the reason underlying a different binding kinetic and open the avenue to a rational structure-based design of residence time to optimize efficacy and minimize toxicology.

14.4   Conclusions

Free energy methods such as metadynamics and its derivatives PCV and PTmetaD can do much in the lead discovery and lead optimization phases of drug discovery. They can be used to understand the mode of action of a known binder, to calculate the free energy profile efficiently (as shown in the case of CDK2), and even to rationalize the differences in binding kinetics in general and of residence time in particular. Clearly, their computational cost still makes their use impractical for large virtual screening experiments. However, the continuous development of more powerful processors and graphic processing units, as well as the concomitant availability of a free easy-to-use open-source plug-in[25] that works with the most common MD codes, will surely help to facilitate the widespread adoption of these methods in drug design.

References

1. J. E. Ladbury, *Chem. Biol.*, 1996, **3**, 973.
2. N. Moitessier, P. Englebienne, D. Lee, J. Lawandi and C. R. Corbeil, *J. Pharmacol.*, 2008, **153**(suppl.), S7.
3. J. R. Schames, R. H. Henchman, J. S. Siegel, C. A. Sotriffer, H. Ni and J. A. McCammon, *J. Med. Chem.*, 2004, **47**, 1879.
4. P. L. Freddolino, F. Liu, M. Gruebele and K. Schulten, *Biophys. J.*, 2008, **94**, L75.
5. B. Hess, C. Kutzner, D. van Der Spoel and E. Lindahl, *J. Chem. Theory Comput.*, 2008, **4**, 435.
6. *Free Energy Calculations: Theory and Applications in Chemistry and Biology*, ed. C. Chipot and A. Pohorille, Springer, Berlin, 2007.
7. P. G. Bolhuis, D. Chandler, C. Dellago and P. L. Geissler, *Annu. Rev. Phys. Chem.*, 2002, **53**, 291.
8. A. Laio and F. L. Gervasio, *Rep. Prog. Phys.*, 2008, **71**, 126601.
9. J. L. Klepeis, K. Lindorff-Larsen, R. O. Dror and D. E. Shaw, *Curr. Opin. Struct. Biol.*, 2009, **19**, 120.
10. M. Shirts and V. S. Pande, *Science*, 2000, **290**, 1903.
11. G. Bussi, F. L. Gervasio, A. Laio and M. Parrinello, *J. Am. Chem. Soc.*, 2006, **128**, 13435.
12. F. Pietrucci, F. Marinelli, P. Carloni and A. Laio, *J. Am. Chem. Soc.*, 2009, **131**, 11811.
13. D. Branduardi, F. L. Gervasio and M. Parrinello, *J. Chem. Phys.*, 2007, **126**, 054103.
14. F. L. Gervasio, A. Laio and M. Parrinello, *J. Am. Chem. Soc.*, 2005, **127**, 2600.
15. D. Branduardi, F. L. Gervasio, A. Cavalli, M. Recanatini and M. Parrinello, *J. Am. Chem. Soc.*, 2005, **127**, 9147.
16. M. Masetti, A. Cavalli, M. Recanatini and F. L. Gervasio, *J. Phys. Chem. B*, 2009, **113**, 4807.
17. C. Domene, M. L. Klein, D. Branduardi, F. L. Gervasio and M. Parrinello, *J. Am. Chem. Soc.*, 2008, **130**, 9474.
18. A. Berteotti, A. Cavalli, D. Branduardi, F. L. Gervasio, M. Recanatini and M. Parrinello, *J. Am. Chem. Soc.*, 2009, **131**, 244.

19. D. E. Shaw, P. Maragakis, K. Lindorff-Larsen, S. Piana, R. O. Dror, M. P. Eastwood, J. A. Bank, J. M. Jumper, J. K. Salmon, Y. Shan and W. Wriggers, *Science*, 2010, **330**, 341.

20. D. C. Swinney, *Curr. Opin. Drug. Discovery Dev.*, 2009, **12**, 31.

21. J. Fidelak, J. Juraszek, D. Branduardi, M. Bianciotto and F. L. Gervasio, *J. Phys. Chem. B*, 2010, **114**, 9516.

22. V. Limongelli, M. Bonomi, L. Marinelli, F. L. Gervasio, A. Cavalli, E. Novellino and M. Parrinello, *Proc. Natl. Acad. Sci. U. S. A.*, 2010, **107**, 5411.

23. D. Cvijovicacute and J. Klinowski, *Science*, 1995, **267**, 664.

24. G. Bussi, A. Laio and M. Parrinello, *Phys. Rev. Lett.*, 2006, **96**, 090601.

25. PLUMED: https://sites.google.com/site/plumedweb/.

26. V. Babin, C. Roland, T. A. Darden and C. Sagui, *J. Chem. Phys.*, 2006, **125**, 204909.

27. M. Ceccarelli, C. Danelon, A. Laio and M. Parrinello, *Biophys. J.*, 2004, **87**, 58.

28. A. Barducci, R. Chelli, P. Procacci, V. Schettino, F. L. Gervasio and M. Parrinello, *J. Am. Chem. Soc.*, 2006, **128**, 2705.

29. G. Fiorin, A. Pastore, P. Carloni and M. Parrinello, *Biophys. J.*, 2006, **91**, 2768.

30. X. Biarnés, A. Ardèvol, A. Planas, C. Rovira, A. Laio and M. Parrinello, *J. Am. Chem. Soc.*, 2007, **129**, 10686.

31. M. Bonomi, F. L. Gervasio, G. Tiana, D. Provasi, R. A. Broglia and M. Parrinello, *Biophys. J.*, 2007, **93**, 2813.

32. K. Kamiya, M. Boero, M. Tateno, K. Shiraishi and A. Oshiyama, *J. Phys. Condens. Matter*, 2007, **19**, 365220.

33. S. Piana, A. Laio, F. Marinelli, M. Van Troys, D. Bourry, C. Ampe and J. C. Martins, *J. Mol. Biol.*, 2008, **375**, 460.

34. S. Piana, *J. Phys. Chem. A*, 2007, **111**, 12349.

35. E. Piccinini, M. Ceccarelli, F. Affinito, R. Brunetti and C. Jacoboni, *J. Chem. Theory Comput.*, 2008, **4**, 173.

36. G. Petraglio, M. Bartolini, D. Branduardi, V. Andrisano, M. Recanatini, F. L. Gervasio, A. Cavalli and M. Parrinello, *Proteins: Struct., Funct., Bioinf.*, 2008, **70**, 779.

37. M. Ceccarelli, R. Anedda, M. Casu and P. Ruggerone, *Proteins: Struct., Funct., Bioinf.*, 2008, **71**, 1231.

38. L. Sutto, M. D'Abramo and F. L. Gervasio, *J. Chem. Theory Comput.*, 2010, **6**, 3640.

39. S. K. Kearsley, *Acta Crystallogr., Sect. A*, 1989, **45**, 208.

40. M. K. Gilson and H.-X. Zhou, *Annu. Rev. Biophys. Biomol. Struct.*, 2007, **36**, 21.

41. Y. Deng and B. Roux, *J. Phys. Chem. B*, 2009, **113**, 2234.

42. M. Malumbres and M. Barbacid, *Nat. Rev. Cancer*, 2009, **9**, 153.

43. G. Kontopidis, C. McInnes, S. R. Pandalaneni, I. McNae, D. Gibson, M. Mezna, M. Thomas, G. Wood, S. Wang, M. D. Walkinshaw and P. M. Fischer, *Chem. Biol.*, 2006, **13**, 201.
44. I. Buch, M. J. Harvey, T. Giorgino, D. P. Anderson and G. De Fabritiis, *J. Chem. Inf. Model.*, 2010, **50**, 397.
45. R. A. Copeland, D. L. Pompliano and T. D. Meek, *Nat. Rev. Drug Discovery*, 2007, **6**, 249.
46. W. L. Smith, P. Borgeat and F. A. Fitzpatrick, *Biochemistry of Lipids, Lipoproteins and Membranes*, Elsevier, Amsterdam, 1991.
47. R. A. Copeland, J. M. Williams, J. Giannaras, S. Nurnberg, M. Covington, D. Pinto, S. Pick and J. M. Trzaskos, *Proc. Natl. Acad. Sci. U. S. A.*, 1994, **91**, 11202.
48. C. A. Lanzo, J. Sutin, S. Rowlinson, J. Talley and L. J. Marnett, *Biochemistry*, 2000, **39**, 6228.
49. M. C. Walker, R. G. Kurumbail, J. R. Kiefer, K. T. Moreland, C. M. Koboldt, P. C. Isakson, K. Seibert and J. K. Gierse, *Biochem. J.*, 2001, **357**, 709.
50. D. Picot, P. J. Loll and R. M. Garavito, *Nature*, 1994, **367**, 243.
51. M. Bonomi, D. Branduardi, F. L. Gervasio and M. Parrinello, *J. Am. Chem. Soc.*, 2008, **130**, 13938.

Computer-Assisted Design of Drug-Like Synthetic Libraries

P. SENECI[a,b,c], V. FRECER[a,d,e,f] AND S. MIERTUS*[a,f]

[a] International Centre for Science and High Technology, UNIDO, AREA Science Park, Padriciano 99, I-34012 Trieste, Italy; [b] Dipartimento di Chimica Organica e Industriale, Università degli Studi di Milano, Via Venezian 21, I-20133 Milan, Italy; [c] CISI scrl, Via Fantoli 16/15, I-20138 Milan, Italy; [d] Cancer Research Institute, Slovak Academy of Sciences, SK-83391 Bratislava, Slovakia; [e] Department of Physical Chemistry of Drugs, Faculty of Pharmacy, Comenius University, SK-83232 Bratislava, Slovakia; [f] International Centre for Applied Research and Sustainable Technology – ICARST, Jamnickeho 19, SK-84104 Bratislava, Slovakia
*E-mail: stanislav.miertus@icarst.org

15.1 Introduction

Once upon a time, more talented members of the human race tried to assist diseased fellows by providing them with natural remedies through empirical knowledge, in a traditional medicine scenario.[1] Today, a wealth of high-throughput technologies and cutting-edge knowledge of molecular mechanisms leading to one or another disease state are available to scientists devoted to drug discovery.[2,3] What has not changed in the thousands of years spanning the two approaches is the unmet medical need, as we call it today, which still causes a significant portion of men and women to suffer, or even to die, due to intractable diseases.[4,5] The higher degree of understanding (causes, risks, incidence, symptoms, and so on) does not

RSC Drug Discovery Series No. 23
Physico-Chemical and Computational Approaches to Drug Discovery
Edited by F. Javier Luque and Xavier Barril
© The Royal Society of Chemistry 2012
Published by the Royal Society of Chemistry, www.rsc.org

minimize the emotional and economic burden on our "civilized" shoulders as patients and/or relatives.

Some intriguing calculations defined, in terms of population, how many drug-like compounds satisfying the Lipinski rules,[6] or even their more modern versions,[7] could theoretically be synthesized using any known chemical reaction on any compound which was previously reported in the literature.[8] Being as high as 10^{80}, or as low as 10^{40}, it surely exceeds even the number of single molecules existing at a given time on our beloved planet. Thus, one cannot think of making all what is theoretically possible; even at an improbable "10^6 molecules per week" synthetic speed, it would take close to 20 years to make a billion molecules, around 350 years to make 10^{12} of them, an impressive 70 centuries to reach a mere 10^{15} mark, and so on.

Virtual compounds are easily created, are cheap, do not require heavy logistics, and have no stability problems. Their assembly into a virtual screening library allows us to execute large-scale virtual high-throughput screening (vHTS) campaigns, to constantly update the library content and diversity, and to easily expand the virtual structure–activity relationships (SAR) around a proprietary chemotype. The "virtual-to-tangible" transition is better done on much smaller numbers of prospective hits, to create small-size high-quality collections of active compounds.[9] In our opinion, such arguments should sufficiently justify in everyone's mind the case for rational drug design, and for heavy virtual filtering of drug-like libraries and library individuals before embarking in their synthesis and biological characterization.

Computational methods are constantly used to assist in library design, focusing and evaluation. Their input is significant at any stage of the drug discovery process: vHTS campaigns provide selection criteria for the initial, activity-based heavy filtering of large libraries, while drug-like filters applied on smaller collections may drive them towards drug-like leads. In our opinion, a mixed iterative cycle-based approach including virtual (computer driven, larger numbers) and tangible (biological assay driven, smaller numbers) screening steps is ideally suited on the one hand to rapidly and inexpensively focus onto prospective compounds, and on the other hand to rapidly refine and validate the virtual hypotheses used to find and structurally optimize high-quality hits and leads.

The next section will provide a brief description of computational methods employed in library design, focusing and evaluation. As those are the main topics of several chapters in this book, the last section, constituting the main part of this chapter, will be devoted to the description of successful applications of computer-assisted library design in drug discovery. Our common affiliation in a supra-national institution devoted to assisting in the technology transfer process to developing countries has focused our efforts against common viral, bacterial and parasitic diseases in such areas, including HIV,[10] avian flu,[11] dengue fever,[12] malaria[13] and tuberculosis.[14]

15.2   Methods and Formats in Computer-Assisted Library Design

There are three main factors influencing everyone's choices in designing libraries for drug discovery projects. The first has to do with the nature of the biological target, and with what is known about it. Structure-based library design[15] approaches require a detailed knowledge about the target's active-binding site, leading to *in silico* target-active site reconstruction, using experimental data from X-ray crystallography,[16] NMR[17] or the like. Ligand-based library design[18] is used when a reliable pharmacophore[19,20] can be extracted from the structure of several (the more, and the more diverse, the better) known ligands of the structurally uncharacterized target. As time goes by, more and more targets (even the historically intractable membrane receptors[21,22] or the multi-functional protein complexes[23]) are structurally elucidated, and consequently structure-based approaches become applicable to a larger set of pathology-related targets. Moreover, if a target cannot be over-produced and crystallized, or otherwise structurally characterized, it may nevertheless be similar to other human, animal or vegetal characterized targets; in such a scenario, a homology model[24] can be built by modifying the congener's structure and can be used as such to select target-focused prospective compounds with an acceptable probability of success. Even if nothing is known or predictable about the selected target (a situation which is more and more unlikely, as structural target and ligand-inhibitor information is piling up day after day), libraries designed as a screening collection against such a target must be drug-like; then, virtual Lipinski rule-derived filters[25] and/or ADMET (adsorption, distribution, metabolism, excretion and toxicity) filters[26] can be used *a priori* to discard unlikely drug-like library members.

The second factor deals with drug-like chemical diversity, *i.e.* with what is known about effectors of a given target of relevance. If no natural or synthetic ligands are known, diversity-based libraries[27] (if synthetic efforts are planned) or collections (if the screening set is to be assembled from commercial vendors) are designed to be respectively synthesized or purchased. Their size may vary from several thousands to even millions of individual library collection members, but invariably their virtual members are selected to span at best the drug-like chemical diversity space; they are as chemotype-rich as possible, to maximize the chance of discovering novel effectors of the target. Their virtual screening (based on target affinity) can be compared to a tangible primary HTS campaign, where a likely result is a subset of prospective hits which need further profiling to assess their drug-like features. If one or more active chemotype scaffolds are known, activity-based focused libraries[28] or collections are designed to be either synthesized or purchased. Their size is typically smaller, ranging between a few hundreds to several thousands of virtual library members, which are selected to span at best the drug-like chemical similarity space around one or more active chemotypes. Their virtual screening can be compared to a secondary, more articulate profiling cascade for confirmed hits, keeping a close eye to primary target QSAR acquisition, but also evaluating other hit features potentially important for downstream development (selectivity *versus* similar targets, virtual ADMET screens, and so on). Both dissimilarity diversity-based and similarity activity-based selection are driven by chemical feasibility tractability of the chemotypes included in the library, so as to have an easy "virtual-to-tangible" transition. Virtual chemotype substitution *via* functional group modification is the standard approach to library expansion, while scaffold hopping[29] is a modern technique granting a more thorough evaluation of equivalent core structures for the spatial distribution of active site interacting side chains, and an easier access to novel and patentable lead series.

Third, and perhaps most important, are the techniques used to build reliable target and library individuals' models, and to evaluate and score their interactions. One may want to build a diversity-based or an activity-based library, may have detailed structural information about the target, or about several of its ligands, but in any case it is mandatory to properly describe the molecular features of each considered structure, to employ a validated selection method to focus on similar or dissimilar compounds, and to rely on a target-library evaluation method capable of scoring each individual in terms of interaction potential and to rank their likelihood to be of relevance. As to molecular descriptors,[30] a large number of them were reported in the past decades: physico-chemical properties[31] such as dipole moments, log *P* valuess and melting points, topological indexes[32] such as atom content and chain–ring sequences, fingerprint-based descriptors derived from 2D connection tables[33] or from 3D conformations.[34] None of them is universally more suitable than others, but rather each descriptor-based chemical entity or library representation has advantages or disadvantages in terms of calculation time, reliability and compound chemical function suitability.

As to library–library individual selection methods, one can rely on less demanding reactant-based[35] or on more accurate and time consuming product-based[36] (dis)similarity evaluation; one may directly use pairwise similarity-based comparison methods, *e.g.* clustering,[37] or pre-selection partitioning[38] to define a low-dimensional space independent of compound numerosity or nature. One may rely on a variety of scoring methods to measure ligand–target interactions, including but not limited to docking,[39] target-specific scoring functions,[40] *etc.* Once again, one's experience in computer-assisted library design and evaluation may guide him or her through the variety of existing (commercial or free publicly available) tools to describe, evaluate and score libraries and their individual members.

Other chapters in this book describe in detail and accurately each of the mentioned descriptors, algorithms and methods. From now on we will focus on our experience in designing medium/small drug-like chemical libraries targeted against relevant drug discovery targets. While we cannot claim to have sampled all approaches and methods, we are confident that our examples may be of relevance, and may contribute to evaluate the impact of such techniques in a drug discovery project.

15.3 Examples of Computer-Assisted Library Design

Between 2005 and 2011 we published several examples of structure-based library design approaches, for which the X-ray structure of the selected biological target, either alone or complexed with one or more of its ligands or inhibitors, was available. In order to significantly reduce the virtual library size we invariably used at first a reagent-based selection method,[35] choosing a relatively small number of commercially available fragments on the basis of their similarity with best substitution motifs on known target ligands. Resulting libraries were then filtered on a product basis,[36] prioritizing their individuals with a higher similarity to known target ligands, and obtaining a smaller library whose members were docked on the target's active site. Then 10 to 20 library individuals showing the highest docking score were identified and carefully evaluated in each project.

As combinatorial chemistry increases its efficiency by synthesizing arrays of compounds containing a few common building blocks,[41] rather than painstakingly assembling a library of extremely different individuals, we finally analyzed the recurrence of fragments in active library members, and designed small/medium libraries (from several tens up to a few hundred discrete entities), which in our mind are the best compromise between limiting the number of needed tangible analogs and synthesizing a reasonable number of them to establish a solid QSAR.

We already mentioned the developing countries-directed nature of our efforts, which has almost invariably selected third world country endemic tropical diseases as therapeutic targets for our projects. All of them have a strong impact also for developed countries' citizens, either because the disease is also spread in Western countries, *e.g.* HIV, or because it is re-manifesting itself after long periods of eradication, *e.g.* tuberculosis, or simply because the ever-increasing globalization implies more frequent contacts among the developing and developed worlds, thus increasing the infection risks for Europeans and Americans.

15.3.1 Viral Diseases

15.3.1.1 HIV

We first focused on HIV[10] and the selected molecular target was the aspartic protease (PR), due to its essential role in the maturation of HIV-1 particles and virus replication.[42] Among known PR inhibitors, we selected as structural inspiration for library design the cyclic urea inhibitor XV-638[40] (Figure 15.1, left). Its structure and chemical synthesis[43] inspired the assembly of the computer-designed cyclic urea library shown in Figure 15.1 (right).[44]

Thus, four groups of easily attainable fragments from commercially available sources[45] were selected: R_1, derived from amino acids, R_2, derived from aldehydes, and R_3 and R_4, both derived from aryl halides. Eleven electronic, physico-chemical and topological descriptors were used to filter the R groups for their resemblance to the corresponding R groups of known,

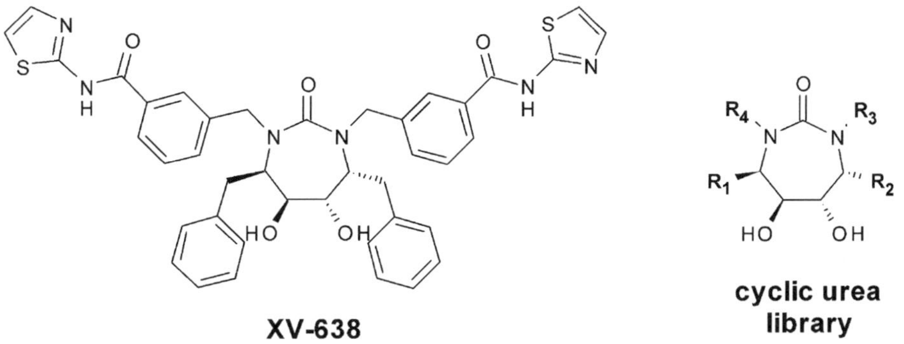

Figure 15.1 *Left*: XV-638, a cyclic urea inhibitor of PR;[40] *right*: generic structure of a computer-designed, substituted cyclic urea library of PR inhibitors.

active inhibitors crystallized into the PR active site; a total of $6R_1 \times 3R_2 \times 18R_3 \times 18R_4 = 5832$ cyclic ureas comprised the first, fragment-derived library. Such a library was further filtered using the same descriptors on whole library members, selecting the most diverse and closer analogs of known PR cyclic urea inhibitors, and a 1000 library size was attained.

Structure-based focusing was then performed using the Monte Carlo ligand fit algorithm of Cerius.[2,46,47] The 10 best binding conformations per virtual compound were generated, made to fit the receptor model generated from the PR–XV-638 complex X-ray structure, and the one showing the highest docking score was selected. The K_i^* (predicted inhibition constant) was calculated using an equation built on the relative rankings and the measured K_i values of a training set made of 12 known cyclic urea inhibitors of PR, and was validated by predicting the K_i values of five structurally related PR inhibitors. This target-specific scoring function was then applied to *in silico* screening to calculate the K_i^* constants for the designed 1000 cyclic urea library. Nine best library compounds from *in silico* screening, shown in Figure 15.2, were characterized virtually in terms of their potency and ADMET properties.

These compounds were predicted to have an overall drug-like profile, and their K_i^* values were up to two orders of magnitude lower than for XV-638, with four of them (labeled [#] in Figure 15.2) expected to be sub-picomolar PR inhibitors. A detailed SAR for any R_1–R_4 fragments was obtained; the predicted potency increase was mostly due to changes in the R_3 and R_4 groups, introducing bulkier substituents. In order to combine the acquired SAR information and chemical feasibility, the most recurrent R_1–R_4 fragments in the 90 highest scoring compounds were identified (darker brown, cyan, blue and green bars, Figure 15.3).

Eventually, a $2R_1 \times 3R_2 \times 4R_3 \times 3R_4 = 72$-membered cyclic urea library L1 (Figure 15.4) was designed and proposed for synthesis and biological characterization.[44]

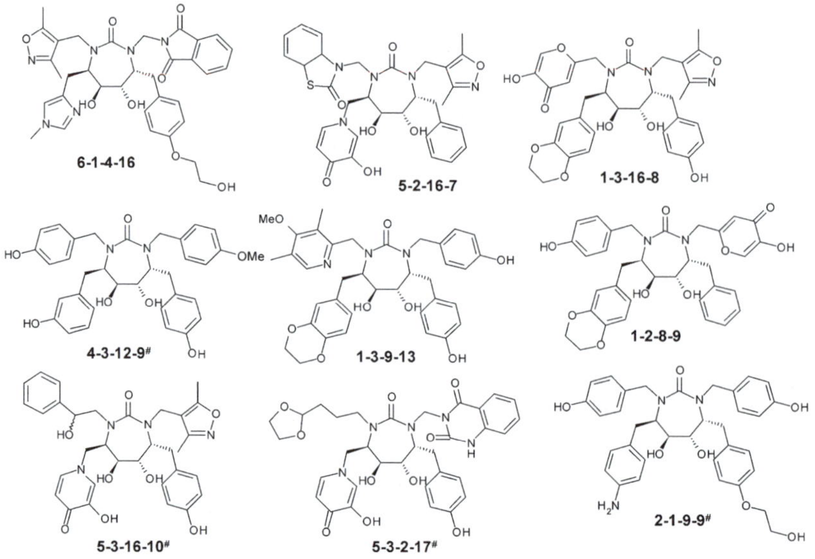

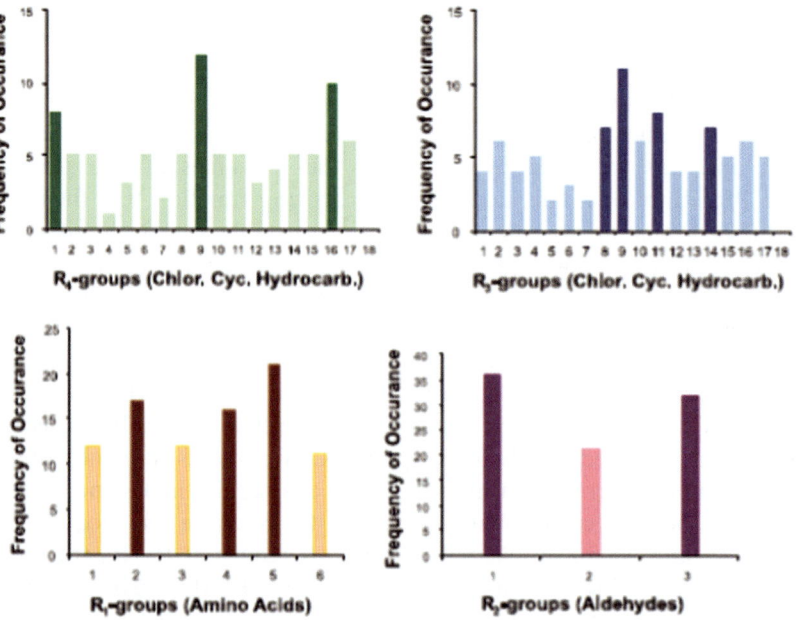

Figure 15.2   Structures of nine highest ranked library members, with lowest K_i^* values against PR. Compounds with sub-picomolar K_i^* are labeled [#]; numbers correspond to the R_1-R_2-R_3-R_4 numbering of the fragments used to build the first 5832-membered library.

Figure 15.3   Frequency of occurrence of R groups in 90 highest ranked library members, with lowest K_i^* values against PR. See Figure 15.4 for the structure of best R_1–R_4 substituents (bar and substituent numbers match).

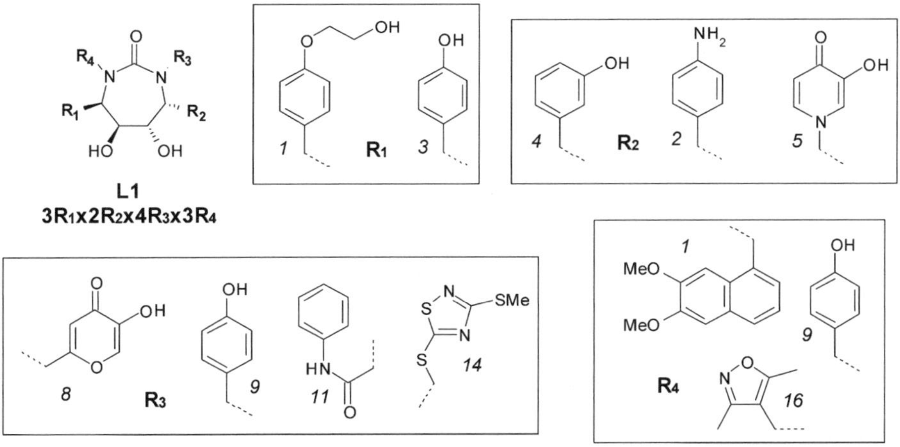

Figure 15.4 Structure of the computer-designed cyclic urea library L1, proposed for synthesis, and of R_1–R_4 substituents used in the library. See Figure 15.3 for the frequency of occurrence of best R_1–R_4 substituents (substituent and bar numbers match).

15.3.1.2 Avian Flu

The avian influenza A virus subtype H5N1,[11] and more specifically its crystallized neuraminidase (NA) membrane glycoprotein,[48] were selected as therapeutic targets for two of our library design projects.

Among the few commercially available NA inhibitors, oseltamivir[49] (Figure 15.5, left) was selected as a structural motif, and a three-point substitution library (R_1, R_2, R_3R_4; Figure 15.5, right) was planned both to

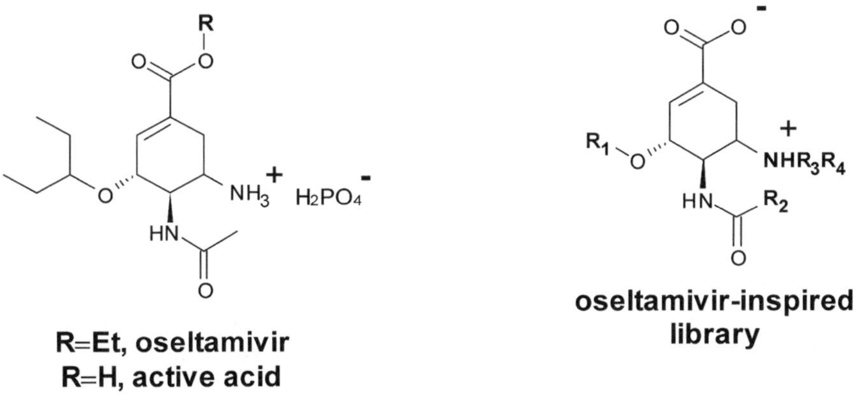

Figure 15.5 *Left*: Oseltamivir and its active acid metabolite, a cyclohexenecarboxylate inhibitor of NA;[49] *right*: generic structure of a computer-designed, substituted cyclohexenecarboxylate library of oseltamivir carboxylate analogs as NA inhibitors.

explore the SAR of these positions and to find active NA inhibitors against oseltamivir-resistant avian influenza strains.[50] According to two reported oseltamivir syntheses,[51,52] three groups of easily attainable fragments from commercially available sources[45] were selected: R_1, derived from alcohols or ketones, R_2, derived from acylating agents, and R_3R_4, derived from either carbonyl compounds or thioureas.

Nine electronic, physico-chemical and topological descriptors were used to filter the R groups for their resemblance to the corresponding R groups of known inhibitors crystallized into the active site of oseltamivir resistance-relevant NA subtype N1; a total of $29R_1 \times 5R_2 \times 22R_3R_4 = 3190$ oseltamivir analogs comprised the first library. The library size was reduced to 2000 members by analog filtering, using the same descriptors on whole library members and selecting the most diverse analogs retaining structural similarity with oseltamivir.

Structure-based focusing was then performed using previously described[46] software, together with known force fields[53] and methods.[54] Oseltamivir was replaced with several conformations for each library member into the binding site of the NA subtype N1–oseltamivir complex X-ray structure. The best fitting conformer for each docked compound was selected. IC_{50}^* (predicted IC_{50}) values were calculated using an equation derived from a training set made up of 14 carbocyclic NA inhibitors, and a validation set was made of three structurally related NA inhibitors. Nine best compounds, following *in silico* screening using the PLP1 scoring function,[55] are shown in Figure 15.6.

Around 200 library members showed an $IC_{50}^* < 1$ nM, *i.e.* a predicted inhibition stronger than oseltamivir. A detailed SAR for any R_1–R_3R_4 fragment was obtained; the predicted potency increase was mostly due to changes in the R_1 groups, where bulky aliphatic and/or aromatic groups were shown to fill a hydrophobic pocket in the binding site. In order to combine the acquired SAR and chemical feasibility, the most recurrent R_1–R_3R_4 fragments in the 208 highest scoring compounds were identified (darker blue, yellow and green bars, Figure 15.7).

Finally, a $7R_1 \times 2R_2 \times 6R_3R_4 = 84$-membered oseltamivir analog library L2 (Figure 15.8) was designed and proposed for synthesis and biological characterization.[50]

A second computer-assisted library design targeted against NA avian flu used as a structural inspiration the pentacyclic compound A-315675[56] (Figure 15.9, left), a more recent NA inhibitor active against several clinically significant oseltamivir-resistant H5N1 strains. In accordance with its reported synthesis,[57] four groups of easily attainable fragments from commercially available sources[58] were selected: R_1 and R_3, derived from alkyl and aryl bromide-derived Grignard reagents, R_2, derived from alkyl halides, and R_4, derived from either Wittig, Grignard, amino or amidino compounds. The generic library is shown in Figure 15.9 (right).[59]

Twenty electronic, physico-chemical and topological descriptors were used to filter the R groups for their resemblance to the corresponding R groups of

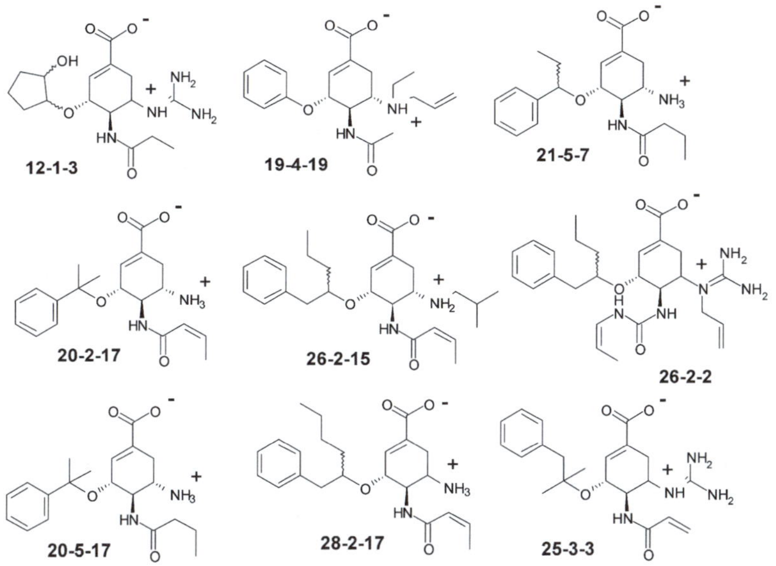

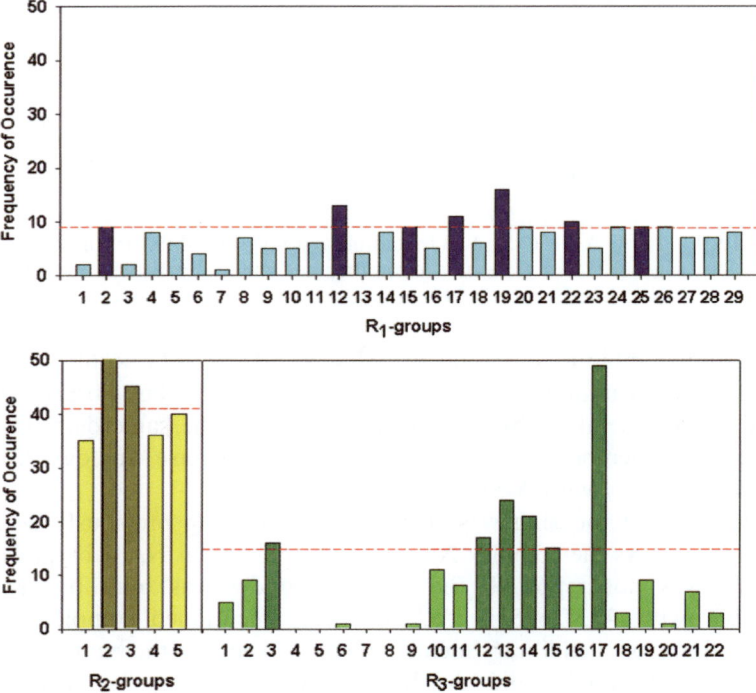

12-1-3   **19-4-19**   **21-5-7**

20-2-17   **26-2-15**   **26-2-2**

20-5-17   **28-2-17**   **25-3-3**

Figure 15.6   Structures of the nine highest ranked library members, with lowest IC_{50}* values against NA. Numbers correspond to R_1-R_2-R_3-R_4 numbering of fragments used to build the first 3190-membered library.

Figure 15.7   Frequency of occurrence of R groups in 208 highest ranked library members, with lowest IC_{50}* values against NA. See Figure 15.8 for the structure of best R_1–R_3R_4 substituents (bar and substituent numbers match).

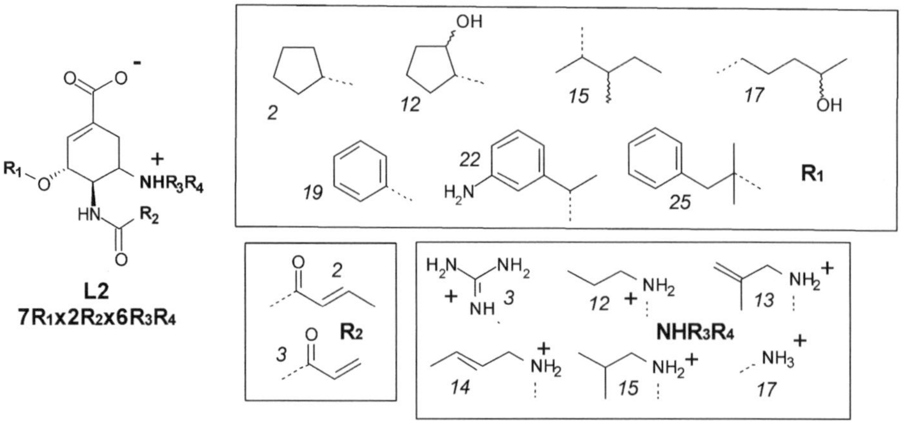

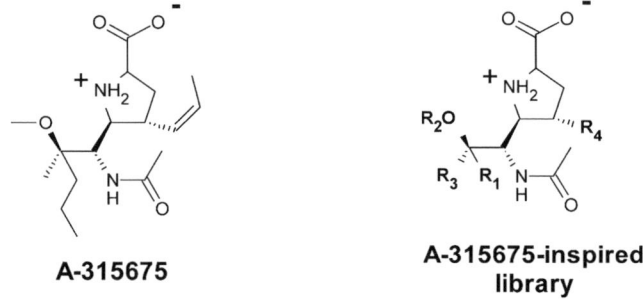

Figure 15.8   Structure of the computer-designed oseltamivir carboxylate analog library L2, proposed for synthesis, and of $R_1–R_3R_4$ substituents used in the library. See Figure 15.7 for the frequency of occurrence of best $R_1–R_3R_4$ substituents (substituent and bar numbers match).

A-315675

A-315675-inspired library

Figure 15.9   *Left*: A-315675, a pyrrolidinecarboxylate inhibitor of NA;[56] *right*: generic structure of a computer-designed, substituted pyrrolidinecarboxylate library of NA inhibitors.

known, similar inhibitors; a total of $48R_1 \times 6R_2 \times 6R_3 \times 6R_4 = 10\,368$ pyrrolidine-based inhibitors comprised the first library. The library size was reduced to 700 members by analog filtering, using the same descriptors on whole library members and selecting the most diverse analogs retaining structural similarity with A-315675.

Structure-based focusing was then performed using previously described methods.[46,53,54] As the complex X-ray structure between NA subtype N1 and A-315675 was not reported, the receptor model was built by replacing oseltamivir with A-315675 in the previously mentioned complex. Several conformations for each library member were docked into the built binding site by replacing A-315675, and the best fitting conformer for each docked compound was selected. $IC_{50}{}^{*}$ values were calculated using an equation derived from a training set made of 13 pyrrolidine-based NA inhibitors, and a validation set was made of three structurally related NA inhibitors. Nine best compounds, following *in silico* screening using the PMF scoring function,[60] are shown in Figure 15.10.

More than 10 library members showed an IC_{50}^* value better than A-315675. A detailed SAR for any $R_1–R_4$ fragment was obtained; the predicted potency increase was likely due to the positively charged secondary amines or amidines in the AM-derived R_4 groups, where a hydrogen bond may be established with the carboxyl side chain of the Asp151 residue. In order to combine the acquired SAR and chemical feasibility, the most recurrent $R_1–R_4$ fragments in the 200 highest scoring pyrrolidine-based NA inhibitors were identified (dark bars, Figure 15.11).

Finally, a $12R_1 \times 2R_2 \times 2R_3 \times 2R_4$ = 96-membered A-315675 analog library L3 (Figure 15.12) was designed and proposed for synthesis and biological characterization.[59]

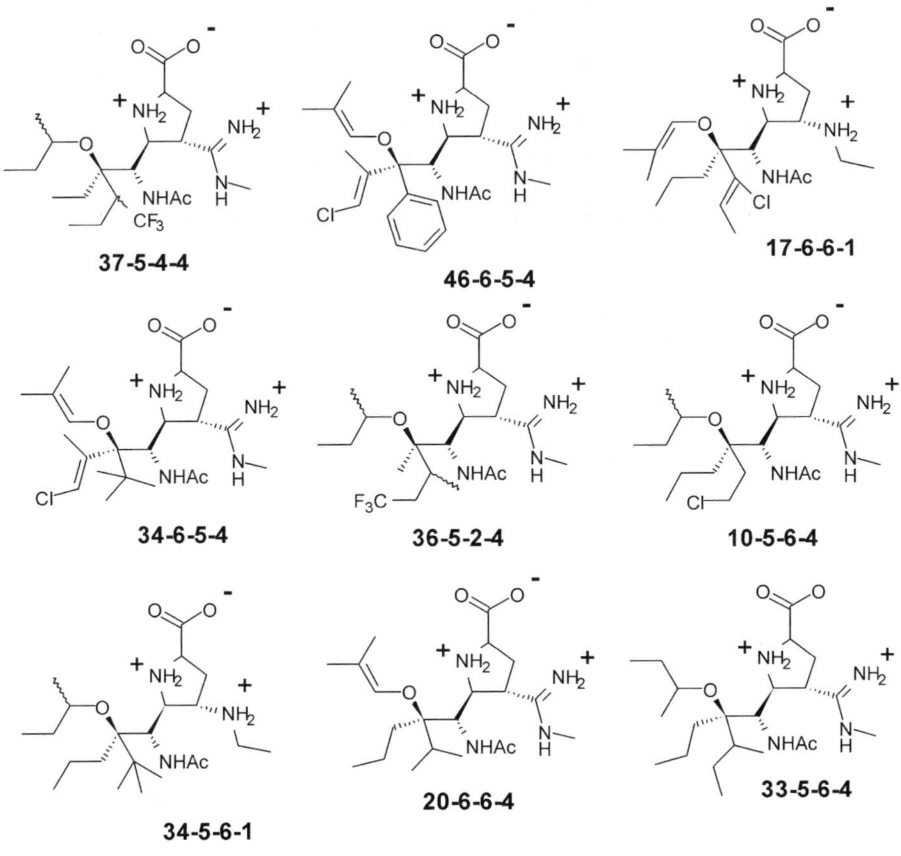

Figure 15.10 Structures of the nine highest ranked library members, with lowest IC_{50}^* values against NA. Numbers correspond to R_1-R_2-R_3-R_4 numbering of fragments used to build the first 10 368-membered library.

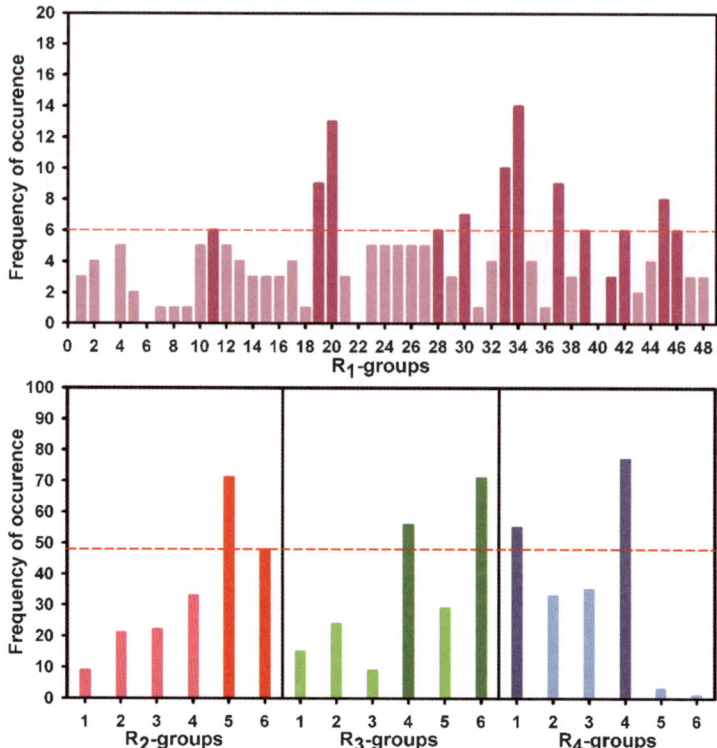

Figure 15.11 Frequency of occurrence of R groups in 200 highest ranked library members, with lowest IC_{50}^{PRE} values against NA. See Figure 15.12 for the structure of best R_1–R_4 substituents (bar and substituent numbers match).

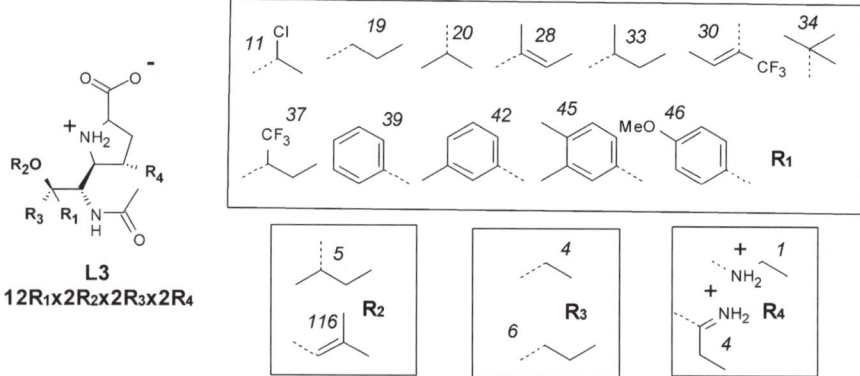

Figure 15.12 Structure of the computer-designed A-315675 carboxylate analog library L3, proposed for synthesis, and of R_1–R_4 substituents used in the library. See Figure 15.11 for the frequency of occurrence of best R_1–R_4 substituents (substituent and bar numbers match).

15.3.1.3   Dengue Fever

Dengue fever,[12] an endemic disease in sub-tropical and tropical regions with no available cure, was targeted through its viral protease target NS3pro,[61] a trypsin-like serine protease vital for viral replication and maturation of infectious Dengue virions.

While no small-molecule inhibitors of NS3pro are known, a tetrapeptidic, N-acylated aldehyde template inhibitor D1[62,63] (Figure 15.13, left) was used as a scaffold for library design. A four-point variation library (R_1–R_4, Figure 15.13, right) was planned both to explore the SAR of the four amino acidic positions and to find more active and proteolitically stable NS3pro inhibitors.[64] According to the published D1 synthesis,[63] we focused on four groups of either commercially available or easily synthesizable amino acid-derived reagents: R_1, derived from natural and unnatural α-aminoaldehydes, the two central R_2 and R_3 groups, derived from natural and unnatural α-amino acids, and R_4, derived from N-acylated natural and unnatural α-amino acids.

A set of electronic, physico-chemical and topological descriptors were used to filter the members of the four building block classes and to focus on more prospective candidates; a total of $21P_1 \times 10P_2 \times 11P_3 \times 4P_4 = 9240$ D1 analogs comprised the first library. The library size was further reduced to 2310 members by simply considering the single, largest P_4 Nle residue, due to both its likely better fit in the corresponding NS3pro S_4 pocket and to the known little effect on inhibitory potency for residue variation in the same pocket.

Structure-based focusing was then performed using previously described methods[46,53,54] and some updates,[65] employing a model built from super-position of the known X-ray structure of the DEN2 serotype NS2B-NS3pro target with the reported X-ray complex between NS2B-NS3pro from highly homologous West Nile virus and the template inhibitor D1.[66] Thus, D1 was replaced in the homology model with several conformations for each library member into the NS3pro binding site. The best fitting conformer for each

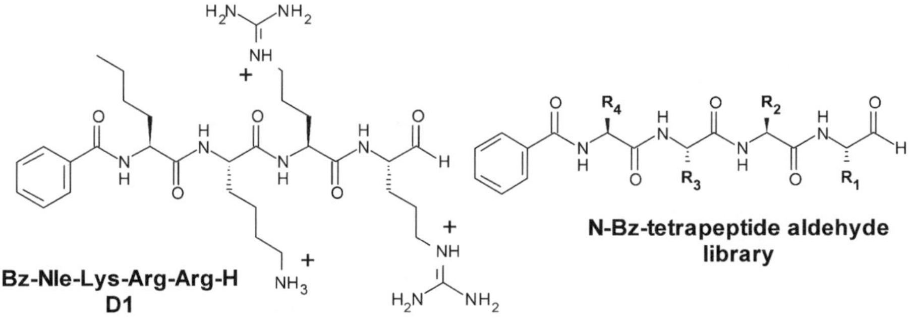

Bz-Nle-Lys-Arg-Arg-H
D1

N-Bz-tetrapeptide aldehyde library

Figure 15.13   *Left*: Template inhibitor D1, an N-acylated tetrapeptide aldehyde inhibitor of NS3pro;[63] *right*: generic structure of a computer-designed, substituted N-benzylated tetrapeptide aldehyde library of NS3pro inhibitors.

docked compound, out of 20 per compound generated by the algorithm, was selected. Then K_i^* values were calculated using an equation derived from a training set made of 12 experimentally tested N-acylated tetrapeptide aldehydes, and a validation set was made of three structurally related compounds. The 16 best compounds, following *in silico* screening using the LUDI scoring function,[67] are shown in Figure 15.14. Ten out of these

Bz-Nle-His-*m*(Gn)Phe-*p*(Gn)Phe-H
1-3-10-9

        Bz-Nle-*p*(Am)Phe-*m*(Gn)Phe-*p*(Am)Phe-H
        1-5-10-5

Bz-Nle-hHis-*m*(Gn)Phe-*p*(Gn)Phe-H
1-4-10-9

        Bz-Nle-His-*m*(Gn)Phe-2Nal-H
        1-3-10-14

Bz-Nle-His-*p*(Gn)Phe-Trp-H
1-3-9-11

        Bz-Nle-*m*(Am)Phe-*m*(Im)Phe-(dMo)Phe-H
        1-6-8-17

Bz-Nle-hHis-*p*(Gn)Phe-*p*(Ph)Phe-H
1-4-9-13

        Bz-Nle-Arg-*m*(Gn)Phe-*p*(Am)Phe-H
        1-1-10-5

Bz-Nle-Lys-Arg-Trp-H
1-2-1-11

        Bz-Nle-His-*m*(Im)Phe-Arg-H
        1-3-8-1

Bz-Nle-*p*(Am)Phe-*m*(Im)Phe-Arg-H
1-5-8-1

        Bz-Nle-His-*p*(Gn)Phe-*p*(Gn)Phe-H
        1-3-9-9

Bz-Nle-*p*(Am)Phe-*p*(Gn)Phe-Trp-H
1-5-9-11

        Bz-Nle-*m*(Gn)Phe-*p*(Gn)Phe-*p*(Gn)Phe-H
        1-10-9-9

Bz-Nle-hHis-*m*(Gn)Phe-Arg-H
1-4-10-1

        Bz-Nle-*p*(Am)Phe-*m*(Gn)Phe-Arg-H
        1-5-10-1

Figure 15.14 16 highest ranked library members, with lowest K_i^* values against NS3pro. Numbers correspond to R_1-R_2-R_3-R_4 numbering of fragments used to build the first 9240-membered library.

compounds did score better than parent D1 in terms of virtual ADMET profiling.

Around 250 library members showed a $K_i^* < 15$ μM, *i.e.* a predicted inhibition similar to, or even stronger than, D1. A detailed SAR for any P_1–P_4 residue was obtained, and in order to combine the acquired SAR and chemical feasibility the most recurrent P_1–P_3 fragments in the 250 highest scoring compounds were identified (dark blue, orange and yellow bars, Figure 15.15). As to P4, it was decided to use only the N-benzoylated Nle residue.

Finally, a $1P_1 \times 8P_2 \times 4P_3 \times 5P_4 = 160$-membered D1 analog library L4 (Figure 15.16) was designed and proposed for synthesis and biological characterization.[64]

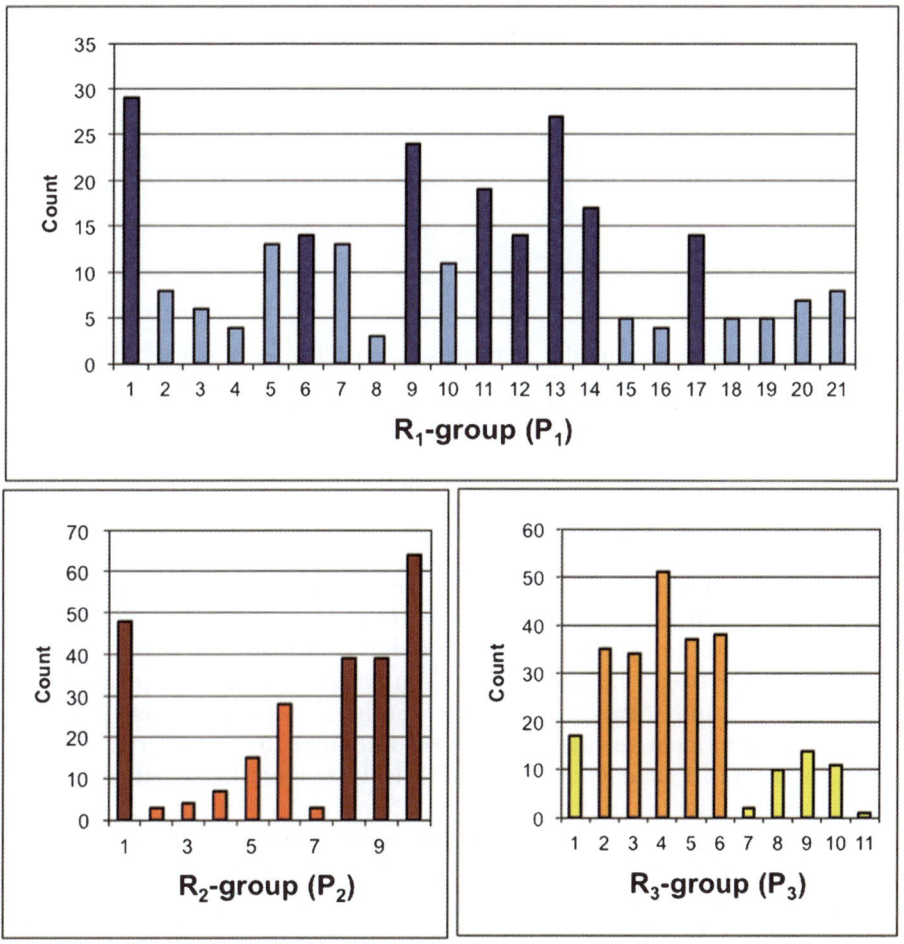

Figure 15.15 Frequency of occurrence of R groups in 250 highest ranked library members, with lowest K_i^* values against NS3pro. See Figure 15.16 for the structure of best R_1–R_4 substituents (bar and substituent numbers match, and N-BzNle is the only P4 selected residue).

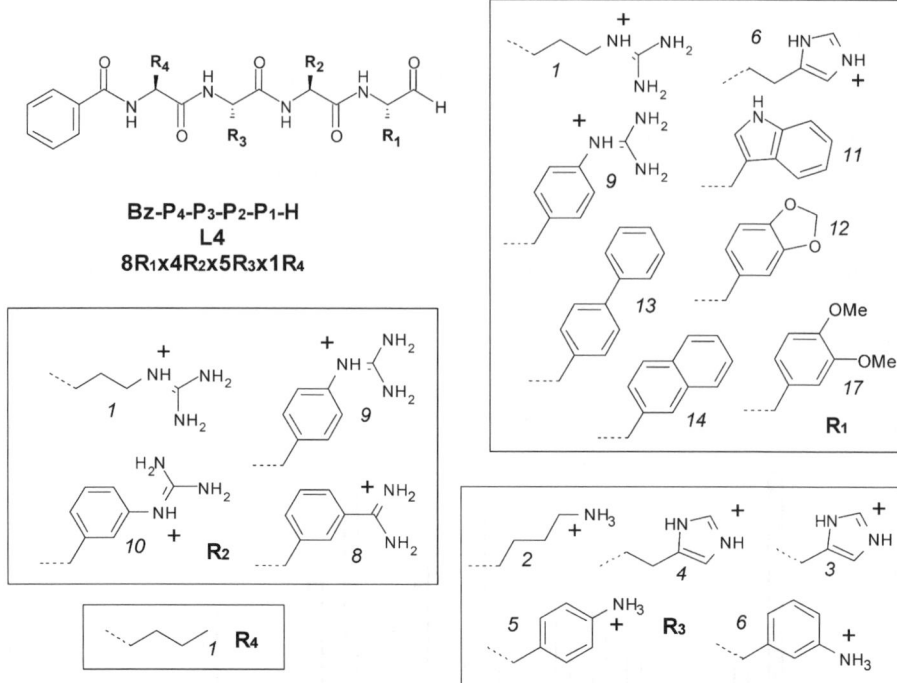

Figure 15.16 Structure of the computer-designed template inhibitor D1 analog library L4, proposed for synthesis, and of R_1–R_4 substituents used in the library. See Figure 15.15 for the frequency of occurrence of best R_1–R_4 substituents (substituent and bar numbers match).

15.3.2 Parasitic Diseases

An anti-malarial library design project was published in 2009, targeting in general the type II fatty acid biosynthesis (FAS-II) pathway[68] in *Plasmodium falciparum*, and in particular its enoyl-acyl carrier protein reductase (PfENR).[69] The tragic impact of malaria in terms of infected population and death count,[13] and the essential role of both the selected pathway and its enzymatic component, ensure a significant relevance for our efforts.

Triclosan[70] (TCL) and its close analog TCL11[71] (Figure 15.17, left and center) are non-competitive and potent PfENR inhibitors (IC$_{50}$ between 50 and 76 nM) which were selected as structural motifs on which to design a library. A three-point substitution library (R_1–R_3, Figure 15.17, right) was planned by substitution of the three Cl atoms on TCL, both to explore the SAR of these positions and to find PfENR inhibitors with higher potencies than TCL.[72] We focused on 40 substitution fragments, which were all compatible for R_1–R_3 substitution, either *via* Grignard reaction on 5-aldehyde precursors[71] (R_1), *via* 4′-nitro reduction and acylation, 4′-carboxy acylation or 4′-nitrile condensation[73] (R_2) and 2′-nitro, cyano or aldehyde functionalization[74] (R_3).

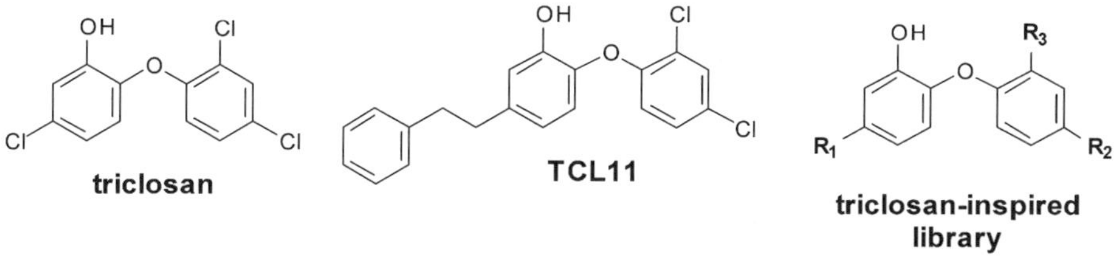

Figure 15.17 *Left*: Triclosan[70] and (*middle*) TCL11,[71] two diaryl ether inhibitors of PfENR; *right*: generic structure of a computer-designed, substituted diaryl ether library of PfENR inhibitors.

Twenty electronic, physico-chemical and topological descriptors were used to filter the R radicals for their resemblance to the corresponding R groups of TCL11 and other known, structurally related inhibitors crystallized into the active site of PfENR; a total of $40R_1 \times 23R_2 \times 11R_3 = 10\,120$ TCL analogs comprised the first library. The library size was reduced to its 1000 most relevant members by analog filtering, using the same descriptors on whole library members, and selecting the most diverse analogs retaining structural similarity with TCL11.

Structure-based focusing was then performed using previously described methods[46,53,54,65] and replacing TCL11 with 20 best conformations for each library member into the binding site of the PfENR–TCL11 complex X-ray structure. The best fitting conformer for each docked compound was selected. IC_{50}^{*} values were calculated using an equation derived from a training set made of 16 TCL analogs, and a validation was set made of four structurally related PfENR inhibitors.

Nine best compounds, following *in vitro* screening using the LUDI scoring function,[67] are shown in Figure 15.18. Eight out of these compounds did score better than parent TCL in terms of virtual ADMET profiling.

Around 260 library members showed an $IC_{50}^{*} < 200$ nM, *i.e.* a predicted inhibition comparable or even stronger (seven compounds labeled # in Figure 15.18) than TCL. A detailed SAR for any R_1–R_3 fragment was obtained and, in order to combine the acquired SAR and chemical feasibility, the most recurrent R_1–R_3 fragments in the 266 highest scoring compounds were identified (darker blue, orange and yellow bars, Figure 15.19).

Finally, an $8R_1 \times 5R_2 \times 3R_3 = 120$-membered TLC analog library L5 (Figure 15.20) was designed and proposed for synthesis and biological characterization.[72]

15.3.3 Bacterial Diseases

Recently we turned our attention to tuberculosis,[14] a widely diffused bacterial disease which is resurfacing through the emergence of traditional treatment-resistant strains. Among the more recently validated molecular targets of *Mycobacterium tuberculosis*, its thymidine monophosphate kinase[75] (TMPK$_{mt}$)

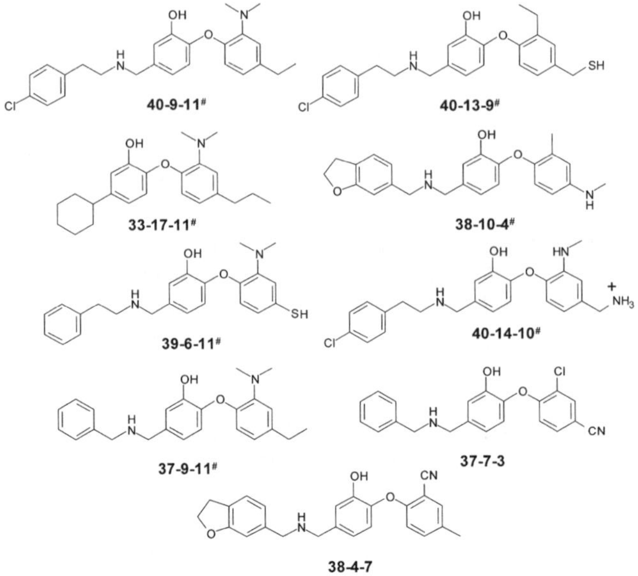

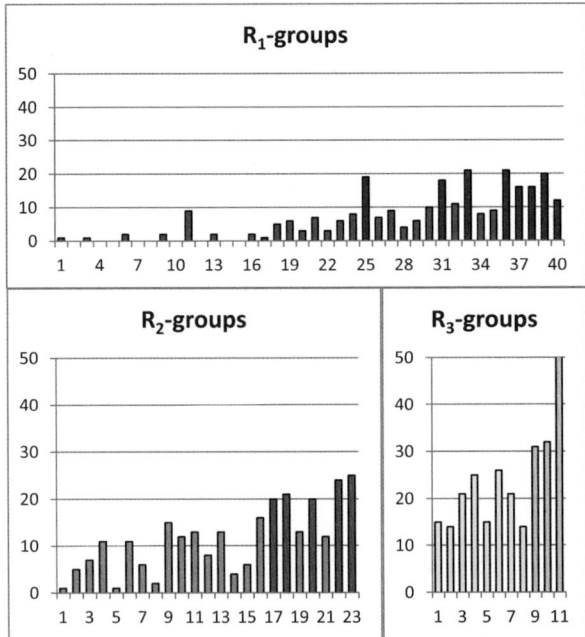

Figure 15.18   Structures of nine highest ranked library members, with lowest IC$_{50}$* values against PfENR. Compounds with < 50 nM IC$_{50}$* are labeled $^\#$; numbers correspond to R$_1$-R$_2$-R$_3$ numbering of fragments used to build the first 10 120-membered library.

Figure 15.19   Frequency of occurrence of R groups in 266 highest ranked library members, with lowest IC$_{50}$* values against PfENR. See Figure 15.20 for the structure of best R$_1$–R$_3$ substituents (bar and substituent numbers match).

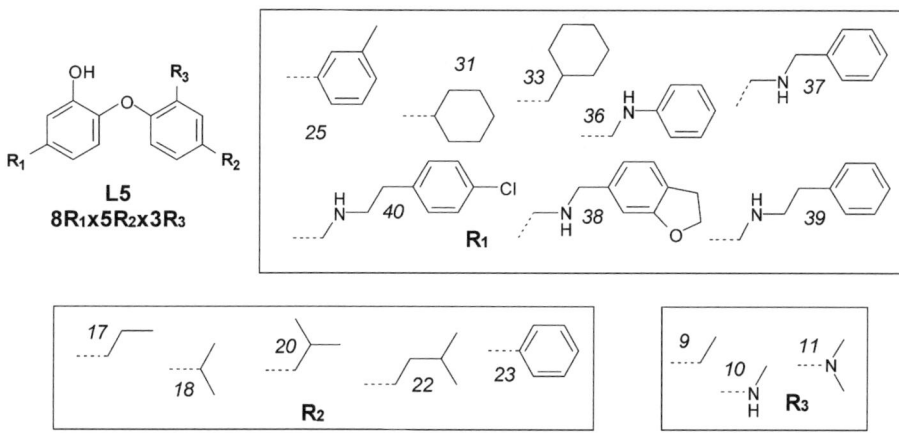

Figure 15.20 Structure of the computer-designed TCL11 analog library L5, proposed for synthesis, and of R_1–R_3 substituents used in the library. See Figure 15.19 for the frequency of occurrence of the best R_1–R_3 substituents (substituent and bar numbers match).

has a crucial role for the pathogen and is largely different from its human counterpart, thus ensuring selectivity and specificity.

Its substrate, deoxythymidine phosphate (dTMP), and two low micromolar-active inhibitors, azido-containing AZTMP[76] and bicyclic VAN1[77] (Figure 15.21, left and center), are known; the X-ray structure of dTMP complexed with TMPK$_{mt}$ was also reported.[78] We selected VAN1 as a structural motif and designed a four-point substitution library (R_1–R_4, Figure 15.21, right) both to explore the SAR of these positions and to find more potent bicyclic TMPK$_{mt}$ inhibitors against rifampicin-resistant *M. tuberculosis* strains;[79] we also replaced the traditional ribose ring with a cyclopentane scaffold, to increase the metabolic stability of designed compounds. We focused on three groups of commercially available[80] fragment functions, assuming that known functional group transformations for the ribose ring-containing dTMP[81] and VAN1[82] analogs could be transferred to our cyclopentane-containing library members:

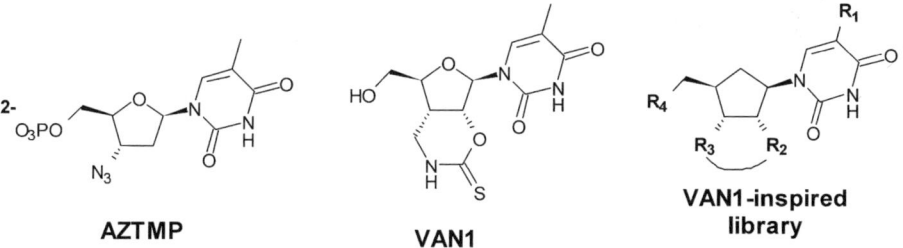

Figure 15.21 *Left*: AZTMP[76] and (*center*) VAN1,[77] respectively a monocyclic and a bicyclic thymidine analog inhibitor of TMPK$_{mt}$; *right*: generic structure of a computer-designed, substituted bicyclic thymidine analog library of TMPK$_{mt}$ inhibitors.

small methyl replacements (R_1), condensed heterocycle replacements (R_2R_3) and primary alcohol substituted replacements (R_4).

Several electronic, physico-chemical and topological descriptors were used to filter the R groups for their resemblance to the corresponding R groups of dTMP, or of VAN1, respectively, crystallized or replaced and minimized into the active site of TMPK$_{mt}$; a total of $6R_1 \times 29R_2R_3 \times 12R_4 = 2088$ VAN1 analogs comprised the first library. Owing to its smaller size, this library was not submitted to analog filtering and was passed as such to structure-based evaluation.

Structure-based focusing was then performed using previously described methods[46,53,54,65] and replacing dTMP-VAN1 with 10 conformations for each library member into the TMPK$_{mt}$ binding site of the refined ligand–TMPK$_{mt}$ complex derived from X-ray structure. The best fitting conformer for each docked compound was selected. The K_i^* values were calculated using an equation derived from a training set made of 18 dTMP-like inhibitors, and a validation set was made of five structurally related compounds. The 16 best compounds, following *in vitro* screening using the PLP1 scoring function,[55] are shown in Figure 15.22.

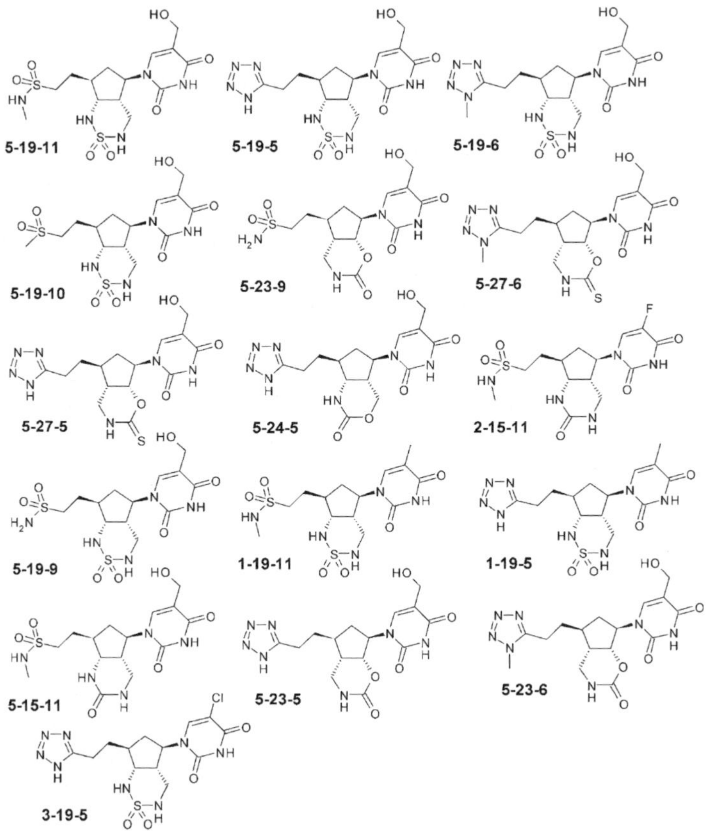

Figure 15.22   Structures of the 16 highest ranking library members, with lowest K_i^* values against TMPK$_{mt}$. Numbers correspond to R_1-R_2R_3-R_4 numbering of fragments used to build the first 2088-membered library.

More than 100 library members showed a $K_i^* < 1$ μM, *i.e.* a predicted inhibition stronger than both AZTMP and VAN1. A detailed SAR for any considered R_1–R_4 substitution was obtained; the predicted potency increase for the best library members was mostly due to the introduction of multiple proton acceptor groups in various parts of the library members. In order to combine the acquired SAR and chemical feasibility, the most recurrent R_1–R_4 fragments in the 100 highest scoring compounds were identified (darker yellow, blue and red bars, Figure 15.23).

Finally, a $1R_1 \times 7R_2R_3 \times 5R_4 = 35$-membered bicyclic VAN1 analog library L6 (Figure 15.24) was designed and proposed for synthesis and biological characterization.[79]

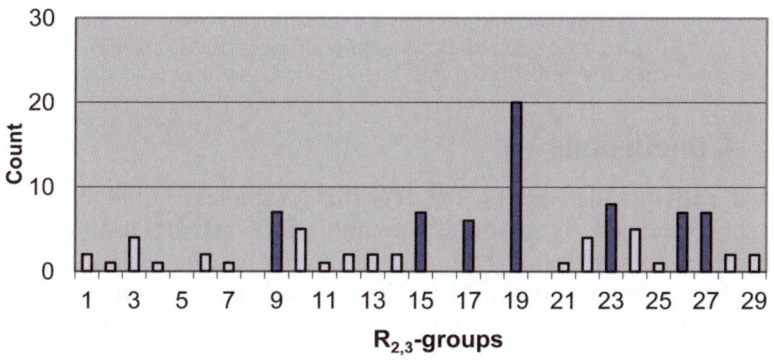

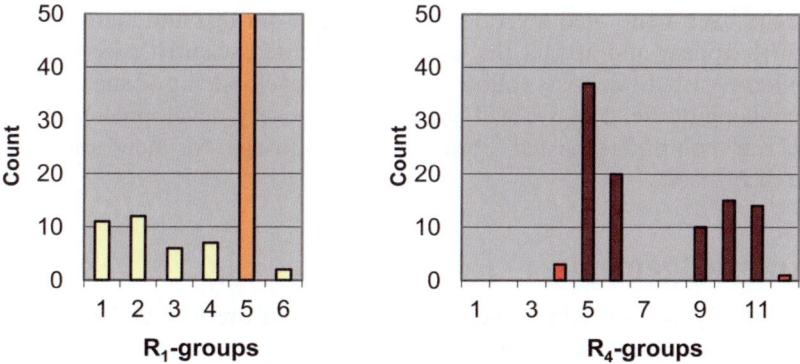

Figure 15.23 Frequency of occurrence of R groups in 100 highest ranked library members, with lowest K_i^* values against TMPK$_{mt}$. See Figure 15.24 for the structure of best R_1–R_4 substituents (bar and substituent numbers match).

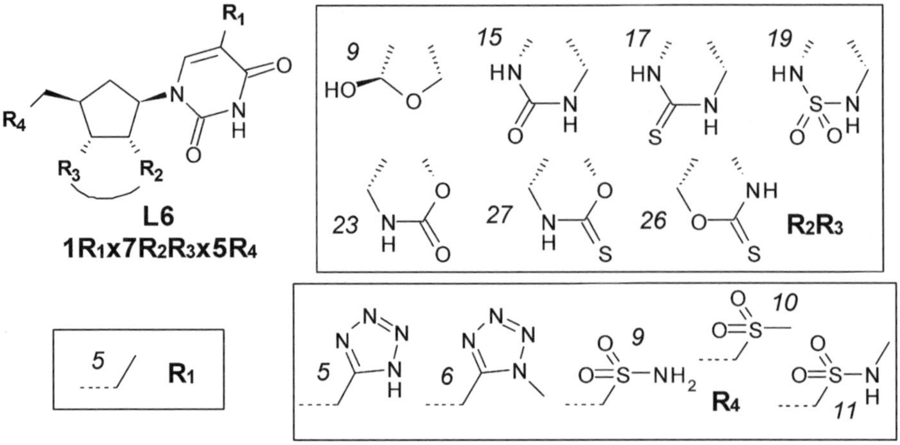

Figure 15.24   Structure of the computer-designed VAN1 analog library L6, proposed for synthesis, and of R_1–R_4 substituents used in the library. See Figure 15.23 for the frequency of occurrence of best R_1–R_4 substituents (substituent and bar numbers match).

15.4   Conclusions

Predicted activities derived by the reported examples should obviously be thoroughly validated by synthesis and inhibitory activity testing, so as to (among other things) refine each and every one of the generated, project-specific library design models. Nevertheless, such studies contain significant value which, in our opinion, should be harvested by skilled medicinal chemists interested in one or another of the therapeutic areas involved in our library design experiments.

On a more general level, computer-assisted library design remains both a relevant component of the drug discovery process and the subject of extensive research.[83] We mentioned several recent trends in our text and many others are bound to appear and attract the interest of talented scientists; we suggest the computer-assisted scientists follow closely the development of such a growing and challenging field, as we will eagerly check every development to further refine our computer-assisted library design strategy for new and exciting research projects.

Acknowledgements

We gratefully acknowledge generous support from the ICS-UNIDO (Trieste, Italy) for multiple parts of these studies.

References

1. C. M. Kaefer and J. A. Milner, *J. Nutr. Biochem.*, 2008, **19**, 347.

2. H. O. Villar and M. R. Hansen, *Expert Opin. Drug Discovery*, 2009, **4**, 1215.
3. Z. Zhu and J. Cuozzo, *J. Biomol. Screen.*, 2009, **14**, 1157
4. http://annualreport.roche.com/10/ar/pharmaceuticals/focus_on_unmet_medical_needs.htm.
5. E. F. Schmidt and D. A. Smith, *Drug Discovery Today*, 2007, **12**, 998.
6. W. P. Walters, A. Ajay and A. M. Murcko, *Curr. Opin. Chem. Biol.*, 1999, **3**, 384.
7. M. Congreve, G. Chessari, D. Tisi and A. J. Woodhead, *J. Med. Chem.*, 2008, **51**, 3661.
8. R. S. Bohacek, C. McMartin and W. C. Guida, *Med. Chem. Res.*, 1996, **16**, 3.
9. D. M. Schnur, *Curr. Opin. Drug Discovery Dev.*, 2008, **11**, 375.
10. P. Zhang, W. Li, H. Chen and X. Liu, *Curr. Med. Chem.*, 2010, **17**, 3393.
11. J. Oxford, R. Lambkin-Williams and A. Mann, *Antiviral Chem. Chemother.*, 2007, **18**, 71.
12. T. Parkinson and D. C. Pryde, *Future Med. Chem.*, 2010, **2**, 1181.
13. T. Wu., A. S. Nagle and K. A. Chatterjee, *Curr. Med. Chem.*, 2011, **18**, 853.
14. A. Koul, E. Arnoult, N. Lounis, J. Guillemont and K. Andries, *Nature*, 2011, **469**, 483.
15. S. Yan and R. Selliah, *Methods Mol. Biol.*, 2011, **685**, 175.
16. Y. Wang, C. Strickland, J. H. Voigt, M. E. Kennedy, B. M. Beyer, M. M. Senior, E. M. Smith, T. L. Nechuta, V. S. Madison, M. Czarniecki, B. A. McKittrick, A. W. Stamford, E. M. Parker, J. C. Hunter, W. J. Greenlee and D. F. Wyss, *J. Med. Chem.*, 2010, **53**, 942.
17. C. Abad-Zapatero, G. F. Stamper and V. S. Stoll, *Methods Princ. Med. Chem.*, 2006, **34**, 249.
18. X. H. Ma, J. Jia, F. Zhu, Y. Xue, Z. R. Li and Y. Z. Chen, *Comb. Chem. High Throughput Screening*, 2009, **12**, 344.
19. C. Acharya, A. Coop, J. E. Polli and A. D. Mackerell, Jr., *Curr. Comput.-Aided Drug Des.*, 2011, **7**, 10.
20. I. Wallach, *Drug Dev. Res.*, 2011, **72**, 17.
21. A. Lange and M. Baldus, *Drug Discovery Ser.* 2006, **4**, 297.
22. J. V. Moller, C. Olesen, A.-M. L. Winther and P. Nissen, *Methods Mol. Biol.*, 2010, **654**, 119.
23. W. Feng, L. F. Pan and M. J. Zhang, *Sci. China: Life Sci.*, 2011, **54**, 101.
24. C. N. Cavasotto and S. S. Phatak, *Drug Discovery Today*, 2009, **14**, 676.
25. S. J. Campbell, A. Gaulton, J. Marshall, D. Bichko, S. Martin, C. Brouwer and L. Harland, *Drug Discovery Today*, 2010, **15**, 3.
26. R. Sistla, C. Ghadiyaram, N. C. Srinivasan and H. S. Subramanya, *Innov. Pharm. Technol.*, 2006, 18.
27. H. O. Villar and M. R. Hansen, *Expert Opin. Drug Discovery*, 2009, **4**, 1215.
28. Y. Fukunishi, and M. Lintuluoto, *Curr. Comput.-Aided Drug Des.*, 2010, **6**, 90.

29. H. Mauser and W. Guba, *Curr. Opin. Drug Discovery Dev.* 2008, **11**, 365.
30. A. Pozzan, *Curr. Pharm. Des.*, 2006, **12**, 2099.
31. A. R. Katritzky, M. Kuanar, S. Slavov, D. C. Hall, M. Karelson, I. Kahn and D. A. Dobchev, *Chem. Rev.*, 2010, **110**, 5714.
32. K. Roy, *Mol. Div.*, 2004, **8**, 321.
33. U. Maran, S. Sild, I. Tulp, K. Takkis and M. Moosus, *Issues Toxicol.*, 2010, **7**, 148.
34. A. Nicholls, N. E. MacCuish and J. D. MacCuish, *J. Comput. Aided Mol. Des.*, 2004, **18**, 451.
35. D. M. Schnur, B. R. Beno, A. J. Tebben and C. L. Cavallaro, *Methods Mol. Biol.*, 2011, **672**, 387.
36. V. J. Gillet, in *Computational Medicinal Chemistry for Drug Discovery*, ed. P. Bultinck, H. de Winter, W. Langenaeker and J. P. Tollenaere, Dekker, New York, 2004, p. 617.
37. D. M. Schnur, A. J. Tebben and C. L. Cavallaro, in *Comprehensive Medicinal Chemistry II*, ed. D. J. Triggle and J. B. Taylor, Elsevier, Oxford, 2006, vol. 4, p. 307.
38. R. Nilakantan, D. S. Nunn, L. Greenblatt, G. Walker, K. Haraki and D. Mobilio, *J. Chem. Inf. Model.*, 2006, **46**, 1069.
39. E. Yuriev, M. Agostino and P. A. Ramsland, *J. Mol. Recognit.*, 2011, **24**, 149.
40. L. Li, M. Khanna, I. Jo, F. Wang, N. M. Ashpole, A. Hudmon and S. O. Meroueh, *J. Chem. Inf. Model.*, 2011, **51**, 755.
41. R. D. Brown, M. Hassan and M. Waldman, *J. Mol. Graphics Model.*, 2000, **18**, 427.
42. R. A. Katz and A. M. Skalka, *Annu. Rev. Biochem.*, 1994, **63**, 133.
43. P. Y. S. Lam, Y. Ru, P. K. Jadhav, P. E. Aldrich, G. V. DeLucca, C. J. Eyermann, C. H. Chang, G. Emmett, E. R. Holler, W. F. Daneker, L. Li, P. N. Confalone, R. J. McHugh, Q. Han, R. Li, J. A. Markwalder, S. P. Seitz, T. R. Sharpe, L. T. Bacheler, M. M. Rayner, R. M. Klabe, L. Shum, D. L. Winslow, D. M. Kornhauser, D. A Jackson, S. Eriksson-Viitanen and C. N. Hodge, *J. Med. Chem.*, 1996, **39**, 3514.
44. V. Frecer, E. Burello and S. Miertus, *Bioorg. Med. Chem.*, 2005, **13**, 5492.
45. Available Chemicals Directory (ACD), version 95.1, MDL Information Systems, San Leandro, CA, 2003; http://cds3.dl.ac.uk/cds/cds.html.
46. Cerius^2 Life Sciences, version 4.5, Accelrys, San Diego, CA, 2000.
47. K. P. Peters, J. Fauck and C. Frommel, *J. Mol. Biol.*, 1996, **256**, 201.
48. J. W. Park and W. H. Jo, *Eur. J. Med. Chem.*, 2010, **45**, 536.
49. L. V. Gubareva, L. Kaiser and F. G. Hayden, *Lancet*, 2000, **355**, 827.
50. T. Rungrotmongkol, V. Frecer, W. De-Eknamkul, S. Hannongbua and S. Miertus, *Antiviral Res.*, 2009, **82**, 51.
51. M. Federspiel, R. Fischer, M. Hennig, H.-J. Mair, T. Oberhauser, G. Rimmler, T. Albiez, J. Bruhin, H. Estermann, C. Gandert, V. Goeckel, S. Goetzo, U. Hoffmann, G. Huber, G. Janatsch, S. Lauper, O. Roeckel-

Staebler, R. Trussardi and A. G. Zwahlen, *Org. Proc. Res. Dev.*, 1999, **3**, 266.

52. P. J. Harrington, J. D. Brown, T. Foderaro and R. C. Hughes, *Org. Proc. Res. Dev.*, 2004, **8**, 86.
53. J. R. Maple, M.-J. Hwang, T. P. Stockfish, U. Dinur, M. Waldman, C. S. Ewing and A. T. Hagler, *J. Comput. Chem.*, 1994, **15**, 162.
54. A. K. Rappé and W. A. Goddard, III, *J. Phys. Chem.*, 1991, **95**, 3358.
55. G. M. Verkhivker, D. Bouzida, D. K. Gehlhaar, P. A. Rejto, S. Arthurs, A. B. Colson, S. T. Freer, V. Larson, B. A. Luty, T. Marrone and P. W. Rose, *J. Comput. Aided Mol. Des.*, 2000, **14**, 731.
56. Y. Abed, B. Nehmé, M. Baz and G. Boivin, *Antiviral Res.*, 2008, **77**, 163.
57. A. C. Krueger, Y. Xu, W. M. Kati, D. J. Kempf, C. J. Maring, K. F. McDaniel, A. Molla, D. Montgomery and W. E. Kohlbrenner, *Bioorg. Med. Chem. Lett.*, 2008, **18**, 1692.
58. ChemBioFinder.com, Scientific Database Gateway, version 2.0.0.26, 2007.
59. T. Rungrotmongkol, T. Udommaneethanakit, V. Frecer and S. Miertus, *Comb. Chem. High Throughput Screening*, 2010, **13**, 268.
60. I. Muegge, *J. Comput. Chem.*, 2001, **22**, 418.
61. B. Falgout, M. Pethel, Y. M. Zhang and C. J. Lai, *J. Virol.*, 1991, **65**, 2467.
62. Z. Yin, S. J. Patel, W. L. Wang, G. Wang, W. L. Chan, K. R. Rao, J. Alam, D. A. Jeyaraj, X. Ngew, V. Patel, D. Beer, S. P. Lim, S. G. Vasudevan and T. H. Keller, *Bioorg. Med. Chem. Lett.*, 2006, **16**, 36.
63. Z. Yin, S. J. Patel, W. L. Wang, W. L. Chan, K. R. Rao, G. Wang, X. Ngew, V. Patel, D. Beer, J. E. Knox, N. L. Ma, C. Erhardt, S. P. Lim, S. G. Vasudevan and T. H. Keller, *Bioorg. Med. Chem. Lett.*, 2006, **16**, 40.
64. V. Frecer and S. Miertus, *J. Comput. Aided Mol. Des.*, 2010, **24**, 195.
65. Cerius² Life Sciences, version 4.6, Accelrys, San Diego, , CA, 2002.
66. P. Erbel, N. Schiering, A. D'Arcy, M. Renatus, M. Kroemer, S. P. Lim, Z. Yin, T. H. Keller, S. G. Vasudevan and U. Hommel, *Nat. Struct. Mol. Biol.*, 2006, **13**, 372.
67. H. J. Boehm, *J. Comput. Aided Mol. Des.*, 1998, **12**, 309.
68. S. A. Ralph, G. G. van Dooren, R. F. Waller, M. J. Crawford, M. J. Fraunholz, B. J. Foth, C. J. Tonkin, D. S. Roos and G. I. McFadden, *Nat. Rev. Microbiol.*, 2004, **2**, 203.
69. C. O. Rock and J. E. Cronan, *Biochim. Biophys. Acta*, 1996, **1302**, 1.
70. N. Surolia and A. Surolia, *Nat. Med.*, 2001, **7**, 167.
71. J. S. Freundlich, F. Wang, H.-C. Tsai, M. Kuo, H.-M. Shieh, J. W. Anderson, I. J. Nkrumah, J.-C. Valderramos, M. Yu, T. R. S. Kumar, S. G. Valderramos, W. R. Jacobs, Jr., G. A. Schiehsher, D. P. Jacobus, D. A. Fidock and J. C. Sacchettini, *J. Biol. Chem.*, 2007, **282**, 25436.
72. V. Frecer, E. Megnassan and S. Miertus, *Eur. J. Med. Chem.*, 2009, **44**, 3009.
73. J. S. Freundlich, J. W. Anderson, D. Sarantakis, H.-M. Shieh, M. Yu, J.-C. Valderramos, E. Lucumi, M. Kuo, W. R. Jacobs, Jr., D. A. Fidock, G. A.

Schiehsher, D. P. Jacobus and J. C. Sacchettini, *Bioorg. Med. Chem. Lett.*, 2005, **15**, 5247.

74. J. S. Freundlich, M. Yu, E. Lucumi, M. Kuo, H.-C. Tsai, J.-C. Valderramos, L. Karagyozov, W. R. Jacobs, Jr., G. A. Schiehsher, D. A. Fidock, D. P. Jacobus and J. C. Sacchettini, *Bioorg. Med. Chem. Lett.*, 2006, **16**, 2163.
75. H. Munier-Lehmann, A. Chafotte, S. Pochet and G. Labesse, *Protein Sci.*, 2001, **10**, 1195.
76. I. Li de la Sierra, H. Munier-Lehmann, A. M. Gilles, O. Barzu and M. Delarue, *J. Mol. Biol.*, 2001, **311**, 87.
77. V. Vanheusden, H. Munier-Lehmann, M. Froeyen, R. Busson, J. Rozenski, P. Herdewijn and S. Van Calenbergh, *J. Med. Chem.*, 2004, **47**, 6187.
78. I. Li de la Sierra, H. Munier-Lehmann, A. M. Gilles, O. Barzu and M. Delarue, *Acta Crystallogr., Sect. D: Biol. Crystallogr.*, 2000, **56**, 226.
79. V. Frecer, P. Seneci and S. Miertus, *J. Comput. Aided Mol. Des.*, 2011, **25**, 31.
80. Available Chemicals Directory (ACD), version 3.0, Symyx Technologies, Santa Clara, CA, 2009.
81. I. Van Daele, H. Munier-Lehmann, M. Froeyen, J. Balzarini and S. Van Calenbergh, *J. Med. Chem.*, 2007, **50**, 5281.
82. I. Van Daele, H. Munier-Lehmann, P. M. S. Hendrickx, G. Marchal, P. Chavarot, M. Froeyen, L. Qing, J. Martins and S. Van Calenbergh, *ChemMedChem*, 2006, **1**, 1081.
83. V. J. Gillet, *Curr. Opin. Chem. Biol.*, 2008, **12**, 372.

Subject Index

Figures are indicated by *italic* page numbers, Tables by **emboldened numbers**, and footnotes by suffix "n"